Hominin Postcranial Remains from Sterkfontein, South Africa, 1936–1995

HUMAN EVOLUTION SERIES

African Biogeography, Climate Change, and Human Evolution
Edited by Timothy G. Bromage and Friedemann Schrenk

Meat-Eating and Human Evolution
Edited by Craig B. Stanford and Henry T. Bunn

The Skull of Australopithecus afarensis
William H. Kimbel, Yoel Rak, and Donald C. Johanson

Early Modern Human Evolution in Central Europe: The People of Dolní Věstonice and Pavlov
Edited by Erik Trinkaus and Jiří Svoboda

Evolution of the Hominin Diet:
The Known, the Unknown, and the Unknowable
Edited by Peter S. Ungar

Genes, Language, & Culture History in the Southwest Pacific
Edited by Jonathan S. Friedlaender

The Lithic Assemblages of Qafzeh Cave
Erella Hovers

Life and Death at the Peştera cu Oase: A Setting for Modern Human Emergence in Europe
Edited by Erik Trinkaus, Silviu Constantin, and João Zilhão

The People of Sunghir: Burials, Bodies, and Behavior in the Earlier Upper Paleolithic
Edited by Erik Trinkaus, Alexandra P. Buzhilova, Maria B. Mednikova, and Maria V. Dobrovolskaya

Hominin Postcranial Remains from Sterkfontein, South Africa, 1936-1995
Edited by Bernhard Zipfel, Brian G. Richmond, and Carol V. Ward

Hominin Postcranial Remains from Sterkfontein, South Africa, 1936–1995

Edited by
Bernhard Zipfel, Brian G. Richmond, and Carol V. Ward

OXFORD
UNIVERSITY PRESS

OXFORD
UNIVERSITY PRESS

Oxford University Press is a department of the University of Oxford. It furthers
the University's objective of excellence in research, scholarship, and education
by publishing worldwide. Oxford is a registered trade mark of Oxford University
Press in the UK and certain other countries.

Published in the United States of America by Oxford University Press
198 Madison Avenue, New York, NY 10016, United States of America.

Library of Congress Cataloging-in-Publication Data
Names: Zipfel, Bernhard, editor. | Richmond, Brian, editor. | Ward, Carol V., editor.
Title: Hominin postcranial remains from Sterkfontein, South Africa, 1936–1995 /
Bernhard Zipfel, co-editor, Brian G. Richmond, co-editor, Carol V. Ward, co-editor.
Description: New York, NY : Oxford University Press, [2020] |
Series: Advances in human evolution series | Includes bibliographical references and index.
Identifiers: LCCN 2019041498 (print) | LCCN 2019041499 (ebook) |
ISBN 9780197507667 (hardback) | ISBN 9780197507681 (epub) |
ISBN 9780197507698 (online) | ISBN 9780197507674 (updf)
Subjects: LCSH: Australopithecines—South Africa—Sterkfontein Caves. |
Fossil hominids—South Africa—Sterkfontein Caves. | Paleoanthropology.
Classification: LCC GN283 .H66 2020 (print) | LCC GN283 (ebook) | DDC 569.0968—dc23
LC record available at https://lccn.loc.gov/2019041498
LC ebook record available at https://lccn.loc.gov/2019041499

1 3 5 7 9 8 6 4 2

Printed by LSC Communications, United States of America

This volume is dedicated to the memory of
Dr. Charles Abram Lockwood III
1970–2008

For his inspiration in initiating this volume and the workshop that led to it,
his thoughtful scholarship of South African fossil hominins, and
his leadership in the field of hominin paleontology.

Photo credit: Photo by Kaye Reed. Courtesy of Institute of Human Origins

Contents

SECTION 3: FUNCTIONAL ANATOMY AND BIOLOGY

SECTION 4: APPENDICES

Preface

Carol V. Ward, Brian G. Richmond, and Bernhard Zipfel

The discovery in 1924 in southern Africa of an early hominin child's skull, referred to *Australopithecus africanus* by Raymond Dart in 1925, was a major event in the history of paleoanthropology, providing the first evidence of early hominins in Africa and overturning conventional ideas about human evolution. Subsequent discoveries over the next several decades, notably from cave deposits at Sterkfontein, yielded the first evidence of postcranial anatomy in *Australopithecus*, and the first evidence that early hominins were habitual bipeds. Nearly 50 years after Dart's report, the discoveries in eastern Africa of a wealth of fossil evidence of the slightly older and craniodentally more primitive taxon, *A. afarensis*, catalyzed debates that continue today about the origin and evolution of human gait and the phylogenetic relationships among early hominins. Since then, *A. afarensis* has formed the main basis of our understanding of early hominin bipedality and paleobiology. However, little attention has been paid to examining variation among species in postcranial anatomy and/or locomotion, although intriguing hints are beginning to appear in the literature. Did multiple varieties of bipedality evolve? Did australopith species differ in positional or manipulative abilities, body proportions, or patterns of sexual dimorphism? These are critical questions for understanding the evolution of australopiths and hominin locomotion, yet research has been limited because only the postcrania of *A. afarensis* have been extensively studied, despite the existence of a large fossil collection from Sterkfontein as well.

The Sterkfontein hominin fossils generally are attributed to the species *Australopithecus africanus*, because most craniodental remains from the site are attributable to that taxon (reviews in Grine, 2013, 2019). However, there may be more than one hominin represented within the sample, even within the most productive member, Member 4, and, given the complex stratigraphy of the site and challenges in dating the deposits, this may or not may be the case. In general, several studies have suggested the presence of two or more australopith taxa within the sample, each citing more morphological variation among the craniodental remains from Sterkfontein than can be attributed to a single species, at least compared to extant hominoid taxa (Clarke, 1988, 1994; Kimbel and White, 1988; Lockwood, 1997, Lockwood and Tobias, 2002). However, it is notable that none of these studies agree on which specimens comprise the different possible taxa or groups, largely due to emphasis on different aspects of morphology varying among the fossils. The likely time depth of the Sterkfontein sample, even within Member 4 (see Pickering and Herries, Chapter 3, this volume), may also complicate assessment of potential taxonomic heterogeneity at the site.

None of the Sterkfontein postcranial fossils can be definitively associated with any craniodental specimens (but see Thackeray et al., 2002) and so cannot be related directly to any of the proposed taxonomic divisions within the sample. However, some studies have cited variation within the postcranial fossils that may also reflect taxonomic variation, although many analyses to date have not tackled this question rigorously. Even though these suggestions have been made occasionally in the literature, no clear or consistent suggestion of two or more taxa has been apparent within the postcranial samples (reviewed in Grine, 2019). Taxonomic variation is one of the key questions that each chapter in this volume addresses (summarized in Ward and Zipfel, Chapter 18, this volume).

Excavations at Sterkfontein cave continued throughout the late 20th century and have yielded one of the largest collections (>150) of postcranial fossils of any hominin taxon. These fossils remain relatively unstudied, and few are published, despite the enormous potential of these fossils for answering questions about *Australopithecus africanus* paleobiology, early hominin variation, and early human evolution. The goal of this volume is to provide descriptions and figures of all of the Sterkfontein fossils and inspire more research on this important collection of hominin fossils.

In 2009, the University of the Witwatersrand held a workshop funded by the Wenner Gren Foundation for Anthropological Research, The Ford Foundation, Palaeontological Scientific Trust (PAST) of South Africa, and the university that brought together experts on early hominin postcranial anatomy to describe and analyze the Sterkfontein fossils, exchange ideas, and foster innovative and internationally collaborative research.

South African students attended the workshop as well, interacting with their more senior colleagues and engaging in stimulating and productive discussions about the Sterkfontein hominins. The participants had a chance to interact with the fossils over several days, describing and comparing the fossils and discussing interpretations and what the fossils tell us about *Australopithecus africanus* and how they inform us about human evolution. Ultimately, these participants contributed chapters to this volume, bringing in other experts as authors as well as their work progressed.

The timing of the workshop helped promote Africa-based paleoanthropology at a crucial time. From a scientific perspective, it is becoming ever more apparent in recent literature and in conferences that variation among hominins is key to understanding the early evolution of bipedality and interpreting how patterns of locomotion relate to the origin of *Homo*, but without more formal work on the Sterkfontein fossils, we are ill-equipped to address differences among *Australopithecus* species.

This volume presents all Sterkfontein hominin postcranial fossils that were available for study when the workshop was convened. Appendix I lists these fossils and what is known about their provenience.

Back row, left to right: Shahed Nalla, Alan Morris, Adam Gordon, J. Michael Plavcan, Matthew Tocheri, William Jungers, Colin Menter, Kristian Carlson, Bernhard Zipfel. Middle row, left to right: Michelle Drapeau, Job Kibii, Roshna Wunderlich, Dipuo Mokokwe, Dominic Stratford, Danielle Vernon, Christine Steininger, David Green, Ronald Clarke. Front row, left to right: Carol Ward, Burt Rosenman, Christopher Ruff, Brian Richmond, Tea Jashashvili, Andrew Gallagher, Brendon Billings, Martin Haeusler.

Since this workshop was held, additional fossils from the Sterkfontein Cave have been recovered and are beginning to be published. The descriptions and analysis presented in this volume should stand as a foundation from which to interpret these and other fossils from Sterkfontein, and from all over Africa, that will be recovered in years to come.

This volume is organized in three parts. The volume opens with a transcript of the opening address for the workshop delivered by the late Professor Dr. Phillip V. Tobias, whose long-standing and indelible contributions to our understanding of the Sterkfontein fossils and to human evolution cannot be overstated. We were all honored to have him open the workshop and this volume. Chapters 1–3 present background and context of the Sterkfontein site and its fossils. J. Francis Thackeray reviews the remarkable history of this important site. Dominic Stratford reviews site formation, and Robyn Pickering and Andrew Herries review the geochronology of the site and temporal context of the fossils. Chapters 4–15 present descriptions and photographs of every postcranial fossil recovered from the site through 1995, along with basic comparative analysis and interpretation of each anatomical region. Chapters 16–17 present more synthetic analyses of anatomy, with consideration of long bone cross-sectional morphology and body proportions. Finally, Chapter 18 synthesizes some of the major conclusions from each chapter to summarize how these analyses have advanced our understanding of the taxonomy, functional anatomy, and paleobiology of the hominins from Sterkfontein.

Many people at the University of the Witwatersrand and beyond supported our efforts in carrying out the workshop. Notable are Belinda Bozzoli, former Deputy Vice Chancellor—Research, Trefor Jenkins, former Acting Director, Institute for Human Evolution (IHE), Bruce Rubidge, Director of the former Bernard Price Institute (BPI) for Palaeontological Research, and Marlize Lombard, former IHE Management Committee. They approved and supported this project without hesitation and allowed us to make use of university resources. Logistical support was kindly given by Rhod McRae-Samuel (Wits Palaeosciences) and Andrea Leenen and Ann Smilkstein from the PAST. Their time and resources in assisting with transport, communication, and catering is very much appreciated. We thank Joe Daley, former Head of the School for Anatomical Sciences at Wits, for graciously hosting us at the school. For access to fossils and comparative skeletal material, we thank the Wits Fossil Access Advisory Committee, Brendon Billings (School of Anatomical Sciences), and Stephanie Potze, Lazerus Kgasi, and Miriam Tawane (Ditsong Museum of Natural History). Roshna Wunderlich spent many hours painstakingly taking photographs of each specimen and made an additional trip to South Africa after the workshop to complete this important task. We are very grateful for her significant contribution. We also thank Mark W. Grabowski for assisting in the editing and checking of fossil descriptions. Our deepest gratitude also extends the Charlie Lockwood family for their support and for providing a grant to the then IHE to support palaeoanthropological activities, including this volume. Finally, we thank all the contributing authors, assistants, and reviewers for their time, enthusiasm, expertise, and patience in completing their respective tasks.

Most importantly, plans for the Workshop on Sterkfontein Postcranial Fossils and for this monograph were initiated by our esteemed colleague Dr. Charlie Lockwood, now deceased. Dr. Lockwood had energized the organization of this meeting and had already made major contributions to our understanding of the paleobiology of *Australopithecus africanus*. As of September 2008, Charles Lockwood was to become the inaugural director of the Institute for Human Evolution at Wits, and he energized this workshop initiative and made it possible. In recognition of his indelible contributions, we are honored to dedicate the Sterkfontein Hominin Postcranial Fossils Workshop and this volume to the memory of Dr. Charles Lockwood.

References

Clarke, R.J., 1988. A new *Australopithecus* cranium from Sterkfontein and its bearing on the ancestry of Paranthropus. Evolutionary History of the "Robust" Australopithecines. Aldine de Gruyter: New York, pp. 285–292.

Clarke, R.J., 1994. On some new interpretations of Sterkfontein stratigraphy. South African Journal of Science 90, 211–214.

Grine, F.E., 2013. The alpha taxonomy of *Australopithecus africanus*. In: Reed, K.E., Fleagle, J.G., Leakey, R.E. (Eds.), The paleobiology of *Australopithecus*. Springer: Dordrecht, pp. 73–104.

Grine, F.E., 2019. The alpha taxonomy of *Australopithecus* at Sterkfontein: the postcranial evidence. Comptes Rendues Paleovol 18, 335–352.

Kimbel, W.H. and White, T.D., 1988. A revised reconstruction of the adult skull of *Australopithecus afarensis*. Journal of Human Evolution 17, 545–550.

Lockwood, C.A., Tobias, P.V., 2002. Morphology and affinities of new hominin cranial remains from Member 4 of the Sterkfontein Formation, Gauteng Province, South Africa. Journal of Human Evolution 42, 389–450.

Thackeray, F., Gommery, D. and Braga, J., 2002. Australopithecine postcrania (Sts 14) from the Sterkfontein Caves, South Africa: the skeleton of' 'Mrs Ples'?: News & Views. South African Journal of science 98, 211–212.

Contributors

Kristian J. Carlson
Department of Integrative Anatomical Sciences
University of Southern California
Los Angeles, CA
USA
Evolutionary Studies Institute
University of the Witwatersrand
Johannesburg
South Africa

Ronald J. Clarke
Evolutionary Studies Institute
School of Anatomical Sciences
University of the Witwatersrand
Johannesburg
South Africa

Jeremy M. DeSilva
Department of Anthropology
Dartmouth College
Hanover, NH
USA
Evolutionary Studies Institute
University of the Witwatersrand
Johannesburg
South Africa

Michelle S. M. Drapeau
Department of Anthropology
University of Montreal
Montreal, Quebec
Canada

Adam D. Gordon
Department of Anthropology
University at Albany
Albany, NY
USA

Mark W. Grabowski
Research Centre in Evolutionary Anthropology and
Palaeoecology
School of Natural Sciences and Psychology
Liverpool John Moores University
Liverpool L3 3AF
United Kingdom
Centre for Ecological and Evolutionary Synthesis
University of Oslo
Norway

David J. Green
Department of Anatomy
Campbell University School of Osteopathic Medicine
South Lillington, NC
USA
Department of Anatomy
Midwestern University
Downers Grove, IL
USA
Evolutionary Studies Institute
University of the Witwatersrand
Johannesburg
South Africa

Martin Haeusler
Institute of Evolutionary Medicine
University of Zurich
Zurich
Switzerland

Andy I. R. Herries
The Australian Archaeomagnetism Laboratory
Department of Archaeology and History
LaTrobe University
Melbourne, VIC
Australia
Centre for Anthropological Research
University of Johannesburg
Johannesburg
South Africa

Ryan W. Higgins
Center for Functional Anatomy and Evolution
Johns Hopkins School of Medicine
Baltimore, MD
USA

Tea Jashashvili
Department of Radiology
University of Southern California
Los Angeles, CA
USA
Department of Geology and Paleontology
Georgian National University
Tblisi
Georgia
Evolutionary Studies Institute
University of the Witwatersrand
Johannesburg
South Africa

William L. Jungers
Department of Anatomical Sciences
Stony Brook University
Stony Brook, NY
USA
Association Vahatra
Antananarivo
Madagascar

Job Kibii
Department of Earth Sciences
National Museums of Kenya
Nairobi
Kenya
Evolutionary Studies Institute
University of the Witwatersrand
Johannesburg
South Africa

Tracy L. Kivell
School of Anthropology and Conservation
University of Kent
Kent
United Kingdom
Evolutionary Studies Institute
University of the Witwatersrand
Johannesburg
South Africa

Michael R. Lague
Department of Biology
The Richard Stockton College of New Jersey
Galloway, NJ
USA

Bruce Latimer
Case Western Reserve University
Cleveland, OH
USA

Colin G. Menter
Department of Biology
University of Florence
Florence
Italy

Shahed Nalla
Faculty of Health Sciences
University of Johannesburg
Johannesburg
South Africa
Evolutionary Studies Institute
University of the Witwatersrand
Johannesburg
South Africa

Kelly R. Ostrofsky
Center for the Advanced Study of Hominid Paleobiology
The George Washington University
Washington, DC
USA

Robyn Pickering
School of Earth Sciences
University of Melbourne
Melbourne, VIC
Australia
Department of Geological Sciences
University of Capetown
Capetown
South Africa

Brian G. Richmond
Product Intelligence at Aura Health
Albany, CA
USA

Burt Rosenman
Department of Biology
Western New England College
Springfield, MA
USA

Christopher B. Ruff
Center for Functional Anatomy and Evolution
Johns Hopkins University School of Medicine
Baltimore, MD
USA

Dominic Stratford
Evolutionary Studies Institute
University of the Witwatersrand
Johannesburg
South Africa

J. Francis Thackeray
Evolutionary Studies Institute
University of the Witwatersrand
Johannesburg
South Africa

Phillip V. Tobias‡
formally at:
School of School of Anatomical Sciences
Johannesburg
South Africa

Matthew W. Tocheri
Department of Anthropology
Lakehead University
Thunder Bay, Ontario
Canada

Human Origins Program, National Museum
of Natural History
Smithsonian Institution
Washington, USA
Australian Research Council Centre of Excellence for
Australian Biodiversity and Heritage
University of Wollongong
Wollongong, NSW
Australia

Carol V. Ward
Department of Pathology and Anatomical Sciences
Department of Anthropology
University of Missouri
Columbia, MO
USA
Evolutionary Studies Institute
University of the Witwatersrand
Johannesburg
South Africa

Roshna Wunderlich
Department of Biological Sciences
James Madison University
Harrisonburg, VA
USA

Bernhard Zipfel
Evolutionary Studies Institute
University of the Witwatersrand
Johannesburg
South Africa

Introductory remarks to the Workshop on Sterkfontein Hominin Postcranial Fossils

Phillip V. Tobias

Opening Remarks to Sterkfontein Hominin Post-cranial Remains Workshop in honor of Charles Lockwood, University of the Witwatersrand, Johannesburg, January 5–9, 2010.

Originally presented to Professor Francis Thackeray, Director of the Institute for Human Evolution, Professor Joe Daly, Head of the School of Anatomical Sciences, The Co-organizers of the Workshop, Colleagues, and Friends from at least three continents.

It is a great pleasure for me to welcome all of you to the School of Anatomical Sciences which you have chosen as the venue for the Workshop. It is of course an obvious choice since the School of Anatomical Sciences houses one of the largest collections of original hominin fossils in the world. The lion's share of these have been recovered from Sterkfontein between 1966 and the present. I shall return to this theme in a moment.

It would be unthinkable for me to continue without dwelling on the man who first proposed that such a workshop be held, Charles Abram Lockwood III. He was due to leave for South Africa to take up, on September 1, 2008, his position as the first director of the new University of the Witwatersrand formation, the Institute for Human Evolution. Six weeks before that, while riding his motor bicycle to work at University College London, he was apparently crushed by a lorry near King's Cross Station—and died on the spot. It was Monday July 14, 2008 ("Bastille Day").

He came to South Africa in 1994 to carry out his doctoral research, with myself as his supervisor. Of 50 or more doctoral and postdoctoral students whom I supervised, Charles stands out as one of the most brilliant, dedicated, and hard-working of all, completing his thesis in close to the minimum time, and obtained the Wits Ph.D. in 1997. He received high plaudits from his examiners, one of whom was Francis Clark Howell. This must have been one of the last such reports that Clark Howell tendered before his own death. Howell wrote of Charles's thesis, "I consider that this fine research accomplishment may well stand at the forefront of a series of contributions focused on ancient Transvaal hominins that will reestablish and reaffirm their extraordinary interest and relevance for a [full] appreciation of humankind's evolutionary diversity, adaptations and history." Lockwood's other examiners were Professor Leslie Aiello and myself.

Charles was one of my most outstanding protégés. He was a fine scholar, a serious and dedicated student. On September 1, 2008, the very date on which he was to take up his new post at Wits, he would have celebrated his 38th birthday—so he was only 37 at the time of his tragic and untimely death.

After completing his Ph.D. here he spent a period as a postdoctoral fellow at the Institute of Human Origins at Arizona State University, working with Bill Kimbel and John Lynch under Don Johanson. There followed a spell at Stony Brook University. Then, when Leslie Aiello left

Phillip V. Tobias, *Introductory remarks to the Workshop on Sterkfontein Postcranial Fossils* In: *Hominin Postcranial Remains from Sterkfontein, South Africa, 1936–1995*. Edited by: Bernhard Zipfel, Brian G. Richmond, and Carol V. Ward, Oxford University Press (2020). © Oxford University Press.
DOI: 10.1093/oso/9780197507667.002.0009

University College London to become president of the Wenner-Gren Foundation in New York City, Charles was appointed to a position in the Anthropology Department she had left. He was doing so remarkably well that only a short time before his lamentable death, he was promoted to the status of Reader at University College London.

In his last letter to me, less than three weeks before he died, he told me with enthusiasm about a workshop on Sterkfontein hominin post-cranial remains, which he was organizing for May or June 2009 as the first such activity of the Institute for Human Evolution. He added, "the final term in UCL recently ended, and I'm starting to prepare for the move." He added, "I'm very much looking forward to getting to Johannesburg in a couple of months."

My last letter to him was on July 9, 2008, five days before his fatal accident. I welcomed his plan for that workshop, suggested seven of my former Ph.D. graduates who might be considered as possible participants in the workshop, and accepted his invitation to open the workshop. He had told me that he had asked two of his American associates, Carol Ward of the University of Missouri and Brian Richmond of George Washington University, to help organize the workshop.

When the IHE and Wits University held a memorial meeting on Charlie's death on August 13, 2008, as part of my address, I stated, "It would be a most fitting tribute to Lockwood's memory if Wits University were to proceed with the holding of the proposed workshop, helped by Ward and Richmond of the USA and by one or two palaeoanthropologists of Wits and neighboring institutions. I foresee a *Charles Lockwood Memorial Meeting* carrying on with his plans for the workshop." Happily this has come to pass, through the conscientious endeavors and time-consuming hard work of Carol Ward and Brian Richmond, joined as a third co-organizer by Bernhard Zipfel.

After the news of Charles's death became known, a shockwave of horror and grief traveled around the world. Letters of sympathy arrived from Professor Nanneke Redclift of University College, London; Fred Grine of Stony Brook; Alan Bilsborough of the University of Durham ("His death is a tragic loss ... particularly to Wits and to early hominid studies and especially the I.H.E. He had done and was doing some outstanding work, was an exemplary teacher and such a pleasant, informal personality, with a good sense of humour and without any pomposity or 'side' to him"); Don Johanson of Arizona State University where Charles had been a star, first as a postdoc. and then as an assistant professor ("This is an especially sad day for Wits. I know that everyone there, especially you, were very pleased that Charlie would be joining the faculty and carrying on the great tradition of paleoanthropology in South Africa.

From my last conversation with Charlie he was thrilled and excited about his move to Wits"); Bernard Wood of George Washington University ("My thoughts are with you. It must be like losing a child. ... At the Nairobi meeting last summer [I] realised just what a very special person [and] scientist he was."). Dr Qian Wang, now of Mercer University, Georgia, who had spent three years with me as a postdoc in Anatomical Sciences, wrote to the American Association of Physical Anthropologists, and proposed that a session of the annual meeting of the AAPA be dedicated to "celebrate Charles's truncated yet very productive and influential life." The vice president of AAPA, Lorena Madrigal, replied to Qian Wang that the Executive Board knew of Charles's passing and was already considering how best to honor him.

Charles had devoted his Ph.D. thesis to the morphology of australopithecine faces, following in the footsteps of Yoel Rak of Tel Aviv.

It is interesting that Charles selected the postcranial remains as the subject for this workshop. Forty years ago, the 1970 Burg Wartenstein Symposium was devoted to the Functional and Evolutionary Biology of Primates. Organized by Russell Tuttle of the University of Chicago, the proceedings appeared as a book of the same name and edited by Tuttle. My chapter in that book included a census of the numbers and anatomical parts of the hominin fossils from each of 14 African sites, 5 in South Africa and 9 in East Africa. Looking over this analysis the other evening, I was interested to note that 138 of the total number of specimens surveyed (1,410) were of postcranial remains. From the South African hominin sites, there were 78 and from the East African sites 60. In a further breakdown of the data I found that 44 postcranial specimens were from Sterkfontein. In preparation for this workshop, Dr. Bernhard Zipfel has recorded over 150 postcranial specimens from Sterkfontein alone. Thus, from 1970 to 2010, the total stockpile of Sterkfontein hominin post-cranial fossils has gone from 44 to 200+. This more than threefold increase is the consequence of a sustained program of systematic excavation of the Sterkfontein Cave deposit that I started in 1966, with Alun R. Hughes, and which continued relentlessly up to the present. In the early 1990s, when Hughes's health failed, the dig at Sterkfontein was continued by Professor Ronald J. Clarke.

These numbers alone justify the need for this workshop. Of course, the minimum numbers of individuals represented by the 200+ postcranial remains are far fewer, as they embrace clusters of carpals, metacarpals, and phalanges, as well as tarsals, metatarsals, and phalanges and partial and virtually complete vertebral columns. Indeed, Sterkfontein is noteworthy and perhaps unique in having yielded at least three partial skeletons of hominins, one of these virtually complete. The taphonomic implications of so high a number of skeletons from this one site, Sterkfontein, are well worthy of exploration.

It is not surprising that with these fair-sized samples of various regional anatomical subsets of postcranial remains, a number of our graduate students have devoted their dissertations and theses to the postcranial bones of australopithecines. These include Daleen Benade on the vertebral column; Lee Berger on the shoulder complex; Colin Menter on the elbow joint; David Ricklan on wrist, hand, and fingers; Ivan Suzman on the hip joint (sadly, not completed); and Peter Christie on the ankle joint.

Nine other postgraduate students devoted their theses to the postcranial elements of proto-historic and recent human populations of South Africa.

All of these postgraduate students and "postdocs" supplemented their fossil and other research materials by studying comparable bones of recent human cadaver-derived skeletons. These are mainly of modern African populations, of which the school has a very large collection. I called it the Raymond Dart Collection of Modern Human Skeletons after Professor Dart, my illustrious predecessor as head of the Department of Anatomy, who started the collection. During your stay here over the next five days, I hope that you will find the time to dip into that collection, which is housed down the corridor in the school.

I shall not be worrying you further during the coming days. I wish you all a most successful workshop and I feel confident that Carol, Brian, and Bernhard will produce a superb volume from the fruits of your collective labors.

Thank you and good luck.
Phillip V. Tobias
January 5, 2010

Section 1

Temporal, geologic, and historical context of the Sterkfontein hominins

1

A summary of the history of exploration at the Sterkfontein Caves in the Cradle of Humankind World Heritage Site

J. Francis Thackeray

The Sterkfontein Caves, situated approximately 70 kilometers southwest of Pretoria, are part of an area declared by UNESCO in 1999 as the Cradle of Humankind World Heritage Site. The caves are important in terms of their relative abundance of Plio-Pleistocene fossils attributed to *Australopithecus*, in addition to rare specimens representing *Paranthropus* and early *Homo,* associated with non-hominin fossils that reflect paleo-environmental changes within the last three million years.

Three historical periods of exploration can be recognized. First (1895–1935), the caves were explored by prospectors interested in the mining of limestone, coinciding with the unsystematic recovery of non-hominin fossils; second (1936–1966), the caves were the subject of scientific research and fieldwork by Robert Broom, his assistant John Robinson, and C. K. (Bob) Brain, all based at the Transvaal Museum within a period when Plio-Pleistocene hominin fossils and artifacts were discovered; and third, a period since 1966 when Professor Phillip Tobias and others from the University of the Witwatersrand (including Alun Hughes, Tim Partridge, Ron Clarke, Kathy Kuman and Dominic Stratford) undertook systematic fieldwork and research and made additional remarkable discoveries of fossils and artifacts under the aegis of the Department of Anatomy, later the School of Anatomical Sciences, and more recently through the Institute for Human Evolution (IHE) which was established in 2007 at the university. In 2013 the IHE was merged with the Bernard Price Institute to form the Evolutionary Studies Institute (ESI).

The early years (1895–1935)

The exploration at Sterkfontein was initially associated with the exploration of caves for limestone which was mined for agricultural purposes as well as for the purification of gold. One of the prospectors was G. Martignalia who used dynamite to open up a "wondergat" in the dolomite at Sterkfontein, *circa* 1895. His son noted, "There are only a few people alive today who saw these caves in the days of their original splendor before they were destroyed by commercial exploitation," as quoted by van Riet Lowe (1947, p. 85). The mining attracted the attention of geologists who noted the presence of fossils in breccia deposits (Draper, 1896).

In 1895 David Draper was appointed as the first secretary of the Geological Society of South Africa, and in the same year he sent samples to the British Museum with a note referring to stalagmites and fossiliferous rock (breccia), including a primate skull and a lion claw (Brain, 1981). Draper's fossiliferous breccia samples were prepared in acetic acid at the British Museum. Oakley (1960) reported the presence of a baboon, a carnivore, an equid, a porcupine, rodents, a lizard and some birds.

J. Francis Thackeray, *A summary of the history of exploration at the Sterkfontein Caves in the Cradle of Humankind World Heritage Site*. In: *Hominin Postcranial Remains from Sterkfontein, South Africa, 1936–1995*. Edited by: Bernhard Zipfel, Brian G. Richmond, and Carol V. Ward, Oxford University Press (2020). © Oxford University Press.
DOI: 10.1093/oso/9780197507667.003.0001

In August 1897, an article appeared in a journal, *English Mechanic and World of Science*, with reference to "some wonderful caves" which had been "discovered recently at a place called Sterkfontein … Limestone had been quarried for some months in a small kopje, and after an explosion after some blasting operations, a cavity of great depth was left." The report referred to the exposure of "magnificent caves. The spectacle was one of great beauty, the light carried by the explorers being reflected from thousands of stalactites," as quoted by Malan (1959, p. 322).

This account, republished in a French journal called *Cosmos*, attracted the attention of a group of Marist brothers who explored the caves and subsequently wrote an article entitled "*Les grottes de Sterkfontein*," also published in *Cosmos*, in 1898. They referred to a fossilized mandible of an antelope, and speculated that it may have been killed by a carnivore.

Draper recognized the need to protect the Sterkfontein Caves from the activities of miners. Through his appeal, the owners of the farm instructed the manager to prevent "wanton destruction" (Malan, 1959, p. 321).

In February 1898, H. Exton (president of the Geological Society) presented an address in which he reported the presence of aragonite in the Sterkfontein cave formations, and suggested that the solution cavities must have been filled with warm water at some time in prehistory. In response to a suggestion by August Prister (1898) that glaciations had occurred in the Gauteng Highveld regions, M. E. Frames expressed interest in the possible relationship between paleoclimatic change and fauna. Frames referred to fossilized remains of horses, antelopes, monkeys, porcupines and bats at Sterkfontein. Without attempting to estimate their geological age, he commented that the geology pointed to some degree of antiquity.

Similar breccia deposits were discovered at Taung in 1924. Following the announcement of the "Taung Child," the type specimen of *Australopithecus africanus* (Dart, 1925), some fossiliferous breccia from Sterkfontein was sent to the University of the Witwatersrand, at a time when mining at Sterkfontein was being managed by George W. Barlow, associated with the Glencairn Lime Company. However, this material did not attract interest since it was considered to be relatively recent (Dart and Craig, 1959).

J. H. S. Gear collected samples of fossils from Sterkfontein, but unfortunately these have not been relocated in the collections at the South Institute for Medical Research, where Gear became Director (Tobias, 1973; Brain, 1981). In 1935, Trevor R. Jones, at that time a student of Professor Raymond Dart at the University of the Witwatersrand, collected a specimen of an extinct baboon (*Parapapio*). Additional fossil primates were collected in 1936 by two other of Dart's students, G. W. H. Schepers and H. le Riche. These fossils were shown to Robert Broom who had been appointed as a paleontologist at the Transvaal Museum in 1934.

The Transvaal Museum years (1936–1966)

The discovery of *Parapapio* at Sterkfontein stimulated Broom to visit the caves in 1936 when he met George Barlow, the mine manager who had previously worked at Taung where fossil baboons had also been discovered. Broom encouraged Barlow to look for anything which resembled the Taung Child. Within a short period, on August 17, 1936, Barlow showed Broom an australopithecine endocranial cast (Sts 60). After more exploration, Broom (1936) found associated parts of the cranium (TM 1511) of the same individual.

Additional fossils were found at Sterkfontein within the next three years, during which time H. B. S. Cooke (1938) prepared a map of the breccia deposits in relation to the dolomite. Broom worked closely with the miners. However, by 1939 the mining was no longer commercially viable, coinciding with the onset of World War II, and no more fossils were discovered until 1947 when the Prime Minister of the Union of South Africa, Jan Smuts, encouraged Broom to continue work at the site. On April 18, 1947, an almost complete cranium of *Australopithecus africanus* (Sts 5, nicknamed "Mrs Ples") was discovered, broken in two pieces as a result of a dynamite explosion (Broom, 1947). Four months later, a partial skeleton (Sts 14) was found close to the locality of "Mrs Ples." Broom continued work at Sterkfontein until his death in 1951. Until his departure for Madison, Wisconsin, John Robinson maintained fieldwork and preparation of fossils. Robinson (1972) published an important book on australopithecine postcrania, with special reference to Sts 14.

In 1956, C. K. (Bob) Brain discovered stone artifacts in the so-called West Pit. Focusing on sedimentology in an attempt to quantify changes in past rainfall, Brain (1958) published a study titled "The Transvaal ape-man-bearing cave deposits."

In 1958, the owners of Sterkfontein (Mr. and Mrs. E. Stegmann) donated an area of 20 morgen to the University of the Witwatersrand, for purposes of conservation and paleontological fieldwork.

The years associated with the University of the Witwatersrand (1966 until present)

In 1966, Professor Phillip Tobias embarked on a long-term systematic excavation of the Sterkfontein deposits (Tobias and Hughes, 1969). He subsequently established the Sterkfontein Research Unit with many collaborators. Alun Hughes supervised fieldwork, and Ian Watt undertook a detailed survey. On the basis of an extensive geological study, Tim Partridge (1978) recognized a sequence of six Members from Sterkfontein. Kathy Kuman recognized stone artifacts from Member 5. She recognized both Oldowan and Acheulian industries. Louis Scott explored the breccia deposits and flowstones for pollen; Marion Bamford studied fossilized wood, including liana (vine) samples; Alan Turner studied carnivores; Vera Eisenmann worked on equids; and Elizabeth Vrba analyzed bovids to asses both chronology and paleoenvironments. It was recognized that Member 4 was associated with periods when tragelaphines (associated with woodland savanna) were more abundant by comparison with fauna from Member 5 associated with relatively high abundances of grassland (alcelaphine) fauna such as wildebeest and hartebeest.

In August 1976, cranial remains of a hominin (StW 53) were discovered from deposits thought to be associated with stone artifacts from Member 5. The hominin fossil was attributed to *Homo habilis* (Hughes and Tobias, 1977). Recently, Ron Clarke has suggested that StW 53 is more likely to be associated with Member 4, and recognizes it as an australopithecine.

Other important discoveries during excavations at Sterkfontein under the direction of Phillip Tobias included an australopithecine cranium (StW 505) and a partial skeleton (StW 431), both from Member 4. Even more remarkable was a complete australopithecine skull and skeleton (StW 573) nicknamed "Little Foot" (Clarke and Tobias, 1995; Clarke, 1998) discovered by Ron Clarke, Stephen Motsumi, and Nkwane Molefe (Figure 1.1) from Member

2 in the Silberberg Grotto. Clarke (2008) believes this specimen to represent a "second species" of *Australopithecus*, distinct from *A. africanus*, but resembling the 3-million-year-old australopithecine fossils from Makapansgat, initially described by Dart as *A. prometheus*.

Provisional attempts to estimate the chronology of cave deposits at Sterkfontein were undertaken using biostratigraphy, based partly on non-hominin primates (Delson, 1988), suids (White and Harris, 1977), bovids (Vrba 1975), and carnivores (Turner, 1997), compared to fauna from radio-metric deposits in East Africa. Cosmogenic, uranium-lead (U-Pb) (Walker et al., 2006), and paleomagnetic dates (Herries and Shaw, 2011), together with faunal data, have been used in efforts to obtain absolute dates of the Sterkfontein cave deposits, ranging from 3.7 million years for Member 2, to about 2 million years for the top of Member 4, with Member 5 at about 1.8 million years.

The age of "Little Foot" from Member 2 is especially challenging, having been dated initially at 3.3 million years on the basis of paleomagnetic and faunal data (Partridge et al., 1999). Cosmogenic dates of about 4 million years (Partridge et al., 2003) were considered to be overestimates. More recently, using the same technique but with a careful study of more than ten samples, Granger et al. (2015) secured a date of 3.67 million years. This was challenged by Kramers and Dirks (2016, 2017) who contended that a younger date should be considered. A rebuttal was published by Stratford et al. (2017). U-Pb dates obtained from flowstones (Walker et al., 2006; Pickering

Figure 1.1 Professor Ronald J. Clarke, Stephen Motsumi, and Nkwane Molefe after the initial discovery of "Little Foot," attributed to *Australopithecus prometheus*. Credit: University of the Witwatersrand.

and Kramers, 2010), at *circa* 2 million years before present, are considered to be too young for the "Little Foot" skeleton itself. They may well be correct for the flowstones adjacent the skeleton, but these flowstones (as infillings) clearly postdate the fossil.

Ron Clarke and his assistants worked meticulously on the StW 573 skeleton which first saw light of day in 2014 when the blocks in which it had been preserved were lifted out of the Silberberg Grotto. Cleaning of the fossil using airscribes continued until 2017 when the skeleton was proudly announced to the world by Clarke in the Phillip Tobias Vault at the ESI. "Little Foot" continues to be the subject of detailed study, including the use of CT scanning. Clarke (2019) published a summary of the historical events relating to the discovery, excavation, and recent research of his extraordinary discovery.

Fieldwork at Sterkfontein is currently directed by Dominic Stratford and his team. One of their projects relates to the spatial and temporal distribution of artifacts and fossils, including hominins. New discoveries can be expected in the decades to come, with greater precision regarding chronology and context, in contrast to the circumstances under which Robert Broom worked almost a century ago.

Conclusion

Having been explored for more than a century, Sterkfontein is one of the most important cave sites in the Cradle of Humankind, recognized by UNESCO as a World Heritage Site. It has yielded more than 700 hominin specimens and a large quantity of non-hominin fossils, including primates, bovids, suids, carnivores, rodents, and insectivores as well as pollen and wood. For historical reasons, part of the collections is curated by the Transvaal Museum (now called the Ditsong National Museum of Natural History in Pretoria), and another part is curated by the University of the Witwatersrand in Johannesburg. Through the efforts of Robert Broom, John Robinson, Bob Brain, Elizabeth Vrba, Phillip Tobias, Alun Hughes, Tim Partridge, Ron Clarke, Dominic Stratford, and many others, the Sterkfontein cave deposits have provided a wealth of paleoanthropological and paleoenvironmental data which continue to be explored in an effort to understand human evolution in the context of Plio-Pleistocene changes in climate, habitats, and faunal species diversity.

Acknowledgments

I thank Brian Richmond, Carol Ward, and Bernhard Zipfel for the opportunity to contribute this historical summary for the Sterkfontein workshop held in honor of the late Charlie Lockwood whom I greatly admired. I wish to express my appreciation to John Robinson for taking me to Sterkfontein in the early 1960s, when as a young schoolboy I was allowed to collect fossils for the Transvaal Museum from breccia dumps, thereby stimulating my interest in paleontology; to Bob Brain and Elizabeth Vrba for the opportunity to prepare Sterkfontein fossils at the Transvaal Museum when I was employed as a laboratory assistant at the Transvaal Museum in 1971; and to Professor Phillip Tobias for his enthusiastic encouragement when I was embarking on a paleontological career. I wish to thank the Lockwood family, the National Research Foundation, the Wenner Gren Foundation, the Ford Foundation, the Andrew Mellon Foundation, and the French Embassy in South Africa for supporting the Institute for Human Evolution of which I was Director from 2009 to 2013, prior to its merger with the Bernard Price Institute to establish the Evolutionary Studies Institute at the University of the Witwatersrand.

References

Brain, C.K., 1981. The Hunters or the Hunted: an Introduction to African Cave Taphonomy. University of Chicago Press: Chicago.

Broom, R., 1936. A new fossil anthropoid skull from South Africa. Nature 138, 486–488.

Broom, R., 1947. Discovery of a new skull of the South African ape-man, *Plesianthropus*. Nature 159, 1–365.

Clarke, R.J., 1998. First ever discovery of a well-preserved skull and associated skeleton of an *Australopithecus*. South African Journal of Science 94, 460–463.

Clarke, R.J., 2008. Latest information on Sterkfontein's *Australopithecus* skeleton and a new look at *Australopithecus*. South African Journal of Science 104, 443–449.

Clarke, R.J., Tobias, P.V., 1995. Sterkfontein Member 2 foot bones of the oldest South African hominin. Science 269, 521–524.

Clarke, R.J., 2019. Excavation, reconstruction and taphonomy of the StW 573 *Australopithecus prometheus* skeleton from Sterkfontein Caves, South Africa. Journal of human evolution 127, 41–53.

Cooke, H.B.S., 1938. The Sterkfontein bone breccia: a geological note. South African Journal of Science 35, 204–208.

Dart, R.A., 1925. *Australopithecus africanus*: the man-ape of South Africa. Nature 115, 195–199.

Dart, R.A., Craig, D., 1959. Adventures with the Missing Link. Harper: New York.

Delson, E. 1988. Chronology of South African australopith site units. In: Grine, F. (Ed.), Evolutionary History of the "Robust" Australopithecines. Aldine de Gruyter: New York, pp. 317–324.

Draper, D., 1896. Report of meeting, 8 April, 1895. Geological Society of South Africa. Transactions of the Geological Survey of South Africa 1, 11.

Frames, M.E., 1898. Remarks on Prof. Prister's paper on glacial phenomena at Pretoria and on the rand. Transactions of the Geological Society of South Africa 3, 91–95.

Granger, D.E., Gibbon, R.J., Kuman, K., Clarke, R.J., Bruxelles, L., Caffee, M.W., 2015. New cosmogenic burial ages for Sterkfontein Member 2 *Australopithecus* and Member 5 Oldowan. Nature 522, 85–88.

Herries, A.I.R., Shaw, J. 2011. Paleomagnetic analysis of the Sterkfontein paleocave deposits; age implications for the hominin fossils and stone tool industries. Journal of Human Evolution 60, 523–539.

Hughes, A., Tobias, P.V., 1977. A fossil skull probably of the genus *Homo* from Sterkfontein, Transvaal. Nature 265, 310–312.

Kramers, J.D., Dirks, H.G.M., 2016. The age of fossil StW 573 ("Little foot"): an alternative interpretation of 26Al/10Be burial data. South African Journal of Science 113, 1–8.

Kramers, J.D., Dirks, H.G.M., 2017. The age of fossil StW 573 ("Little Foot"): reply to comments by Stratford et al. (2017). South African Journal of Science 113, 1–3.

Malan, B.D., 1959. Early references to the Sterkfontein caves. South African Journal of Science 55, 321–324.

Oakley, K.P., 1960. The history of Sterkfontein with a comment by B.D. Malan. South African Journal of Science 56, 110.

Partridge, T.C., 1978. Re-appraisal of lithostratigraphy Sterkfontein hominin site. Nature 275, 282–287.

Partridge, T.C., Granger, D.E., Caffee, M.W., Clarke, R.J., 2003. Lower Pliocene hominin remains from Sterkfontein. Science 300, 607–612.

Partridge, T.C., Shaw, J., Heslop, D., Clarke, R.J., 1999. The new hominin skeleton from Sterkfontein, South Africa: age and preliminary assessment. Journal of Quaternary Science 14, 293–298.

Pickering, R., Kramers, J.D., 2010. Re-appraisal of the stratigraphy and determinations of new U-Pb dates for the Sterkfontein hominin site, South Africa. Journal of Human Evolution 59, 70–86.

Robinson, J.T., 1972. Early Hominin Posture and Locomotion. University of Chicago Press: Chicago.

Stratford D., Granger, D.L., Bruxelles, L., Clarke, R.J., Kuman, K., Gibbon, R.J., 2017. Comments on "The age of fossil StW 573 ('Little Foot'): an alternative interpretation of 26Al/10Be burial data." South African Journal of Science 113, 1–3.

Tobias, P.V., 1973. A new chapter in the history of the Sterkfontein early hominin site. Journal of the South African Biological Society 14, 30–44.

Tobias, P.V., Hughes, A.R., 1969. The new Witwatersrand University excavation at Sterkfontein. South African Archaeological Bulletin 24, 158–169.

Turner, A., 1997. Further remains of Carnivora (Mammalia) from the Sterkfontein hominin site. Paleontologia africana 34, 115–126.

Van Riet Lowe, C., 1947. Die ontdekking van die Sterkfontein grotte. South African Journal of Science 1, 85–86.

Vrba, E.S., 1975. Some evidence of chronology and paleoecology of Sterkfontein, Swartkrans and Kromdraai from the fossil Bovidae. Nature 254, 301–304.

Walker, J., Cliff, R.A., Latham, A.G., 2006. U-Pb isotopic age of the StW 573 hominin from Sterkfontein, South Africa. Science 314, 1592–1594.

White, T.D., Harris, J.M., 1977. Suid evolution and correlation of African hominin localities. Science 198, 13–21.

2

The geological setting, cave formation, and stratigraphy of the fossil-bearing deposits at Sterkfontein Caves

Dominic Stratford

The importance of understanding the spatial and stratigraphic context of the caves and their interred huge faunal assemblages has not gone un-noticed, as is evident at Sterkfontein from stratigraphic interpretations spanning over 76 years (e.g., Cooke, 1938; Wilkinson, 1983; Partridge & Watt, 1991; Pickering & Kramers, 2010; Herries & Shaw, 2011; Stratford et al., 2014; Bruxelles et al., 2019). Indeed, stratigraphic research was one of the research priorities when Phillip Tobias started directing excavations in 1966 (Tobias & Hughes, 1969). In recent years, under the direction of Professor Clarke, and since the discovery of the StW 573 "Little Foot" skeleton, the need for more focused stratigraphic work has intensified, leading to several new stratigraphically focused works (Pickering & Kramers, 2010; Herries & Shaw 2011; Bruxelles et al., 2014; Stratford et al., 2014; Bruxelles et al., 2019). The impetus for a renewed dedication to stratigraphic research has grown from continuing controversy over a basic chronological framework for the deposits and specific hominin specimens, like StW 573 (e.g. Granger et al., 2015; Kramers & Dirks, 2017a,b; Stratford et al., 2017; Bruxelles et al., 2019). The lack of contextual and chronological control has been a major challenge regarding palaeoanthropological work at Sterkfontein. As the richest repository of *Australopithecus* fossils in the world, improvements in our understanding of context and chronology are essential if South African hominin evolutionary histories are going to be understood. This chapter summarizes the geological and geomorphological context of the Sterkfontein karst system, and gives a brief description of the current stratigraphic interpretation of the Sterkfontein deposits.

Sterkfontein geological and geomorphological context

The Sterkfontein cave system has formed within a stromatolitic dolomitic limestone (the Malmani Subgroup) deposited during the Late Archaean (2.5–2.6 billion years ago) (Eriksson et al., 2001). At this time, an inland sea occupied the Transvaal Basin, depositing the limestone and chert beds into five formations (Eriksson et al., 1993). Uplift of the Johannesburg Dome to the southeast and intracratonic sag to the north (Eriksson et al., 2001) of the Cradle of Humankind area resulted in the dipping of the dolomites to the north to northwest and the development of abundant vertical faulting. The planation of the landscape resulted in the subaerial exposure of all five dolomitic members in the cradle. The Sterkfontein Caves occupy the boundary between the two basal formations, the Oaktree and Monte Christo (Figure 2.1), which are differentiated based on relative chert bed abundance and thickness (Eriksson et al., 2006). The Oaktree Formation is poor in chert beds whereas the overlying Monte Christo Formation is relatively rich in chert beds. Dominant faults are orientated roughly east-west (a result of uplift to the south) and subordinate faults are orientated roughly north-south. The activity of the faults and compound fractures has led to a complex history of speleogenesis, allogenic infilling, and rekarstification.

There have been a number of different speleogenetic models proposed for the Sterkfontein Caves (e.g., Partridge, 1973; Wilkinson, 1973, 1983; Partridge and Watt, 1991; Martini et al., 2003; Klimchouk, 2007; Bruxelles

Dominic Stratford, *The geological setting, cave formation, and stratigraphy of the fossil-bearing deposits at Sterkfontein Caves*. In: *Hominin Postcranial Remains from Sterkfontein, South Africa, 1936–1995*. Edited by: Bernhard Zipfel, Brian G. Richmond, and Carol V. Ward, Oxford University Press (2020). © Oxford University Press.
DOI: 10.1093/oso/9780197507667.003.0002

Geology

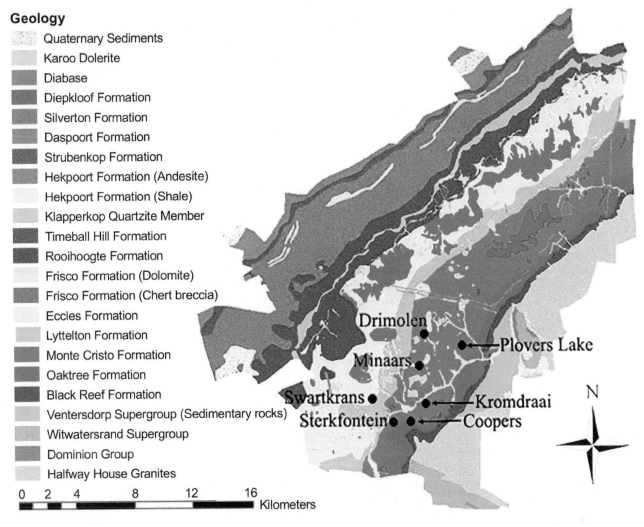

- Quaternary Sediments
- Karoo Dolerite
- Diabase
- Diepkloof Formation
- Silverton Formation
- Daspoort Formation
- Strubenkop Formation
- Hekpoort Formation (Andesite)
- Hekpoort Formation (Shale)
- Klapperkop Quartzite Member
- Timeball Hill Formation
- Rooihoogte Formation
- Frisco Formation (Dolomite)
- Frisco Formation (Chert breccia)
- Eccles Formation
- Lyttelton Formation
- Monte Cristo Formation
- Oaktree Formation
- Black Reef Formation
- Ventersdorp Supergroup (Sedimentary rocks)
- Witwatersrand Supergroup
- Dominion Group
- Halfway House Granites

0   2   4   8   12   16
Kilometers

Figure 2.1   Geological map of the Cradle of Humankind (adjusted from Obbes, 2000).

et al., 2009; Dirks and Berger, 2013). Each model has specific implications on the timing of the opening of the caves to the landscape surface (i.e., the maximum age of the deposits) and the order in which deposition occurred underground (i.e., where the oldest sediment in the caves will be found). While the specific model is still to be determined with certainty, in the most basic terms, primary formation of the chambers and passages occurred under phreatic conditions (below the groundwater level) where dissolution was focused along vertical faults and unconformities in the dolomite. The influence of a near-static groundwater, low meteoric recharge (Martini et al., 2003), and fault distribution led to the formation of a complex network of tall, narrow, subparallel, and perpendicular passages. Major passages bear a general east-west trend relating to the dominant fault regime but show no hierarchical passage development relating to water flow direction. Larger passages and chambers formed through a combination of phreatic dissolution in areas where faults

are particularly dense or intersect (Wilkinson, 1983), as well as vadose (above the groundwater level) collapse (Osborne, 2002) (e.g., Elephant Chamber; Figures 2.2, 2.3). A good example of this is the Name Chamber, Silberberg Grotto, and Milner Hall chamber complex near the center of the system where a particularly complex system of laterally and vertically articulated chambers and passages formed as a result of the close proximity of six faults (Figure 2.2; Wilkinson, 1983; Stratford et al., 2014).

The vadose cave network can be described as a two-level system comprising a lower, intact subterranean network of chambers and passages and an upper level, represented by a large, deroofed chamber exposed on the landscape surface, named the Fossil Cavern (Robinson, 1962). The same combination of passage and chamber forming factors resulted in the development of the 60m long, 25m wide east-west oriented Fossil Cavern whose north-south extent has been controlled by the major east-west trending faults. The exposed cave infills in the

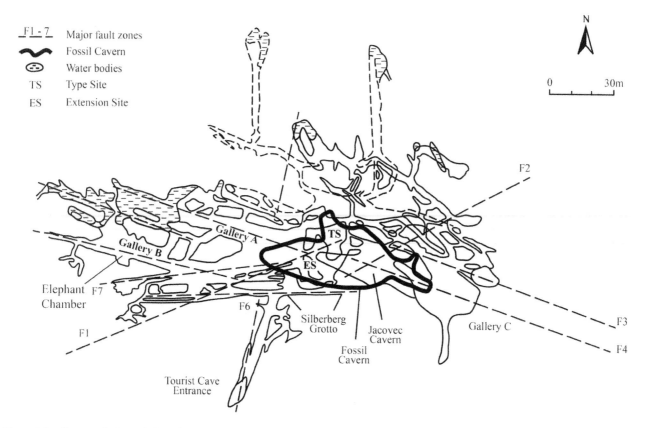

Figure 2.2 Geomorphological plan of the Sterkfontein Cave system. Major controlling faults are shown in relation to the upper level (Fossil Cavern; bold line) and the lower level (subterranean area; thin lines). Adjusted from Wilkinson (1983), with permission from Elsevier.

Fossil Cavern were the focus of extensive excavations carried out by Alun Hughes (Tobias and Hughes, 1969).

The opening of the caves to the landscape

The change of the cave to a vadose system and the associated removal of the buoyant support caused internal collapse throughout the network (e.g., Figure 2.3). This too was controlled by existing vertical faults in the dolomite, facilitating both an increase in passage height and an enlargement of chambers where fault systems and passage density are high. In some cases, this breakdown led to the development of openings to the surface, thereby allowing externally derived (allogenic) sediment to enter the caves. The timing of the opening of the caves in relation to the groundwater level has been debated (Partridge, 1978; Wilkinson, 1983, 1985; Partridge & Watt, 1991; Pickering & Kramers, 2010; Stratford et al., 2014; Bruxelles et al., 2019) and has significant implications for the maximum age of the Sterkfontein deposits (Wilkinson, 1973).

Most openings at Sterkfontein are represented by apertures high in the roof of passages or chambers. Where landscape erosion has intersected with the uppermost

passages, or the roof has collapsed due to surface erosion, long deep gullies are opened. Many openings are very steep sided and represent a serious natural trap danger to animals, and thus "death trap" assemblages represent a common bone accumulation agent (Pickering, 1999; Kibii, 2004; Pickering et al., 2004a).

At Sterkfontein, vertical shafts have developed between upper and lower galleries (e.g., the Name Chamber), where a 12m vertical shaft articulates the western end of the overlying Fossil Cavern with the deeper Name Chamber (Clarke, 1994; Stratford et al., 2012). Connections also exist between the landscape surface and lower gallery—depositing sediments directly into the deepest areas of the caves (e.g., the Milner Hall). There are also examples of single chambers articulated to other chambers and the landscape through multiple conduits—creating coeval infilling processes from multiple sources (e.g., Milner Hall and the Jacovec Cavern). The chambers located on the southern boundary of the cave network (Silberberg Grotto, Fossil Cavern, Milner Hall, and Jacovec Cavern) (Figure 2.2), have particularly complex infilling histories directed by dynamic activity of openings near areas of high fault density (Wilkinson, 1973).

Figure 2.3 Sterkfontein Elephant Chamber. Note the sloping dolomite and interbedded chert (A) and evidence of the vadose collapse on the cave floor (B). View looking west along "Gallery B" in Figure 2.2. Railing in background is 1 m high and roof is 13 m high (Photo: D. Stratford).

Figure 2.4 An image showing localized erosion of Member 4 stony gray breccia (upper right and lower left) that formed deep, vertical pockets subsequently filled with artefact-bearing red, matrix-supported sediments of Member 5 (center). When the pockets were excavated by Alun Hughes in the early 1970s, were filled with a mixture of decalcified Member 5 (Oldowan-bearing) and modern sediments. Total station box for scale (Photograph by L. Bruxelles with permission).

More recent openings, or reactivation of previous openings and faults, have caused localized erosion and rekarstification of more ancient breccias and led to collapses of undercut deposits, inverted stratigraphic sequences where new sediments are washed into cavities, and vertical deposit contacts. An excellent example of this is the western end of the surface exposed deposits of the Fossil Cavern where artefact-bearing Member 5 has filled into deep, vertical (east-west oriented) cavities in *Australopithecus*-bearing Member 4 (Figure 2.4). Opening shape (White, 2007), receptacle topography (Kidwell et al., 1986), and infilling sediment properties (Bertran et al., 1997; Bertran & Texier, 1999) all play dynamic but identifiable roles in the morphological development of the deposit and the accumulation, distribution, and preservation of archaeological and palaeontological material. These patterns are, however, greatly complicated by multigenerational diagenetic processes acting over long periods of time.

Most of the allogenic deposits in Sterkfontein have developed through colluvial sediment accumulation, forming characteristic talus cones (Martini et al., 2003) with varying contributions of flowing water. The formation of talus slope deposits (referred to as sediment gravity flows) involve intricate processes (e.g., Lowe, 1982; Postma, 1986; Major, 2000). These are complicated more so in karst

environments where chemical and physical diagenesis is temporally and spatially highly variable (Karkanas et al., 2000; Goldberg and Sherwood, 2006).

Current stratigraphic interpretations

The complexities of Sterkfontein stratigraphy are aptly demonstrated by the number of interpretations and refinements (e.g., Cooke, 1938; Brain, 1958; Wilkinson, 1973; Partridge, 1978; Clarke, 1994; Pickering and Kramers, 2010; Herries & Shaw, 2011; Bruxelles et al., 2014; Stratford et al., 2014; Bruxelles et al., 2019; Table 2.1). Most works have focused on providing a macro-scale stratigraphic interpretation in an attempt to describe the entire site in the simplest chronological sequence possible (e.g., Partridge, 1978; Wilkinson, 1983; Partridge and Watt, 1991; Clarke, 2006; Pickering and Kramers, 2010) and some (e.g., Clarke, 1994a; Kuman and Clarke, 2000; Stratford et al., 2014; Bruxelles et al., 2014, 2019) have focused on specific deposits and areas. Generally, interpretations have proposed progressively more complex scenarios as researchers have tried to account for the dynamic nature of karst cave depositional processes. Macro-scale interpretations of the stratigraphy have tended to simplify the complex depositional processes acting within the caves.

The exception to this trend in increasingly intricate interpretations is the site-scale stratigraphic interpretation by Pickering and Kramers (2010), who attempt to simplify the member system through the re-examination of the Partridge and Watt (1991) sediment cores, combined with uranium-series dating of speleothems identified as bounding sedimentary units.

All recent works adopt the member system to classify bodies of sediments, a system applied to Sterkfontein by Partridge (1978). The established system of Partridge and Watt (1991) is described here for two reasons. First, because this book is focused on the hominin fossils, most readers will be familiar with the established member system nomenclature. Departure from that system here is likely to confuse more than clarify interpretations of hominin provenience. Second, debate is ongoing over many aspects of the deposits including extent and geometry, sediment flow direction, opening position, and deposit contact presence and morphology. This brief overview is not the stage for a detailed discussion of these debates but recently proposed adjustments are presented where pertinent. As dedicated stratigraphic investigations continue to challenge the Partridge and Watt (1991) model, however, it will need to be refined, adjusted, and reassessed.

The Sterkfontein deposits (incorporated within the "Sterkfontein Formation; SACS (1980)") are separated into six members, Member 1 to Member 6 (M1–M6) (Figure 2.5). The members are numbered in proposed chronological order of infilling. Members 2 and 3 were described from their exposures underground in the Silberberg Grotto, while Members 4, 5, and 6 are exposed on the landscape surface and are contained in the deroofed Fossil Cavern (Partridge, 1978). It should be noted that several fossiliferous deposits found at Sterkfontein are not incorporated into the member system. The sediments in the Jacovec Cavern are a good example of these more remote infillings of potentially significant age with as yet unknown stratigraphic correlations to the centrally located members (Partridge et al., 2003). Other remote fossiliferous deposits can be found in the extreme west and east of the site and in the adjacent Lincoln Cave.

Member 1 (M1) represents a mass of sediments accumulated prior to the opening of the caves to the surface, during early breakdown of the caves in the vadose period. These autogenic sediments are comprised of "occasionally voluminous blocks set in a dark brown manganiferous matrix" (Martini et al., 2003, p. 57).

Member 2 (M2) represents one of the earliest known fossil-bearing deposits to have accumulated in the caves. M2 sediments were originally described from exposures in the Silberberg Grotto (Partridge, 1978) and then further identified in sediment cores examined by Partridge and Watt (1991). The M2 sediments within the Silberberg Grotto contain two facies (Clarke, 2006; Bruxelles et al., 2014) and represent the proximal portion of a gradually accumulated extensive talus deposit that spread into the Milner Hall through multiple entrances (Stratford et al., 2014; Bruxelles et al., 2019). Famously, M2 has yielded the most complete single *Australopithecus* skeleton yet found (StW 573) (Clarke, 1998; Clarke, 2019). This discovery has

Table 2.1 **Major stratigraphic works conducted at Sterkfontein.**

Author	Year	Interpretations
Cooke	1938	Recognized the deposits as a single breccia body exposed at the surface and "lower cave."
Brain	1958	Recognized the deposits as a single breccia exposed in two underground areas.
Robinson	1962	Divided the breccias into "lower," "middle," and "upper." Underground sediments were a single breccia except for a collapse of "middle" breccia into the Name Chamber.
Wilkinson	1973	Proposed a "deep phreatic" karstification model from the continuous exposures of breccias through all depths of the caves.
Partridge	1978	Described the 6 Members of the "Sterkfontein Formation" from breccia exposures in the Silberberg Grotto and surface excavations. Silberberg Grotto is considered the base of the sequence.
Wilkinson	1983, 1985	Reinterpretation of the full depth of deposits in relation to the formation of the caves. Proposes the deepest deposits may be the oldest.
Partridge and Watt	1991	Supports Partridge's initial interpretation that the Silberberg Grotto contains the oldest sediments from descriptions of sediment cores.
Clarke	1994	Proposed three phases of formation in the M5 deposit and clarifies Robinson's (1962) hypothesis of collapsed M5 sediments in the Name Chamber.
Kuman and Clarke	2000	Refined M4 and M5 stratigraphy and proposed distribution of hominin specimens in relation to the M4/M5 boundary.
Clarke	2006	Proposed the erosion and collapse of the StW 573 torso and subsequent intrusion of speleothem in the resultant void.
Reynolds et al.	2007	Proposed the erosion and movement of fossils and artifacts from the western Member 5 areas through an articulating tunnel into the Lincoln Cave.

continued

Table 2.1 Continued

Author	Year	Interpretations
Ogola	2009	Proposed the continuation of M4 below the Member 5 west.
Pickering and Kramers	2010	Proposed a re-assignment of Member 3 as distal Member 4; suggests deepest deposits in the Milner Hall and Jacovec are younger than M2 in the Silberberg.
Herries and Shaw	2011	Generally proposed younger dates for the deposits and an intermediate age for the StW 53 infill between M4 and M5.
Stratford et al.	2012	Suggested a rapid and gradual re-working of the Oldowan-bearing M5 sediments into the Name Chamber and Milner Hall.
Stratford et al.	2014	Suggested M2 originally accumulated down to close to the current base level suggesting the current depth of the caves were formed when the caves opened to the surface.
Bruxelles et al.	2014	Proposed a refined sequence of depositional and erosional processes active in the taphonomy of StW 573 and identified instrusive flowstone characters in speleothems around the skeleton.

Author	Year	Major data resources
Cooke	1938	Faunal representation; sediment description
Brain	1958	Faunal representation; sediment description
Robinson	1962	Faunal and stone tool representation; sediment description
Wilkinson	1973	Cave geomorphology; deposit distribution
Partridge	1978	Sedimentological analysis of exposed breccias
Wilkinson	1983, 1985	Cave geomorphology; deposit distribution
Partridge and Watt	1991	Sedimentological analysis of sediment core samples
Clarke	1994	Sediment description and stone tool, hominin representation
Kuman and Clarke	2000	Hominin and stone tool representation and spatial distribution, sediment description
Clarke	2006	Faunal representation; taphonomy; deposit and sediment feature description
Reynolds et al.	2007	Artifact, hominin, faunal representation; U-Pb speleothem dates
Ogola	2009	Artifact, hominin and faunal representation; taphonomy
Pickering and Kramers	2010	U-Pb seriation of speleothem, longitudinal facies identification from exposed breccias and Partidge's 1989 sediment cores
Herries and Shaw	2011	Palaeomagnetic seriation of siltstones and associated speleothems; faunal and artifact representation and ESR dates
Stratford et al.	2012	Deposit and sediment description, macro- and micro-fauna and artifact a representation
Stratford et al.	2014	Sedimentological and geochemical analysis; artifact and fauna representation; taphonomy; spatial analysis of deposits
Bruxelles et al.	2014	Micromorphology, geochemistry, sedimentology

prompted a great deal of research into the stratigraphy and age of this deposit, the latest accounts of which can be found in Bruxelles et al., 2019. The StW 573 specimen represents the only hominin found within the M2 deposit and Clarke suggests the extraordinary preservation of the entire skeleton is due to the hominin's entrance into the cave close to the end of the M2 deposition (Clarke, 2006; Bruxelles et al., 2019; Clarke, 2019). Widespread post-depositional erosion of the calcified deposit caused the formation of extensive voids around and through the body of the skeleton, which were later filled with flowstones and associated siltstones (Bruxelles et al., 2014, 2019; Granger et al., 2015) making dating the specimen problematic (e.g., Partridge et al., 1999, 2003; Berger et al., 2002; Clarke, 2002; Walker et al., 2006, Pickering and Kramers, 2010; Pickering et al., 2011; Herries and Shaw, 2011; Granger et al., 2015; Kramers & Dirks, 2017a,b; Stratford et al., 2017; Bruxelles et al., 2019).

A covering flowstone (unit 3A in Partridge, 1978) subsequently formed on top of the M2 deposit, before M3 was deposited directly on top of the stalagmite body (Clarke, 2002; Martini et al., 2003). Mining of the

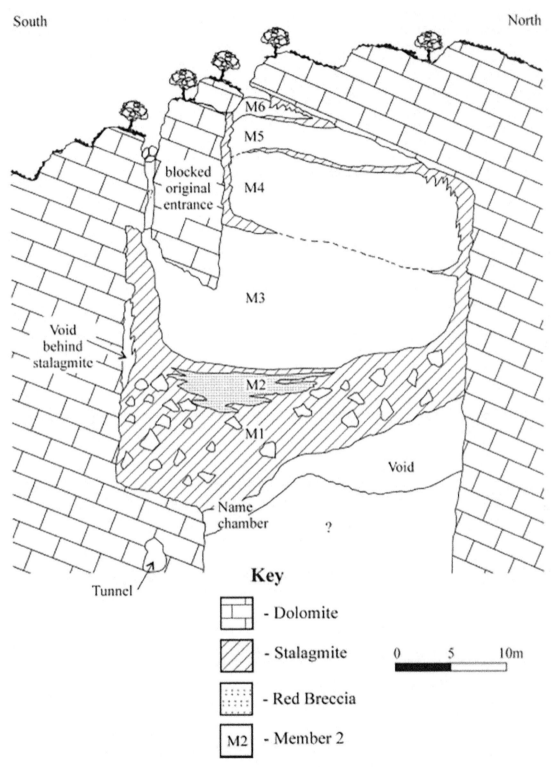

South

North

M6
M5
M4
M3

blocked
original
entrance

Void
behind
stalagmite

M2

M1

Void

Name
chamber

?

Tunnel

Key

☐ - Dolomite

▨ - Stalagmite

⋮ - Red Breccia

M2 - Member 2

0   5   10m

Figure 2.5 Composite stratigraphic column of the main sedimentary members described initially by Partridge (1978). Various adjustments have been made since the Partridge work but the major sediment body nomenclature remains the same (Clarke, 2006).

large 'boss' stalagmite that formed on top of Member 1 in the central eastern area of the Silberberg Grotto produced a large amount of fossiliferous rubble on the floor of the Silberberg Grotto. This rubble provided the first faunal evidence from the Silberberg Grotto M2 sediments (Broom, 1945; Tobias, 1979) and may also have included a component of M3 sediments (Pickering et al., 2004a).

Member 3 (M3) has been described as a colluvial talus deposit with potential inclusion of an aeolian component (Partridge, 1978). Partridge describes the deposit as approximately eight metres thick and underlying the eastern area of Member 4 (Partridge, 1978; Martini et al., 2003). The association of M3 and M4 (the oldest of the surface exposed deposits) is subject to some debate (Berger et al., 2002; Clarke, 2002, 2006; Pickering and Kramers, 2010; Bruxelles et al., 2019). Pickering and Kramers (2010) used deposit separating flowstones and longitudinal sediment facies identification in their reanalysis of Partridge's 1989 sediment core samples to propose that Member 3 represents a distal portion of the Member 4 deposit and not a distinct deposit as proposed by Partridge (1978) and then Partridge and Watt (1991). As yet, Member 3 has not been systematically sampled, and its associations with the other members and its sedimentological properties will remain unclear until a representative sampling of the sediments and fauna is carried out.

Member 4 (M4) is the largest of the currently sampled Sterkfontein members and is exposed across the majority of the surface excavation. The northern, western, and lateral boundaries of M4 are still unclear, and work by Ogola (2009) suggests that M4 may stretch west, across and underneath Member 5 (M5) (Figure 2.6). Given the size of the M4 deposit, it is not surprising that the deposit may have accumulated over 600,000 years (Pickering and Kramers, 2010; Herries and Shaw, 2011) and that a variety of taphonomic accumulation agents is present (Kibii, 2004; Pickering et al., 2004b). It is also not surprising that a great deal of variation is seen in the faunal assemblages and hominin morphology within the M4 deposit if it does indeed represent landscape, ecological, and hominin evolution over a period exceeding half a million years.

M4 represents the main repository of *Australopithecus*, and has yielded hundreds of hominin bones (MNI of 87; Pickering et al., 2004b) and a possible second species of *Australopithecus* (Clarke, 2008; Clarke, 2013; Clarke & Kuman, 2019). The deposit also contains the only fossil wood yet discovered in the Cradle of Humankind, providing a rare palaeoenvironmental perspective (Bamford, 1999). Although four 'beds' (3 clastic and 1 speleothemic) were identified within M4 (Partridge, 1978), the central, decalcified area of the was excavated as a single unit and so the extent and associations of Partridge's beds

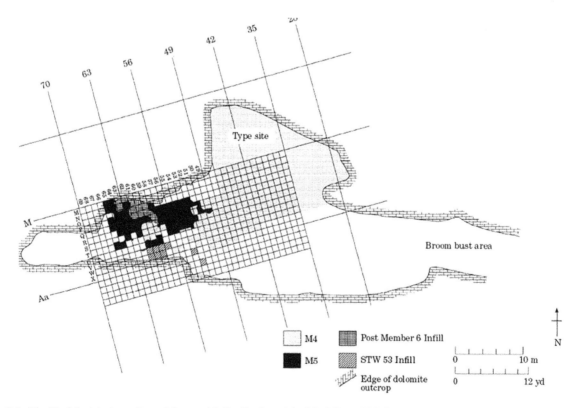

Figure 2.6 The Sterkfontein type site and the spatial distribution of the M4, M5, and M6 breccias. From Kuman and Clarke (2000).

are unclear, making inferences of assemblage associations based on faunal and hominin evidence problematic. This poor stratigraphic resolution also means the infilling dynamics and internal morphology of the deposit are also unclear (Partridge and Watt, 1991; Wilkinson, 1983; Pickering and Kramers, 2010). A number of collapse events may have also evidently affected parts of the M4 deposit (Robinson, 1962; Partridge and Watt, 1991).

In the southern area of the extension site close to the preserved boundary between M4 and M5 is the breccia known as the StW 53 infill (Figure 2.6). This small area of the site has been heavily affected by solution cavities but yielded the StW 53 cranium (Hughes and Tobias, 1977). Originally, this area was associated with M5 and the stone tool assemblages nearby (Partridge, 1978), but Kuman and Clarke's (2000) refined stratigraphic work in the area of Member 5 demonstrates that the specimen was yielded from an irregular remnant of M4 cemented to the southern wall of the cave with M5 forming around it. Herries and Shaw (2011), who date the area to an intermediate period between M4 and M5, also support its distinction from M5.

Partial collapses and localized erosion of deposits are common, and the spaces created are often filled with sediments from distinct sources, as can be seen in the case of M5. At the western end of the surface excavation, M5 has formed unconformably against the side, and in pockets of, the eroded M4 deposit (Figure 2.4). Mixing of the two deposits (M4 and M5) was first recognized by Robinson (1962), who found blocks of M4 breccia within the M5 deposit. Unfortunately, the location of these blocks was not recorded. Due to the heavy-duty excavation methods used to extract the heavily calcified sediments, provenience data from past excavations is limited in resolution. Also, during the excavation of the western M4 and eastern M5 area, no records were kept of pertinent stratigraphic features. Given the irregularity of residual deposit contacts in this area, it is extremely difficult to predict the pre-excavation morphology of the deposits and their contacts. The M4 and M5 boundary is a good example of ongoing stratigraphic interpretative issues exacerbated by previous research and excavation methodologies. The problem of the M4/M5 boundary, from a paleoanthropological perspective, arises when attempting to provenience hominin fossils that have been excavated from the area and associate them to a specific deposit. In order to understand intra- or inter-species variation, solid stratigraphic associations are necessary and deserve dedicated studies (Karkanas, et al., 2000; Farrand, 2001).

Kuman and Clarke (2000) proposed a spatial distribution for the M4 and M5 deposits (Figure 2.6) based on faunal and artifactual contents. This does, however, assume that the samples are representative of the respective squares and that certain taxa are exclusive to certain deposits and have experienced minimal localized mixing. Kuman and Clarke (2000) have determined the hominin taxa based on morphology, not provenience (Clarke, pers. comm., 2010). Appendix 1 (this volume) presents the most recent member allocation for the hominin material (also see Kuman and Clarke, 2000). The reconstruction of previous deposit contacts from poorly provenienced fauna, without stratigraphic support, is problematic and as Harris (1979) points out, classification of deposit based on contents alone is risky, and deposit differentiation should be supported by a host of stratigraphic data. Unfortunately, much of this supportive data (bulk sediment samples, for instance) has been lost and so deposit classification through contents is one of the only options available. Dedicated studies utilizing micromorphology and geochemistry of deposit contact remnants and detailed 3D modeling with fabric analysis of exposed deposit profiles are currently underway to improve the stratigraphic resolution of this area.

Member 6 is the smallest of the Sterkfontein deposits and formed on top of the flowstone that capped M5 West filling the remainder of the cavern to the roof only at the western end of the site (Kuman and Clarke, 2000; Ogola, 2009). The underside of this breccia and a portion of the uppermost M5 East Acheulean deposit were eroded and filled with both Middle Stone-Age-bearing sediments and older reworked Acheulean artifacts and is classified as post-Member 6 (Kuman and Clarke, 2000; Ogola, 2009).

Summary

It has been over 70 years since the first stratigraphic interpretation of the deposits exposed in the Sterkfontein Caves by Basil Cooke (1938). Since this first attempt to describe the physical and chronological extent of the fossiliferous sediments exposed on the Sterkfontein hill and "lower cave," each work (Table 2.1) has proposed a more detailed account of the formation and physical and temporal extent of the interred deposits. It is not surprising that many of these studies have challenged previous hypotheses. The complex nature of the karst, coupled with the long deposit accumulation times and pervasive diagenetic processes, has obscured those proxies usually used to delineate depositional entities. In this manner, these processes have complicated attempts to understand the cultural, taxonomic, morphological, landscape, and environmental evolution at Sterkfontein. This complexity starts with the formation of the dolomites 2.6 billion years ago and its burial, early karstification, uplift, and fracturing over huge time scales. These processes have affected the size and shape of the cave system, as well as the position, nature, and timing of the opening to the landscape surface. Many of these processes, such as the process of cave formation, are still debated and have important implications for the age and distribution of the fossils. As the caves have gradually accumulated sediments, bones, and stone tools over hundreds of thousands of years, the dynamic processes of erosion, collapse, and redistribution have modified many aspects of the original assemblages. The six members making up the Sterkfontein Formation represent a macro-level chronostratigraphic reconstruction of the filling of the cave and are commonly differentiated by the presence or absence of certain fauna, stone tool types, and taphonomic agents; the identification of interbedding flowstones; or the description of basic sedimentological characters. The sequence starts with Member 1, formed before the cave was open, followed by the first allogenic sediment accumulation (Member 2) *c.* 3.76 million years ago (Granger et al., 2015). On top of Member 2, Member 3 and the *Australopithecus*-rich Member 4 accumulated within the relative confines of the Fossil Cavern. The Oldowan- and early Acheulean-bearing Member 5 formed around the heavily eroded M4 surface, followed by Member 6, filling the last space near the western roof and finally by Post-Member 6, a mix of younger and older reworked sediments filling an eroded pocket. With the exception of the flowstone that separates M2 and M3, deposit boundaries and their morphology are difficult to identify and correlate across 3D space from the available sediment cores and exposed profiles. Both sources of evidence are problematic from a theoretical and methodological perspective. Most of the deposit contacts are erosional and irregular, leading to difficulty in differentiating deposit and assemblage contact areas. Even more difficult is retrospectively modeling deposit contact morphology in areas that have been excavated. Characterizing taxonomic, morphological, or cultural variability from excavated deposits where no stratigraphy was recognized is difficult but is the subject of ongoing work. This is exactly the case within the important *Australopithecus*-bearing M4 and at its contact with the Oldowan-bearing M5.

As more recent dedicated stratigraphic research has identified complications inherent to cave sites, increasingly intricate site formation histories have been proposed. As Goldberg and Sherwood (2006) point out, the processes influencing the formation of fauna or artifact-bearing deposits do not often receive the attention they deserve, with the role of sediment accumulation being simplified, and perhaps overly simplified, in an effort to facilitate interpretations. Current research at Sterkfontein aims to apply new mutlidisciplinary stratigraphic analyses to new stratigraphically sensitive excavations, thereby providing high-resolution contextual support to recovered assemblages and attempts to increase our understanding of excavated deposit morphology through identification and study of sedimentary and stratigraphic features preserved in the remnants of previous excavations.

Acknowledgments

I would like to thank the editors of this volume for inviting me to contribute to this important monograph. I appreciate their inclusion of this subject in a work dedicated to postcranial hominin material. I would also like to thank the reviewers and editors for their suggestions which have significantly improved this chapter. I would also like to thank Palaeontological Scientific Trust (PAST), the National Research Foundation, The Department of Science and Technology, The Center of Excellence Palaeosciences, The Leakey Foundation, The African Origins Platform for their support of the Sterkfontein research program. I would also like to thank the Sterkfontein research team for their continuing dedicated work at the site.

References

Bamford, M., 1999. Pliocene woods from an early hominid cave deposit, Sterkfontein, South Africa. South African Journal of Science 95, 231–237.

Berger, L.R., Lacruz, R., de Ruiter, D.J., 2002. Brief communication: revised age estimates of *Australopithecus*-bearing deposits at Sterkfontein, South Africa. American Journal of Physical Anthropology 119, 192–197.

Bertran, P., Hetu, B., Texier, J-P., van Steijn, H., 1997. Fabric characteristics of subaerial slope deposits. Sedimentology 44, 1–16.

Bertran, P., Texier, J-P., 1999. Facies and microfacies of slope deposits. Catena 35, 99–121.

Brain, C.K., 1958. The Transvaal Ape-Man-Bearing Cave Deposits. Transvaal Museum Memoir No. 11. Pretoria.

Broom, R., 1945. A new primitive hyaena from Sterkfontein. South African Museum Association Bulletin 3, 273.

Bruxelles, L., Clarke, R.J., Maire, R., Ortega, R., Stratford, D.J., 2014. Stratigraphic analysis of the Sterkfontein StW 573 *Australopithecus* skeleton and the implications for its age. Journal of Human Evolution 70, 36–48.

Bruxelles, L., Stratford, D.J., Maire, R., Pickering, T.R., Heaton, J.L., Beaudet, A., Kuman, K., Crompton, R., Carlson, K.J., Jashashvili, T. and McClymont, J., 2019. A multiscale stratigraphic investigation of the context of StW 573 'Little Foot' and Member 2, Sterkfontein Caves, South Africa. Journal of Human Evolution 133, 78–98.

Bruxelles, L., Quinif, Y., Wienin, M., 2009. How can Ghost Rocks help in karst development? 15th International Congress of Speleology Proceedings 1, 814–818.

Clarke, R.J., 1994. On some new interpretations of Sterkfontein stratigraphy. South African Journal of Science 90, 211–214.

Clarke, R.J., 1998. First ever discovery of a well-preserved skull and associated skeleton of *Australopithecus*. South African Journal of Science 94, 460–463.

Clarke, R.J., 2002. On the unrealistic "Revised age estimates" for Sterkfontein. South African Journal of Science 98, 415–419.

Clarke, R.J., 2006. A deeper understanding of the stratigraphy of Sterkfontein fossil hominid site. Transactions of the Royal Society of South Africa 61, 111–120.

Clarke, R.J., 2008. Latest information on Sterkfontein's *Australopithecus* skeleton and a new look at *Australopithecus*. South African Journal of Science 104, 443–449.

Clarke, R.J., 2019. Excavation, reconstruction and taphonomy of the StW 573 *Australopithecus prometheus* skeleton from Sterkfontein Caves, South Africa. Journal of Human Evolution 127, 41–53.

Clarke, R.J., 2013. *Australopithecus* from Sterkfontein Caves, South Africa. In: Reed, K.E., Fleagle, J.G., Leakey, R.E. (Eds.), The Paleobiology of *Australopithecus*. Springer: Dordrecht, pp. 105–123.

Clarke, R.J., Kuman, K., 2019. The skull of StW 573, a 3.67 Ma *Australopithecus prometheus* skeleton from Sterkfontein Caves, South Africa. Journal of Human Evolution 134, p.102634.

Clarke, R.J., 2013. Australopithecus from Sterkfontein Caves, South Africa. In: Reed, K.E., Fleagle, J.G., Leakey, R.E. (Eds.), The Paleobiology of *Australopithecus*. Springer: Dordrecht, pp. 105–123.

Clarke, R.J., Kuman, K., 2019. The skull of StW 573, a 3.67 Ma Australopithecus prometheus skeleton from Sterkfontein Caves, South Africa. Journal of Human Evolution 134, p.102634.

Cooke, H.B.S., 1938. The Sterkfontein bone breccia: a geological note. South African Journal of Science 35, 204–208.

Dirks, P.H.G.M., Berger, L.R., 2013. Hominin-bearing caves and landscape dynamics in the Cradle of Humankind, South Africa. Journal of African Earth Sciences 78, 109–131.

Eriksson, P.G., Altermann, W., Catuneanu, O., van der Merwe R., Bumby, A.J., 2001. Major influences on the evolution of the 2.76–2.1 Ga Transvaal basin, Kaapvaal craton. Sedimentary Geology 141–142, 205–231.

Eriksson, P.G., Altermann, W., Hartzer, F.J., 2006. The Transvaal Supergroup and its precursors. In: Johnson, M.R., Anhaeusser, C.R., Thomas, R.J., (Eds.), The geology of South Africa, Pretoria. Geological Society of South Africa and Council for Geoscience Publication: Pretoria, pp. 237–260.

Eriksson, P.G., Schweitzer, J.K., Bosch, P.J.A., Schereiber, U.M., Van Deventer, J.L., Hatton, C.J., 1993. The Transvaal Sequence: an overview. Journal of African Earth Sciences 16, 25–51.

Farrand, W.R., 2001. Sediments and stratigraphy in rockshelters and caves: a personal perspective on principles and pragmatics. Geoarchaeology: An International Journal 16, 537–557.

Goldberg, P., Sherwood, S.C., 2006. Deciphering human prehistory the geoarchaeological study of cave sediments. Evolutionary Anthropology 15, 20–36.

Granger, D.E., Gibbon, R.J., Kuman, K., Clarke, R.J., Bruxelles, L., Caffee, M.W., 2015. New cosmogenic burial ages for Sterkfontein Member 2 *Australopithecus* and Member 5 Oldowan. Nature 522, 85–88.

Harris, E.C., 1979. Principles of Archaeological Stratigraphy. Academic Press: London.

Herries, A.I.R., Shaw, J., 2011. Palaeomagnetic analysis of the Sterkfontein palaeocave deposits: implication for the age of the hominin fossils and stone tool industries. Journal of Human Evolution 60, 523–539.

Hughes, A.R., Tobias, P.V., 1977. A fossil skull probably of the genus *Homo* from Sterkfontein, Transvaal. Nature 265, 310–312.

Karkanas, P., Bar-Hosef, O., Goldberg, P., Weiner, S., 2000. Diagenesis in prehistoric caves: the use of minerals that form *in situ* to assess the completeness of the archaeological record. Journal of Archaeological Science 27, 915–929.

Kibii, J., 2004. Comparative taxonomic, taphonomic and palaeoenvironmental analysis of 4-2.3 million year old australopithecine cave infills at Sterkfontein. Ph.D. Dissertation, University of the Witwatersrand.

Kidwell, S.M., Fürsich, F.T., Aigner, T., 1986. Conceptual framework for the analysis and classification of fossil concentrations. PALAIOS 1, 228–238.

Klimchouk, A.B., 2007. Hypogene speleogenesis: hydrogeological and morphogenetic perspective. Special Paper no. 1. National Cave and Karst Research Institute: Carlsbad.

Kramers, J.D., Dirks, P.H.G.M., 2017. The age of fossil StW573 ('Little Foot'): An alternative interpretation of 26Al/10Be burial data. South African Journal of Science 113, 1–8.

Kramers, J.D., Dirks, P.H.G.M., 2017. The age of fossil StW573 ('Little Foot'): Reply to comments by Stratford et al. (2017). South African Journal of Science, 113, 1–3.

Kuman, K., Clarke, R.J., 2000. Stratigraphy, artefact industries and hominid associations, Member 5. Journal of Human Evolution 38, 827–847.

Lowe, D.R., 1982. Sediment gravity flows: II. Depositional models with special reference to the deposits of high density turbidity currents. Journal of Sedimentary Petrology 52, 279–297.

Major, J.J., 2000. Gravity-driven consolidation of granular slurries-implications for debris-flow deposition and deposit characteristics. Journal of Sedimentary Research 70, 64–83.

Martini, J., Wilinger, P.E., Moen, H.F.G., Keyser, A., 2003. Contribution to the speleology of Sterkfontein cave,

Gauteng Province, South Africa. International Journal of Speleology 32, 43–49.

Ogola, C., 2009. The Sterkfontein western breccias: stratigraphy, fauna and artefacts. Ph.D. Dissertation, University of the Witwatersrand.

Osborne, R.A.L., 2002. Cave breakdown by vadose weathering. International Journal of Speleology 31, 37–53.

Partridge, T.C., 1973. Geomorphological dating of cave openings at Makapansgat, Sterkfontein, Swartkrans and Taung. Nature 246, 75–79.

Partridge, T.C., 1978. Re-appraisal of lithostratigraphy of Sterkfontein hominid site. Nature 275, 282–286.

Partridge, T.C., Granger, D.E., Caffee, M.W., Clarke, R.J., 2003. Lower Pliocene hominid remains from Sterkfontein. Science 300, 607–612.

Partridge, T.C., Shaw, J., Heslop, D., Clarke, R.J., 1999. The new hominid skeleton from Sterkfontein, South Africa: age and preliminary assessment. Journal Quaternary Science 14, 293–298.

Partridge, T.C., Watt, I.B., 1991. The stratigraphy of the Sterkfontein hominid deposit and its relationship to the underground cave system. Palaeontologia Africana 28, 5–40.

Pickering, T.R., 1999. Taphonomic interpretations of the Sterkfontein early hominid site (Gauteng, South Africa) reconsidered in the light of recent evidence. Ph.D. Dissertation, University of Wisconsin.

Pickering, T.R., Clarke, R.J., Heaton, J.L., 2004a. The context of Stw 573, an early hominid skull and skeleton from Sterkfontein Member 2: Taphonomy and paleoenvironment. Journal of Human Evolution 46, 277–295.

Pickering, T.R., Clarke, R.J., Moggi-Cecchi, J., 2004b. The role of carnivores in the accumulation of the Member 4 hominid assemblage. American Journal of Physical Anthropology 125, 1–15.

Pickering, R., Kramers, J.D., 2010. Re-appraisal of the stratigraphy and determination of new U-Pb dates for the Sterkfontein hominin site, South Africa. Journal of Human Evolution 59, 70–86.

Pickering, R., Kramers, J.D., Hancox, P.J., de Ruiter, D.J., Woodhead, J.D., 2011. Contemporary flowstone development links early hominin bearing cave deposits in South Africa. Earth and Planetary Science Letters 306, 23–32.

Postma, G., 1986. Classification for sediment gravity-flow deposits based on flow conditions during sedimentation. Geology 14, 291–294.

Robinson, J.T., 1962. Sterkfontein stratigraphy and the significance of the Extension Site. South African Archaeological Bulletin 17, 87–107.

Stratford, D.J., Bruxelles, L., Clarke, R.J., Kuman, K., 2012. New interpretations of the fossil and artefact-bearing deposits of the Name Chamber, Sterkfontein. South African Archaeological Bulletin 67, 159–167.

Stratford, D.J., Grab, S., Pickering, T.R., 2014. The stratigraphy and formation history of fossil- and artefact-bearing sediments in the Milner Hall of Sterkfontein Cave, South Africa: new interpretations and implications for palaeoanthropology and archaeology. Journal of African Earth Science 96, 155–167.

Stratford, D.J., Granger, D.L., Bruxelles, L., Clarke, R.J., Kuman, K., Gibbon, R.J., 2017. Comments on 'The age of fossil StW573 ('Little Foot'): An alternative interpretation of 26Al/10Be burial data'. South African Journal of Science 113, 1–3.

Tobias, P.V., 1979. The Silberberg Grotto, Sterkfontein, Transvaal, and its importance in palaeo-anthropological research. South African Journal of Science 75, 161–164.

Tobias, P.V., Hughes, A.R., 1969. The new Witwatersrand University excavation at Sterkfontein. South African Archaeological Bulletin 24, 158–69.

Walker, J., Cliff, R.A., Latham, A.G., 2006. U-Pb isotopic age of the StW 573 hominid from Sterkfontein, South Africa. Science 314, 1592–1594.

White, W.B., 2007. Cave sediments and palaeoclimate. Journal of Cave and Karst Studies 69, 76–93.

Wilkinson, M.J., 1973. Sterkfontein cave system: evolution of a karst form. M.A. thesis, University of the Witwatersrand.

Wilkinson, M.J., 1983. Geomorphic perspectives on the Sterkfontein australopithecine breccias. Journal of Archaeological Science 10, 515–529.

Wilkinson, M.J., 1985. Lower-lying and possibly older fossiliferous deposits at Sterkfontein. In: Tobias, P.V. (Ed.), Hominid Evolution: Past, Present and Future. Alan Liss: New York, pp. 165–170.

3

A new multidisciplinary age of 2.61–2.07 Ma for the Sterkfontein Member 4 australopiths

Robyn Pickering and Andy I.R. Herries

The Sterkfontein Caves are situated in the "Cradle of Humankind" (Cradle) UNESCO World Heritage Site, some 40km northwest of Johannesburg, in the Gauteng Province of South Africa (Figure 3.1). The palaeokarst deposits at Sterkfontein Caves have yielded the largest global collection of *Australopithecus africanus* fossils, consisting of over 600 fragments (Wood and Richmond, 2000) and contains one of the most complete hominin skeletons ever found (StW 573) (Clarke, 1998), which Clarke (2013) has attributed, along with certain other fossils from MB4, to *Australopithecus prometheus*. Sterkfontein Caves are a complex fossil site with a complicated depositional history, which today is poorly exposed and partially mined away but has nonetheless been the subject of much ongoing investigation (Bruxelles et al., 2014; Stratford et al., 2014; Granger et al., 2015; Stratford, Chapter 2 this volume).

Unlike in East Africa, the South African cave sites do not contain any material suitable to potassium-argon or argon-argon dating, and until recently no equivalent direct radiometric technique had been proved suitable or successful. Calcium carbonate layers, or flowstone, in caves can be dated using uranium-series dating, and advances in the U-Pb method have allowed the dating of Pliocene and Pleistocene speleothems (Woodhead et al., 2006; Woodhead and Pickering, 2012). This is beginning to change the approach to the chronology of the Cradle cave sites (Walker et al., 2006; Pickering and Kramers, 2010; Pickering et al., 2010; Pickering et al., 2011 a,b; Pickering et al., 2019). However, while U-Pb dating can provide precise ages for the flowstone layers above and

below fossil-bearing deposits, these data are only half the story, often providing large, hundreds of thousands of years, time windows during which sedimentation could take place. Here we address this issue by combining the Sterkfontein U-Pb ages with the detailed magnetostratigraphy of the deposits (Herries and Shaw, 2011), where the pattern of normal and reversed signals can be pinned down using U-Pb ages and thereby provide a narrower time interval for the sediments and fossils (Pickering et al., 2011a; Herries et al., 2013). The addition of Electron Spin Resonance (ESR) dating of the fossils themselves (Schwarcz et al., 1994; Grine et al., 1996; Curnoe, 1999; Curnoe et al., 2001; Herries and Shaw, 2011; Herries and Adams, 2013) provides additional age data, meaning that we are now able to construct a robust, direct, and independent chronology by dating the speleothem, sediments, and inter-bedded fossils. In this chapter we use the ESR ages of Curnoe (1999) that have good spatial association, published in Herries and Shaw (2011), rather than previous ESR ages (Schwarcz et al., 1994) whose context is uncertain. The implications of such a combined approach for the early hominin bearing cave sites across the Cradle are broad, but in this chapter we focus just on the Member 4 (MB4), australopith-bearing deposits at Sterkfontein.

Sterkfontein Caves and Member 4

The caves at Sterkfontein have acted as receptacles for surface sediments over the last few million years and,

Robyn Pickering and Andy I.R. Herries, *A new multidisciplinary age of 2.61–2.07 Ma for the Sterkfontein Member 4 australopiths*. In: *Hominin Postcranial Remains from Sterkfontein, South Africa, 1936–1995*. Edited by: Bernhard Zipfel, Brian G. Richmond, and Carol V. Ward, Oxford University Press (2020). © Oxford University Press. DOI: 10.1093/oso/9780197507667.003.0003 given the abundance of early hominin and other fossils preserved in these sediments, have been the subjects of much investigation. The most enduring interpretative description of the sediments is the Member (MB) system proposed by Partridge (Partridge, 1978; Partridge and Watt, 1991; Partridge, 2000) where the deposits are classified into six Members, 1 to 6, oldest to youngest. Today, only a few fossiliferous sediments occur underground, notably in the Jakovec Cavern, Silberberg Grotto, and the Name Chamber (Clarke, 1998, 2008; Partridge et al., 2003), with the bulk of the fossils coming from the surface exposures of MB4.

Sterkfontein has been under more or ess constant excavation since 1966, and the "Type Site" and "Extension Site" as originally distinguished by Robinson (1952) have subsequently merged into one large excavation (Figure 3.1). MB4 is exposed in the "Type Site" as a massive breccia, in the true sense, referring only to angular, coarse-grained material, dominated by angular clasts in a fine-grained matrix. The MB4 deposits closest to the

remaining cave walls are capped by a flowstone layer, while the exposures in the middle of the Type Site end in a number of large dolomite blocks, suggesting some type of roof collapse and the end of sedimentation (Pickering and Kramers, 2010; Herries and Shaw, 2011).

A series of five boreholes were drilled through the surface exposures of MB4 and Member 5 (MB5) in an attempt to link these deposits with those in the underground chambers of the cave system (Partridge and Watt, 1991) (Figure 3.2). These cores were re-examined in detail by Pickering and Kramers (2010) and briefly summarized here. Of the five cores, Boreholes (BH) 1, 4 and 5 are the most similar, each consisting of a basal layer of residium (heavily weathered dolomite), a flowstone, a thick package of sediments, another flowstone layer, and a second package of sediments. BH1 and BH4 were clearly drilled through the surface exposures of MB4 and we interpret the second, uppermost package of sediments in these cores as MB4, with the thick flowstone layer at the base. Using the middle flowstone layer as a marker horizon,

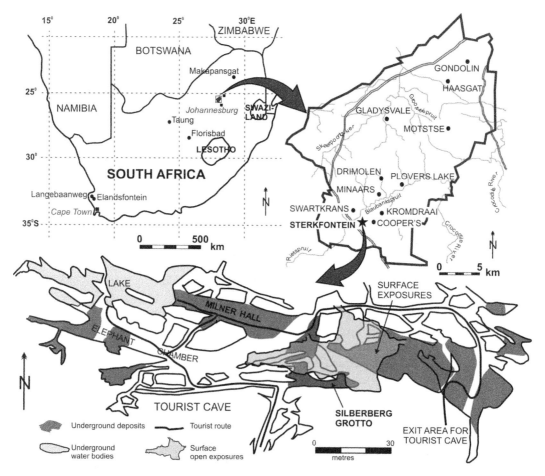

Figure 3.1 A map of southern Africa showing the location of Sterkfontein Caves within the Cradle of Humankind and the position of the deposits exposed at surface in relation to the underground sections of the cave system (map from Pickering and Kramers, 2010).

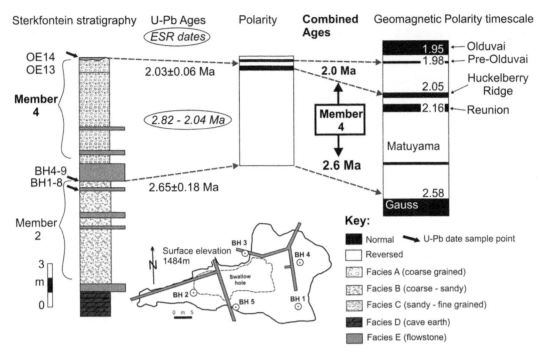

Figure 3.2 A composite stratigraphic log for the Sterkfontein boreholes, updated from Partridge and Watt (1991), compared to magnetostratigraphy of Sterkfontein, as per Herries (2003), using the U-Pb dates of Pickering et al. (2010) and Pickering and Kramers (2010) (recalculated by Pickering et al., 2018) as pinpoints to provide new, combined age range of 2.61–2.07 Ma for Member 4.

BH5 can be correlated to BH1 and BH4, and again the upper package of sediments interpreted as MB4 (Figure 3.2). In all three cores the MB4 sediments consist of a mixture of massive, course-grained, reddish-brown sands, with abundant dolomite blocks and highly calcified (Facies A) and reddish-brown sands, with some dolomite clasts, less well calcified but bone rich (Facies B, Figure 3.2) (see Pickering and Kramers, 2010, for more detailed discussion).

Beneath the MB4 deposits are an older series of breccia layers separated by flowstones that have been referred to as Member 3 (MB3) by Partridge (Partridge and Watt, 1991; Partridge, 2000). However, these deposits do not necessarily represent the same deposits as those classified as MB3 by Partridge (1978) in the Silberberg Grotto. As such, Pickering and Kramers (2010) considered the underlying deposits exposed in the BHs to represent Member 2 (MB2) and that the former MB3 deposits in the Silberberg Grotto consist of the distal portions of both MB4 and MB2. Despite diagrams by Partridge and Watt (1991) and Partridge (2000) showing the various borecores through the entire Sterkfontein Formation, they did not reach the depth of the MB2 and MB3 deposits as exposed in the Silberberg Grotto, meaning that the link made between these exposures is, in fact, tenuous. A detailed account of the stratigraphy of the sediments surrounding StW 573 is given by Bruxelles et al. (2014), who argue against a simple layer-cake stratigraphy for the deposits.

U-Pb dating flowstone layers

Samples for U-Pb dating were collected from flowstone at the base of MB4, exposed in BH1, BH4, and BH5 (samples BH4/9 and BH1/8) (Figure 3.1) and at the top of MB4 from the surface exposures of the Type Site (samples OE13 and OE14). Samples were given names depending on their location, so sample BH4/9 is the 9th sample collected from borehole 4 and OE13 is the 13th sample collected from the Open Excavations. A major limiting factor on the success of U-Pb dating of speleothems is the concentration and distribution of U; material with less than ~1 part per million (ug/g) of U will not have produced a datable quantity of Pb in the few millions of years since formation (Pickering et al., 2010). This problem can be solved by the careful pre-screening of samples prior to dating, using phosphor imaging to produce a "map" of U distributions (Cole et al., 2002; Walker et al., 2006) and picking out U-rich layers for dating (Pickering et al., 2010). Samples were prepared for U-Pb dating and measured on a double focusing Nu Instruments® multicollector ICP-Mass spectrometer following the methods detailed in Pickering et al. (2010). Ages were calculated using Tera-Wasserberg plots of $^{238}U/^{206}Pb$ against $^{207}Pb/^{206}Pb$ and final ages take into account initial $^{234}U/^{238}U$ ratios and full analytical uncertainties (Pickering et al., 2011a). The ages originally reported in Pickering and Kramers (2010) were calculated using $^{238}U/^{204}Pb$ ratios; here we present the same data, but using the

^{238}U/^{206}Pb ratio instead, which produces a much more precise age, as the uncertainly associated with the ^{204}Pb measurements is eliminated. There is no substantial change in the ages themselves, just a marked improvement in the precision (Pickering et al., 2018). Table 3.1 shows all U-Pb ages, as well as paleomagnetic orientations and ESR ages.

The boreholes reveal a substantial flowstone layer underlying the MB4 sediments, dated to 2.65 ± 0.183 Ma in BH4. In BH1 the basal section of the massive flowstone has been exhaustively sampled and could not be dated. However, a thin flowstone layer below this, was dated to 2.80 ± 0.140 Ma (BH1/8) indicating a maximum age for the overlying thick flowstone. The errors on these two ages are around 5%, which is typical for U-Pb dating in this time range, and puts the error for these age estimates close to the ~1% error on Ar-Ar ages from East African hominin sites (McDougall et al., 2012). U-Pb errors over 10% are not uncommon in this ~2.5-3.0 Ma range, where there is still residual initial (^{234}U/^{238}U), which needs to be taken into account in the age calculations in order to not overestimate the ages, but the very low levels of ^{234}U make it difficult to measure this ratio precisely. The substantial

Table 3.1. Summary of all the published ages for Sterkfontein Member 4. U-Pb ages from Walker et al. (2006), Pickering and Kramers (2010), and Pickering et al. (2010, 2018). Palaeomagnetic data from Herries and Shaw (2011). ESR ages originally from Curnoe (1999), published in Herries and Shaw (2011).

U-Pb	Sample name	Stratigraphic Unit	Age	± 2SE	% error
	OE-13	Top Member 4	1.784	0.090	5.0
	OE-14	Top Member 4	2.030	0.061	3.0
	SB-1	Silberberg Grotto	2.275	0.172	7.7
	STA15	Silberberg Grotto	2.240	0.080	3.6
	BH4-9	Base Member 4	2.645	0.140	6.9
	BH1-8	Base Member 2	2.800	0.140	5.0
	BH1-15	Base Member 2	2.747	0.172	6.3
Palaeomag	**Sample name**	**Stratigraphic Unit**	**Orientation**	**Depth**	
	STER-A01U	Mrs. Ples flowstone (MPFS), top MB4	R	8.7	
	STER-16U	MPFS	I	8.75	
	STER-16L	MPFS	I	8.8	
	STER-A01L	MPFS	I	8.85	
	STER-A02	MPFS	N	8.9	
	STER-A03	MPFS	N	8.95	
	STER-A04	MPFS	N	9.0	
	STER-A05U	MPFS	I	9.05	
	STER-A05L	MPFS	R	9.1	
	STER-A06	MPFS	I	9.15	
	STER-A07	MPFS	I	9.2	
	STER-12U	MPFS	N	9.25	
	STER-15U	MPFS	N	9.3	
	STER-12M	MPFS	I	9.35	
	STER-15L	MPFS	R	9.4	
	STER-A08	MPFS	R	9.45	
	STER-12L	MPFS	R	9.5	
	STER-15C	MPFS	R	9.55	
	STER-14C	Member 4B	R	12.8	
	STER-14S	Member 4B	R	13.0	

Table 3.1 Continued

U-Pb	Sample name	Stratigraphic Unit	Age	± 2SE	% error
	STER-13	Member 4A	R	14.7	
	STER-11U	Member 4A	R	15.9	
	STER-11L	Member 4A	I	16.1	
	STER-10C	Member 4A	R	17.5	
	STER-10S	Member 4A	R	17.6	
ESR	**Sample name**	**Stratigraphic Unit**	**Age**	**± 2SE**	**% error**
	1348a	Member 4	2.06	0.18	8.7
	1348b	Member 4	2.32	0.28	12.0
	1347a	Member 4	2.60	0.22	8.5
	1352a	Member 4	3.09	0.29	9.3
Average:			2.42	0.38	15.7

error on the initial (^{234}U/^{238}U) goes on to contribute to the large errors on the final U-Pb ages. The flowstone sample OE14 is clearly capping the MB4 deposits and we use this date of 2.030 ± 0.061 Ma to provide a minimum age for the sediments. The nearby flowstone dated to 1.784 ± 0.090 Ma (OE13) provides an upper age estimate for a period of flowstone development at the end of MB4 and so the maximum age for the early *Homo, Paranthropus*, and stone tool-bearing Member 5. Paleomagnetic analysis and ESR ages for the Member 5 deposits confirm such a post-1.78 Ma age for Member 5 (Herries and Shaw, 2011). A cosmogenic burial on a single manuport from the MB5 Oldowan infill of 2.18 ± 0.21 Ma (Granger et al., 2015) falls outside this range and is yet another example of cosmogenic burial ages being substantially older than all other estimates for the same deposits (see also Pickering, 2015, for discussion on Wonderwerk Cave).

The U-Pb dated flowstone above and below MB4 therefore gives an age bracket of ~2.6 – ~2.0 Ma. However, the full errors associated with the ages must be taken into account, and this gives a maximal age range of 2.83-1.97 Ma and a minimal age range of 2.47-2.09 Ma (Figure 3.1). A few more U-Pb dates for each flowstone could help narrow down this period of around 1 million years—however, no more suitable exposures are visible, and of the limited flowstones exposed in the boreholes, only the flowstones dated here contained sufficient U for dating. This is a good example of the limitations of the sole use U-Pb dating.

Paleomagnetic analysis

Herries and Shaw (2011) outlined paleomagnetic methodology, which was primarily conducted on the speleothem portion of the cave deposits because of concerns over the stability of remanence within the sediments of the MB4 deposits from earlier studies. However, a growing body of research (McFadden et al., 1979; Thackeray et al., 2002; Herries et al., 2006 a,b; Adams et al., 2007; Adams et al., 2010; Dirks et al., 2010; Herries and Shaw, 2011; Pickering et al., 2011a) is demonstrating that the siltstone and speleothem deposits from the South African are suitable for paleomagnetic analysis as the remanence in speleothems are locked in almost instantaneously as the calcite is precipitated and are thus adequate recorders of the magnetic field at the time of their formation (Lascu and Feinberg, 2011). The poor early results from Sterkfontein, and other sites, as noted by Brock et al. (1977) and Jones et al. (1986) are likely due to the fact that collapse breccias were sampled. The nature of their formation means that there is potentially no mechanism by which the remanence can be set into the samples. A further advantage of the South African speleothems is that they form slowly enough to record short excursions or reversals in the Earth's magnetic field (Dirks et al., 2010; Herries and Shaw, 2011; Pickering et al., 2011a) that would not normally be recorded in fluvial depositional settings, as is most often the context of hominin-bearing sites in East Africa.

The speleothem and clastic deposits from MB4 all record a reversed polarity attributed to the early part of the Matuyama Chron based on the faunal age estimates and ESR dating of the deposits (Herries and Shaw, 2011), so younger than the Gauss Matuyama boundary at 2.61 Ma (Figure 3.2; Table 3.1) and older than the beginning of the Olduvai Chron at 1.95 Ma (Singer, 2014). The flowstone associated with Sts 5 (referred to here as the "Mrs. Ples" flowstone) at the top of MB4 records two episodes of normal and intermediate polarities, followed by reversed polarities (Figure 3.1). The nature of the reversals within the speleothem suggest they represent a short reversal of the magnetic field and Herries (2003) initially correlated them to the Réunion events between ~2.21 and ~2.14 Ma. At the time this was the most parsimonious correlation, as the Réunion events were then the best resolved magnetic events. Herries and Shaw (2011) later correlated these to the Réunion and Huckleberry Ridge Events at ~2.16 and ~2.05 Ma, respectively, although they note that this could also result from complexity of the magnetic field during one of these magnetic reversal events not yet documented in the magnetostratigraphy.

Combining the U-Pb ages and palaeomagenetic sequence: New age ranges

Since all the previous publications concerning the age of Sterkfontein Member 4 the age and sequencing of magnetic reversals for the Quaternary (last ~2.61 Ma) has been refined in the Geomagnetic Instability Timescale (Singer, 2014) and the ages from this new timescale are now used here. Alone the U-Pb ages can only give a broad time range of a maximal of 2.83-1.97 Ma and a minimal of 2.47-2.09 Ma. By combining the U-Pb with the paleomagnetic analysis, the MB4 deposits cannot be older than the Gauss-Matuyama boundary at 2.61 Ma based on the reversed polarity of both the clastic deposits and speleothem. The fact that these both have the same polarity (reversed) suggests that sedimentation of MB4 began shortly after the lower flowstone formed.

The flowstone capping MB4 in which Herries and Shaw (2011) identified the short geomagnetic field event(s) is the same one dated by Pickering and Kramers (2010) to 2.03 ± 0.06 Ma (2.09-1.97 Ma), the so-called Mrs. Ples flowstone. Given that the U-Pb age of this flowstone from its very base, the flowstone could not have started growing before 2.09 Ma and this part of the flowstone is slightly older than the magnetic field events. As such they cannot represent the older Réunion event at 2.2 Ma, or the Feni SubChron at 2.16-2.12 Ma (Singer, 2014). Given the U-Pb age, the magnetic field event(s) likely represent variation during the Huckleberry Ridge Event at 2.07 Ma or this event and the later Pre-Olduvai events at ~1.98 Ma (Lai and Channell, 2009). Both these events have been identified at the Malapa, from which *Australopithecus sediba* was recovered (Pickering et al., 2011a). At Malapa the Huckleberry Ridge event has been directly dated to between 2.062 ± 0.021 Ma through the combination of the paleomagnetic analysis of the and U-Pb dating the flowstones underlying the hominin fossils (Dirks et al., 2010). The dates for this event at Sterkfontein and Malapa correlate well both with each other and as well the suggested ages based on Argon-Argon dating (Singer, 2014). This gives a minimum age of 2.07 Ma for MB4 and suggests

that the entire deposit must have formed sometime between 2.61 and 2.07 Ma (Figure 3.1).

While this is the first, chronometric age estimate for MB4, 2.61 to 2.07 Ma provides a long ~540 ka time window; the fossils themselves provide several lines of evidence, in the form of ESR dates on fossil tooth enamel and using specific species as biochronological markers, which can be used together in an attempt to narrow down the age of the deposits. In isolation, such an approach becomes circular, but here we start the U-Pb ages and paleomagnetic analysis to provide the ages framing the fossil deposits.

Schwarcz et al. (1994) undertook the first attempt at directly dating the MB4 fossils using ESR, but this study used teeth already excavated, meaning that their exact provenience and surrounding dosimetry was not clear. Here we use the ESR ages of Curnoe (1999) that were undertaken on in situ recovered teeth four of which can confidently be assigned to MB4 (Herries and Shaw, 2011) (Table 3.1). Three of the four ESR ages fall within the 2.61-2.07 Ma age bracket for MB4, with an even spread of ages during this period: from 2.06 ± 0.18 Ma as the youngest, 2.32 ± 0.28 in the middle and 2.60 ± 0.22 as the oldest. These ages hint that the MB4 fossils accumulated throughout the 2.61-2.07 Ma time window. Recent advances in ESR dating at other similar aged sites in the region, such as the ~2.61 Ma Drimolen Makondo, shows that ESR correlates well with U-Pb and paleomagnetism, and works well over this time range (Herries et al., 2019). The fourth age is older, 3.09 ± 0.29 Ma and is similar in age to the deposits beneath MB4 in the boreholes, dated by Pickering and Kramers (2010) to post-2.80 ± 0.14 Ma. All the other chronological evidence presented here (U-Pb ages, reversed paleomagnetic signal) indicate that MB4 is not older than 2.61 Ma, so in order to explain the presence of this older tooth, we suggest that some older material was exposed at the surface, this tooth was eroded out, and then it was incorporated into MB4. Alternatively, and we believe more likely, this tooth originates from the older deposits but became mixed in with the MB4 deposits in the large area of decalcified sediment in the middle of the Sterkfontein open excavations known as the "Swallow Hole."

This issue of mixing and intrusion of older and/or younger material into the MB4 deposits at Sterkfontein is not new. The presence of *Equus* fossils, which can be used as a strong biochronological marker as this genus only appears in Africa after 2.3 Ma (Geraads et al., 2004), has been contested by Clarke (2007) who argues that these fossils are from a younger part of the Sterkfontein deposits and are intrusive into MB4. Within the 2.61–2.07 Ma age we propose here for MB4, there is room for *Equus* to be present without being intrusive and the youngest ESR age of 2.06 ± 0.18 Ma suggests that fossils this young are present in MB4. *Equus* is certainly present at Malapa by 1.98 Ma (Dirks et al., 2010) and at Haasgat prior to 1.95 Ma (Herries et al., 2014). Given the presence of the much older ESR age (3.09 ± 0.29 Ma) we cannot rule out the mixing of deposits of different ages, although it is more likely for older fossil material to be mixed into younger deposits by the formation of younger cavities within the older deposits, or possibly during early excavations of the decalcified deposits prior to the current stratigraphic framework. Reynolds et al. (2007) has documented the mixing of older hominin material into much younger deposits in Lincoln Cave at Sterkfontein and the same has been seen at Swartkrans (Herries and Adams, 2013). Stratford et al. (2014) have also shown how material from these surface exposures of palaeokarst have been reworked into lower chambers at Sterkfontein,

While the bulk of the fossils recovered from Sterkfontein are from MB4, hominins and other fauna are also preserved in two underground chambers, the Jakovec Cavern (Partridge et al., 2003; Clarke, 2008) and the Silberberg Grotto (Clarke, 1998, 2007, 2008). How these deposits and fossils relate to MB4 is not immediately clear and has been the subject of much debate. The presence of *Equus* in the Jakovec Cavern deposits (Kibii, 2004) suggests an age of <2.3 Ma, hinting that these deposits are coeval with MB4. This is in stark contrast with the cosmogenic burial ages of quartz with the Jakovec Cavern sediments of ~4 Ma (Partridge et al., 2003). Given the suggestion by Clarke (2008) that the Jakovec Cavern hominin fossils are *A. africanus*, a similar age for the Jakovec Cavern and MB4 does not seem unreasonable despite its greater depth in the cave. Multiple entrances through to different caverns at Sterkfontein could easily have been open at the same time and so material could have been deposited synchronously.

A similar situation is suggested by the U-Pb and paleomagnetic age assessment of the Silberberg Grotto deposits and *Australopithecus prometheus* fossil StW 573 (Clarke, 2013), which we argue is also dated to between 2.6 and 2.2 Ma (Pickering and Kramers, 2010; Pickering et al., 2010; Herries and Shaw, 2011). While an older age of 3 Ma has been hypothesized for this fossil (Bruxelles et al., 2014), the only evidence to support this age is an erosional hiatus that occurs between the deposition of the fossil and the formation of the U-Pb dated capping speleothem, something other authors have agreed on for some time (Walker et al., 2006; Pickering and Kramers, 2010; Herries and Shaw, 2011). The basal speleothem has reversed polarity, as opposed to the normal polarity of the capping speleothem (Herries and Shaw, 2011), indicating that these flowstones did not form synchronously. Indeed within the 400 ka duration of the 2.6–2.2 Ma age estimated based on the paleomagnetic

analysis, there is scope for a significant hiatus. Granger et al. (2015) use an isochron approach to refine their cosmogenic nuclide age for the sediments surrounding StW 573 to 3.67 ± 0.16 Ma. This has been questioned by Kramers and Dirks (2017), who, using the same data, argue that an age of 2.8 Ma is more likely, given the strong possibility of recycling sediment within the cave system. We prefer this 2.8 Ma age, although the reversed signal in the basal flowstone suggests an even younger age of more like 2.61 Ma. Thus, we argue that the occurrence of *Australopithecus prometheus* fossils in MB4 (Clarke, 2013) and the similar reversed polarity of the two deposits, together with the U-Pb ages, suggests that the Silberberg Grotto and MB4 deposits likely formed synchronously.

The U-Pb ages and paleomagnetic data presented here combine to give an age estimate for MB4 of between 2.61 and 2.07 Ma. If two species of *Australopithecus* do occur in MB4 as Clarke (2008, 2013) and other authors (Kimbel and Rak, 1993; Kimbel and White, 1988) have suggested, this may reflect two species existing at the same time or temporal variation over half a million years or more of deposition. Fossil Sts 5 represents the youngest *A. africanus* fossil from the MB4 deposit, being deposited around the time of the capping flowstone at ~2.07 Ma and is only slightly older than *Australopithecus sediba* from Malapa at 1.98 Ma (Pickering et al., 2011a). This temporal variation needs to be taken into account when assessing the variation in anatomy seen within the MB4 hominins, as well as in other studies such as palaeodietary analyses. For example, Balter et al. (2012) have suggested that *A. africanus* had a much wider dietary breadth than other species such as *Paranthropus robustus* or early *Homo*. However, the *Paranthropus robustus* specimens sampled were primarily from Swartkrans Member 1, a temporally more restricted site (2.0–1.8 Ma; Pickering et al., 2011; Herries and Adams, 2013), while those representing *Australopithecus africanus* could cover more than 0.54 Ma of deposition and perhaps also two species. Such variation would not be surprising with this age range in mind.

The minimum age for MB4 of 2.07 Ma and the 1.98 Ma age for Malapa (Pickering et al., 2011) indicate that in South Africa *Australopithecus* survived for a much longer period of time than previously suggested or as seen in East Africa, where *A. garhi* represents the last of the genus at ~2.5 Ma (Asfaw et al., 1999). If the Stw53 fossil also represents *Australopithecus* (Clarke, 2013), rather than early *Homo* (Curnoe and Tobias, 2006; Smith and Grine, 2008), then *Australopithecus* may have survived to less than 1.8 Ma (Kuman and Clarke, 2000). The Stw53 deposit, otherwise known as Member 5a (Partridge, 2000), has a reversed polarity attributed to the Matuyama Chron post the Olduvai event (<1.78 Ma) based on ESR ages (between 1.71 ± 0.35 and 1.63 ± 0.16 Ma; Herries and

Shaw, 2011) and is underlain by a flowstone U-Pb dated to 1.81 ± 0.06 (Pickering and Kramers, 2010). Clarke (2007) suggested that StW 53 comes from a younger extension of MB4, which would place the upper age for MB4 at ~1.8 Ma and not 2.0 Ma. We prefer to view the StW 53 deposit as an intermediate deposit separate from both MB4 and MB5.

Conclusions

We present a summary of published U-Pb and ESR dating, as well as paleomagnetic analysis of the MB4 deposits at Sterkfontein. Each of these methods represents great advances in understanding the chronology of the South African hominin-bearing cave deposits but in isolation has its limitations. The U-Pb and ESR ages come with necessarily large age uncertainties, especially on the ages in the ~3.0–2.5 Ma time range, and the paleomagnetic sequences require some kind of direct dating in order to pin the sequences to an absolute time scale. Here we have combined the data sets, each compensating for the other's weaknesses and provide a new, direct age estimate of 2.61–2.07 Ma for MB4 and suggest that it likely formed over a ~540 ka-year period, which may explain some of the variability seen in the hominin fossils from the deposit.

Acknowledgments

Andy Herries learned about Charlie Lockwood's death via sms while surveying for fossil sites in the Karoo and could not believe that such a young talented mind could be gone in such a manner; we fondly remember many great moments round the campfire at Makapansgat for many years, as well as many times of being beaten at ultimate Frisbee. Robyn Pickering met Charlie briefly at a Smithsonian workshop in 2005 but was struck at both his ability as a researcher and generosity as a person. Charlie leaves a gaping hole in palaeoanthropology, particularly in South Africa, but the many people he taught and influenced over the years remain to honor him. We acknowledge the late Tim Partridge and Andre Keyser, who both strove to better understand the geological context of the South African fossils. We thank Ron Clarke, Steven Motsumi, Able Molepolle, and Lucas Sekowe for providing access to the deposits at Sterkfontein. We would like to thank two anonymous reviewers for their constructive and helpful comments on the manuscript. U-Pb analysis was funded by the Swiss National Research Foundation (grant 20-113658) and the University of Melbourne (McKenzie fellowship); paleomagnetic analysis by the British Academy and the Australian Research Council's Future Fellowship FT120100399.

References

Adams, J.W., Herries, A.I.R., Hemingway, J., Kegley, A.D.T., Kgasi, L., Hopley, P., Reade, H., Potze, S., Thackeray, J.F., 2010. Initial fossil discoveries from Hoogland, a new Pliocene primate-bearing karstic system in Gauteng Province, South Africa. Journal of Human Evolution 59, 685–691.

Adams, J.W., Herries, A.I.R., Kuykendall, K.L., Conroy, G.C., 2007. Taphonomy of a South African cave: geological and hydrological influences on the GD 1 fossil assemblage at Gondolin, a Plio-Pleistocene paleocave system in the Northwest Province, South Africa. Quaternary Science Reviews 26, 2526–2543.

Asfaw, B., White, T.D., Lovejoy, C.O., Latimer, B., Simpson, S., Suwa, G., 1999. *Australopithecus garhi*: a new species of early hominid from Ethiopia. Science 284, 629-635.

Balter, V., Braga, J., Telouk, P., Thackeray, J.F., 2012. Evidence for dietary change but not landscape use in South African early hominins. Nature 489, 558–560.

Brock, A., McFadden, P.L., Partridge, T.C, 1977. Preliminary palaeomagnetic results from Makapansgat and Swartkrans. Nature 266, 249.

Bruxelles, L., Clarke, R.J., Maire, R., Ortega, R., Stratford, D., 2014. Stratigraphic analysis of the Sterkfontein StW 573 *Australopithecus* skeleton and implications for its age. Journal of Human Evolution 70, 36–48.

Clarke, R.J., 1998. First ever discovery of a well-preserved skull and associated skeleton of *Australopithecus*. South African Journal of Science 94, 460–463.

Clarke, R.J., 2007. A deeper understanding of the stratigraphy of Sterkontein fossil hominid site. Transactions of the Royal Society of South Africa 61, 111–120.

Clarke, R.J., 2008. Latest information on Sterkfontein's *Australopithecus* skeleton and a new look at *Australopithecus*. South African Journal of Science 104, 443–449.

Clarke, R.J., 2013. *Australopithecus* from Sterkfontein Caves, South Africa. In: Reed, K.E., Fleagle, J.G., Leakey, R.E. (Eds.), The Paleobiology of *Australopithecus*. Springer: Dordrecht, pp. 105–123.

Cole, J.M., Nienstedt, J., Spataro, G., Rasbury, E.T., Lanzirotti, A., Celestian, A.J., Nilsson, M., Hanson, G.N., 2002. Phosphor imaging as a tool for *in situ* mapping of ppm levels of uranium and thorium in rocks and minerals. Chemical Geology 193, 127–136.

Curnoe, D., 1999. A contribution to the question of early *Homo* in southern Africa: researchers into dating, taphonomy and phylogeny reconstruction. Ph.D. Dissertation, Australian National University.

Curnoe, D., Grün, R., Taylor, L., Thackeray, F., 2001. Direct ESR dating of a Pliocene hominin from Swartkrans. Journal of Human Evolution 40, 379–391.

Curnoe, D., Tobias, P.V., 2006. Description, new reconstruction, comparative anatomy, and classification of the Sterkfontein StW 53 cranium, with discussions about the taxonomy of other southern

African early *Homo* remains. Journal of Human Evolution 50, 36–77.

Dirks, P., Kibii, J.M., Kuhn, B.F., Steininger, C., Churchill, S.E., Kramers, J.D., Pickering, R., Farber, D.L., Meriaux, A.S., Herries, A.I.R., King, G.C.P., Berger, L.R., 2010. Geological setting and age of *Australopithecus sediba* from Southern Africa. Science 328, 205–208.

Geraads, D., Eisenmann, V., Petter, G., 2004. The large mammal fauna of the Oldowan sites of Melka Kunture. In: Chavaillon, J., Piperno, M. (Eds.), Studies on the Early Paleolithic Site of Melka Kunture, Ethiopia. Istituto Italiano di Preistoria e Protostoria: Florence. pp. 169–192.

Granger, D.E., Gibbon, R.J., Kuman, K., Clarke, R.J., Bruxelles, L., Caffee, M.W., 2015. New cosmogenic burial ages for Sterkfontein Member 2 *Australopithecus* and Member 5 Oldowan. Nature 522, 85.

Grine, F.E., Jungers, W.L., Schultz, J., 1996. Phenetic affinities among early *Homo* crania from East and South Africa. Journal of Human Evolution 30, 189–225.

Herries, A.I., Kappen, P., Kegley, A.D., Patterson, D., Howard, D.L., de Jonge, M.D., Potze, S., Adams, J.W., 2014. Palaeomagnetic and synchrotron analysis of 1.95 Ma fossil-bearing palaeokarst at Haasgat, South Africa. South African Journal of Science 110, 1–12.

Herries, A.I.R., 2003. Magnetostratigraphy of the South African hominid palaeocaves. American Journal of Physical Anthropology S36, 113.

Herries, A.I.R., Adams, J.W., 2013. Clarifying the context, dating and age range of the Gondolin hominins and *Paranthropus* in South Africa. Journal of Human Evolution 65, 676–681.

Herries, A.I.R., Adams, J.W., Kuykendall, K.L., Shaw, J., 2006a. Speleology and magnetobiostratigraphic chronology of the GD 2 locality of the Gondolin hominin-bearing paleocave deposits, North West Province, South Africa. Journal of Human Evolution 51, 617–631.

Herries, A.I.R., Pickering, R., Adams, J.W., Curnoe, D., Warr, G., Latham, A.G., Shaw, J., 2013. A multidisciplinary perspective on the age of *Australopithecus* in southern Africa. In: Reed, K.E., Fleagle, J.G., Leakey, R. (Eds.), Paleobiology of *Australopithecus*. Springer: Dordrecht, pp. 21–40.

Herries, A.I.R., Reed, K.E., Kuykendall, K.L., Latham, A.G., 2006b. Speleology and magnetobiostratigraphic chronology of Buffalo Cave fossil site, Makapansgat, South Africa. Quaternary Research 66, 233–245.

Herries, A.I.R., Shaw, J., 2011. Paleomagnetic analysis of the Sterkfontein palaeocave deposits: implications for the age of the hominin fossils and stone tool industries. Journal of Human Evolution 60, 523–539.

Jones, D.L., Brock, A., McFadden, P.L., 1986. Palaeomagnetic results from the Kromdraai and Sterkfontein hominid sites. South African Journal of Science 82, 160–163.

Kibii, J.K., 2004. Comparative taxonomic, taphonomic and palaeoenvironmental analysis of 4–2.3 million year old australopithecine cave infills at Sterkfontein. Ph.D. dissertation, University of the Witswatersrand.

Kimbel, W.H., Rak, Y., 1993. The importance of species taxa in paleoanthropology and an argument for the phylogenetic concept of the species category. In: Kimbel, W.H., Martin, L.B. (Eds.), Species, Species Concepts, and Primate Evolution. Plenum Press: New York, pp. 461–484.

Kimbel, W.H., White, T.D., 1988. Variation, sexual dimorphism and the taxonomy of *Australopithecus*. In: Grine, F.E. (Ed.), Evolutionary History of the "Robust" Australopithecines. Aldine de Gruyter: New York, pp. 175–198.

Kramers, J.D., Dirks, P.H.G.M., 2017. The age of fossil StW 573 ("Little Foot"): an alternative interpretation of 26Al/10Be burial data. South African Journal of Science 113, 1–8.

Kuman, K., Clarke, R.J., 2000. Stratigraphy, artefact industries and hominid associations for Sterkfontein, Member 5. Journal of Human Evolution 38, 827–847.

Lai, C., Channell, J.E.T., 2009. Geomagnetic excursions. In: Kono, M. (Ed.), Treatise on Geophysics: Geomagnetism. Elsevier: Amsterdam, pp. 373–416.

Lascu, I., Feinberg, J.M., 2011. Speleothem magnetism. Quaternary Science Reviews 30, 3306–3320.

McDougall, I., Brown, F.H., Vasconcelos, P.M., Cohen, B.E., Thiede, D.S., Buchanan, M.J., 2012. New single crystal 40Ar/39Ar ages improve time scale for deposition of the Omo Group, Omo–Turkana Basin, East Africa. Journal of the Geological Society 169, 213–226.

McFadden, P.L., Brock, A., Partridge, T.C., 1979. Paleomagnetism and the age of the makapansgat hominid site. Earth and Planetary Science Letters 44, 373–382.

Partridge, T.C., 1978. Re-appraisal of lithostratigraphy of Sterkfontein hominid site. Nature 275, 282–287.

Partridge, T.C., 2000. Hominid-bearing cave and tufa deposits. In: Partridge, T.C., Maud, R.R. (Eds.), The Cenozoic in Southern Africa. Oxford University Press: Oxford, pp. 100–125.

Partridge, T.C., Granger, D.E., Caffee, M.W., Clarke, R.J., 2003. Lower Pliocene hominid remains from Sterkfontein. Science 300, 607–612.

Partridge, T.C., Watt, I.B., 1991. The stratigraphy of the Sterkfontein hominid deposit and its relationship to the underground cave system. Palaeontologica Africana 28, 35–40.

Pickering, R., 2015. U-Pb dating small buried stalagmites from Wonderwerk Cave, South Africa: a new chronometer for Earlier Stone Age cave deposits. African Archaeological Review 32, 645–668.

Pickering, R., Dirks, P.H.G.M., Jinnah, Z., de Ruiter, D.J., Churchill, S.E., Herries, A.I.R., Woodhead, J.D., Hellstrom, J.C., Berger, L.R., 2011a. *Australopithecus sediba* at 1.977 Ma and implications for the origins of the genus *Homo*. Science 333, 1421–1423.

Pickering, R., Herries, A.I.R., Woodhead, J.D., Hellstrom, J.C., Green, H.E., Paul, P., Ritzman, T., Strait, D.S., Schoville, B.J., Hancox, P.J., 2019. U-Pb dated flowstones restrict South African early hominin record to dry climate phases. Nature 565, 226–229.

Pickering, R., Kramers, J.D., 2010. A re-appraisal of the stratigraphy and new U-Pb dates at the Sterkfontein hominin site, South Africa. Journal of Human Evolution 59, 70–86.

Pickering, R., Kramers, J.D., Hancox, P.J., de Ruiter, D.J., Woodhead, J.D., 2011b. Contemporary flowstone development links early hominin bearing cave deposits in South Africa. Earth and Planetary Science Letters 306, 23–32.

Pickering, R., Kramers, J.D., Partridge, T., Kodolanyi, J., Pettke, T., 2010. U–Pb dating of calcite–aragonite layers in speleothems from hominin sites in South Africa by MC-ICP-MS. Quaternary Geochronology 5, 544–558.

Reynolds, S.C., Clarke, R.J., Kuman, K.A., 2007. The view from the Lincoln Cave: mid-to late Pleistocene fossil deposits from Sterkfontein hominid site, South Africa. Journal of Human Evolution 53, 260–71.

Robinson, J.T., 1952. The australopithecine-bearing deposits of the Sterkfontein area. Annals of the Transvaal Museum 22, 1–19.

Schwarcz, H.P., Grün, R., Tobias, P.V., 1994. ESR dating studies of the australopithecine site of Sterkfontein, South Africa. Journal of Human Evolution 26, 175–181.

Singer, B.S., 2014. A Quaternary geomagnetic instability time scale. Quaternary Geochronology 21, 29–52.

Smith, H.F., Grine, F.E., 2008. Cladistic analysis of early *Homo* crania from Swartkrans and Sterkfontein, South Africa. Journal of Human Evolution 54, 684–704.

Stratford, D., Grab, S., Pickering, T.R., 2014. The stratigraphy and formation history of fossil- and artefact-bearing sediments in the Milner Hall, Sterkfontein Cave, South Africa: new interpretations and implications for palaeoanthropology and archaeology. Journal of African Earth Sciences 96, 155–167.

Thackeray, J.F., Kirschvink, J.L., Raub, T.D., 2002. Paleomagnetic analyses of calcified deposits from the Plio-Pleistocene hominid site of Kromdraai, South Africa. South African Journal of Science 98, 537–540.

Walker, J., Cliff, R.A., Latham, A.G., 2006. U-Pb isotopic age of the StW 573 hominid from Sterkfontein, South Africa. Science 314, 1592–1594.

Wood, B., Richmond, B.G., 2000. Human evolution: taxonomy and paleobiology. Journal of Anatomy 197, 19–60.

Woodhead, J., Hellstrom, J., Maas, R., Drysdale, R., Zanchetta, G., Devine, P., Taylor, E., 2006. U-Pb geochronology of speleothems by MC-ICPMS. Quaternary Geochronology 1, 208–221.

Woodhead, J., Pickering, R., 2012. Beyond 500ka: progress and prospects in the U-Pb chronology of speleothems, and their application to studies in palaeoclimate, human evolution, biodiversity and tectonics. Chemical Geology 322–323, 290–299.

Section 2

Postcranial anatomy of the
Sterkfontein hominins

4

The partial skeletons

Carol V. Ward, Martin Haeusler, and Bernhard Zipfel

Sterkfontein is notable not only for its wealth of isolated bones but also for the fortuitous preservation of two partial skeletons, both attributed to *Australopithecus africanus*, Sts 14 and StW 431. Both of these skeletons have been discussed extensively in the literature, as noted throughout the chapters in this volume. A third and most complete skeleton, StW 573 "Little Foot" from Member 2 in the Sterkfontein Formation (Clarke, 1998), has also been recovered. This skeleton, however, is not included here, as it was still under study at the time of writing of this volume.

This monograph describes the individual bones from Sts 14 and StW 431 in separate chapters by element to place the morphology of each in comparative context with the isolated bones also found at Sterkfontein. In this brief chapter, we present these specimens as complete skeletons so that the reader has a complete reference for each when reading about the individual bones described throughout this monograph.

Sts 14

Sts 14 is a skeleton that was recovered in a single block of breccia from Sterkfontein Member 4 in late 1947 by Robert Broom and John T. Robinson (Broom and Robinson, 1947; Broom et al., 1950) (Figures 4.1 and 4.2). The right hipbone was chiseled out of the block described at the time, but the rest of the specimen remained unprepared because the bones were thought to too fragile to survive mechanical preparation. In the late 1960s, acid preparation became available, so John Robinson made a cast of the block and then prepared it at that time using acid to dissolve the breccia surrounding the bones (Robinson, 1972), although

the exact type and proportions of acid(s) used in the preparation are undocumented.

As a result of Robinson's efforts, the first partial skeleton known for any australopith was revealed (Robinson, 1972) (Figure 4.2). Sts 14 still forms an integral part of our understanding of the postcranial morphology of *Australopithecus africanus*. No craniodental remains from this individual are known. Thackeray et al. (2002) explored the possibility of Sts 14 being associated with the cranium known as "Mrs. Ples" (Sts 5); however, there is no direct evidence to suggest that these specimens were from the same individual. Further, perhaps because of the preparation method, the bones remain quite fragile, and most are too fragile to be molded and cast.

When Robinson prepared the skeleton, he reconstructed missing parts of some of the vertebrae and proximal femur with a dense plaster. Most of these reconstructions are appropriate and likely quite accurate. However, one of the vertebrae, Sts 14B, the penultimate lumbar element, was reconstructed incorrectly, and Ward et al. (Chaper 10 in this volume) present a corrected version. The reconstruction of this vertebra affects some metrics, but not the overall functional or phylogenetic interpretations of this skeleton (see Chapter 10, this volume). The proximal femur was reconstructed by following the general contours of the partly preserved neck, and by estimating femoral head size and position using the acetabulum, and does not preserve any original morphology (see Desilva and Grabowski Chapter 12, this volume).

The Sts 14 skeleton preserves 19 partial vertebrae, some fragmentary ribs, the sacrum, right hipbone, much of the left hipbone, and the proximal two-thirds of the right femur, all from the same individual. Robinson identified the vertebrae, pelvis, and femur but did not ascribe

Carol V. Ward, Martin Haeusler, and Bernhard Zipfel, *The partial skeletons*. In: *Hominin Postcranial Remains from Sterkfontein, South Africa, 1936–1995*. Edited by: Bernhard Zipfel, Brian G. Richmond, and Carol V. Ward, Oxford University Press (2020). © Oxford University Press.
DOI: 10.1093/oso/9780197507667.003.0004

Figure 4.1 Robert Broom (1866–1951) holding breccia block containing the Sts 14 skeleton. Photo courtesy of the Ditsong Museum of Natural History.

individual ribs to particular levels (Robinson, 1972). In 2015, the ribs were studied carefully and attributed to particular levels by Carol Ward (Ward et al., Chapter 10, this volume). They were then accessioned into the collections of the Ditsong Museum of Natural History by Stephanie Potze, who was the curator of these collections at the time.

Given the small size of this specimen Sts 14 is mostly likely from a female. All physeal plates are fully fused, indicating that she was fully adult at the time of death.

StW 431

The StW 431 partial skeleton (Figure 4.3) was discovered at Sterkfontein Member 4 in 1987 by Alun Hughes and his team under the direction of Philip Tobias (Tobias, 1992) and published originally by Toussaint et al. (2003). Four additional pieces of the skeleton were later identified by Kibii and Clarke (2003) among the faunal remains from Sterkfontein. All parts of the skeleton were found deep below the top of Member 4 in partially decalcified breccia, in an area that extends horizontally only three square yards, but vertically over a range between 21'11" and 26'6" below the surface (Appendix 4). They are

identifiable as deriving from a single individual based on the fact that they share a similar color and texture, state of preservation, and developmental age, and were all recovered in close proximity within the deposit. Some craniodental remains were also recovered in the area, but these remains include duplicated elements and variable states of age and wear, suggesting that more than one individual is represented. Several of the teeth are relatively worn, however, and could have come from the same individual as the postcranial elements, but cannot be firmly attributed as such. In addition, a fragment of temporal squama and two metacarpals were also found in the same part of the grid, and although they cannot be definitively excluded from belonging to StW 431, they also cannot be firmly attributed to the same individual (Toussaint et al., 2003).

Originally, each pelvic and vertebral element now attributed to and accessioned as StW 431 were assigned individual accession numbers (Benade, 1990; Kibii and Clarke, 2003; Toussaint et al., 2003). Drapeau and Menter (Chapter 7) and Tocheri and Kibii (Chapter 8) describe the reassignments of numbers.

Given the relatively large size of this specimen and distinct muscle markings, it likely was from a male

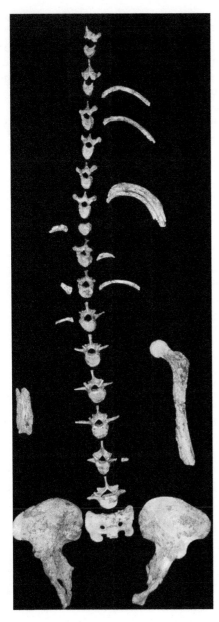

Figure 4.2 The Sts 14 partial skeleton housed at the Ditsong Museum of Natural History.

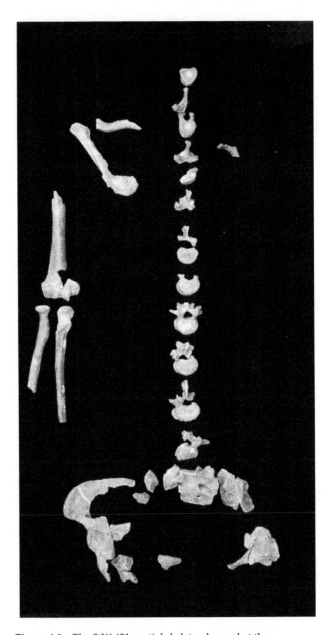

Figure 4.3 The StW 431 partial skeleton housed at the Evolutionary Studies Institute, University of the Witwatersrand.

individual. All physeal plates are obliterated, indicating adult status. Osteophytic growths along some of the vertebral bodies suggest that this individual was not a young adult, although no precise age estimate is possible.

Final note

Descriptions and discussion of each element attributed to these two remarkable skeletons are found within the chapters in Section 2 of this volume. Section 3 includes consideration of body size and proportions.

References

Benade, M.M., 1990. Thoracic and lumbar vertebrae of African hominids ancient and recent: morphology and functional aspects with specific reference to upright posture. M.A. Thesis, University of the Witwatersrand.

Broom, R., Robinson, J.T., 1947. Further remains of the Sterkfontein Ape-Man, *Plesianthropus*. Nature 160, 430–431.

Broom, R., Robinson, J.T., Schepers, G.W.H., 1950. Sterkfontein ape-man, *Plesianthropus*. Transvaal Museum Memoir 4, 1–117.

Clarke, R.J., 1998. First ever discovery of a well-preserved skull and associated skeleton of *Australopithecus*. South African Journal of Science 94, 460–463.

Kibii, J.M., Clarke, R.J., 2003. A reconstruction of the Stw 431 *Australopithecus* pelvis based on newly discovered fragments. South African Journal of Science 99, 225–227.

Robinson, J.T., 1972. Early hominid posture and locomotion. University of Chicago Press: Chicago.

Thackeray, F., Gommery, D., Braga, J., 2002. Australopithecine postcrania (Sts 14) from the Sterkfontein Caves, South Africa: the skeleton of "Mrs Ples"?: news & views. South African Journal of Science 98, 211–212.

Tobias, P.V., 1992. New researches at Sterkfontein and Taung with a note on Piltdown and its relevance to the history of palaeo-anthropology. Transactions of the Royal Society of South Africa, 1–14.

Toussaint, M., Macho, G., Tobias, P., Partridge, T., Hughes, A., 2003. The third partial skeleton of a late Pliocene hominin (StW 431) from Sterkfontein, South Africa. South African Journal of Science 99, 215–223.

5

Scapula, clavicle, and proximal humerus

David J. Green

Among the debates about the earliest stages of human evolution is whether or not the early hominin upper limb continued to serve a locomotor function following the advent of bipedalism. The Sterkfontein scapulae, clavicles, and proximal humeri represent an important resource in this debate, and we discuss them in detail here. Appendix 1 presents member, locus with grid reference, and date of discovery of each specimen where known.

Descriptions

Scapulae

Sts 7b—Right partial scapula

Preservation. The specimen is the lateral portion of an adult right scapula preserving the glenoid fossa and coracoid process, the base of the spine, about 71 mm of the axillary border, and small portions of the blade near the scapular neck (Figure 5.1). Its surface texture is weathered. The specimen is preserved in several pieces that have been glued together along large cracks, the most significant of which traverses the glenoid horizontally. The inferior half of the posterior margin of the glenoid is missing. The coracoid is missing its distal tip and appears depressed inferiorly and anteriorly, possibly as a result of post-depositional deformation. The spinoglenoid notch is traversed by the large crack that divides the glenoid. The spine is broken about 11 mm posteriorly from its base. Both the supraglenoid and infraglenoid tubercles are preserved. Some pieces of bone are missing along the axillary border, but the ventral "bar" of bone is preserved on the anterior aspect. Virtually none of the supraspinous

fossa remains, but small portions of the infraspinous fossa are preserved inferior and posterior to the spinoglenoid notch. The anterior side of the fragment is more concave than is typical for hominins, indicating some postmortem deformation likely occurred (Vrba, 1979).

Morphology. The coracoid projects about 23 mm anterosuperiorly relative to the glenoid and there is 17.5 mm between the broken coracoid and the anterior margin of the glenoid fossa. The coracoid is about 16 mm thick superoinferiorly at the point where it turns laterally (Figure 5.1). The spine projects dorsally in lateral view (Figure 5.1); at its base, the spinoglenoid notch is about 12 mm from the medial margin of the glenoid cavity (in posterior view). The glenoid measures 36.1 mm superoinferiorly and is estimated to have been 23 mm in mediolateral width, given the slightly damaged posterior margin. The infraglenoid tubercle is about 22.5 mm from the spinoglenoid notch and 8 mm from the inferior glenoid margin.

StW 162—Left partial scapula

Preservation. This specimen is the lateral portion of a left scapula including most of the glenoid fossa and about 38 mm of the bone medial to it (Figure 5.1). The superior and posterior margins are broken, giving the glenoid a pronounced "keyhole" appearance. A small piece of bone is missing, leaving a notch along the posterior border. The coracoid is broken off at its base. The infraglenoid tubercle is preserved, but none of the axillary border remains medially; a small portion of the ventral bar is preserved on the anterior surface. The lateral surface of spinoglenoid notch is preserved. The base of the spine is present throughout the width of the fragment

David J. Green, *Scapula, clavicle, and proximal humerus.* In: *Hominin Postcranial Remains from Sterkfontein, South Africa, 1936–1995.* Edited by: Bernhard Zipfel, Brian G. Richmond, and Carol V. Ward, Oxford University Press (2020). © Oxford University Press.
DOI: 10.1093/oso/9780197507667.003.0005

Sts 7b

StW 162

StW 366

StW 431d

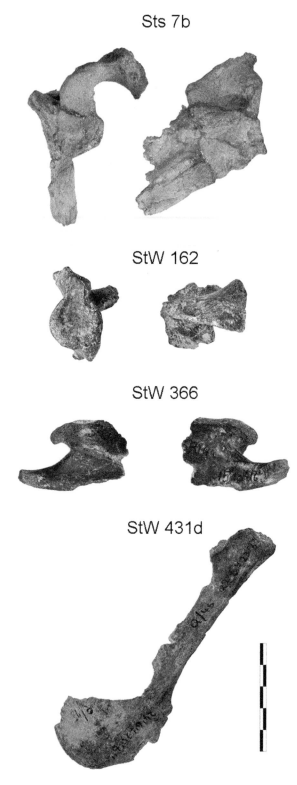

Figure 5.1 Sts 7b in lateral and posterior views, StW 162 in lateral and inferior views, StW 366 in superior and inferior views, and StW 431d in posterior view. Scale bar indicates 5 mm.

but is broken 17 mm posterior to its root. A small portion of the supraspinous fossa remains superior to the spine. The subscapular fossa is preserved at the scapular neck as is a correspondingly small portion of the infraspinous fossa.

Morphology. The spine projects posteriorly from its base, somewhat more superiorly than in Sts 7b. The spinoglenoid notch is 15.5 mm from the medial margin of the glenoid cavity (in posterior view). The glenoid measures 36.5 mm superoinferiorly and 25.9 mm mediolaterally (the broken posterior margin is superior to the point where this measurement is taken). The scapular neck is broad, with the infraglenoid tubercle about 29 mm from the spinoglenoid notch. The infraglenoid tubercle is 15 mm from the inferior glenoid margin.

StW 366—Left scapular fragment

Preservation. This fragment preserves 39 mm of the blade (Figure 5.1). An incipient process representing the medial margin of the glenoid is visible, but the glenoid and coracoid process are missing. The adjoining portion of the spine is preserved laterally as far as the medial edge of the acromion, 30 mm lateral to the spinoglenoid notch. Only 12 mm of the posterosuperior margin of the glenoid remains. Starting at the medial edge of the fragment, the posterior margin of the spine is broken off, then is intact for 15 mm, then is abraded away. A portion of the supraspinous fossa remains superior to the spine, including a nutrient foramen, but little else; a correspondingly small portion of the infraspinous fossa remains. None of the axillary border remains.

Morphology. A low, roughened ridge for the attachment of deltoid runs along inferior margin of the preserved section of the spine. The spinoglenoid notch is about 7 mm wide and is 9 mm from the inferior margin of the spine above it. The medial aspect of the acromion rises about 24.5 mm above the blade. The spine is 9.2 mm thick at its lateral margin.

StW 431d—Right partial scapula

Preservation. This specimen is most of the axillary border of a right scapula associated with the StW 431 partial skeleton (Figure 5.1). The bone is preserved from the infraglenoid tubercle to the inferior angle, and is about 138 mm long. What remains of the scapula is well preserved. The axillary border is complete, save for a small portion near the inferior angle. On the anterior surface, the ventral bar is fully preserved, but little of the scapular blade remains. About 20 mm of the vertebral border and adjacent part of the blade are also preserved.

Morphology. The distance between the infraglenoid tubercle and the inferior angle is 133 mm. There is a distinct ridge between the well-defined teres major and minor attachment sites. An elevated ridge of bone on the ventral side is evident, which is broken just superior to the inferior angle. Often, the bone of the scapular blade is translucent, save for the scapular border and visible thickenings of bone that run parallel to the muscular line of the subscapularis; this ridge is continuous with the most inferior of these thickened bony "crests."

Clavicles

StW 431g—Right partial clavicle

Preservation. This specimen—associated with the StW 431d scapula and partial skeleton—is 73 mm of the central region of a right clavicular diaphysis that is missing both articular ends (Figure 5.2). The medial break is through the medial curvature; most of the lateral curvature is preserved. The conoid tubercle is present at the point where the bone curves posteriorly toward the acromial end. The medial aspect shows a jagged, but likely fresh, break.

Morphology. This clavicle measures 8.9 mm superoinferiorly by 13.2 mm anteroposteriorly at the medial break and 7.5 mm by 14.8 mm laterally. A prominent crest for deltoid is visible along the anterior margin of the superior surface of the bone, which also gives the conoid tubercle a somewhat more prominent appearance. Posterior and medial to it, the bone surface is barely concave anteroposteriorly. The inferior surface of the bone is concave anteroposteriorly toward the lateral break as the bone flattens toward its acromial end.

StW 582—Right partial clavicle

Preservation. This specimen is 70 mm of a right clavicle preserving neither articular end; more of the medial portion of the shaft is preserved than in StW 431g (Figure 5.2). Most of the medial and lateral curvatures are preserved, as is part of the conoid tubercle.

Morphology. The diaphysis is 8.2 mm superoinferiorly by 10.5 mm anteroposteriorly at the medial break, and 7.6 mm by 11.4 mm laterally. The conoid tubercle is weak, as is the deltoid crest which lies along the ventral margin of the superior surface. At the lateral end, the inferior surface becomes anteroposteriorly concave as it approaches the acromial end.

Humeri

Sts 7a—Right proximal humerus

Preservation. This is more than half of a proximal humerus and diaphysis—associated with the Sts 7b scapula—that is 215 mm in maximum length (Figure 5.3). The head is well

StW 431g

StW 582

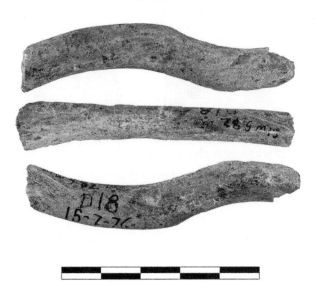

Figure 5.2 StW 431g and StW 582 in superior, anterior, and inferior views.

preserved, as are the tubercles, but the inferior region of the greater tubercle is crushed. The diaphysis distal to this point is also severely damaged, having been crushed anteroposteriorly and mediolaterally and broken into many fragments. The inferior portion of the shaft (medial border length about 50 mm) has no apparent distortion other than a large crack (1.5 mm wide) running the length of the medial cortex.

Morphology. A prominent crest separating the supraspinous, infraspinous, and teres minor attachment sites is clearly visible, and the lesser tubercle—with the attachment site for the subscapularis muscle—is preserved.

Sts 7a

Sts 328

Sts 517

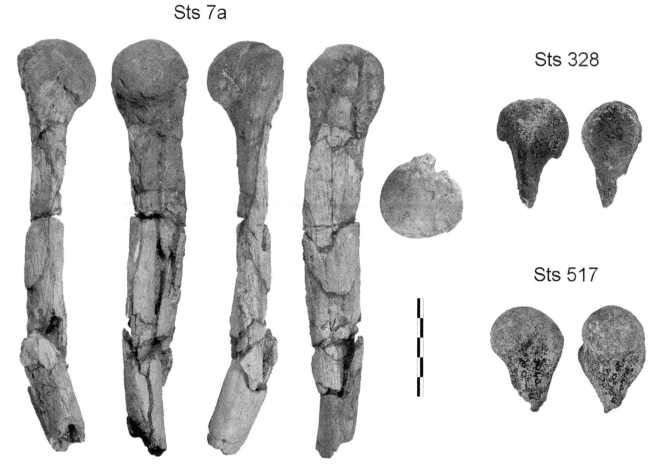

Figure 5.3 Sts 7a humeral fragment in anterior, medial, posterior, lateral, and superior views. StW 328 in anterior and medial views, StW 517 in anterior and medial views.

The humeral head is undistorted, despite the damage to the diaphysis, and measures 40 mm superoinferiorly by 39 mm anteroposteriorly.

StW 328—Right proximal humerus

Preservation. This is the anterior half of a proximal humerus and about 25 mm of the medial portion of the diaphysis (Figure 5.3). The head is mediolaterally narrow owing to considerable surface abrasion and exfoliation, which have exposed a good deal of cancellous bone. Most margins of the preserved portion are badly abraded except for the superior margin, which is only lightly abraded, and an intact section 13 mm long inferiorly. The greater tubercle is preserved for 23 mm of its superoinferior length, though its posterior margin is abraded away. The surface of the tubercle is pitted, obscuring surface morphology. The lesser tubercle remains but is missing bone along its posterior margin. The intertubercular groove is preserved throughout the length of the fragment, but its surface is also pitted. *Morphology.* The head projects proximally above the greater tubercle and measures ~33.5 mm superoinferiorly

and 28.5 mm anteroposteriorly. No distinct muscle insertion sites are preserved on the greater tubercle, but the subscapularis insertion site is clear on the lesser tubercle and measures about 18 mm superoinferiorly by 10 mm mediolaterally. Two nutrient foramina are visible just anterior to this point. The medial margin of the diaphysis forms a distinct, rounded ridge flanked by flatter areas of the bone that are set at about a right angle to one another.

StW 517—Left proximal humerus

Preservation. This is the head and about 22 mm of the medial aspect of adjacent shaft (Figure 5.3). About two-thirds of the lesser tubercle is present, but the greater tubercle and intertubercular groove are absent. The head is missing its superoanterior margin and there is pitting across the articular surface, but this does not obscure its contours.
Morphology. The head measures ~36 mm superoinferiorly and 34.5 mm anteroposteriorly. The subscapularis insertion face is present, but damaged superiorly; its mediolateral width is 9.2 mm.

Comparative morphology

Materials and methods

Metric data

An Immersion MicroScribe G2 digitizer was used to collect three-dimensional landmark points to calculate scapular lengths, indices, and angles on comparative samples of adult modern humans (n = 108), chimpanzees (110), gorillas (138), and orangutans (53); other measurements were taken on the anterior aspect of the scapula and proximal humerus with plastic sliding digital calipers (Figure 5.4; Table 5.1). Available landmarks were recorded for the Sterkfontein material, in addition to comparative fossil material of *Australopithecus afarensis* (A.L. 288-1), *A. sediba* (MH2), and early *Homo* (KNM-WT 15000)[1]. Angle measurements were derived in three dimensions with the software package "R" (Ihaka and Gentleman, 1996). Angles were calculated in space by first defining a plane represented by three of four points that also represented the two lines of

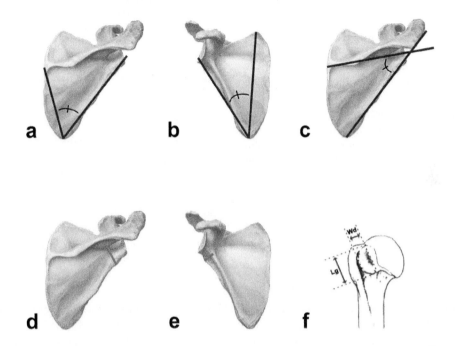

Figure 5.4 Depictions of measurements used in this study (Table 5.1). (a) axillary/infraspinous medial border angle, (b) ventral bar/glenoid angle, (c) axillary border/spine angle, (d) infraspinous neck width, (e) lateral expansion of the subscapularis fossa, and (f) length and width of the subscapularis insertion facet on the lesser tubercle (images 'a–e' modified from Schuenke et al., 2007, 'f' is from Larson, 1995).

Table 5.1 List of measurements used in this study.

Measurement name	Description
Axillary/infraspinous medial border angle	The angle formed by the medial border and infraspinous breadth line
Ventral bar/glenoid angle	The angle formed by the "ventral bar" of the subscapularis fossa and glenoid height line
Axillary border/spine angle	The angle formed by the axillary border and the spine
Spine/glenoid angle	The angle formed by the spine and glenoid height line
Infraspinous neck width	The distance between the infraglenoid tubercle and the spinoglenoid notch
Lateral expansion of subscapularis fossa	The perpendicular distance from the infraglenoid tubercle to the ventral bar
Subscapularis insertion facet height	The total height of the subscapularis insertion facet on the lesser tubercle of the proximal humerus
Subscapularis insertion facet width	The total width of the insertion facet, perpendicular to the height

See also Figure 5.4 for depictions of above measurements.

1 The subadult KNM-WT 15000 specimen is being considered here since it is comparable in size to the adult fossils.

interest and then projecting those lines against the plane so that they might intersect to calculate a given angle.

Measurements requiring further estimation

Since some of the specimens are missing landmarks, adjustments were necessary to estimate certain measurements. In lieu of the original lines between landmark points to define angles in the extant samples, lines were estimated as closely as possible in order to measure the comparable angles.

Sts 7b

Because the blade of Sts 7b is especially curved, the line used to estimate spine orientation went from the spinoglenoid notch to a point along the line of the base of the spine just below the medial extent of the preserved spine (Figure 5.1). This line was used to estimate both spinoglenoid and axillospinal angles. To account for the curvature of the axillary border, this line was deemed to be between the infraglenoid tubercle and a point about midway and down the preserved border and slightly posterior to the bone to account for the damage. This line was also used to estimate axillospinal angle. Ventral bar/glenoid angle was estimated without issue, though the more lateral portion of the ventral bar was used to define that line so the angle measurement would not be affected by the deformation of the fossil.

StW 162

As in Sts 7b, the spinoglenoid notch and the available portion of the spine medial to it were used as the second point of the line to estimate spinoglenoid angle (Figure 5.1). Very little of the ventral bar is preserved, and what

is preserved medial to the infraglenoid tubercle is damaged. Just enough of the bar is present to estimate two points for measuring ventral bar/glenoid angle and the lateral expansion of the subscapularis fossa.

StW 431d

This specimen (Figure 5.1) preserves very little of the vertebral border superior to the inferior angle. Although there are no landmarks superior to this point, one can extrapolate several possible paths of the vertebral border, rendering estimates of blade shape speculative at best. Thus, the estimate presented below is tentative, but in line with Berger's (1994) assessment.

Expanded discussions of how comparative fossil measurements were taken may be found in Green (2010) and Green and Alemseged (2012).

Results

Bivariate scapular shape characteristics

Scapular blade shape

The angle formed by the axillary border and the infraspinous portion of the medial border broadly describes the overall mediolateral breadth of the scapula. *Pan* has the narrowest axillary/infraspinous medial border angle, while *Gorilla* is broader, with a larger average angle than both *Homo* and *Pongo* (Table 5.2). Of the Sterkfontein material, only an estimate of the StW 431d axillary/infraspinous medial border angle is available; this relatively high value is nearest to *Gorilla*, but outside its 95% confidence limit (Table 5.2). KNM-WT 15000 is mediolaterally narrower than StW 431d, most like comparably aged *Pan*

Table 5.2 Scapular angular measures; mean values (and confidence intervals) are reported for extant taxa.

	Axillary/infraspinous medial border angle (°)	Ventral bar/glenoid angle (°)	Axillary border/spine angle (°)	Spine/glenoid angle (°)
Pan troglodytes	42.9[a]	127.8[a]	22.8[a]	91.5[d]
	(42.1–43.6)	(126.8–128.7)	(22.0–23.7)	(90.4–92.5)
Gorilla gorilla	54.9[d]	130.0[b]	30.1[b]	89.6[c]
	(54.2–55.7)	(129.1–131.0)	(29.4–30.8)	(88.8–90.3)
Pongo pygmaeus	46.9[b]	131.9[b]	34.9[c]	82.3[a]
	(45.7–48.1)	(130.1–133.7)	(33.4–36.6)	(80.5–84.2)
Homo sapiens	51.5[c]	142.3[c]	45.6[d]	87.6[b]
	(50.9–52.2)	(141.1–143.5)	(44.6–46.6)	(86.6–88.5)
Sts 7b	–	128.0	29.8	83.4
StW 162	–	124.3	–	81.7
StW 431d	59.3	–	–	–
A.L. 288-1	–	132.2	38.2	83.7
MH2	55.8	131.3	40.0	83.5
KNM-WT 15000	47.5	137.6	69.6	68.2

Differences between extant taxa were assessed by a one-way ANOVA (analysis of variance). Mean values with different superscript letters represent taxa that were significantly different at the α ≤ 0.05 level and in ascending order (e.g., *Pan* had a significantly smaller ventral bar/glenoid angle than *Homo*); taxa with the same superscript were not significantly different from one another (e.g., *Gorilla* and *Pongo* for ventral bar/glenoid angle).

(0.85 [0.82–0.88]), just less than that of *Homo* (0.92 [0.91–0.94]) and greater than *Gorilla* (0.85 [0.83–0.86]) (Green and Alemseged, 2012).

Lateral expansion of the subscapularis fossa

Pan and *Gorilla* have the broadest lateral expansions of the subscapularis fossa relative to glenoid size, followed by *Pongo* and then *Homo*, which is significantly smaller than all three great ape taxa (Figure 5.7; Table 5.4). All the australopiths all have relatively broad lateral expansions; StW 162, A.L. 288-1, and MH2 all fall within the *Pan* confidence limit, while Sts 7b is just above it (Figure 5.7; Table 5.4). The KNM-WT 15000 lateral expansion is also within the confidence limit of comparably aged *Pan* (0.54 [0.53–0.56]) (Green and Alemseged, 2012).

Clavicle

With both articular ends missing, it is difficult to measure clavicular curvature in the Sterkfontein fossils for reconstructing pectoral girdle position relative to the upper thorax (Ohman, 1986; Voison, 2006). In superior view, both Sterkfontein clavicles share an anteriorly concave curvature laterally, with incipient medial curvatures (anteriorly convex); this morphology is present in modern human clavicles but also shared by many great apes. It is difficult to assess superior and inferior curvatures when the clavicle is viewed anteroposteriorly, given the limited preservation. As preserved, the Sterkfontein clavicles do not show marked superior curvature near the medial end of the bone. Modern humans, with lower and broader shoulders, display similarly low levels of anteroposterior curvature, while the more cranially oriented clavicles of African apes generally show pronounced medial curvature (Voison, 2006). The conoid tubercles of StW 431g and 582 are not

prominent and the deltoid origin site has a more anterior position along the lateral curve than in apes, similar to the condition seen in humans. These traits also align the Sterkfontein clavicles more closely with *Homo* than the African apes (Ohman, 1986; Partridge et al., 2003).

Proximal humerus

The relative height and width of the subscapularis insertion facet are considered together and also relative to humeral head size (Tables 5.5 and 5.6). *Pongo* and *Homo* have the smallest ratios of subscapularis insertion facet height and width, while *Pan* and *Gorilla* have the largest ratio values, with insertion facet height being nearly twice the width (Table 5.6). Sts 7a, KNM-ER 1473, and Omo 119-73-2718 have ratios that were more similar to the African apes, while StW 517 has the largest ratio of all sampled fossils and extant taxa (Table 5.6). Both StW 328 and KNM-BC 1745 fall intermediate between the African apes and *Pongo* and *Homo*. A.L. 288-1 has the lowest subscapularis insertion height:width ratio of all the fossils, it being just lower than the *Pongo* lower bound (Table 5.6). A.L. 333-87 and A.L. 333-107 are most similar to *Homo* in subscapularis insertion height:width.

Discussion and conclusions

Apelike characteristics of the australopith shoulder are commonly cited as evidence of arboreal activity (Vrba, 1979; Stern and Susman, 1983; Susman et al., 1984; Alemseged et al., 2006; Larson, 2007; Green and Alemseged, 2012). Hunt (1991: 529) proposed that "a cranially oriented glenoid fossa may be an adaptation to

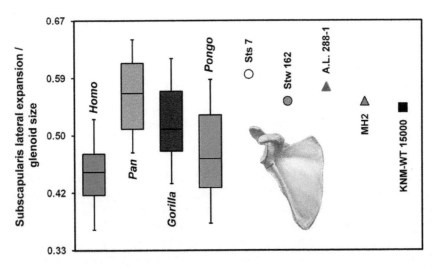

Figure 5.7 Box plots of subscapularis lateral expansion breadth (divided by glenoid size) in extant taxa as compared with fossil individuals.

Table 5.5 Raw and relative proximal humerus dimensions; mean values (and confidence intervals) are reported for extant taxa.

	Humeral head height (superoinferior)	Humeral head width (anteroposterior)	Humeral head size [(ht. × wd.)^0.5]
Pan troglodytes	38.5[a]	39.6[a]	39.0[a]
	(37.9–39.1)	(39.0–40.2)	(38.4–39.6)
Gorilla gorilla	54.8[c]	56.8[b]	55.8[c]
	(53.5–56.1)	(55.5–58.1)	(54.5–57.0)
Pongo pygmaeus	42.4[b]	40.9[a]	41.6[b]
	(41.0–43.7)	(39.6–42.1)	(40.3–42.9)
Homo sapiens	43.4[b]	41.0[a]	42.1[b]
	(42.6–44.1)	(40.3–41.6)	(41.5–42.8)
Sts 7a	40.3	39.2	39.7
StW 328	33.4	28.5	30.8
StW 517	36.2	34.6	35.4
A.L. 288-1	28.3	26.8	27.5
A.L. 333-107	41.3	35.0	38.0
KNM-BC 1745	30.0	28.5	29.2
KNM-ER 1473	43.3	43.7	43.5
Omo 119-73-2718	40.0	36.0	37.9

Differences between extant taxa were assessed by a one-way ANOVA. Mean values with different superscript letters represent taxa that were significantly different at the $\alpha \leq 0.05$ level and in ascending order.

Table 5.6 Raw and relative subscapularis insertion site (lesser tubercle) dimensions; mean values (and confidence intervals) are reported for extant taxa.

	Subscapularis insertion height	Subscapularis insertion height/humeral head size	Subscapularis insertion width	Subscapularis insertion width/humeral head size	Subscapularis insertion height/ width
Pan troglodytes	22.2[a]	0.57[c]	11.4[a]	0.29[b]	1.95[c]
	(21.7–22.6)	(0.56–0.58)	(11.2–11.7)	(0.288–0.30)	(1.91–1.99)
Gorilla gorilla	31.4[b]	0.56[b]	15.8[d]	0.28[a]	1.99[d]
	(30.4–32.3)	(0.55–0.57)	(15.3–16.2)	(0.278–0.289)	(1.95–2.04)
Pongo pygmaeus	21.5[a]	0.52[a]	13.2[c]	0.32[c]	1.65[a]
	(20.6–22.4)	(0.50–0.53)	(12.6–13.7)	(0.31–0.33)	(1.57–1.72)
Homo sapiens	21.3[a]	0.50[a]	12.4[b]	0.29[b]	1.70[b]
	(20.5–21.7)	(0.55–0.57)	(12.1–12.8)	(0.289–0.30)	(1.67–1.74)
Sts 7a	23.4	0.59	10.9	0.27	2.15
StW 328	17.7	0.57	9.8	0.32	1.81
StW 517	22.0	0.62	9.2	0.26	2.39
A.L. 288-1	14.2	0.52	9.1	0.33	1.56
A.L. 333-87	24.5	-	14.3	-	1.71
A.L. 333-107	25.5	0.67	14.4	0.38	1.77
KNM-BC 1745	22.0	0.75	11.8	0.40	1.86
KNM-ER 1473	23.0	0.53	11.5	0.26	2.00
Omo 119-73-2718	25.2	0.66	12.6	0.33	2.00

Differences between extant taxa were assessed by a one-way ANOVA. Mean values with different superscript letters represent taxa that were significantly different at the $\alpha \leq 0.05$ level and in ascending order.

distributing strain more evenly over the glenohumeral joint capsule during unimanual arm-hanging, since a laterally or ventrally oriented glenoid fossa concentrates strain on the caudal aspect of the glenohumeral joint capsule." Sts 7b and StW 162's glenohumeral joints are more cranially oriented than A.L. 288-1 and MH2, which are both more cranially oriented than modern humans (Table

5.2). All arboreal hominoids share a cranially oriented shoulder joint, supporting its characterization as a feature related to climbing (Shea, 1986; Hunt, 1991).

Mensforth et al. (1990), Inouye and Shea (1997), and Haile-Selassie et al. (2010) challenged the assertion that the glenohumeral joint of A.L. 288-1 should garner functional attention, instead attributing its cranial orientation to that individual's small size. Mensforth et al. (1990) and Inouye and Shea (1997) used ontogenetic considerations, not to understand how the joint position changed throughout development but to find extant individuals of a comparable size to A.L. 288-1. They contended that there was overlap between juvenile modern humans and apes and that the differences only manifested later in development. Stern (2000) criticized this approach because ontogenetic allometry is not necessarily equivalent to intraspecific allometry among adults. Furthermore, the results presented here and elsewhere (Green and Alemseged, 2012; Green, 2013) show both Sts 7b and StW 162 specimens to be larger (based on glenoid dimensions) than A.L. 288-1, and also possess more cranially oriented glenohumeral joints (Table 5.2). Additionally, a scapula attributed to the A.L. 288-1-sized *Homo floresiensis* (LB6/4) has a "hyperhuman", laterally facing glenohumeral joint (Morwood et al., 2005; Larson et al., 2007: 725). Put together, this suggests that hominin glenohumeral joint orientation cannot be explained by allometry.

The orientation of the scapular spine is an important indicator of dorsal scapular blade shape. Larson et al. (1991) updated models proposed by Oxnard (1963, 1967) and Ashton and Oxnard (1964) with regard to the functional importance of spine orientation. Larson et al. (1991) contended that the functionally relevant aspect of spine orientation related to its base (as opposed to the dorsal crest), which determined the line of action of the dorsal members of the rotator cuff. This is of particular relevance for locomotor considerations, as Larson and Stern (1986) previously determined the infraspinatus muscle to be principally involved in glenohumeral joint stabilization during suspensory activities. The Sts 7b axillary border/spine angle measure was most similar to *Gorilla*, while A.L. 288-1 and MH2 were intermediate between *Pongo* and *Homo* (Table 5.2).

Larson and Stern (1986, 2013) and Larson (1988) argued that the subscapularis muscle plays an important role in vertical climbing and that relative expansion of the inferolateral portion of the fossa might be advantageous for suspensory primates (Larson, 1995). The results presented above support this hypothesis—the subscapularis lateral expansion is widest in the great apes and narrowest in *Homo* (Table 5.4). All of the fossils showed particularly broad lateral expansions, within the chimpanzee and gorilla range. The australopiths' lateral expansions accord with previous assertions of their primitive scapulae, but the broader region in KNM-WT 15000 is at odds with its otherwise humanlike appearance.

Larson and Stern (1986) also suggested that shoulder mobility might be enhanced by functional differentiation of the upper and lower portions of the subscapularis muscle (see also Omi et al., 2010). Larson (1995) extended this hypothesis by proposing that this would be evident in the relative shape (height and width) of the subscapularis' insertion site on the lesser tubercle of the proximal humerus. As hypothesized, the modern human insertion height:width ratio was significantly lower than *Pan* and *Gorilla*, but *Pongo* had the lowest average value of all taxa (Table 5.6). Furthermore, this trait might be most relevant for broadly distinguishing suspensory taxa from non-suspensory taxa, as might also have been the case for infraspinous neck breadth (Table 5.4). The average East African *Australopithecus* subscapularis insertion height:width ratio was 1.82, while the South African (Sterkfontein, *A. africanus*) average was 2.12 (Table 5.6). The East African average was intermediate between the African apes and *Pongo* and *Homo*, while the South African average was greater than all of the extant taxa. As with glenohumeral joint and spine orientation, this might be indicative of slight differences in scapular shape between *A. africanus* and *A. afarensis*, which accords with variation in postcranial anatomy and limb proportions thought to reflect differences in the role of arboreal behaviors in their locomotor repertoires (McHenry and Berger, 1998; Richmond et al., 2002; Green et al., 2007; Harmon, 2009). More fossils are needed to confirm this hypothesis, however, as a discriminant function analysis of traits shared by the two species did not find a significant difference between them.

Acknowledgments

I wish to thank B. Richmond, C. Ward, and B. Zipfel for asking me to participate in this workshop and contribute to this volume, and for their helpful comments. D. Hunt, L. Gordon, E. Westwig, I. Tattersall, G. Garcia, J. Chupasko, M. Omura, Y. Haile-Selassie, L. Jellema, M. Harman, A. Gill, E. Mbua, S. Muteti, M. Yilma, P.V. Tobias, B. Zipfel, S. Potze, and T. Perregil were invaluable to me in helping to coordinate museum visits. I greatly appreciate A. Gordon's analytical assistance with R code, R. Wunderlich's photography, and B. Seitelman's help in preparing the figures. Finally, I wish to acknowledge the National Science Foundation IGERT grant (9987590), NSF Doctoral Dissertation Improvement Grant (BCS-0824552), NSF (BCS-0914687), The Leakey Foundation, and the Wenner-Gren Foundation.

References

Alemseged, Z., Spoor, F., Kimbel, W.H., Bobe, R., Geraads, D., Reed, D., Wynn, J.G., 2006. Juvenile early hominin skeleton from Dikika, Ethiopia. Nature 443, 296–301.

Ashton, E.H., Oxnard, C.E., 1964. Functional adaptations in the primate shoulder girdle. Proceedings of the Zoological Society of London 142, 49–66.

Berger, L.R., 1994. Functional morphology of the hominoid shoulder, past and present. Ph.D. Dissertation, University of the Witwatersrand.

Green, D.J., 2010. Shoulder functional anatomy and development—implications for interpreting early hominin locomotion, Ph.D. Dissertation, George Washington University.

Green, D.J., 2013. Ontogeny of the hominoid shoulder: the influence of locomotion on morphology. American Journal of Physical Anthropology 152, 239–260.

Green, D.J., Alemseged Z., 2012. *Australopithecus afarensis* scapular ontogeny, function, and the role of climbing in human evolution. Science 338, 514–517.

Green, D.J., Gordon, A.D., Richmond, B.G., 2007. Limb-size proportions in *Australopithecus afarensis* and *Australopithecus africanus*. Journal of Human Evolution 52, 187–200.

Haile-Selassie, Y., Latimer, B.M., Alene, M., Deino, A.L., Gibert, L., Melillo, S.M., Saylor, B.Z., Scott, G.R., Lovejoy, C.O., 2010. An early *Australopithecus afarensis* postcranium from Woranso-Mille, Ethiopia. Proceedings of the National Acadamy of Sciences 107, 12121–12126.

Harmon, E.H., 2009. The shape of the early hominin proximal femur. American Journal of Physical Anthropology 139, 154–171.

Hunt, K.D., 1991. Mechanical implications of chimpanzee positional behavior. American Journal of Physical Anthropology 86, 521–536.

Ihaka, R., Gentleman, R., 1996. R: a language for data analysis and graphics. Journal of Computational and Graphical Statistics 5, 299–314.

Inouye, S.E., Shea, B.T., 1997. What's your angle? Size correction and bar-glenoid orientation in "Lucy" (A.L. 288-1). International Journal of Primatology 18, 629–650.

Larson, S.G., 1988. Subscapularis function in gibbons and chimpanzees: implications for interpretation of humeral head torsion in hominoids. American Journal of Physical Anthropology 76, 449–462.

Larson, S.G., 1995. New characters for the functional interpretation of primate scapulae and proximal humeri. American Journal of Physical Anthropology 98, 13–35.

Larson, S.G., 2007. Evolutionary transformation of the hominin shoulder. Evolutionary Anthropology 16, 172–187.

Larson, S.G., Jungers, W.L., Morwood, M.J., Sutikna, T., Jatmiko, Saptomo, E.W., Due, R.A., Djubiantono, T., 2007. *Homo floresiensis* and the evolution of the hominin shoulder. Journal of Human Evolution 53, 718–731.

Larson, S.G., Stern, J.T. Jr., 1986. EMG of scapulohumeral muscles in the chimpanzee during reaching and "arboreal" locomotion. American Journal of Anatomy 176, 171–190.

Larson, S.G., Stern, J.T., Jr., 2013. Rotator cuff muscle function and its relation to scapular morphology in apes. Journal of Human Evolution 65, 391–403.

Larson, S.G., Stern, J.T. Jr., Jungers, W.L., 1991. EMG of serratus anterior and trapezius in the chimpanzee: scapular rotators revisited. American Journal of Physical Anthropology 85, 71–84.

McHenry H.M., Berger, L.R., 1998. Body proportions in *Australopithecus afarensis* and *A. africanus* and the origin of the genus *Homo*. Journal of Human Evolution 35, 1–22.

Mensforth, R.P., Latimer, B., Senturia, S., 1990. A review of the functional significance of the AL-288 axilloglenoid angle. American Journal of Physical Anthropology 81, 267–268.

Morwood, M.J., Brown, P., Jatmiko, Sutikna, T., Saptomo, E.W., Westaway, K.E., Due, R.A., Roberts, R.G., Maeda, T., Wasisto, S., 2005. Further evidence for small-bodied hominins from the Late Pleistocene of Flores, Indonesia. Nature 437, 1012–1017.

Ohman, J.C., 1986. The first rib of hominoids. American Journal of Physical Anthropology 70, 209–229.

Omi, R., Sano, H., Ohnuma, M., Kishimoto, K.N., Watanuki, S., Tashiro, M., Itoi, E., 2010. Function of the shoulder muscles during arm elevation: an assessment using positron emission tomography. Journal of Anatomy 216, 643–649.

Oxnard, C.E., 1963. Locomotor adaptations in the primate forelimb. Symposia of the Zoological Society of London 10, 165–182.

Oxnard, C.E., 1967. The functional morphology of the primate shoulder as revealed by comparative anatomical, osteometric and discriminant function techniques. American Journal of Physical Anthropology 26, 219–240.

Partridge, T.C., Granger, D.E., Caffee, M.W., Clarke, R.J., 2003. Lower Pliocene hominid remains from Sterkfontein. Science 300, 607–612.

Richmond, B.G., Aiello, L.C., Wood, B.A., 2002. Early hominin limb proportions. Journal of Human Evolution 43, 529–548.

Shea, B.T., 1986. Scapula form and locomotion in chimpanzee evolution. American Journal of Physical Anthropology 70, 475–488.

Stern, J.T., 2000. Climbing to the top: a personal memoir of *Australopithecus afarensis*. Evolutionary Anthropology 9, 113–133.

Stern, J.T., Susman, R.L, 1983. The locomotor anatomy of *Austalopithecus afarensis*. American Journal of Physical Anthropology 60, 279–317.

Susman, R.L, Stern, J.T., Jungers, W.L., 1984. Arboreality and bipedality in the Hadar hominids. Folia Primatologia 43, 113–156.

Voison, J-L., 2006. Clavicle, a neglected bone: morphology and relation to arm movements and shoulder architecture in primates. Anatomical Record 288A, 944–953.

Vrba, E.S., 1979. A new study of the scapula of *Australopithecus africanus* from Sterkfontein. American Journal of Physical Anthropology 51, 117–130.

Walker, A.C., Leakey, R.E.F., 1993. The postcranial bones. In: Walker A.C., Leakey, R.E.F. (Eds.), The Nariokotome *Homo erectus* Skeleton. Harvard University Press: Cambridge, pp. 95–160.

6

Distal humerus

Michael R. Lague and Colin G. Menter

All six of the distal humeri described here have been described by Menter (2002); four (StW 38, 124, 150, 182) were originally described by Senut and Tobias (1989), and one (StW 431c) was described by Toussaint et al. (2003). With the exception of StW 150 and 182, the specimens unambiguously derive from Member 4. StW 150 and 182 were originally described by Senut and Tobias (1989) as having derived from "Member 5" and "Member 4 or 5," respectively; some subsequent studies have indicated Member 4 as the source for both specimens (Pickering et al., 2004; Clarke, 2013; see review by Grine, 2019). Additional Sterkfontein distal humeri that are not described below (due to unavailability) include a specimen from Jacovec Cavern (StW 602; Partridge et al., 2003) and the humeri associated with the largely undescribed StW 573 skeleton from the Silberberg Grotto (Clarke, 2008). Appendix I presents Member, locus with grid reference, and date of discovery (where known) of each specimen.

Descriptions

StW 38—Right distal humeral shaft fragment

Preservation. StW 38 is broken proximal to the articular surface and measures only 46 mm long, with a maximum anteroposterior dimension of 16.8 mm (Figure 6.1). It is mediolaterally widest at the distal end (35.7 mm), where only the most proximal parts of the olecranon fossa (~8 mm deep) and the adjacent medial and lateral pillars are preserved. Although the anterolateral side of the specimen is badly crushed and missing bone, its lateral border is mostly preserved, albeit with an irregular edge. Spalling

of cortical bone is apparent on the distal two-thirds of the posterolateral surface.

Morphology. At the proximal break (19 mm wide mediolaterally), the diaphysis is triangular in section and preserves a rounded medullary canal (~3 mm diameter). Anteroposterior cortical thickness is 4.8 mm on the posterior side but cannot be measured with certainty elsewhere. The posterior surface proximal to the olecranon fossa is flat. The well-preserved medial border is rounded with a faint crest and is mildly concave in anterior view.

StW 124—Left distal humeral shaft fragment

Preservation. This 41.4 mm-long shaft fragment is broken proximal to the distal articular surface and has a maximum anteroposterior width of 13.8 mm (Figure 6.1). The distal fracture is oriented obliquely (proximolateral to distomedial) such that the only preserved aspect of the olecranon fossa is its proximomedial margin. The adjacent medial pillar extends further distally than the rest of the specimen and measures 12 mm mediolaterally at the distal break. The most distal point on the specimen was likely just proximal to the most proximal point of the medial trochlear crest (both anteriorly and posteriorly) and lateral to the missing medial epicondyle. No lateral pillar is present and the entire lateral diaphyseal border has been sheared away and likely would have extended 3–4 mm further laterally. Pitting marks the fragment anteriorly and medially, and the posterior surface shows fine-lined cracks laterally.

Morphology. The medial border is marked by a faint crest on an otherwise rounded border that is mildly abraded. The proximal break is roughly triangular in section, with a rounded medullary canal (~5 mm in diameter),

Michael R. Lague and Colin G. Menter, *Distal humerus*. In: *Hominin Postcranial Remains from Sterkfontein, South Africa, 1936–1995*. Edited by: Bernhard Zipfel, Brian G. Richmond, and Carol V. Ward, Oxford University Press (2020). © Oxford University Press.
DOI: 10.1093/oso/9780197507667.003.0006

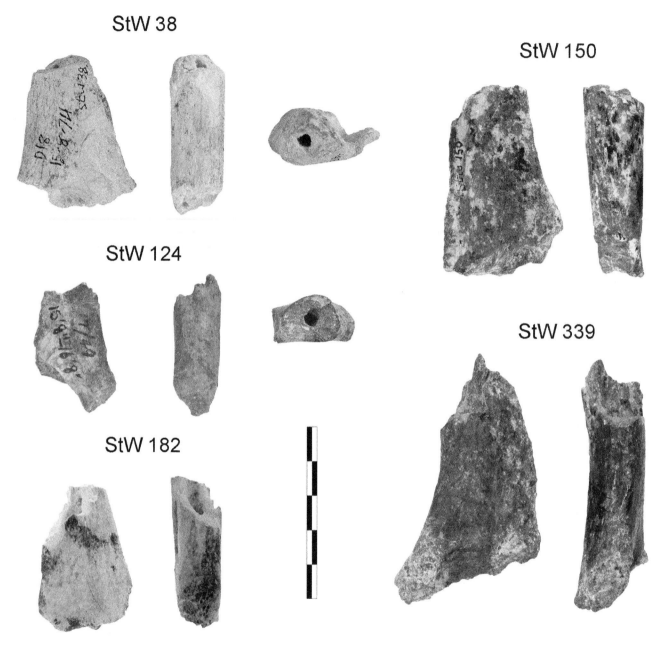

StW 38

StW 150

StW 124

StW 339

StW 182

Figure 6.1 Six distal humeri from Sterkfontein. Most fossils shown in posterior (left photo in each case) and medial views.

surrounded by cortical bone of varying thickness (4–5). The posterior surface exhibits a mild, centrally placed, longitudinal concavity that becomes subtly deeper and wider as it approaches the proximal edge of the olecranon fossa.

StW 150—Left distal humeral shaft fragment

Preservation. StW 150 is a 58 mm-long fragment that is broken proximal to the distal articular surface, with a maximum mediolateral width of 37 mm (across the distal break) (Figure 6.1). It preserves virtually all the radial

fossa, the proximal half of the coronoid fossa, the most proximal border and floor of the olecranon fossa (~6 mm deep), and the most proximal portions of the medial and lateral pillars. The surface is weathered and a longitudinal crack traverses the proximal three-quarters of the anterior shaft. A deeper gash (16 mm long) extends obliquely from the proximal break on the posterior surface. The lateral border of the fragment is abraded and preserves some adhering breccial remnants.

Morphology. Distinct lateral and medial supracondylar ridges on the edges of the specimen are nearly in the same plane as the flattened posterior surface. The lateral supracondylar ridge (up to 3 mm wide) is rounded and flanked anteriorly along its length by a narrow groove. An obliquely oriented crest borders the radial fossa laterally that would have extended to form the proximal aspect of the lateral epicondyle. The posterior diaphysis is mildly concave due to a shallow longitudinal furrow (~16 mm in its maximum width just proximal to the olecranon fossa) that extends proximally from the olecranon fossa; the furrow becomes less pronounced (flatter) and narrower as it extends proximally such that posterior surface appears relatively flat at its most proximal extent, though the morphology here is partly obscured by the aforementioned surface damage. At the proximal end, the roughly triangular diaphysis measures 17.8 mm anteroposteriorly and 24.5 mm mediolaterally (measured distal to a small region of missing bone on the lateral edge). Breccia obscures the medullary cavity, but cortical thickness can be measured anteriorly (4.7 mm).

StW 182—Right distal humeral shaft fragment

Preservation. This 44 mm-long shaft fragment preserves the proximal margin of the olecranon fossa but is missing everything distal to this point (Figure 6.1). Due to oblique breaks on either side of the olecranon fossa, neither pillar is preserved. The specimen measures 13.9 mm anteroposteriorly and 23.1 mm mediolaterally at the proximal break, widening to 29 mm just proximal to the olecranon fossa. The proximal break runs proximolateral to distomedial such that the medial border is about one-third shorter than the lateral border. A piece of cortex is spalled off the anteromedial surface and the medial border is cracked and pitted.

Morphology. Distinct lateral and medial supracondylar ridges are present that are nearly in the same plane as the flattened posterior surface. The rounded lateral supracondylar ridge (3 mm wide) is more pronounced than its medial counterpart and is flanked anteriorly along its length be a narrow groove. The posterior surface is flat except in the region just proximal to the olecranon fossa, where it is mildly concave.

StW 339—Right distal humeral shaft fragment

Preservation. StW 339 is a 76 mm-long shaft fragment that is fractured obliquely (proximolateral to distomedial) through the floor of the olecranon fossa such that virtually all of the medial pillar is preserved (9.9 mm wide), but most of the lateral pillar is missing (Figure 6.1). The proximomedial halves of the coronoid fossa (15 mm wide) and the olecranon fossa (16 mm wide, 7 mm deep) are preserved. The proximal part of the medial epicondyle is preserved, but its medial extent is missing and its projection can only be estimated. The lateral epicondyle

and radial fossa are not preserved. The specimen is comprised of two pieces that have been glued together. The smaller proximolateral shaft fragment (32 mm proximodistal, 23 mm mediolateral) joins tightly to the rest of the specimen, albeit with some areas of missing bone along the crack (particularly anteriorly). The specimen widens distally to a maximum mediolateral width of 42 mm, though mediolateral shaft dimensions are not accurate due to missing bone along the entire lateral border of the specimen.

Morphology. The diaphysis is triangular in section at the proximal break (17.1 mm anteroposterior, 25.0 mm mediolateral). Cortical thickness at the proximal break varies (4.5 mm to 5.5 mm). The shaft is slightly anteriorly concave (posteriorly convex) in lateral view. The posterior surface of the shaft is transversely convex and heavily abraded. From the most anteriorly projecting region of the shaft, the anteromedial surface slopes obliquely back to meet a rounded medial supracondylar ridge. In contrast, the anterolateral surface begins by sloping back from the same region but forms a plateau in the coronal plane that extends laterally into a damaged lateral supracondylar crest, thereby creating a concavity that runs longitudinally along the anterolateral surface. The lateral supracondylar ridge would extend further laterally if not for the aforementioned missing bone.

StW 431c—Right distal humerus

Preservation. This distal half of a right humerus (162 mm long) (Figure 6.1) is part of the StW 431 partial skeleton from Member 4 (Menter, 2002; Toussaint et al., 2003; Ward et al., Chapter 4, this volume). The distal articular surface is largely intact. The specimen was originally extracted in five pieces that fit well together, with some bone missing along a transverse break through the proximal diaphysis (anteriorly) and along a longitudinal break through the posterior trochlea. With all pieces joined together, the diaphysis extends ~18 mm further proximally on the anterior side than on the posterior side. A portion of the medial shaft (30 mm in maximum proximodistal height when measured posteriorly) is missing proximal to the trochlea such that the specimen lacks the proximomedial part of the olecranon fossa and the anteroproximal surface of the medial epicondyle. The lateral pillar is present (17 mm wide), but the medial pillar is largely missing. The lateral trochlear crest is mildly abraded and the medial trochlear crest is chipped along both its anterodistal and posterior margins.

Morphology. The diaphysis is oval in the cross-section at its proximal break (22.5 mm anteroposterior, 20.7 mm mediolateral). Although the exposed cortical bone at the

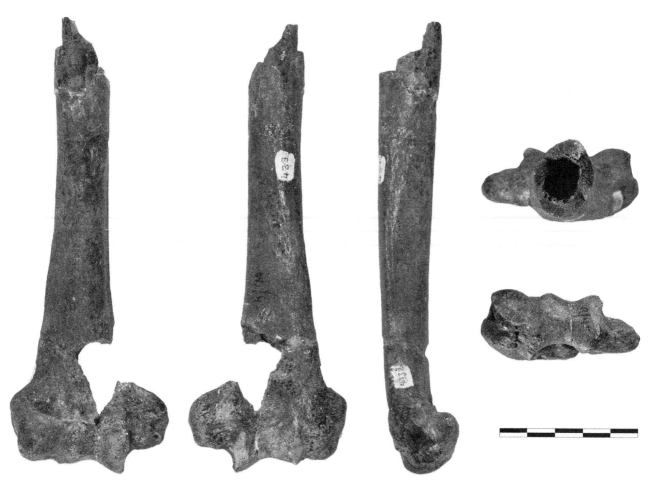

Figure 6.2 The StW 431c humerus (bottom) shown in (from left to right) anterior, posterior, lateral, and proximal (top) and distal (bottom) views.

proximal fracture is not all in the same transverse plane, cortical thickness is fairly uniform on all sides (roughly 4.5 mm). The diaphysis flattens to 16.7 mm in anteroposterior width further distally at the margin of the large missing shaft piece, where the section is more triangular. The diaphysis is faintly curved in the sagittal plane (posteriorly convex). A blunt lateral supracondylar ridge is apparent from the lateral epicondyle up to a point about 65 mm proximal to the lateral epicondyle, where it becomes flush with the diaphyseal surface. It is flanked anteriorly along much of its length by a mild vertical concavity. Although the distal shaft is missing on the medial side, the medial supracondylar ridge is detectable for a span of about 10 mm proximal to the broken section.

The lateral epicondyle is located proximal to the capitulum and projects about 5 mm laterally from the margin of the coronoid fossa. The medial epicondyle (largely intact apart from the missing bone on its anteroproximal surface) projects 17 mm medially from the trochlear margin, resulting in a biepicondylar width of 59.4 mm. The globular capitulum has a squared lateral edge and its posterior extent is marked by a ridge of nonarticular bone, as is the proximal margin of the trochlea. The trochlea is bounded by distinct medial and lateral ridges, the former of which extends only slightly more distally than the latter. The prominent lateral trochlear ridge divides the articular surface into radial and ulnar components that are almost equal in width (21.5 mm and 18.0 mm, respectively). Posteriorly, the ridge extends about 5 mm along the lateral margin of the olecranon fossa, which measures 21 mm wide and 7 mm deep. The posterior surface of the shaft just proximal to the olecranon fossa is slightly concave. McHenry and Brown (2008) present additional measurements of this specimen.

Comparative anatomy and analysis

Materials and methods

The intent of this study is to examine the morphometric affinities of the Sterkfontein humeri among fossil hominins and extant hominids, and to assess the possibility of taxonomic heterogeneity in the Member 4 assemblage. To quantify diaphyseal shape, two-dimensional coordinates were collected from the surface outline of a transverse section through the distal diaphysis at about 19% of total humeral length, including two "Type 2" landmarks (the most medial and lateral points of the section) and 58 sliding semilandmarks (see details in Lague, 2015). The taxonomic utility of distal humeral diaphyseal cross-sectional shape has been demonstrated by previous studies (Senut, 1981; Susman et al., 2001; Lague, 2015), though none included specimens from Sterkfontein other than StW 431c. Although our analyses involve landmark-based techniques (i.e., geometric morphometrics), Table 6.1 provides linear measurements of the same diaphyseal section for the Sterkfontein humeri, along with summary statistics for four extant hominid samples.

The comparative sample of fossils consists of a wide variety of Plio-Pleistocene hominins (Table 6.2). We include three fossils not previously examined in the diaphyseal shape analysis by Lague (2015): (1) IB 7594 (Gomboré I, Ethiopia),[1] (2) D4507 (Dmanisi, Georgia), and (3) MLD 14 (Makapansgat, South Africa). IB 7594 exhibits elbow joint morphology that significantly differs from that of specimens allocated to Paranthropus (Lague and Jungers, 1996; Lague, 2015) and has been attributed to Homo sp. (Chavaillon et al., 1977; Di Vincenzo et al., 2015). The results presented here strongly support the suggestion by Lague (2015) that this specimen represents Homo

erectus. D4507 is associated with a large adult H. erectus skeleton from Dmanisi (Lordkipanidze et al., 2007). In the absence of an available surface scan, coordinates for D4507 were collected from the ~20% section published by Di Vincenzo et al. (2015).[2] Although the humeral fragment from Makapansgat (MLD 14) has been suggested to belong to a baboon (Wolpoff, 1973), it is likely from a hominin (Boné, 1955; Senut, 1978). Given the possibility that MLD 14 represents A. africanus (or another hominin at Makapansgat), we explore its morphometric affinities with the Sterkfontein humeri here.

Two of the specimens from Sterkfontein (StW 38, 124) required virtual reconstruction due to missing material on the lateral side. Locations of missing coordinates on these specimens were estimated via thin-plate spline (TPS) interpolation based on a reference group comprised of the entire set of intact fossils (Gunz et al., 2009; Adams and Otarola-Castillo, 2013; Adams et al., 2015). The resulting geometric reconstructions of StW 38 and StW 124 (Figure 6.2) strongly resemble those based on manual estimation of missing outline information (done prior to TPS interpolation). As the choice of estimation technique does not substantially alter the analytical results, we provide only those results based on geometric reconstructions.

Via tpsRelw software (Rohlf, 2010), we used orthogonal least-squares generalized Procrustes analysis (GPA) to superimpose raw landmark configurations into the same shape space (Gower, 1975; Rohlf and Slice, 1990).

Table 6.1 Linear measurements of the distal humeral diaphysis (in millimeters).[a]

	Sex	n	Mediolateral Width				Anteroposterior Width			
			Mean	SD	Min	Max	Mean	SD	Min	Max
Homo sapiens	F	11	25.3	1.74	21.9	27.9	15.9	1.30	14.4	18.6
	M	10	28.2	2.52	23.2	32.5	17.5	1.49	15.5	20.1
Pan troglodytes	F	11	28.4	2.90	25.2	34.2	17.8	1.74	15.2	20.6
	M	10	30.3	1.59	28.1	33.2	18.9	1.18	17.2	20.8
Gorilla gorilla	F	7	33.0	2.00	30.6	36.8	21.4	1.46	19.1	23.0
	M	5	45.0	3.55	40.7	48.8	27.1	0.71	26.0	27.9
Pongo pygmaeus	F	8	25.9	2.23	21.3	28.5	16.4	1.46	14.4	18.4
	M	4	34.8	1.60	32.6	36.4	21.3	1.41	19.2	22.4
StW 38			27.0[b]				15.6			
StW 124			23.8[b]				13.9			
StW 150			27.2				17.2			
StW 182			24.1				13.2			
StW 339			30.8[b]				16.5			
StW 431c			26.9				16.8			

[a] Measurements taken approximately 19% of humerus length from the distal end (as in Lague, 2015). All extant specimens are from the Cleveland Museum of Natural History; *Pongo* and *Homo* specimens also derive from the National Museum of Natural History (USA).

[b] Reconstructed.

1 Gomboré IB 7594 is also known in the literature as MK3.

2 A section of IB 7594 published by the same authors (in their Supplementary Online Material) was also digitized for comparison to our own section; as the two sections are almost identical, use of the published section of D4507 in the current data set was deemed reasonable.

Table 6.2 Fossil hominin humeri used in this study (arranged by species).

Specimen	Site	Country	Species[a]	elbow[b]
StW 38	Sterkfontein	South Africa	*A. africanus??*	●
StW 124	Sterkfontein	South Africa	*A. africanus??*	
StW 150	Sterkfontein	South Africa	*A. africanus??*	
StW 182	Sterkfontein	South Africa	*A. africanus??*	
StW 339	Sterkfontein	South Africa	*A. africanus??*	
StW 431c	Sterkfontein	South Africa	*A. africanus??*	
StW 602	Sterkfontein	South Africa	*A. africanus??*	●
MLD 14	Makapansgat	South Africa	hominin?	
KNM-KP 271	Kanapoi	Kenya	*A. anamensis*	●
A.L. 322-1	Hadar	Ethiopia	*A. afarensis*	●
A.L. 137-48a	Hadar	Ethiopia	*A. afarensis*	●
A.L. 288-1m	Hadar	Ethiopia	*A. afarensis*	●
A.L. 288-1s	Hadar	Ethiopia	*A. afarensis*	●
MH1 (UW88-88)	Malapa	South Africa	*A. sediba*	●
MH2 (UW88-57)	Malapa	South Africa	*A. sediba*	●
TM 1517e	Kromdraai	South Africa	*P. robustus*	●
SKX 10924	Swartkrans	South Africa	*P. robustus*	●
OH 80-10	Olduvai Gorge	Tanzania	*P. boisei*	
KNM-ER 739	Koobi Fora	Kenya	*P. boisei*	●
KNM-ER 1591	Koobi Fora	Kenya	*P. boisei*	
KNM-ER 6020	Koobi Fora	Kenya	*P. boisei*	●
KNM-ER 1504	Koobi Fora	Kenya	*P. boisei*	●
SKX 19495	Swartkrans	South Africa	*H. aff. habilis*	
SK 2598	Swartkrans	South Africa	*H. aff. habilis*	
SK 24600	Swartkrans	South Africa	*H. aff. habilis*	●
OH 62	Olduvai Gorge	Tanzania	*Homo habilis*	
KNM-ER 3735a	Koobi Fora	Kenya	*Homo habilis*	●
SKX 34805	Swartkrans	South Africa	*H. erectus*	
KNM-WT 15000f	West Turkana	Kenya	*H. erectus*	●
KNM-ER 1808t	Koobi Fora	Kenya	*H. erectus*	
IB 7594	Gomboré I	Ethiopia	*H. erectus*	●
D4507	Dmanisi	Georgia	*H. erectus*	●

[a] Species allocations based on established craniodental associations, conventional taxonomy, and/or the results of Lague (2015).

[b] The black dot (●) indicates preservation of elbow joint morphology.

Semilandmarks (designated via a "sliders" file) were allowed to slide along the diaphyseal outline using the criterion of minimized bending energy (Bookstein, 1997). Principal Components Analysis (PCA) was used to provide a visual summary (via ordination) of morphometric affinities among the aligned specimens. Procrustes distances (D_p) were calculated as a measure of shape dissimilarity between specimen pairs (Rohlf, 1999; Zelditch et al., 2012).

We used a bootstrap resampling procedure (Efron and Tibshirani, 1986) to assess the magnitude of diaphyseal shape variation represented by different fossil test samples, including: (1) the six specimens from Sterkfontein and (2) a mixed sample of seven specimens from Sterkfontein and Makapansgat. Overall variation in diaphyseal shape for a given test sample was measured as the sum of all possible squared pairwise Procrustes distances (SSD; cf. Baab, 2008). Bootstrapping was used to generate an empirical distribution of SSD in each extant reference taxon, against which the test sample SSD could be evaluated. For each of 10,000 iterations, SSD was calculated for a sample of n specimens drawn randomly (and with replacement) from the given reference taxon. In essence, the null hypothesis of the bootstrap procedure is that the variation in the given test sample (of size n) is not significantly higher than one would expect for a random sample of n specimens drawn from a single species. Reported probability values equal the frequency with which a bootstrap SSD was found to be equal to or higher than that of the test sample.

To run additional bootstrap tests using fossil variation for reference, we used a group of 11 australopith fossils representing four species (*Australopithecus anamensis*, *A. afarensis*, *Paranthropus robustus*, *P. boisei*). Although these specimens represent a good deal of temporal/geographic/taxonomic diversity, they share very similar diaphyseal morphology that is distinct from that of *A. sediba* and early *Homo* (Lague, 2015). This group therefore represents the largest possible sample of fossil humeri that is still relatively homogeneous in shape. Using this combined australopith group allows us to test whether the variation observed at Sterkfontein is significantly greater than that observed among non-*sediba* australopiths in general.

Since the Jacovec humerus (StW 602) was not available for this study, we could not include it in the diaphyseal shape analysis. This specimen is of major interest, as it may be derived from older sediments than Member 4 (Partridge et al., 2003; but see Herries et al., 2013; Pickering and Herries, Chapter 3, this volume), and is also one of the few Sterkfontein humeri to preserve distal articular anatomy. To provide some comparative context for this specimen, we quantified its anterior joint surface (18 landmarks) using the photograph published by Partridge et al. (2003) and combined it with the data set from Lague (2014), along with two additional fossils noted above (IB 7594 and D4507). The Procrustes distance between StW 602 and StW 431c was evaluated against distributions of all possible pairwise distances within fossil and extant reference groups.

Results

Diaphyseal size and shape comparison

With respect to overall size, the six Sterkfontein specimens exhibit a range of variation similar to that observed for *H. erectus* (Figure 6.3). The two smallest specimens (StW 124, 182) are similar in size to an average *A. afarensis* humerus and are just slightly smaller than the smallest *H. erectus* specimen (SKX 34805). The largest Sterkfontein specimen (StW 339) approaches the size of the largest (presumed) *H. erectus* (IB 7594) and is beyond the size of all sampled modern humans except one (but well within the chimpanzee size range). Inclusion of MLD 14 extends the upper size range of the South African specimens; among the fossils, its centroid size is exceeded only by the three largest *P. boisei* specimens (KNM-ER 1591, KNM-ER 739, KNM-ER 6020). Most of the shape variation among the fossils (> 62%) is accounted for by the first two principal components (Figure 6.4); PC3 (18.8%) is largely uninformative with respect to taxonomic differences and is not shown. The arrangement of specimens on the PCA plot differs somewhat from that of Lague (2015), largely due to the addition of two specimens (IB 7594, D4507) that have expanded the range of

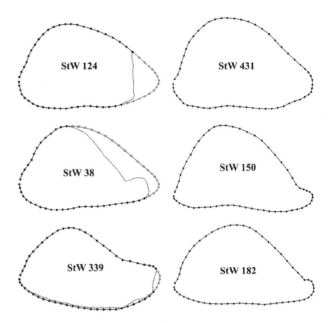

Figure 6.3 Landmark configurations for distal diaphyseal sections (~19% humerus length) of Sterkfontein humeri. Sections are scaled to similar mediolateral width. Orange landmarks are those estimated via TPS interpolation (see text) due to missing bone; dashed lines correspond to actual preserved contours. All left humeri were mirrored to create "right" humeri. Anterior toward the top, medial to the left.

H. erectus shape variation away from KNM-WT 15000. Given the subadult status of KNM-WT 15000, it is likely that the other four *H. erectus* specimens are, on average, a better representation of adult *H. erectus* morphology. (As an interesting contrast, however, the subadult MH1 humerus is highly similar to the adult MH2 humerus.) Although KNM-WT 15000 bears some similarity to *A. afarensis* with respect to those aspects of diaphyseal shape variation captured by PC1 and PC2, it deviates from all fossil taxa (including *Australopithecus*) along PC3 (not shown).

Procrustes distances to species means for the Sterkfontein/Makapansgat fossils (Table 6.3) accord well with the PCA results. Correlations among the Procrustes distance vectors (Table 6.4), along with morphometric affinities apparent from PCA, suggest three morphological groups: (1) StW 38, 124, 431c, (2) StW 150, 182, and (3) StW 339, MLD 14. For each group, all within-group distance vectors are significantly correlated with one another ($P < 0.05$), but not with any distance vector outside the group. "Group 1" fossils are extremely similar to one another and most closely resemble humeri attributed to *P. robustus* and *A. afarensis* (especially the former), as well as those of modern humans. They generally exhibit greater similarity to modern species than do members of the other two groups. "Group 2" fossils are most similar

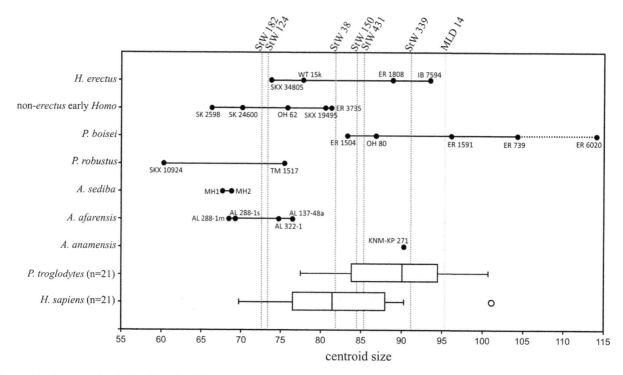

Figure 6.4 Cross-sectional size of the distal humeral diaphysis (based on centroid size of 60 two-dimensional coordinate points). Boxes for extant species contain 50% of the data (median at horizontal line) and "whiskers" represent min./max. values (excluding an unusually large modern human). The diaphyseal size for KNM-ER 6020 is likely inflated due to expansion cracks on its posterior surface.

Table 6.3 Procrustes distances of Sterkfontein and Makapansgat specimens to various species means based on distal diaphyseal shape.

	Homo habilis	*Homo erectus*[a]	*A. sediba*	*A. afarensis*	*P. robustus*	*P. boisei*	*Homo sapiens*	*Pan trog.*	*Gorilla gorilla*	*Pongo pyg.*
StW 38	0.0951	0.0765	0.0617	0.0479	0.0327	0.0594	0.0449	0.0705	0.0896	0.0629
StW 124	0.0888	0.0939	0.0652	0.0434	0.0317	0.0494	0.0562	0.0685	0.0858	0.0648
StW 150	0.1758	0.0521	0.1241	0.1395	0.1157	0.1573	0.1071	0.1479	0.1719	0.1295
StW 182	0.1351	0.0536	0.0794	0.1116	0.0852	0.1213	0.0904	0.1245	0.1422	0.1007
StW 339	0.1436	0.1286	0.0583	0.1122	0.1133	0.1182	0.1316	0.1328	0.1613	0.1357
StW 431	0.1207	0.0790	0.0928	0.0584	0.0440	0.0792	0.0329	0.0775	0.0880	0.0590
MLD 14	0.1219	0.1065	0.0355	0.0832	0.0781	0.0915	0.1045	0.1019	0.1312	0.1025

[a] Based on group mean that does not include the subadult KNM-WT 15000.

Table 6.4 Pearson product moment correlation coefficients (below diagonal) and associated *P* values (above) for the distance vectors of Sterkfontein and Makapansgat fossils.[a]

	StW 38	StW 124	StW 150	StW 182	StW 339	StW 431	MLD 14
StW 38	–	**0.000**	0.316	0.201	0.177	**0.002**	0.090
StW 124	**0.914**	–	0.942	0.804	0.245	**0.030**	0.124
StW 150	0.354	−0.027	–	**0.000**	0.458	0.200	0.493
StW 182	0.442	0.090	**0.951**	–	0.131	0.262	0.148
StW 339	0.464	0.405	0.266	0.512	–	0.916	**0.000**
StW 431	**0.851**	**0.680**	0.443	0.392	0.038	–	0.683
MLD 14	0.563	0.519	0.246	0.493	**0.985**	0.148	–

[a] Distance vectors consist of ten Procrustes distances to species mean diaphyseal shapes (see Table 6.3). Significant correlations (*P* < 0.05) are in bold.

to *H. erectus* (even if KNM-WT 15000 is used in calculation of the *H. erectus* mean) and relatively dissimilar from all other taxa. "Group 3" fossils bear a strong similarity to the humeri of *A. sediba* (MH1, MH2). None of the Sterkfontein specimens have the unusual anteroposteriorly flattened shape of non-*erectus* early *Homo*.

Since the PCA results are largely driven by early *Homo*, we ran an additional PCA based solely on australopiths that also excluded the two *erectus*-like "Group 2" specimens (which also strongly influence the PCA). Most of the shape variation (> 58%) of this more exclusive sample is accounted for by the first two PCs (Figure 6.5). *A. sediba* and the "Group 3" specimens differ from other australopiths (PC1) in presenting a somewhat concave anterolateral surface combined with a more evenly rounded (symmetrical) posterior surface related to more central (less medial) placement of the most posterior point. The three "Group 1" specimens (StW 38, 124, 431c) overlap with a mixed-species set of australopiths (*A. anamensis*, *A. afarensis*, and *P. robustus*) that are differentiated from *P. boisei* along PC2 by virtue of a flatter posterior surface.

Bootstrap analyses of diaphyseal shape variation

Diaphyseal shape variation within the Sterkfontein sample (SSD = 0.1451, *n* = 6) is somewhat higher than the average SSD observed in each of the extant reference bootstrap distributions (Table 6.5). Nevertheless, for all four extant species distributions, the Sterkfontein SSD is below the most extreme 5% of values; i.e., there is a reasonable probability (0.090 < *P* < 0.401) of sampling six specimens with a higher SSD than that observed for Sterkfontein. Hence, with reference to intraspecific variation in modern hominids, we cannot reject the null hypothesis that shape variation at Sterkfontein is consistent with that of a single species. Inclusion of MLD 14 with the Sterkfontein sample (*n* = 7) increases SSD (0.1973) but does not result in rejection of the null hypothesis (0.091 < *P* < 0.421).

Bootstrap tests reveal that the same null hypothesis (i.e., single species) is also not rejected for a sample consisting of all 13 non-Sterkfontein australopith fossils (representing five different species). For either reference taxon of sufficient sample size (*Homo*, *Pan*), the large majority of bootstrap SSD values are higher than the SSD (0.4842) for the mixed australopith sample. In contrast, the degree

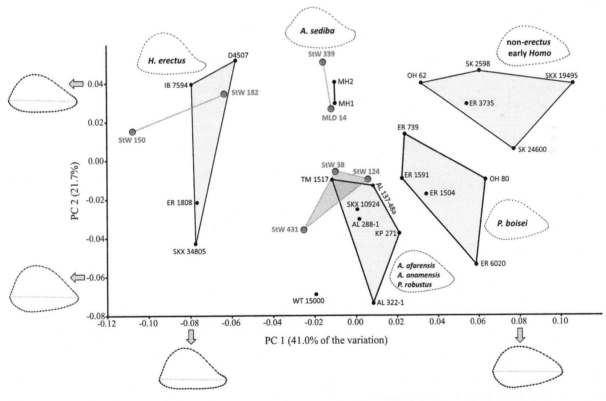

Figure 6.5 PCA results for early hominin diaphyseal shape. Sterkfontein/Makapansgat fossils are in red; dotted red lines connect fossils with significantly correlated distance vectors (Table 6.4). Group mean shapes are indicated on the plot. Shapes outside the plot depict variation along each principal component. The Procrustes distance matrix for the fossils is significantly correlated with the matrix of Euclidean distances based solely on PC1 and PC2 scores (Mantel test; *R* = 0.91, *P* < 0.002), indicating that the first two PCs are a good representation of the overall variation.

Table 6.5 Results of bootstrap analyses based on four extant reference samples.[a,b]

Test Sample	n	Sample SSD	Homo avgSSD	Homo P	Pan avgSSD	Pan P	Gorilla avgSSD	Gorilla P	Pongo avgSSD	Pongo P
Sterkfontein	6	0.1451	0.1251	0.291	0.1176	0.202	0.1011	0.090	0.1369	0.401
Sterkfontein + MLD 14	7	0.1973	0.1759	0.323	0.1637	0.207	0.1414	0.091	0.1896	0.421
australopith (5 species)	13	0.4842	0.6514	0.929	0.6081	0.868	N/A[c]	N/A[c]	N/A[c]	N/A[c]
early Homo	9	0.6528	0.3010	0	0.2806	0	0.2422	0	0.3269	0

[a] P indicates the frequency with which a SSD value greater than or equal to the sample SSD was observed in 10,000 bootstrap samples; 'avgSSD' is the mean of the given bootstrap distribution.

[b] The "early Homo" sample does not include KNM-WT 15000. The "australopith" sample does not include Sterkfontein; left and right humeri of A.L. 288-1 were averaged.

[c] Size of test sample exceeds size of reference sample.

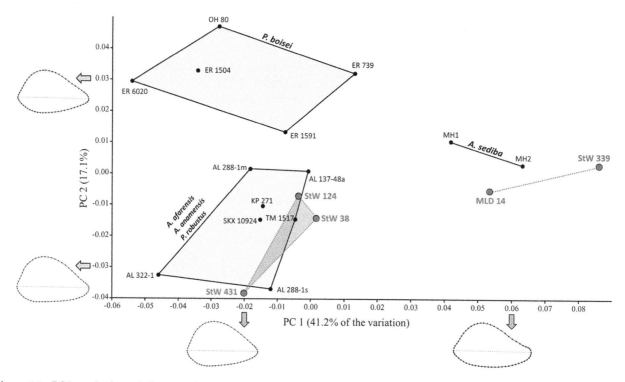

Figure 6.6 PCA results for early hominin diaphyseal shape, excluding early Homo and the H. erectus-like Sterkfontein specimens (StW 150 and 182). Sterkfontein/Makapansgat fossils are in red; dotted red lines connect fossils with significantly correlated distance vectors (Table 6.4). Shapes depict variation along each principal component.

of shape variation produced by a mixed-species sample of early Homo (n = 9; SSD = 0.6528) is higher than that for every randomly sampled subset of nine specimens for all four reference groups (i.e., the probability of randomly sampling nine specimens with equal or higher SSD in an extant species equals zero).

Shape variation among the Sterkfontein specimens appears unusually high when all 11 non-sediba australopiths (four species) are used as a reference group (Figure 6.6). In particular, none of the bootstrap SSD values for the mixed sample of non-sediba australopiths equal or exceed the SSD (0.1451) observed for the six Sterkfontein specimens. Removal of either StW 339 or the two erectus-like

specimens (StW 150, 182) from the Sterkfontein sample still yields SSD in the upper 5% of the non-sediba australopith bootstrap distribution (P = 0.021 and P = 0.006, respectively). In contrast, the SSD value for a group consisting of only the "Group 1" specimens (StW 38, 124, 431c) is exceeded by the large majority of bootstrap values (P = 0.957).

Elbow joint morphology

Based on Procrustes distances, StW 431c and StW 602 are more similar to one another in articular morphology than either specimen is to any other fossil. The distance between

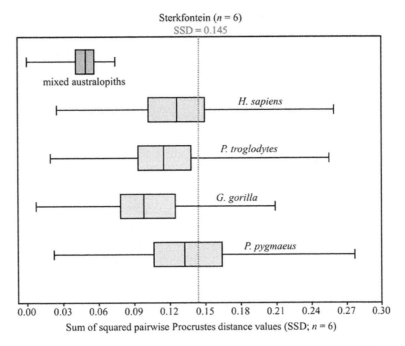

Sterkfontein (*n* = 6)
SSD = 0.145

mixed australopiths

H. sapiens

P. troglodytes

G. gorilla

P. pygmaeus

0.00 0.03 0.06 0.09 0.12 0.15 0.18 0.21 0.24 0.27 0.30

Sum of squared pairwise Procrustes distance values (SSD; *n* = 6)

Figure 6.7 SSD bootstrap distributions for four extant species and a mixed-species group of 11 non-*sediba* australopiths (four species). Bootstrap distributions are based on 10,000 random samples of six individuals. Boxes contain 50% of the data (median at horizontal line) and "whiskers" represent min./max. values. The degree of humeral shape variation at Sterkfontein (red dotted line) can be sampled with reasonable probability (*P* > 0.05) in all four extant taxa, but is extremely high with reference to the combined australopith (non-Sterkfontein) sample.

them (D_p = 0.0645) is less than: (1) the distance between the two non-*erectus* early *Homo* specimens (0.0730), (2) the distance between the two *P. robustus* specimens (0.0931), (3) all possible pairwise distances in the *P. boisei* sample (0.1044, 0.0769, 0.0797), (4) all possible pairwise distances in the *H. erectus* sample (0.0647, 0.0779, 0.0911), and (5) every pairwise distance in the *A. afarensis* sample except that between the left and right humeri of A.L. 288-1 (0.0449). In the context of extant species variation, the D_p between StW 431c and StW 602 is in the lower 3% of all possible pairwise values for all five extant species (Table 6.6).

Discussion and conclusions

Sterkfontein postcranial material has perhaps most often been compared to homologous elements of *Australopithecus afarensis*, a potential ancestor of *A. africanus*. Numerous papers have noted postcranial similarity between the two species (McHenry, 1983, 1986; Senut and Tobias, 1989; Abitbol, 1995; Häusler and Berger, 2001; Dobson, 2005; Green and Gordon, 2008; Berge and Goularas, 2010), suggesting relative stasis in postcranial evolution and similarity in habitual limb use. On the other hand, differences between *A. afarensis* and *A. africanus* have been noted with respect to limb-size proportions (McHenry and Berger, 1998; Green et al., 2007), size of the

Table 6.6 Extant species sample sizes and percentage of pairwise D_p values exceeding that between StW 602 and StW 431 (based on 18 2D landmarks of the elbow joint).[a]

Taxon	F	M	Total	% >
Homo sapiens	30	36	66	98.6
Pan troglodytes	18	24	42	97.3
Pan paniscus	8	8	16	97.5
Gorilla gorilla	19	15	34	97.3
Pongo pygmaeus	14	10	24	98.2

[a] Museum sources as in Lague (2014).

lumbosacral joint (Dobson, 2005), and the morphology of the hand (Marzke et al., 1992), femur (Harmon, 2009b), and tibia (Berger and Tobias, 1996). In some ways, the more craniodentally derived *A. africanus* may possess a more primitive "ape-like" skeleton (Berger and Tobias, 1996; McHenry and Berger, 1998).

Accurate characterization of the *A. africanus* postcranium is potentially confounded by taxonomic heterogeneity within the Member 4 "Type Site" assemblage. Despite the historical tendency to assume that hominin material from Member 4 represents *A. africanus*, extensive craniodental variation may indicate that more than one species is represented (e.g., Clarke, 1988; Kimbel and White, 1988; Kimbel and Rak, 1993; Clarke, 1994; Moggi-Cecchi et al., 1998; Lockwood and Tobias, 2002; Moggi-Cecchi and

Boccone, 2007; Clarke, 2008; Grine, 2019). Quantitative studies of craniodental variation have not produced unequivocal support for the multiple species hypothesis, nor a consistent answer as to how the sample should be divided (Grine, 2013). Postcranial evidence for multiple species at Sterkfontein (e.g., Clarke and Tobias, 1995; Partridge et al., 2003; Harmon, 2009a) is also limited, partly due to the need for multiple, sufficiently preserved fossils of the same informative skeletal element.

Quantitative approaches to questions of taxonomic heterogeneity in fossil samples typically engender statistical analysis of variation in modern reference species, especially those with a close phylogenetic relationship to the fossil organisms under study. The use of modern taxa to model intraspecific variation in paleospecies is not without its problems, including the potential for variation in fossil taxa to be inflated due to temporal mixing. Alternatively, in cases where multiple related fossil species exhibit very similar skeletal anatomy, statistical assessment of variation based on modern reference taxa may be useless for fossil species delimitation.

Based on modern hominid shape variation, the null hypothesis that the Sterkfontein humeri sample a single species is not rejected, even with the inclusion of the specimen from Makapansgat. On the other hand, the degree of diaphyseal shape variation associated with a mix of multiple australopith species can also be sampled with high probability within modern hominid species. This phenomenon says as much about variation within modern hominids as it does about variation among the fossils. Clearly, acceptance of the "single-species" null hypothesis does not necessarily indicate that all of the Sterkfontein specimens are conspecific. In fact, although the magnitude of shape variation among the Sterkfontein humeri appears low relative to modern hominids, it is extremely high when assessed relative to variation in a diverse sample of australopith humeri from multiple species across Africa. In essence, the variation at Sterkfontein is significantly greater than one would expect based on sampling six non-Sterkfontein australopith humeri at random.

The pattern of shape variation is suggestive of three distinct humeral morphs at Sterkfontein and Makapansgat, grouped as follows: (1) StW 38, 124, 431c, (2): StW 150, 182, and (3) StW 339, MLD 14. The perception of three groups may simply be an artefact of randomly sampling a single paleospecies at a relatively low sample size. Nevertheless, as the diaphyseal region of the humerus appears to be a useful taxonomic indicator (Lague, 2015), it is plausible that these group boundaries are biologically meaningful, especially in light of the fact that humeral variation at Sterkfontein is greater than that for multiple hominin sites combined.

The assumption that all of the humeral specimens from Sterkfontein/Makapansgat represent *A. africanus*

precludes a simple description of the morphology of this species, since shape variation within *A. africanus* would then overlap considerably with multiple Plio-Pleistocene hominin morphologies. We therefore summarize humeral morphology based on the above group divisions, with the recognition that the apparent morphological boundaries may not correspond to taxonomic boundaries.

Member 5 *Homo erectus*? (StW 150, StW 182)

The results of this study suggest that two of the Sterkfontein humeri (StW 150, StW 182) represent *Homo erectus*. These two specimens strongly resemble other *H. erectus* specimens in having a strongly projecting anterior surface, a flattened posterior surface, and a distinct and anteroposteriorly narrow lateral supracondylar ridge adjacent to a shallow anterior longitudinal groove. It is possible that this "*erectus*-like" morphology is encompassed within the natural variation of some form of *Australopithecus* at Sterkfontein (*A. africanus* or otherwise), though inclusion of StW 150 and 182 in the *A. africanus* hypodigm dramatically increases the degree of implied intraspecific variation.

Although Member 4 is not known to contain *H. erectus* fossils, the geological provenience of these two specimens is muddled (Grine, 2016); in effect, it is not clear from the relevant literature (e.g., Senut and Tobias, 1989; Kuman and Clarke, 2000; Partridge, 2000; Pickering et al., 2004; Clarke, 2013) whether they originate from Member 4 or Member 5. Hence, allocation of StW 150 and StW 182 to something other than *A. africanus* does not necessarily imply species heterogeneity in Member 4. Indeed, their strong similarity to *H. erectus* suggests that Member 5 is their true origin, as these deposits contain fossils of *H. erectus* as well as Acheulean stone tools (Kuman and Clarke, 2000; Granger et al., 2015). The other four Sterkfontein specimens are more clearly derived from Member 4 and have their closest morphological affinities among australopiths.

Australopithecus africanus (StW 38, StW 124, StW 431c)

Three of the specimens (StW 38, 124, 431c) are extremely similar to the humeri of most non-*sediba* australopiths (particularly *A. afarensis* and *P. robustus*) with respect to distal diaphyseal morphology. Although lacking associated craniodental material, Toussaint et al. (2003) attributed the StW 431 skeleton to *A. africanus* on the basis of "Occam's razor" and its general morphological similarity to *A. afarensis*. Multiple researchers have also noted that the StW 431 elbow joint, in particular, is similar to that of other known australopiths (Lague and Jungers, 1996; Bacon, 2000; Menter, 2002; Toussaint et al., 2003; McHenry and Brown, 2008; Lague, 2014). Given their morphological affinities, and the assumption that most (if not all) of the hominin postcranial elements of Sterkfontein Member 4 belong to *A. africanus*, it is reasonable to allocate StW 38, StW 124, and StW 431 to this species.

Although we lack the benefit of knowing its diaphyseal morphology, there is no evidence to suggest that the potentially older Jacovec specimen (StW 602) should be attributed to a different species than StW 431. Based on joint morphology, StW 602 bears a greater resemblance to StW 431c than it does to any other Plio-Pleistocene fossil specimen. Indeed, the two fossils are far more similar to one another than one would expect from randomly sampling two conspecific specimens from a modern hominid sample. Partridge et al. (2003) also noted the similarity between these two specimens; although StW 602 reportedly has a "more flattened shaft" than StW 431c (p. 611), the extent of this difference is unclear. The unusually strong similarity between StW 431c and StW 602 supports suggestions that sediments from Jacovec Cavern and Member 4 are similar in age (Herries et al., 2013), though it is also true that early hominins are generally homogeneous with respect to elbow joint shape. In short, StW 602 is characterized by the same unusual elbow joint morphology that differentiates virtually all Plio-Pleistocene hominins from modern humans and apes (Lague and Jungers, 1996; Lague, 2014).

The cross-sectional shape of the distal humeral diaphysis of *A. africanus* fits within a pattern of long-term stability within the genus *Australopithecus* and does not differ appreciably from that of either *A. afarensis* (a potential ancestor) or *P. robustus* (a potential descendent). *A. africanus* morphology is also distinctly different from that of other early hominin species (i.e., *H. habilis*, *H. erectus*, *A. sediba*, and *P. boisei*). The morphology shared by *P. robustus* and most species of *Australopithecus* is similar to that characterizing modern hominids in general (Lague, 2015). Indeed, Procrustes distances between group averages indicate that *H. sapiens* is more similar to certain nonhuman hominids (*Pongo*, *Pan*) than to some early hominins (*H. habilis*, *A. sediba*, *P. boisei*). The relative similarity observed among modern hominids makes it difficult to associate the cross-sectional shape of *Australopithecus* with any particular pattern of upper limb use. This is not to say that chimpanzees, modern humans, and most australopiths are identical in cross-sectional morphology, though the differences are subtle compared to the more extreme variation observed in early *Homo*. Future investigations will hopefully clarify to what degree (if any) distal diaphyseal shape is useful in reconstructing upper limb function.

Australopithecus sp. indet. (StW 339, MLD 14)

The cross-sectional morphology of StW 339 is somewhat unusual among the Sterkfontein humeri and uncommon among most hominids. On the other hand, it bears a striking resemblance to MLD 14 and the two specimens of *A. sediba* (MH1, MH2). The similarity of MLD 14 to the other three specimens suggests that it is indeed a hominin (Boné, 1955; Senut, 1978) and not a baboon (*contra* Wolpoff, 1973), an assertion that is supported by qualitative comparisons with available baboon humerus scans. All four fossils differ from other australopiths in having the combination of a concave anterolateral surface (somewhat reminiscent of *H. erectus*), small anteroposterior width relative to mediolateral width, and a more symmetrically rounded posterior surface (due to more central placement of the greatest point of posterior convexity).

Despite the morphological resemblance of StW 339 and MLD 14 to the geologically younger *A. sediba* specimens, the former are considerably larger in cross-sectional size. Indeed, StW 339 and MLD 14 are the two largest humeri in the present South African sample. It may be entirely coincidental that among the Sterkfontein/Makapansgat humeri, the two specimens that share similarly unusual cross-sectional morphology also happen to be the largest (by a good margin), though this seems improbable. Sexual dimorphism is also an unlikely explanation for this dichotomous pattern of size-shape differences at Sterkfontein/Makapansgat, as StW 431c (i.e., the next largest specimen after StW 339) is *least* similar to *A. sediba* among the remaining australopith-like specimens (and is also thought to be male; Toussaint et al., 2003). It is possible that the variation observed here reflects the mixing of temporally discrete (and morphologically distinct) populations of *A. africanus* over the course of Sterkfontein's depositional history (Kimbel, 2007). Nevertheless, in comparison to the degree of shape variation among multiple australopith species, sample variation is unusually high when StW 339 and MLD 14 are combined with the other three definitive Member 4 specimens (StW 38, StW 124, StW 431c).

Interestingly, the pattern of humeral shape variation observed across the two sites of Sterkfontein and Makapansgat is similar to the pattern observed for proximal femora. Harmon (2009a) found that the overall degree of "*A. africanus*" femoral shape variation is unusually high compared to variation within extant taxa (and within *A. afarensis*). Two specimens from Member 4 (StW 522, StW 99) are particularly dissimilar with respect to neck length (Partridge et al., 2003); in fact, the shape difference between StW 522 and StW 99 is unlikely to be sampled within modern hominid taxa (Harmon, 2009a). A femoral specimen from Makapansgat (MLD 46) is similar to StW 522 but differs significantly from StW 99 when neck length is included as a shape variable (Harmon, 2009a). Hence, at least two femoral morphs are apparent at Sterkfontein, one of which is similar to the one observed at Makapansgat. Indeed, much like the MLD 14 humerus, the Makapansgat femur is relatively large in overall size (Reed et al., 1993).

We cannot exclude the possibility that StW 339 and MLD 14 belong to the same species as the specimens described above. Nevertheless, given the potential presence of two *Australopithecus* species at Sterkfontein and Makapansgat, the high variability among the Sterkfontein Member 4 humeral specimens, and the observation of similar patterns of size/shape variation among humeri and femora, we consider it prudent to limit allocation of StW 339 and MLD 14 to the genus level (i.e., *Australopithecus*). In doing so, we recognize the possibility that these specimens derive from a second species of *Australopithecus* that bears a greater resemblance to *Australopithecus sediba* than to *A. afarensis* with respect to distal humeral shape. Future investigations of Sterkfontein postcrania should incorporate homologous elements from Malapa to explicitly test whether any other regions of the skeleton are characterized by distinctive *sediba*-like morphology.

Acknowledgments

We thank Carol Ward, Bernhard Zipfel, and Brian Richmond for the invitation to contribute to the present volume. Tremendous thanks are owed to J. Michael Plavcan and Carol Ward for providing scans of most of the original fossil specimens used in this study. Scans of additional specimens were obtained with the help and generosity of Manuel Domínguez-Rodrigo (OH 80), William Kimbel (OH 62 cast), Fabio Di Vincenzo (IB 7594), Ashley Kruger and Bernhard Zipfel (StW 339), and Yohannes Haile-Selassie and Lyman Jellema (casts of Hadar material). Casts of the Malapa specimens were purchased from University of the Witwatersrand with the kind assistance of Bonita de Klerk. Thanks also to Adam Van Arsdale for providing the photograph of D4507 used to quantify its joint morphology. This research was supported by grants from the National Science Foundation (SBR 9712585 and BCS 0647557) and grants from the Leakey Foundation and the Wenner-Gren Foundation (to JM Plavcan & CV Ward). To Charlie … you are sorely missed as both a friend and a valued colleague.

References

Abitbol, M,M., 1995. Reconstruction of the STS 14 (*Australopithecus africanus*) pelvis. American Journal of Physical Anthropology 96, 143–158.

Adams, D.C., Collyer, M.L., Sherratt, E., 2015. *geomorph*: software for geometric morphometric analyses. R package version 2.1.5 ed.

Adams, D.C., Otarola-Castillo, E., 2013. *geomorph*: an R package for the collection and analysis of geometric morphometric shape data. Methods in Ecology and Evolution 4, 393–399.

Baab, K.L., 2008. The taxonomic implications of cranial shape variation in *Homo erectus*. Journal of Human Evolution 54, 827–847.

Bacon, A-M., 2000. Principal components analysis of distal humeral shape in Pliocene to recent African hominids: the contribution of geometric morphometrics. American Journal of Physical Anthropology 111, 479–487.

Berge, C., Goularas, D., 2010. A new reconstruction of Sts 14 pelvis (*Australopithecus africanus*) from computed tomography and three-dimensional modeling techniques. Journal of Human Evolution 58, 262–272.

Berger, L.R., Tobias, P.V., 1996. A chimpanzee-like tibia from Sterkfontein, South Africa and its implications for the interpretation of bipedalism in *Australopithecus africanus*. Journal of Human Evolution 30, 343–348.

Boné, E.L., 1955. Quatre fragments post-crâniens du gisement à Australopithèques de Makapansgat (N. Transvaal). L'Anthropologie 59, 462–469.

Bookstein, F.L., 1977. Morphometric tools for landmark data: geometry and biology. Cambridge University Press: Cambridge.

Chavaillon, J., Chavaillon, N., Coppens, Y., Senut, B., 1977. Présence d'hominidé dans le site oldowayen de Gomboré I à Melka Kunturé, Éthiopie. Comptes Rendus de l'Académie des Sciences 285, 961–963.

Clarke, R.J., 1988. A new *Australopithecis* cranium from Sterkfontein and its bearing on the ancestry of *Paranthropus*. In: Grine, F. (Ed.), Evolutionary History of the "Robust" Australopithecines. Aldine de Gruyter: New York, pp. 285–292.

Clarke, R.J., 1994. Advances in understanding the cranial anatomy of South African early hominids. In: Corruccini, R.S., Ciochon, R.L. (Ed.), Integrative paths to the past. Prentice Hall: Englewood Cliffs, pp. 205–222.

Clarke, R.J., 2008. Latest information on Sterkfontein's *Australopithecus* skeleton and a new look at *Australopithecus*. South African Journal of Science 104, 443–449.

Clarke, R.J., 2013. *Australopithecus* from Sterkfontein Caves, South Africa. In: Reed, K.E., Fleagle, J.G., Leakey, R.E.F. (Ed.), The Paleobiology of *Australopithecus*. Springer: Dordrecht, pp. 105–123.

Clarke, R.J., Tobias, P.V., 1995. Sterkfontein Member 2 foot bones of the oldest South African hominid. Science 269, 521–524.

Di Vincenzo, F., Rodriguez, L., Carretero, J.M., Collina, C., Geraads, D., Piperno, M., Manzi, G.. 2015. The massive fossil humerus from the Oldowan horizon of Gombore I, Melka Kunture (Ethiopia, >1.39 Ma). Quaternary Science Reviews 122, 207–221.

Dobson, S,D., 2005. Are the differences between Stw 431 (*Australopithecus africanus*) and A.L. 288–1 (*A. afarensis*) significant? Journal of Human Evolution 49, 143–154.

Efron, B., Tibshirani, R., 1986. Bootstrap methods for standard errors, confidence intervals, and other measures of statistical accuracy. Statistical Science 1, 54–77.

Gower, J.C., 1975. Generalized Procrustes Analysis. Psychometrika 40, 33–51.

Granger, D.E., Gibbon, R.J., Kuman, K., Clarke, R.J., Bruxelles, L., Caffee, M.W., 2015. New cosmogenic burial ages for Sterkfontein Member 2 *Australopithecus* and Member 5 Oldowan. Nature 522, 85–88.

Green, D.J., Gordon, A.D., 2008. Metacarpal proportions in *Australopithecus africanus*. Journal of Human Evolution 54, 705–719.

Green, D.J., Gordon, A.D., Richmond, B.G., 2007. Limb-size proportions in *Australopithecus afarensis* and *Australopithecus africanus*. Journal of Human Evolution 52, 187–200.

Grine, F.E., 2013. The alpha taxonomy of *Australopithecus africanus*. In: Reed, K.E., Fleagle, J.G., Leakey, R.E.F. (Eds.), The Paleobiology of *Australopithecus*. Springer: Dordrecht, pp. 73–104.

Grine, F.E., 2019. The alpha taxonomy of *Australopithecus* at Sterkfontein: the postcranial evidence. Comptes Rendues Paleovol 18: 335–352.

Gunz, P., Mitteroecker, P., Neubauer, S., Weber, G.W., Bookstein, F.L., 2009. Principles for the virtual reconstruction of hominin crania. Journal of Human Evolution 57, 48–62.

Harmon, E., 2009a. Size and shape variation in the proximal femur of *Australopithecus africanus*. Journal of Human Evolution 56, 551–559.

Harmon, E.H., 2009b. The shape of the early hominin proximal femur. American Journal of Physical Anthropology 139, 154–171.

Häusler, M., Berger, L., 2001. Stw 441/465: a new fragmentary ilium of a small-bodied *Australopithecus africanus* from Sterkfontein, South Africa. Journal of Human Evolution 40, 411–417.

Herries, A.I.R., Pickering, R., Adams, J.W., Curnoe, D., Warr, G., Latham, A.G., Shaw, J., 2013. A multidisciplinary perspective on the age of *Australopithecus* in southern Africa. In: Reed, K.E., Fleagle, J.G., Leakey, R.E.F. (Eds.), The Paleobiology of *Australopithecus*. Springer: Dordrecht, pp. 21–40.

Kimbel, W.H., 2007. The species and diversity of australopiths. In: Henke, W., Tattersall, I. (Ed.), Handbook of Paleoanthropology. Springer: Berlin, pp. 1539–1573.

Kimbel, W.H., Rak, Y., 1993. The importance of species taxa in paleoanthropology and an argument for the phylogenetic concept of the species category. In: Kimbel, W.H., Martin, L.B. (Eds.), Species, Species Concepts and Primate Evolution. Plenum Press: New York, pp. 461–484.

Kimbel, W.H., White, T. 1988. Variation, sexual dimorphism and the taxonomy of *Australopithecus*. In: Grine, F. (Ed.), Evolutionary History of the "Robust" Australopithecines. Aldine de Gruyter: New York, pp. 175–192.

Kuman, K., Clarke, R.J., 2000. Stratigraphy, artefact industries and hominid associations for Sterkfontein, member 5. Journal of Human Evolution 38, 827–847.

Lague, M.R., 2014. The pattern of hominin postcranial evolution reconsidered in light of size-related shape variation of the distal humerus. Journal of Human Evolution 75, 90–109.

Lague, M.R., 2015. Taxonomic identification of Lower Pleistocene fossil hominins based on distal humeral diaphyseal shape. PeerJ 3, e1084.

Lague, M.R., Jungers, W.L., 1996. Morphometric variation in Plio-Pleistocene hominid distal humeri. American Journal of Physical Anthropology 101, 401–427.

Lockwood, C.A., Tobias, P.V., 2002. Morphology and affinities of new hominin cranial remains from Member 4 of the Sterkfontein Formation, Gauteng Province, South Africa. Journal of Human Evolution 42, 389–450.

Lordkipanidze, D., Jashashvili, T., Vekua, A., Ponce de Leon, M.S., Zollikofer, C.P.E., Rightmire, G.P., Pontzer, H., Ferring, R., Oms, O., Tappen, M., Bukhsianidze, M., Agusti, J., Kahlke, R., Kiladze, G., Martinez-Navarro, B., Mouskhelishvili, A., Nioradze, M., Rook, L., 2007. Postcranial evidence from early *Homo* from Dmanisi, Georgia. Nature 449, 305–310.

Marzke, M.W., Wullstein, K.L., Viegas, S.F. 1992. Evolution of the power ("squeeze") grip and its morphological correlates in hominids. American Journal of Physical Anthropology 89, 283–298.

McHenry, H.M., 1983. The capitate of *Australopithecus afarensis* and *A. africanus*. American Journal of Physical Anthropology 62, 187–198.

McHenry, H.M., 1986. The first bipeds: a comparison of the *A. afarensis* and *A. africanus* postcranium and implications for the evolution of bipedalism. Journal of Human Evolution 15, 177–191.

McHenry, H.M., Berger, L.R., 1998. Body proportions in *Australopithecus afarensis* and *A. africanus* and the origin of the genus *Homo*. Journal of Human Evolution 35, 1–22.

McHenry, H.M., Brown, C.C., 2008. Side steps: the erratic pattern of hominin postcranial change through time. Journal Human Evolution 55, 639–651.

Menter, C.G., 2002. Hominid distal humeri, proximal radii and proximal ulnae from Sterkfontein, in comparison with the elbow joints of other Plio-Pleistocene fossil hominids. Ph.D. Dissertation, University of the Witwatersrand.

Moggi-Cecchi, J., Boccone, S. 2007. Maxillary molar cusp morphology of South African australopithecines. In: Bailey, S.E., Hublin, J-J. (Eds.), Dental Perspectives on Human Evolution. Springer: Dordrecht, pp. 53–64.

Moggi-Cecchi, J., Tobias, P.V., Beynon, A.D., 1998. The mixed dentition and associated skull fragments of a juvenile fossil hominid from Sterkfontein, South Africa. American Journal of Physical Anthropology 106, 425–465.

Partridge, T.C., 2000. Hominid-bearing cave and tufa deposits. In: Partridge, T.C., Maud, R.R. (Eds), The Cenozoic of Southern Africa Oxford Monographs on Geology and Geophysics. Oxford University Press: Oxford, pp. 100–133.

Partridge, T.C., Granger, D.E., Caffee, M.W., Clarke, R.J., 2003. Lower Pliocene hominid remains from Sterkfontein. Science 300, 607–612.

Pickering, T.R., Clarke, R.J., Moggi-Cecchi, J., 2004. Role of carnivores in the accumulation of the Sterkfontein Member 4 hominid assemblage: a taphonomic reassessment of the complete hominid fossil sample (1936-1999). American Journal of Physical Anthropology 125, 1–15.

Reed, K.E, Kitching, J.W., Grine, F.E., Jungers, W.L., Sokoloff, L.. 1993. Proximal femur of *Australopithecus africanus* from Member 4, Makapansgat, South Africa. American Journal of Physical Anthropology 92, 1–15.

Rohlf, F.J., 1999. Shape statistics: Procrustes superimpositions and tangent spaces. Journal of Classification 16, 197–223.

Rohlf, F.J., 2010. tpsRelw, relative warps analysis. version 1.49 ed. Department of Ecology and Evolution, State University of New York at Stony Brook.

Rohlf, F.J., Slice D.E., 1990. Extensions of the Procrustes method for the optimal superimposition of landmarks. Systematic Zoology 39, 40–59.

Senut, B., 1978. Révision de quelques pièces humérales plio-pléistocène Sud-Africaines. Bulletins et Mémoires de la Société d'anthropologie de Paris, 223–229.

Senut, B., 1981. Humeral outlines in some hominoid primates and in Plio-Pleistocene hominids. American Journal of Physical Anthropology 56, 275–283.

Senut, B., Tobias, P., 1989. A preliminary examination of some new hominid upper limb remains from Sterkfontein (1974-1984). Comptes Rendues de l'Academie des Sciences, Paris 308, 565–571.

Susman, R.L., de Ruiter, D., Brain, C.K., 2001. Recently identified postcranial remains of *Paranthropus* and Early *Homo* from Swartkrans Cave, South Africa. Journal of Human Evolution 41, 607–629.

Toussaint, M., Macho, G.A., Tobias, P.V., Partridge, T.C., Hughes, A.R., 2003. The third partial skekleton of a late Pliocene hominin (StW 431) from Sterkfontein, South Africa. South African Journal of Science 99, 215–223.

Wolpoff, M.H., 1973. Posterior tooth size, body size, and diet in South African gracile Australopithecines. American Journal of Physical Anthropology 39, 375–393.

Zelditch, M.L, Swiderski, D.L., Sheets, H.D., 2012. Geometric morphometrics for biologists: a primer. Elsevier: San Diego.

7

Ulna and radius

Michelle S. M. Drapeau and Colin G. Menter

The hominin ulnae and radii from Stekfontein are described in detail, followed by the comparative anatomy and analysis. Appendix I presents Member, locus with grid reference, and date of discovery of each specimen, where known.

Descriptions

Ulnae

StW 108—Left proximal ulna

Preservation. This is a partial left ulna 96.9 mm long (Figure 7.1) first described by Senut and Tobias (1989). The dorsal half of the bone is missing for the proximal 40 mm of the fragment, so that only the distal portion of the fragment preserves the full circumference of the diaphysis. Anteriorly at the proximal end of the fossil, the inferior 11.3 mm of the trochlear notch is preserved. The radial notch is present, only missing its proximo-posterior margin, and with its margins slightly eroded. The matrix is preserved on the tip of the trochlear keel at the coronoid.

Morphology. The distal part of the trochlear notch is keeled. The minimum distance between the keel and the radial notch is 10.6 mm. The medial side of the trochlear notch measures 11.9 mm mediolaterally as preserved, but damage obscures its original dimension. The radial notch is 12 mm anteroposteriorly by 9.8 mm proximodistally, and is oriented laterally. The brachialis insertion is well defined but not particularly rugose or deep. Its lateral margin is a round boss that arises from the anterior margin of the radial notch. It is likely that this boss is also the medial margin for the attachment of *m. supinator*. The preserved part of the supinator crest forms a well-developed crest projecting about 2.2 mm from the shaft at its highest, 40 mm from the distal margin of the radial notch. Distal to this, the supinator crest merges rapidly into a more blunt, although well-marked, interosseous crest. In distal view, the shaft is roughly triangular. At the break, the shaft measures 15.7 mm anteroposteriorly and 14.1 mm mediolaterally.

StW 113—Left proximal ulna

Preservation. This is about half of a proximal ulna 137 mm long (Figure 7.1), first described by Senut and Tobias (1989). There is some damage that has removed some of the trochlear notch articular surface proximomedially and medially. Also, there is damage posterior to the supinator crest that may exaggerate the apparent depth of the *anconeus* insertion. The diaphyseal fragment StW 340 might fit loosely to this specimen.

Morphology. The trochlear notch is keeled. The coronoid beak projects slightly further anteriorly than the olecranon beak. Proximally, the medial portion extends further proximally than the lateral, so that the proximal margin of the articular surface angles from proximal and medial to distal and lateral. The mediodistal quadrant of the notch does not have a rounded profile but instead is formed of almost straight anterior and medial borders set about 110° from each other transversely. The articular surface of the notch is continuous proximodistally, but waisted (even when accounting for damage, narrowing at its center to 11.8 mm mediolaterally). Proximally, the trochlear articular surface measures 12.7 mm medially and 13.0 mm laterally from the keel, and distally measures 8.7 mm and 12.6 mm, respectively. All quadrants of the trochlear notch are concave proximodistally and slightly so mediolaterally, except the proximolateral quadrant that is flat mediolaterally.

Michelle S. M. Drapeau and Colin G. Menter, Ulna and radius. In: *Hominin Postcranial Remains from Sterkfontein, South Africa, 1936–1995.* Edited by: Bernhard Zipfel, Brian G. Richmond, and Carol V. Ward, Oxford University Press (2020). © Oxford University Press.
DOI: 10.1093/oso/9780197507667.003.0007

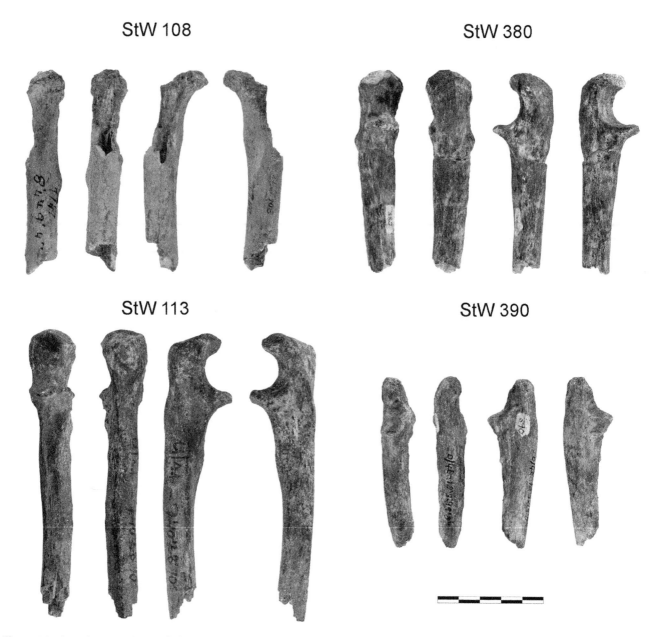

Figure 7.1 Anterior, posterior, medial, and lateral views of Sterkfontein proximal ulnar fragments.

The olecranon extends beyond the articular surface medially as well as proximally. In the lateral or medial view, the profile of the olecranon is angled posteriorly for the most part with a small anterior segment that is angled anteriorly. The insertion for *m. triceps brachii* is rugose and measures 21.4 mm mediolaterally by 15.5 mm anteroposteriorly. Anterior to the insertion at the proximal extent of the bone, but posterior to the area for the capsular attachment, there is a small area for the deep olecranon bursa about 7.4 mm wide anteroposteriorly.

The radial notch is very small, only 10.7 mm transversely by 7.6 mm proximodistally, and faces laterally. The *m. brachialis* insertion is about 27.0 mm proximodistally,

8.3 mm mediolaterally, and 0.9 mm deep. It is rugose between its clear lateral and medial margins, and probably includes the attachment of *m. pronator teres* as well. Its medial margin is a fine crest that arises proximally from the sublime tubercle. Laterally, its margin is formed by a rounded boss running distally from the anterior margin of the radial notch to become strongest 24 mm distal to the trochlear notch and then gradually to merge with the anterior shaft.

The supinator crest projects about 2.8 mm from the shaft surface and is about 14 mm long proximodistally. There are some rugosities along its posterior side. A small crest marking the posterior limit of the *m. anconeus* insertion descends from the olecranon and merges with posterior diaphysis at the level of the radial notch. Here the posterior margin forms a sharp crest that becomes rounder at the level of the ulnar tuberosity. The projection of the crest may have been exaggerated by preparation damage to the floor of the *m. anconeus* insertion (see section "Preservation").

The interosseous crest is well defined, starting about 48 mm distal to the radial notch/trochlear surface junction and extending to the distal break. Some rugosities posterior to it may correspond to the *m. abductor pollicis longus* origin.

The diaphysis is mildly concave anteriorly and even less so medially. Distally the diaphyseal cross-section forms an equilateral triangle, with a flat lateral side and two more convex ones. It is 14.5 mm wide anteroposteriorly and 13.2 mm mediolaterally at the distal break, with measurable cortical thickness of 4.4 mm posterolaterally and 4.6 mm posteromedially.

StW 380—Right proximal ulna

Preservation. This is a right proximal ulna, preserving approximately one-third to one-quarter of the total length of the bone, measuring 96.6 mm (Figure 7.1). It was recovered in two pieces that refit perfectly with each other just below the radial notch (Menter, 2002), with some surface bone missing at that break posteriorly. The tip of the olecranon beak and the proximolateral margin of the trochlear notch are eroded, exposing cancellous bone. The distomedial margin of the trochlear notch is also truncated just anterior to the sublime tubercle. There is some cortex missing in the depth of the trochlear notch.

Morphology. The medial trochlear articular surface extends more proximally than the lateral. The notch is moderately keeled and slightly waisted, narrowing to 14.9 mm at its midpoint. All quadrants of the trochlear notch are concave proximodistally. The two medial quadrants are also concave mediolaterally, while the lateral ones are convex. Although the olecranon beak is eroded, it would not have projected as far anteriorly as the coronoid beak.

The radial notch is small, circularly shaped, 10.4 mm mediolaterally by 10.2 mm proximodistally. It faces anterolaterally and is flush with the adjacent bone surface. The olecranon projects proximally from the articular surface. It is 21.1 mm wide mediolaterally. Its anteroposterior dimension cannot be measured. In the lateral view, the olecranon presents a long posteriorly sloped surface, which accommodates the insertion for *m. triceps brachii*, and a shorter anterior segment that is anteriorly sloped toward the olecranon beak, which is similar to what is observed on other Sterkfontein ulnae.

The crest flanking the tuberosity posteriorly is sharp. The tuberosity presents weak rugosities on its floor, and is about 24.9 mm proximodistally by 9.0 mm mediolaterally and is 0.7 mm deep. The supinator crest projects about 2.0 mm from the shaft at its maximum point. Just posteriorly another crest probably for the posterior extent of the *mm. supinator* and *flexor digitorum profundus* muscle attachments descends distally about 32 mm on the posterior border of the diaphysis. This crest curves anteriorly and terminates in a large tubercle at the proximal end of the interosseous crest, which continues until the distal break.

A small ridge ascends proximally from the supinator crest to the *m. triceps brachii* insertion area on the olecranon probably representing the anterior margin of the *m. anconeus*. A crest demarcating its posterior surface runs distally and posteriorly to meet a faint crest for attachment of *m. flexor carpi ulnaris* at the level of the distal margin of the radial notch. Distal to that point, a faint crest runs down the middle of the posteromedial surface of the bone. The sublime tubercle projects moderately, leaving only a shallow groove posteriorly, and extends into the posterior margin of the ulnar tuberosity.

At the distal break, the diaphyseal cross-section is triangular and measures 14.2 mm anteroposteriorly by 13.1 mm mediolaterally. The posterolateral surface of the diaphysis is flattened, but other sides are more convex transversely. Cortical thicknesses at the break are 4.2 mm nteriorly, 3.6 mm posteromedially, and 4.0 mm posterolaterally. The bone is straight anteroposteriorly and mediolaterally, but the articular surface is slightly tilted medially.

StW 390—Right proximal ulna

Preservation. This is a very weathered, fragmentary portion of proximal right ulna measuring 80 mm as preserved (Figure 7.1). It extends from near the center of the trochlear notch to a point just distal to the supinator crest. Only a small portion of the trochlear surface remains along the radial notch. The posterior and anterior margins of the radial notch are slightly abraded. The ulnar tuberosity is present but very weathered. The distal break is also very weathered.

Morphology. The radial notch faces laterally and measures 12.6 mm anteroposteriorly and 9.2 mm proximodistally as preserved with slight abrasion along the margins. The supinator crest is well delineated but is not very robust, projecting only 0.4 mm from the shaft. Weathering has obscured its distal extent. The margins of the attachment for *m. anconeus* and *m. flexor carpi ulnaris* are visible posteriorly and merge at the level of the radial notch. The diaphysis is very weathered at the break, so its shape may not be an accurate reflection of the morphology, and cortical thickness cannot be measured.

StW 326—Right distal ulna

Preservation. This is the distal 80–85% of a right ulna that is 180 mm long as preserved (Figure 7.2). It was recovered in three fragments (proximal fragment 72 mm, middle 45 mm, distal 69 mm) that fit well (Menter, 2002). At the proximal break, the diaphysis has an uncharacteristic depression on the anterolateral surface that may represent perimortem damage. Distally, the styloid process is missing its anteromedial surface.

Morphology. This can be estimated to have been 221–232 mm long originally. The diaphysis is slender and straight except for a mild anterior concavity near midshaft. Proximally, two faint crests on the dorsal surface merge into the dorsal margin of the bone that may correspond to the posteroinferior margin of the *m. anconeus* attachment and the end of the supinator crest. Proximally, the diaphysis is flat posterolaterally, with well-marked posterior and anterolateral margins, the latter corresponding to the interosseous crest. This flat surface becomes more convex distally. Anterior to the anterolateral margin,

the diaphysis is transversely convex toward its proximal extent but flattens distally, particularly at the level of the pronator quadratus crest. The medial diaphyseal surface is transversely convex for the length of the bone. Although the shaft fragments have been glued, not allowing for direct observation of the midshaft cross-sectional shape, it is roughly triangular. Its approximate-midshaft dimensions are 11.0 mm anteroposterior and 10.4 mm mediolateral. At the distal break, the cross-section is teardrop shaped, with the interosseous crest as the tip and located anteriorly. Figure 7.2 shows the fossil ulna.

Just proximal to the head, the diaphysis narrows to 10.2 mm mediolateral by 8.1 mm anteroposterior. The *m. pronator quadratus* crest is well delineated, extending proximally about 41 mm from the ulnar head. The head measures 14.0 mm mediolateral and 6.9 mm anteroposterior and is crescent shaped in distal view. The area for insertion of the triangular articular disc is 2.6 mm deep from the distal-most point on the head. The groove for the tendon of *m. extensor carpi ulnaris* is apparent but shallow, possibly due to erosion and the truncated styloid process.

StW 326   StW 349   StW 568

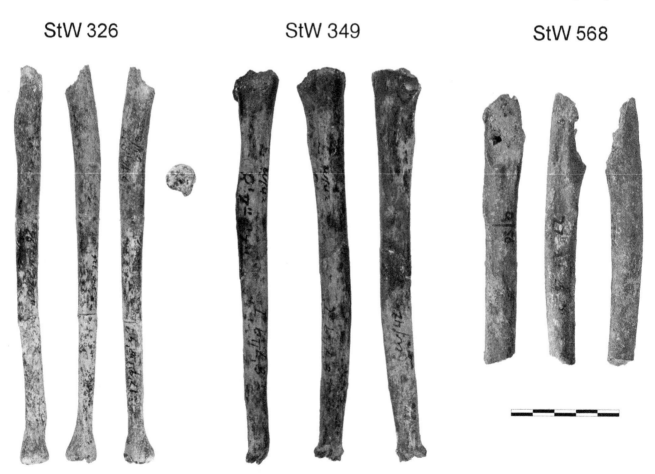

Figure 7.2 Anterior, medial, lateral and distal views of StW 326. Anterior, medial and lateral views of StW 349. Anterior, lateral, and medial views of StW 568.

StW 349—Right ulnar diaphysis

Preservation. This is 179 mm of a right ulnar diaphysis missing its epiphyses (Figure 7.2). It was recovered in two pieces that fit together along a diagonal break near its midpoint (Menter, 2002). Proximally the bone is broken just proximal to the incomplete radial notch, which is eroded on all its margins and preserves only an area about 4 mm by 5 mm of the articular surface of adjacent trochlear notch. Distally, about 7 mm of the diaphysis is crushed just proximal to the pronator quadratus crest, deforming the cross-section. There is a longitudinal crack along the middle third of the anterior surface with a bit of cortex is raised by about 1 mm above it along the junction of the two fragments. This specimen may belong to the same individual as fragmentary radius StW 348 (Tobias, 1985; Menter, 2002).

Morphology. The radial notch is oriented laterally but its original dimensions are not preserved. In the posteromedial view, a truncated sublime tubercle projects strongly along the proximal break. The ulnar tuberosity is demarcated by a small crest medially that arises from the sublime tubercle and laterally by the anterior margin of the bone. The ulnar tuberosity measures about 23 mm proximodistally and 8.2 mm mediolaterally and is about 1.4 mm deep. The supinator crest is sharp, projecting about 1 mm from the shaft proximally and merging distally with the posterior margin of the bone. A strong crest for the attachment of *m. anconeus* extends about 27 mm distally.

The interosseous border begins 50 mm distal to the trochlea with a sharp edge that becomes blunter distally. At about midpoint of the preserved fragment, the shaft is rugose posterior to the interosseous crest. A small crest probably representing the junction of the origins of *mm. abductor* and *extensor pollicis longus* can be felt on the posterolateral surface.

The posterolateral surface of the shaft is flat, the anterior surface is transversely convex, and the posteromedial surface is convex proximally and flattens distally. The diaphysis is triangularly shaped particularly at midshaft where it is 14.6 mm mediolateral and 11.9 mm anteroposterior. Overall the shaft is mildly concave anteriorly and only very slightly so mediolaterally.

StW 568—Right proximal ulnar diaphysis

Preservation. This is a proximal half of a right ulnar diaphysis that measures 124 mm long as preserved (Figure 7.2). Proximally, it is broken diagonally from just distal to the insertion of the *m. triceps* to halfway through the ulnar tuberosity. The distal break is probably just slightly distal to midshaft. Breccia seems to still be present around the supinator crest.

Morphology. The anterior crest of the diaphysis is well marked, but the medial margin of the tuberosity is not well delineated. The supinator crest is partially masked by breccia but was not very pronounced. The interosseous crest begins about 16 mm from the distal break, projecting about 1.3 mm from the diaphysis proximally but becoming blunter distally. The posterolateral margin of the diaphysis is flat along the proximal end of this crest. At the distal break, bone measures 12.9 mm anteroposteriorly by 14.1 mm mediolaterally. Breccia obscures measurement of cortical thickness.

StW 340—Left (?) distal ulna

Preservation. The piece appears to be a distal left shaft extending from a point just distal to midshaft and to one just proximal to the pronator quadratus crest, measuring 58 mm in total length (Figure 7.3). However, this identification is tentative given the fragmentary nature of the specimen. This fragment might fit loosely with StW 113, and if so, it would definitively be a left ulna.

Morphology. The proximal cross-sectional shape is an equilateral triangle, with what is probably the posterolateral aspect being a bit flatter than the other surfaces. At the proximal break, the diaphysis is 12.7 mm anteroposterior and 12.6 mm mediolateral, with cortical thickness of 3.6 mm anteromedially and the 4.3 mm on the other sides. Distally, the cross section is missing some cortex, but it has the shape of a more anteroposteriorly elongated triangle.

StW 398a—Left proximal ulna

Preservation. This is the proximal end of a left ulna measuring 21 mm as preserved (Figure 7.3). Distally, it is broken through the middle of the ulnar tuberosity and supinator crest. The medial and lateral margins of the olecranon are broken off. Medially, the damage is more extensive and part of the trochlear notch articular surface is also missing proximomedially. The articular surface has some surface pitting at its deepest point.

Morphology. The notch has a small keel and is waisted, narrowing to 13.0 mm. The notch appears to be tilted laterally because it extends further proximally on the medial side than the lateral one. The coronoid beak projects more anteriorly than the olecranon. The radial notch faces mostly laterally and very slightly anteriorly and measures 10.7 mm mediolaterally by 9.1 mm proximodistally.

The olecranon projects proximally beyond the olecranon beak by 8 mm. Only the proximal part of the ulnar tuberosity is preserved, but it is well delineated by an anterior ridge and a medial crest. It is 1.8 mm deep, with steep walls. The *m. supinator* crest is truncated, but the

StW 340

StW 398a

StW 571

StW 398b

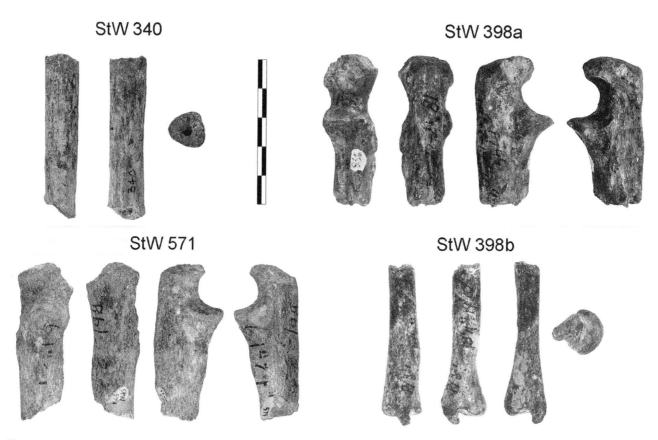

Figure 7.3 Anterior, posterior, and distal views of StW 340. Anterior, posterior, medial, and lateral views of StW 398a. Posterior, medial, lateral, and distal views of StW 398b. Anterior, posterior, lateral, and medial views of StW 571.

preserved part is well marked, projecting about 0.7 mm from the shaft. The fragment is broken too proximally to observe the interosseous crest. Posteriorly, on the lateral margin of the olecranon, the insertion of the *m. anconeus* is well marked, while the attachment for *m. flexor carpi ulnaris* is also well marked but less sharp, and they meet at a level just distal to the radial notch. At the distal break the bone measures 17.9 mm anteroposteriorly by 13.9 mm mediolaterally but is too weathered to measure cortical thickness.

StW 398b—Left distal ulna

Preservation. This specimen is a distal end of a left ulna measuring 22 mm in maximum length (Figure 7.3). It was found with StW 398a and originally given the number StW 399 (Tobias, 1986) but renumbered to StW 398b later (Menter, 2002). It preserves the head, styloid, and pronator quadratus crest.

Morphology. The styloid process is rather gracile and curves slightly medially toward its tip. The head is crescent shaped, measuring 14.1 mm mediolaterally by 6.8 mm anteroposteriorly, with a fairly deep area for attachment of the triangular disc. The groove for *m. extensor carpi ulnaris*

is deep and well defined. The insertion for *m. pronator quadratus* is also well marked extending 47 mm proximal to the ulnar head. The interosseus crest is strong and curves laterally and flanked by a faint longitudinal depression posteriorly. The cross-sectional shape at the break is triangular, measuring 9.7 mm anteroposteriorly by 9.3 mm mediolaterally, with cortical thickness of 2.1 mm anteriorly and posterolaterally and 2.7 mm posteromedially.

StW 571—Right proximal ulna

Preservation. This is a right proximal ulna measuring 60 long (Figure 7.3). Proximally, the olecranon and proximal portion of the trochlea are broken off (Menter, 2002). The margins of the remaining articular surfaces are eroded, making any articular measurements impossible. The sublime tubercle is broken away. Distally, it is broken through the distal portion of the ulnar tuberosity. The supinator crest is well preserved. The floor of the ulnar tuberosity appears to have been "excavated" during preparation, which seems to have exaggerated the depth of the fossa and erased the floor morphology.

Morphology. The trochlear notch is moderately keeled. It is difficult to ascertain the orientation of the radial notch because very little is preserved, but it appears to be mostly lateral with a slight anterior tilt. The ridge flanking the ulnar tuberosity anteriorly is relatively sharp, and the crest bounding it posteriorly is large and sharp. The supinator crest is strong, projecting about 1.8 mm from the shaft and fading away about 14 mm from the radial notch where it is continuous with a low, rounded ridge. On the lateral aspect of the olecranon, the proximal margin of the *m. anconeus* attachment is well marked, while medially the margin of the *m. flexor carpi ulnaris* is difficult to see but can be felt. The two merge posteriorly about 6.6 mm distal to the radial notch. The distal break is irregular so dimensions of the whole bone and cortical thickness cannot be measured reliably.

StW 431b—Right proximal ulna

Preservation. This is a right proximal ulna (Figure 7.4), part of a partial skeleton (Tobias, 1992; Menter, 2002; Toussaint et al., 2003; Ward et al., Chapter 4, this volume). About the proximal two-thirds are preserved and measure 180 mm in length. They were recovered in three pieces that join well at transverse breaks (Menter, 2002). The proximal piece measures 74 mm and includes all of the proximal articulation and major muscles attachments. At that proximal break, there is some cortex missing on the posteromedial margin. The middle piece measures 71 mm and the distal piece 34 mm (Menter, 2002). At the break between the middle and distal piece, there is some cortex missing anteriorly, but no other damage to the diaphysis is apparent. Proximally, the trochlear notch is truncated lateroanteriorly to posteromedially from the anterior tip of the radial notch to the just distal to the waisting of the notch medially, leaving nothing of the coronoid process except for the posterior margin of the sublime tubercle. The olecranon is abraded proximolaterally, and the rim of the trochlear notch articular surface is eroded there as well. There are some very minor abrasions on the anterior surface of the olecranon beak. The anterior rim of the radial notch presents some erosion.

Morphology. The notch appears to be slightly tilted laterally. It has a moderate central keel and is strongly waisted. At the junction of the olecranon and coronoid process, slightly raised, non-articular bone ridges visible on the medial and lateral margins delineate the articular surface. The three preserved quadrants of the trochlear notch present proximodistal concave surfaces. While mediolaterally, the proximomedial quadrant is also concave, the proximolateral is convex (a convexity that might be exaggerated because of the erosion on the lateral margin of that surface) and the distolateral is flat. Although most of the coronoid beak is missing, it clearly projects more anteriorly than the olecranon beak. The radial notch is oriented laterally, with only very little anterior rotation. The notch is large compared to the other Sterkfontein specimens, measuring 14.4 mm anteroposteriorly by 11.4 mm proximodistally.

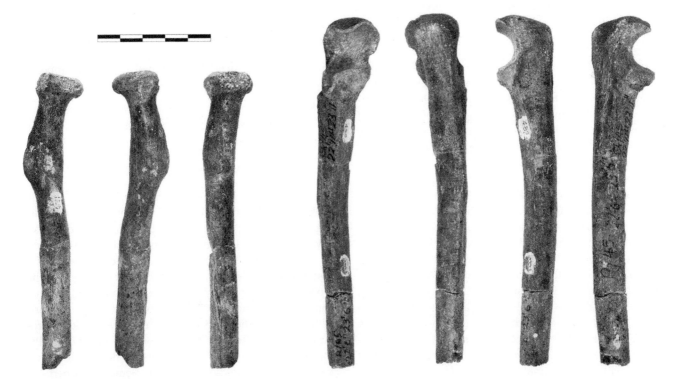

Figure 7.4 Right proximal radius (left) and ulna (right) of StW 431. The radius is in posterior, anterior, and medial views. The ulna is in anterior, posterior, medial, and lateral views.

Because some of the articular margin is missing anteriorly, the anteroposterior value may be an underestimate of the actual anatomical value. All the articular surfaces that are preserved have rims that project above the non-articular bone instead of being continuous with it, except for the waisted surface of the trochlear notch, which has projections of non-articular bone. Because of this lipping, there are clear grooves medial and lateral to the proximal trochlear notch surfaces and the radial notch is raised on a bony plateau instead of being continuous with the shaft.

The olecranon projects proximal to the trochlear notch by about 8 mm. It is also strongly projecting medially pass the articular surface by 3.3 mm, a morphology also observed in StW 113 but to a lesser degree. The olecranon is also much wider than the articular surface at the coronoid/olecranon junction, the articular surface being 12.3 mm wide while the olecranon is 20.0 mm. Distal to that point, it is impossible to evaluate if non-articular bone projects pass the articular surface since too much of the coronoid is missing. In proximal view, at the *m. triceps brachii* insertion site, there is a round depression of non-cortical bone, which exposes trabecular bone that does not appear to be peri- or postmortem damage. The hole is about 2.6 mm by 3.5 mm wide and 1.8 mm deep. The anterior limit of the *m. triceps* insertion is not clearly marked and cannot be accurately separated from the deep bursa surface. In the lateral view, and similarly to other Sterkfontein specimens, the olecranon profile is mostly posteriorly slopped, with a much shorter segment that slops anteriorly.

The ulnar tuberosity is well demarked by an anterior boss that originates from the truncated coronoid and merges with the anteromedial surface 52.6 mm distal to the trochlear/radial notches junction. Medially, the tuberosity is limited by a well-marked, but modest crest that originates from the truncated sublime tubercle. The tuberosity is about 1.0 mm deep. The rugosity of the floor of the insertion is difficult to observe because it is partially covered by a catalogue number. What is visible, suggests that the floor was not very rugose.

As with most other Sterkfontein specimens, the supinator crest is well marked and 6.0 mm wide, projecting some 1.8 mm from the shaft, and is located just distal to the radial notch. The posterior aspect of this crest descends and tapers distally into a very faint line that ends 45.4 mm istal to the trochlear/radial notches junction. The total length of the supinator crest is 30.2 mm.

The interosseous crest is blunt but well defined, marked by a rugose tubercle proximally. It starts 50.8 mm distal of the trochlear/radial notches junction. On the shaft, some rugosities are observed at the posterior margin of the diaphysis and posterior to the interosseous crest at the level of its proximal extremity.

On the lateral aspect of the olecranon, there is a small tubercle separated from the proximolateral trochlear notch by a small groove 0.8 mm deep. This groove can be traced distally, where it widens and deepens to 1.4 mm as the articular surface becomes more waisted. From the proximal tubercle, a small crest runs inferiorly and slightly posteriorly until it becomes a fairly well-marked ridge at the level of the radial notch. It then merges into the posterior surface. This crest likely marks the posterior attachment limit of *m. anconeus* while the tubercle and area anterior to the groove mark its anterior attachment. A bit distally to this ridge, the posterior margin becomes visible with its rugosities described above.

On the medial aspect of the olecranon, the limit of the *m. flexor carpi ulnaris* attachment can be seen without being well crested. Distally, a line marking the attachment of *m. flexor carpi ulnaris* is faint, but it merges very laterally with the ridge for *m. anconeus* at the level of the radial notch, about 29.7 mm distal to the olecranon process. More anteriorly, there is a deep sulcus just posterior to the sublime tubercle. This depression is part of the origin for *m. flexor digitorum profondus*.

In the middle third of the fragment, the diaphysis is flattened laterally. At that level, the diaphysis is transversally triangular, with the two other sides slightly bulging. At the distal break, the cross-sectional shape is an anteroposteriorly elongated triangle, with the anterior lateral margin being the interosseous crest and the posterior margin being the distolateral vertex. At that level, the lateral surface of the diaphysis has ceased to be flat, but has become a bit more bulging. The diaphyseal curvature is moderate anteroposteriorly and more pronounced mediolaterally, with the lateral side convex. Because of this mediolateral curve, the proximal articulation appears slightly displaced medially relative to the long axis of the diaphysis.

Radii

Sts 68—Left proximal radius

Preservation. This is a proximal fragment of a left radius, broken just distal of the radial tuberosity and measuring 60.6 mm long (Figure 7.5). It is eroded and looks like it was originally in two pieces that were glued together in a fairly good fit through its midsection. Because of a distal diagonal break and the distortion of the shaft, the cross-sectional anatomy at the break is not in its original condition. The head is eroded posteriorly. The margins of the edge of the head are also mildly abraded anterolaterally and medially, but the fovea is intact. Distally, the shaft is missing numerous cortical flakes, mostly on the medial and posterior surfaces. Some are superficial while others are deep. The cortex on the anteromedial aspect of the neck is heavily eroded. The medial portion

Sts 68 StW 46 StW 105

StW 139 StW 516

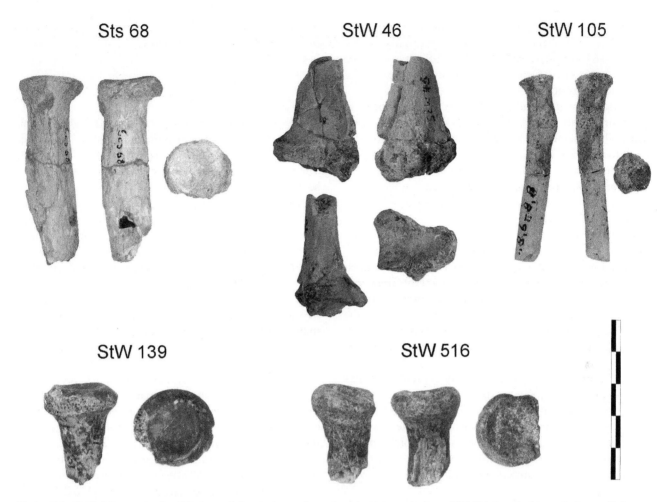

Figure 7.5 Radial fragments. Sts 68 is in medial, anterior, and proximal (anterior up) views. StW 46 is in anterior, posterior, medial, and distal (anterior up) views. StW 105 is in anterior, medial, and proximal (anterior down) views. StW 139 is in posterior and proximal (anterior up) views. StW 516 is in posterior, medial, and proximal (anterior up) views.

of the tuberosity is damaged and on the posterior margin of the tuberosity two pieces of cortex, the proximal one measuring 10.2 mm by 5.1 mm and the distal one measuring 14.6 mm by 9.2 mm, are displaced distoanteriorly. Proximal to the tuberosity, some cortex is missing measuring 11.2 mm by 5.2 mm. For this reason, the posterior and distal margins of the tuberosity are distorted and the proximal margin is eroded, making reliable measurements of that structure difficult. Posterolaterally, toward the distal margin of the fragment, the cortex is missing on an area 6.1 mm by 3.6 mm, creating a hole.

The specimen has been coated with a varnish, giving the specimen a yellow tint. On the anterior and posterior surfaces, some recent flaking has occurred, peeling off the varnish and some surface bone. None of this more recent damage seems to have modified the shape of the bone in any significant way. Given the damaged nature of this specimen, few reliable measurements are possible.

Morphology. The hominin status of Sts 68 has been questioned (Tobias et al., 1977; Brain, 1981), but its morphology suggests that it is indeed hominin (Susman and Grine, 1989; Grine and Susman, 1991; Menter, 2002). It presents a head that is rather short proximodistally, unlike other Sterkfontein specimens, measuring about 6.7 mm at its shortest point. In the proximal view, the head, although eroded, presents a capitular fovea that is fairly large (maximum diameter 16.0 mm) relative to the total diameter of the head (maximum preserved 21.4 mm). This seems to distinguish this specimen from other Sterkfontein specimens that usually have a relatively small fovea. The depth of the fovea is 2.1 mm. As observed on other fossil specimens, the head projects laterally past the shaft for 5.8 mm, which is the point of maximum projection. Medially and anteriorly, the head projects only mildly past the shaft, about 2.4 mm anteromedially where it can be reliably measured. Maximum neck diameter is 14.9 mm, measured anteroposteriorly. The tuberosity does not project anteriorly but is marked by a faint, blunt crest. The preserved area of the posterior margin suggests that the tuberosity may have been delineated posteriorly by a fairly sharp crest and presents a shallow (1.0 mm deep) depression at its center measuring 12.9 mm proximodistally by 4.6 mm anteroposteriorly. It is difficult to evaluate how postmortem deformation might have influenced this particular morphology.

Angulation of the neck relative to the shaft appears to be slight, but little is preserved of the shaft past the tuberosity, which makes the evaluation of the angulation difficult to ascertain with confidence. Overall, from the neck to the distal break, the shaft does not fluctuate in diameter much and, therefore, is somewhat tubular.

Sts 2198b—Right (?) proximal radius

Preservation. This fragment includes the head and the neck and is 23 mm long (not figured). It is the same specimen as the Sts 2198a humerus, which is a right. The entire floor and most of the width of the capitular fovea are preserved, but the rim of the fovea is damaged around the entire specimen except from posteromedial to anterior. Most of the periphery of the articular circumference is damaged except posteriorly. The neck is fairly well preserved anteriorly, but posterolaterally a small piece has been displaced obliquely proximoposteriorly.

Morphology. This is one of the smallest radial heads in the Sterkfontein collection. It seems to have little beveling which corresponds to the shallow zona conoidea of the humerus. Its neck is narrow anteromedially to anterolaterally. Due to its poor preservation, little can be said about this specimen.

StW 46—Left distal radius

Preservation. This is a damaged and broken distal end to a left radius of total length 40.8 mm (Figure 7.5). The bone is comprised of several pieces joined together. Most joints are missing only tiny amounts of bone along them, but a large one just proximal to the distal articular surface has a large wedge missing along its lateral edge that continues posteriorly for most of the length of the fragment. The distal articulation is abraded along the edges of the distal articular surface. The ulnar notch is well preserved. The styloid process is not preserved and looks to have been abraded away. Most of the anterior rim of the distal articulation has been broken away.

Morphology. At the proximal break, the bone measures 10.2 mm anteroposteriorly and 15.5 mm mediolaterally. Cortical bone is 2.0 mm thick anteriorly and 2.3 mm posteriorly. The maximum anteroposterior width across the distal articular surface is 19.9 mm. The estimated maximum mediolateral width across the distal articular surface is 26.8 mm, but this is less than it would have been originally, as the styloid process is not preserved. The dorsal tubercle is very prominent. The ulnar notch is mainly flat and not deeply concave transversely. The maximum width of the ulnar notch is 14.6 mm and height is 4.8 mm.

StW 105—Right juvenile proximal radius

Preservation. This is a fragment of a juvenile right proximal radius (Figure 7.5). The specimen consists of two pieces joined at its midsection that is 60 mm in maximum length. The radial epiphysis is missing but the proximal diaphyseal surface is well preserved. On the posterior part of the proximal diaphysis there is a small adhering piece of breccia. A number of long cracks run longitudinally on both pieces, with a few small chips of cortex missing from the proximal piece. Distally the fossil is broken obliquely, and proximal to the break on the posterolateral side, 6 mm of cortex is missing.

Morphology. Very little morphology is present in this juvenile specimen as the proximal epiphysis is absent. The maximum diameter of the proximal diaphyseal plate is 13.3 mm taken anteroposteriorly. The neck is very short, roughly 14.1 mm from center of tuberosity to the margin of the epiphyseal surface, and with the minimum breadth of the neck is 7.7 mm. Immediately distal to the radial tuberosity the diaphysis is round in section with a diameter of 8.4 mm. A mild interosseus crest is apparent near the distal end of the fossil where the diaphysis is 8.0 mm anteroposteriorly and 8.4 mm mediolaterally and has a cortical thickness of 2.1 mm on all sides. Assuming that the specimens preserves approximately 40 to 50% of the diaphysis, the estimated age would be ~5 years of age using a modern human standard.

StW 139—Radial head and partial neck

Preservation. This is a proximal fragment of a left radius that preserves the head and part of the neck and is only 30.9 mm long (Figure 7.5). Menter (2002) sided this as right radius, but it is actually from the left side (see also Senut and Tobias, 1989; Grine and Susman, 1991). The circumference of the radial head is preserved only from the posteromedial to the anterolateral aspect. The capitular fovea is almost entirely intact, except for a segment from the posteromedial to posterolateral rim of the fovea. Up to 22 mm of the neck is preserved, but only on the posterior side; hence, the radial tuberosity is not preserved.

Morphology. The head is circular and the articular circumference of the radial head is beveled presenting a three-part articulation as described by Grine and Susman (1991). The beveling is broad and sloped to fit a deep capitulotrochlear sulcus (zona conoidea). The depth of the fovea is relatively deep, about 2.6 mm, while the mediolateral of the capitular fovea is 15.6 mm. The head is about 23 mm in diameter; anteromedially the head projects slightly past the shaft for 2.1 mm. The preserved portion of neck measures 12.8 mm by 13.4 mm in its minimum and maximum dimensions, respectively.

StW 431a—Right proximal radius

Preservation. This is a proximal half of a right radius (Figure 7.4) that is part of the Sterkfontein Member 4 partial skeleton (Menter, 2002, Toussaint et al., 2003; Ward et al., Chapter 4, this volume). The specimen is well preserved and consists of two pieces reunited in fairly good apposition. The proximal piece (length 80 mm) includes the head, neck, radial tuberosity, and part of the proximal shaft. The distal piece is broken transversely and this break is approximately 25 mm distal to the distal border of the radial tuberosity. The distal piece is 54 mm proximodistally and, distally, is broken obliquely. The radial head is in good condition anteriorly, except for damage that starts on the medial side of the head and continues to the posteromedial aspect for about 15.2 mm. Posteriorly, there is a small 5.0 mm by 3.2 mm chip of articular bone missing just posterior to the fovea. The remainder of the head is well preserved. Remnants of casting material are visible inside the broken areas of the radial head. The specimen is a dark brown and in places stained black by manganese. At the union of the two pieces just anterior to the interosseous crest a chip of bone is missing on both sides of the break.

Morphology. The head of the radius shows the African ape and australopithecine three-part articular relationship. The capitular fovea is deep and the articular circumference is beveled indicating a deep capitulo-trochlear sulcus, as observed on the matching humerus, StW 431c. The head measures 22.2 mm mediolaterally and 22 mm anteroposteriorly, while the fovea is 14.4 mm in both axes. The minimum proximodistal height of the head is 8.4 mm. The radial neck appears more robust than in African apes but less so than in humans. The radial neck is short as in humans and just distal to it is a large and heavily rugose radial tuberosity. The radial tuberosity is very prominent and raised above the shaft with very prominent insertions for the posterior fibers of *m. biceps brachii*. On the anteromedial aspect of the tuberosity it is a smooth, rounded border. Distal to the tuberosity is the anterior oblique line, which is rugose. In its proximal portion, the interosseous border is sharp while the distal segment is rounded. On the medial border just below the sharp part of the interosseous border, is what is most likely the nutrient foramen. The anteroposterior diameter at the level of the foramen is 13.0 mm and the mediolateral width is 13.8 mm. In distal view, the shaft is circular to oval. Laterally the radial head projects 7.7 mm from the shaft while medially the head projects 2.2 mm from the shaft.

StW 516—Left proximal radius

Preservation. This is a fragmentary proximal end of a left radius that preserves parts of the head and neck (Figure 7.5). Proximodistally only 31.8 mm is preserved while the greatest breadth is 22.8 mm, being the preserved portion of the head from the posteromedial to anterolateral. The specimen is heavily calcified and is a gray to gray-black piece of bone that is heavily manganese stained. In addition, the internal part of this bone as viewed from the distal aspect is filled with calcite crystals. The radial head is preserved from the medial side around posteriorly to the posterolateral side. From the posterolateral side around anteriorly the side of the radial head and the articular circumference are missing or heavily damaged. The capitular fovea is preserved posteriorly and on the lateral side but is slightly damaged posteromedially and to a greater degree anteromedially. Little of the radial neck is present, the longest preserved portion (20 mm in length) being the lateral side of the radial neck. The posterior side of the radial neck has a large fragment of cortex missing, 16 mm proximodistally and 9.7 mm mediolaterally. Distally the shaft is broken obliquely from medial to lateral. Medially the shaft is broken proximal to the radial tuberosity.

Morphology. From what is preserved the radial head is circular, but as the anterior portion is poorly preserved it is not possible to determine whether the radial head is beveled. The greatest breadth of the preserved portion of the radial head is 22.8 mm while mediolaterally the head is estimated at 22.4 mm. The mediolateral neck width is 12.6 mm. The posterior rim of the capitular fovea is very high proximally and creates a steep slope from the floor to the posterior rim. The medial side of the radial head gives the impression of being proximodistally larger than in StW 139 and StW 431a, but damage on all three specimens precludes taking precise measurements. Laterally the head projects 5.6 mm past the shaft while medially it projects 3.5 mm past the shaft.

StW 125—Right radial diaphyseal fragment

Preservation. This is a midshaft fragment of a right radius 117 mm long (Figure 7.6) that was originally described by Senut and Tobias (1989). The proximal end is broken obliquely and is missing most of the anterior surface. The shaft is mildly abraded across most of its length, especially on the posterior and lateral surfaces and along the interosseus crest. There are a number of mild longitudinally cracks running across all surfaces. Distally, the shaft is broken transversely. A large nutrient foramen just anterior to the interosseous border is evident, 42.7 mm distal to the most proximal point of the shaft.

Morphology. Very little morphology is present as only the shaft is preserved. Proximally, the diaphysis is triangular

StW 125   StW 348   StW 354   StW 528

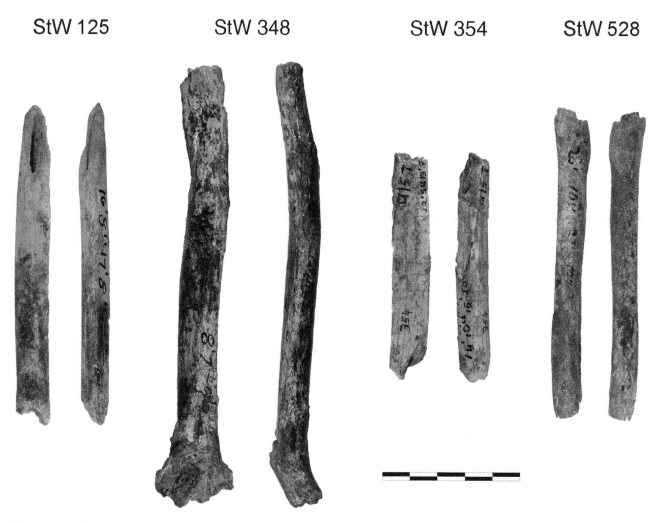

Figure 7.6   Radial diaphysis fragments. All views are anterior and medial.

in section measuring 9.4 mm anteroposteriorly by 11.7 mm mediolaterally, but the break is too oblique to measure cortical thickness. At the level of the nutrient foramen, it is 10.7 mm by 13.3 mm. Distally it is oval in section and measures 10.4 mm by 13.5, mm with its cortical thicknesses of 3 mm anteriorly and posteriorly and 3.5 mm medially and laterally. A rugose insertion for *m. pronator teres* extends 15.2 mm from the proximal break. Where the interosseous crest is not abraded, it forms a sharp and small crest starting from the distal part of the nutrient foramen.

StW 348—Right radial diaphysis

Preservation: This is a right radial diaphysis with a total length of 157 mm (Figure 7.6). The specimen is badly crushed and broken at both the proximal and distal ends. This specimen preserves the distal end of the radial tuberosity and most of the shaft, but it is missing about a quarter of its distal end. The distal third of the radial tuberosity is preserved but it is crushed and displaced.

The bone appears undistorted and largely intact from the level of the crest for *m. pronator teres* distally for about 80 mm. The interosseous border is evident but its proximal part is slightly abraded.

Morphology. The radial tuberosity seems to be as large and robust, although damaged. The anteroposterior width at the level of the nutrient foramen is 12.2 mm and mediolaterally is 15.7 mm. The interosseous border is strong, and anterior to it is a shallow groove for *m. flexor pollicis longus*, running for approximately 40 mm along the middle of this fossil. The attachment for *m. pronator teres* is rugose and 24.7 mm proximodistally. The oblique line is only mildly evident.

StW 354—Left radial diaphyseal fragment

Preservation. This is a fragment of left radial midshaft 83 mm long (Figure 7.6). This specimen has been assembled from three main fragments in fairly good opposition, and cracks traverse the specimen but do not disturb the bone contours appreciably. In several places preparation damage is evident on the posterior aspect of the proximal half obscuring surface contours.

Morphology. The diaphysis is triangular in cross-section throughout the length of the fragment. Near the proximal end, the bone measures 10.6 mm anteroposteriorly by 13.5 mm mediolaterally. A nutrient foramen is present adjacent to the distal break. At this point, the diaphysis is 12.4 mm anteroposteriorly and 14.2 mm mediolaterally.

StW 528—Right radial diaphyseal fragment

Preservation. This is a well-preserved shaft fragment of a right radius extending from the distal half of the radial tuberosity to the distal part of the *m. pronator teres* insertion (Figure 7.6). The preserved length is 115.5 mm. The shaft is mottled by manganese staining. The nutrient foramen is visible at the interosseous crest, on the distal third of the bone. Proximally the bone is broken transversely while distally it is broken by a jagged and oblique break.

Morphology. The shaft is one of the smallest in the collection. It is smaller than, for example, the shafts of StW 348 and StW 431a. Its widths at the level of the nutrient foramen are as follows: mediolateral width, 13.0 mm and anteroposterior width, 10.4 mm. The radial tuberosity does not have pronounced posterior lipping as observed in StW 431a and the anterior is very smooth and rounded. It has a slight furrow in its middle as has StW 431a. The interosseous border is not sharply defined; just anterior to it is a slight concavity for the attachment of *m. flexor pollicis longus*. The pronator teres insertion is present on its lateral surface, but it is not rugose and not very evident. The oblique line is not evident. Viewed distally, the shaft is an anteroposteriorly flattened oval.

Comparative anatomy

Materials and methods

The comparative sample includes ulnae and radii from *Homo sapiens*, and wild-shot specimens of *Pan paniscus*, *Pan troglodytes*, *Gorilla gorilla*, and *Pongo pygmaeus*. Because the subspecies of gorillas are variable morphologically, eastern (*G. g. gorilla*) and western (*G. g. berengei* and *G. g. graueri*) gorillas were separated in the univariate and bivariate analyses. To insure adequate sample sizes for all taxa in the multivariate analyses, the gorilla subspecies were lumped into one group (*Gorilla*) as were bonobos and chimpanzees (*Pan*). Tables 7.1 and 7.2 lists the comparative sample and the fossil specimens for the univariate and bivariate analyses; Tables 7.3 and 7.4 lists the comparative sample and the fossil specimens for the discriminant and principal component analyses.

In order to compare taxa of widely different sizes, variables are presented as ratios or controlled for size before comparison. For the distal ulna, a size surrogate was calculated as the geometric mean of four measurements: maximum anteroposterior and mediolateral diameters of the distal end and anteroposterior and mediolateral minimum diameters of the diaphysis proximal to the head. These measurements were selected because they describe the size of distal ulna and were measurable on two Sterkfontein distal specimens (StW 326 and StW 398b).

Trochlear notch orientation (UNO) and olecranon orientation (UOO) and length (UOL) were measured following the methods presented in Drapeau (2004) and Drapeau et al. (2005), modified to include incomplete specimens. Following Lovejoy et al. (2009, sup. information), the orientation of the trochlear notch (UNO) and the olecranon (UOO) were measured relative to the proximal dorsal profile of the diaphysis (Figure 7.7). To measure keeling of the trochlea (UKL), we calculated the ratio of the maximum articular trochlear diameter to the notch diameter (UMTD/UTD; Figure 7.7). UMTD is measured as the maximum diameter of the trochlear notch including the articular surfaces measured parallel to the axis of trochlear notch orientation in lateral view (Figure 7.7). UTD is measured as the diameter of the circle best-fitted (NIH Image J software) to the trochlear notch margin excluding the articular surfaces (Figure 7.7). A large ratio indicates that a large portion of the surfaces is oriented laterally (rather than more anteriorly), which indicates a higher keel. Waisting of the trochlear notch was measured as the mediolateral width of the proximal notch divided by the mediolateral width of the center of the notch (UNMP/UTW; Figure 7.7), with larger values indicating greater waisting.

On the radius, neck constriction was calculated as the anteroposterior diameter of the head divided by the anteroposterior diameter of the neck (RHAP/RBAP; Figure 7.8). Relative head size was calculated as the anteroposterior diameter of the head divided by the anteroposterior diameter of the diaphysis immediately distal to the tuberosity (RHAP/RTAP; Figure 7.8). Relative fovea size is measured as the anteroposterior diameter of the fovea divided by the anteroposterior diameter of the head (RFAP/RHAP; Figure 7.8). All measurements are in the anteroposterior axis in order to include a maximum number of fossil specimens. However, a few fossils could not be measured in that axis, so mediolateral values are presented in the text when possible (relative fovea size in Sts 68 and A.L. 333x-14).

Comparisons among taxa were made with a one-way ANOVA. If the ANOVA was significant, a post hoc test was performed to compare taxa using t-tests with Bonferroni corrections when the among-groups variance was homogeneous (Levene's test) or with a Tamhane T2 test when variance was heterogeneous.

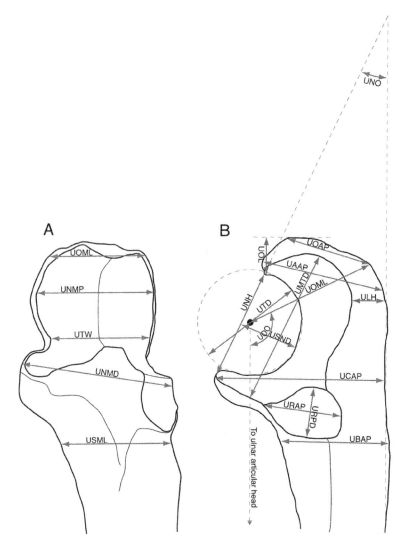

Figure 7.7 Ulnar measurements presented in this chapter. Anterior (A) and lateral (B) views of a left ulna.

Table 7.1 Ulnar measurements used in the univariate and bivariate analyses (mean, standard deviation, n = sample size). All measurements in mm unless otherwise noted.

Taxon/ Specimen[a]	Sex	Stat.	UOL	UNH	UNMP	UNMD	UKL	URAP	URPD	UCAP	UAAP	UMML	UMAP	UDML	UHT	UOO[b]	UNO[b]
H. sapiens	M	Mean	4.1	26.2	26.8	33.4	1.38	18.8	12.3	36.9	26.7	15.2	16.2	18.9	11.9	110.5	155.9
		St. dev.	1.7	2.3	1.8	2.8	0.06	2.3	2.1	2.2	2.4	1.1	1.7	1.8	1.4	21.9	16.5
		n	15	56	57	51	14	15	15	15	15	15	15	54	55	14	14
	F	Mean	4.2	23.3	23.2	28.8	1.37	15.9	9.9	32.8	23.6	12.6	14.0	15.7	10.0	114.2	161.6
		St. dev.	1.2	1.9	1.6	2.6	0.08	2.3	1.5	2.1	1.3	1.7	2.0	1.5	1.2	21.2	11.3
		n	15	43	40	38	15	15	15	15	15	15	15	41	42	15	15
P. paniscus	M	Mean	2.8	20.3	21.0	24.2	1.63	14.4	9.8	33.5	27.7	12.6	14.1	17.8	7.3	97.9	151.0
		St. dev.	0.6	1.2	1.7	2.5	0.44	1.7	1.6	1.3	2.8	1.5	1.8	1.5	1.2	12.5	4.5
		n	6	11	11	11	8	11	11	11	11	11	11	10	10	8	8
	F	Mean	2.7	20.1	19.4	21.6	1.49	14.1	10.2	31.6	25.5	12.1	13.2	16.6	6.7	102.1	151.5
		St. dev.	1.4	2.1	1.4	3.1	0.05	1.0	1.4	2.4	2.4	1.4	1.2	1.0	0.9	10.6	6.8
		n	9	9	9	9	7	9	9	9	9	9	9	9	9	7	7

Table 7.1 Continued

Taxon/Specimen[a]	Sex	Stat.	UOL	UNH	UNMP	UNMD	UKL	URAP	URPD	UCAP	UAAP	UMML	UMAP	UDML	UHT	UOO[b]	UNO[b]
P. troglodytes	M	Mean	3.2	23.5	23.4	32.5	1.46	17.1	12.5	39.0	32.6	15.1	15.8	19.3	8.4	100.7	155.2
		St. dev.	2.7	3.2	2.2	2.9	0.08	1.9	1.4	3.3	2.3	1.9	1.8	1.4	0.8	27.8	24.9
		n	22	23	22	22	11	22	22	22	22	22	22	22	21	11	11
	F	Mean	3.5	21.2	21.4	28.7	1.56	15.1	11.1	34.8	28.8	14.2	14.8	18.1	7.9	100.7	156.6
		St. dev.	2.2	3.1	1.4	3.1	0.19	1.9	1.4	2.3	2.4	1.4	2.0	1.4	0.7	19.7	28.2
		n	32	34	32	32	13	32	32	32	32	33	33	34	33	11	11
G. g. gorilla (western)	M	Mean	2.1	33.8	36.1	56.6	1.37	28.3	20.1	52.3	41.2	22.6	22.5	29.4	13.2	92.9	148.9
		St. dev.	1.9	3.3	3.1	5.1	0.12	3.0	2.1	4.4	4.7	2.4	2.7	2.4	1.5	25.9	17.8
		n	30	30	29	29	7	29	30	29	30	30	30	29	29	9	9
	F	Mean	1.6	26.4	27.7	41.8	1.48	22.3	16.8	41.3	30.9	17.5	17.5	22.3	10.3	96.0	147.1
		St. dev.	2.1	2.5	2.6	3.8	0.13	1.9	1.6	2.8	2.8	1.6	2.1	1.8	0.9	25.4	12.5
		n	26	27	27	27	14	27	27	27	27	27	27	27	27	14	14
G. g. beringei *G. g. graueri* (eastern)	M	Mean	3.3	28.8	36.9	56.3	1.58	28.3	21.5	53.0	41.3	23.8	23.0	27.8	14.8	89.2	142.3
		St. dev.	4.0	1.6	2.3	4.1	0.15	2.9	1.8	3.7	4.7	2.3	2.8	1.7	2.0	19.1	17.3
		n	6	10	9	10	6	10	10	10	9	10	10	9	9	6	6
	F	Mean	2.7	23.1	27.2	42.0	1.49	22.7	17.6	39.5	30.0	18.5	16.5	21.8	11.1	88.2	143.2
		St. dev.	1.7	1.8	1.9	3.0	0.10	1.8	1.2	1.9	2.0	2.1	1.6	1.5	1.5	22.5	16.5
		n	5	9	8	9	8	9	9	9	9	9	9	8	8	8	8
P. pygmaeus	M	Mean	2.3	23.2	24.5	40.0	1.44	21.8	15.6	38.3	26.7	16.2	19.4	24.0	10.6	90.8	145.6
		St. dev.	1.6	3.2	3.3	3.9	0.12	2.0	1.2	3.7	3.2	2.3	2.2	2.6	1.6	17.0	14.3
		n	14	14	14	14	8	14	14	14	14	14	14	14	14	8	8
	F	Mean	2.5	18.2	20.3	33.0	1.55	18.0	13.5	29.7	21.9	11.6	15.9	19.7	9.4	98.0	148.6
		St. dev.	1.3	2.6	1.3	2.3	0.15	1.7	1.5	2.2	1.8	0.7	1.4	1.8	1.4	26.4	22.4
		n	16	16	16	16	10	16	16	16	16	16	16	16	16	11	11
A. africanus																	
StW 108									9.3								
StW 113			7.4	22.9	22.0	22.8	1.5	10.0	8.0	30.3	27.5	13.9	14.1			114.0	166.8
StW 326												10.4	11.0	14.0	6.9		
StW 340												12.6	12.7				
StW 349												11.9	14.6				
StW 380				20.0*	22.0	18.1		9.2	10.2	29.2	23.5*					117.6	168.0
StW 390								12.1	8.3								
StW 398a&b			7.9	18.0		21.6		11.2	9.5	27.1	22.8			14.3	7.2		
StW 431b			7.7		22.8*			14.4	11.7		27.0	15.1	12.5			111.2	
StW 568												14.5	14.6				
StW 571								12.3	8.8*	25.6							
A. afarensis																	
A.L. 438-1			7.7	23.2	25.7	23.0	1.3	16.4	10.2	31.8	26.2	14.8	17.6	17.5	8.9	122.1	169.9
A.L. 288-1[c]			3.6	15.7	16.1	16.0	1.50	9.8	6.8	22.5	18.4	11.6	10.2	12.2	6.0	122.6	170.5
A.L. 333w-36								10.2	9.8	25.1							
A.L. 333-11								15.3									
A.L. 333-12														18.3	7.7		
P. robustus																	
SKX 8761[d]			10.6	13.8	21.3	20.2		11.0	8.7	25.6	23.3	12.2	13.1				

[a] The ape skeletal collection is from the Musée Royal de l'Afrique Centrale, Tervuren, Belgium, the Anthropological Institute of the University of Zurich, Switzerland, the Cleveland Museum of Natural History, and the National Museum of Natural History, Washington, D.C. The human collection is from the Cleveland Museum of Natural History, the Canadian Museum of History, Gatineau, and the Département d'anthropologie, Université de Montréal, Canada.

[b] Measurements are in degrees.

[c] Measured on the right side and, if unavailable, was measured on the left.

[d] Attributed to *P. robustus* by Susman (1989) and Susman et al. (2001).

* = Estimated.

UOL: Proximodistal olecranon height

UNH: Proximodistal height of trochlear notch from coronoid to anconeal tips

UNMP: Mediolateral width of proximal trochlear notch

UNMD: Mediolateral width of distal trochlear notch and radial notch

continued

Table 7.1 Continued

UKL: Trochlear keeling measured as the ratio of the proximodistal height of the trochlear articular surface divided by the height of the trochlear notch (UMTD/UTD) in lateral view (see Figure 7.7)

URAP: Anteroposterior width of radial notch

URPD: Proximodistal height of radial notch

UCAP: Anteroposterior length of coronoid process

UAAP: Anteroposterior length of anconeal process

UMML: Midshaft mediolateral width

UMAP: Midshaft anteroposterior width

UDML: Head mediolateral articular width

UHT: Head anteroposterior articular length

UOO: Orientation of the olecranon mechanical lever (in degrees)

UNO: Orientation of the trochlear notch (in degrees)

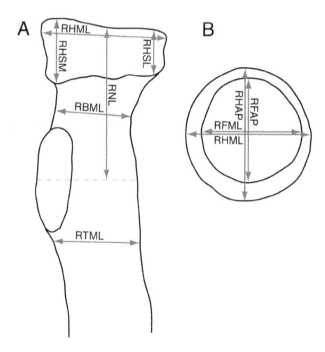

Figure 7.8   Radial measurements presented in this chapter. Anterior (A) and superior (B) views of a left radius (lateral is to the right in both views).

A discriminant function was performed on the proximal ulna, while on the distal ulna, a Principal Component Analysis (PCA) was performed since the covariance matrices between taxa were unequal. For both types of analyses, variables were controlled for size by dividing each variable by the geometric mean of all the variables from the analysis. The analysis of the proximal ulna included 12 variables (Table 7.3), and that of the distal ulna included 5 variables (Table 7.4). The jackknife method is used to assess the classification performance of the discriminant function. The fossils are not included in the discriminant function, but their likeliness of belonging to each extant taxon is evaluated using Mahalanobis distances (following McHenry et al., 2007). For the PCA, the taxa differences for each component are evaluated by an ANOVA, and the similarity of the fossils is evaluated graphically.

Results

Ulnae

The sizes of the 11 Sterkfontein ulnae are easily accommodated within the *A. afarensis* size distribution (n = 6) (Table 7.1). The largest *A. afarensis* specimen (A.L. 438-1) is larger than the largest Sterkfontein specimen (StW 108), while the smallest Hadar specimen (A.L. 288-1) has a smaller head than the smallest Sterkfontein specimen (StW 326), although it is broadly comparable in diaphyseal size (see Table 7.1). The lower size variation is consistent with lower variation in body mass estimates for *A. africanus* than for *A. afarensis* (Grabowski et al., 2015). All Sterkfontein fossils are small compared to large-bodied extant hominoids.

Proximal ulna

The Sterkfontein specimens have olecranon processes that are proximally oriented (Table 7.1; Figure 7.9). As shown previously, humans are also characterized by a proximally oriented olecranon (Drapeau, 2004; Table 7.1), which provides the most efficient leverage for the *m. triceps brachii* when the elbow is flexed (Drapeau, 2004, 2012; Drapeau et al., 2005). This is interpreted as reflecting the use of the triceps and of the forearm in general in a flexed position, as it is the case during manipulation. Apes, in contrast, have an olecranon that is more posteriorly oriented, not projecting proximally to the elbow joint (Figure 7.9; Table 7.1). The Sterkfontein specimens have human-like values, suggesting habitual use of the upper limb when the forearm is flexed (Drapeau, 2004, 2012), possibly for manipulatory activities, consistent with hand proportions (Green and Gordon, 2008) and hand bone trabecular bone structure (Skinner et al., 2015).

The StW 431b as well as the Hadar specimen A.L. 438-1 present resorption lacunae at the *m. triceps brachii* insertion site. In modern humans, resorptions and enthesophytes (the latter also observed on A.L. 438-1) at the olecranon do not appear to be related to greater activity or contraction of the muscle but instead correlate with age (Villotte, 2008). This suggests possibly that both StW 431 and A.L. 438-1 represent older individuals (in the latter

Table 7.2 Radius measurements used in the univariate and bivariate analyses (mean, standard deviation, n = sample size). All measurements in mm.

Taxon/specimen	Sex	Stat.	RHML	RHAP	RFML	RFAP	RHSA	RHSL	RHSP	RHSM	RNL	RBML	RBAP	RTML	RTAP
H. sapiens	M	Mean	23.0	23.2	18.9	19.5	8.6	7.7	9.2	9.7	34.6	14.9	15.4	14.3	13.9
		St. dev.	1.6	1.5	0.3	0.4	0.4	0.7	0.03	0.9	4.0	1.7	1.1	1.4	1.3
		n	43	59	3	3	3	3	2	3	18	18	18	18	18
	F	Mean	19.6	20.1							31.3	12.8	13.1	12.0	12.4
		St. dev.	1.5	1.5							3.3	1.5	1.7	1.3	1.2
		n	43	43							15	15	15	15	15
	I[a]	Mean	20.7	21.2	17.6	18.7	9.0	7.8	8.8	8.9	31.8	13.8	14.2	13.4	13.7
		St. dev.	2.3	2.0	4.5	1.8	1.2	1.0	1.8	1.4	2.8	1.8	1.5	1.4	1.6
		n	21	19	13	15	18	16	12	18	18	18	18	18	18
P. paniscus	M	Mean	21.6	22.6							36.9	12.3	12.5	13.2	12.9
		St. dev.	1.0	1.2							2.8	1.1	1.5	1.3	0.9
		n	11	11							12	11	11	11	11
	F	Mean	20.4	21.0							34.3	12.0	12.2	12.6	12.7
		St. dev.	2.3	2.6							4.3	1.0	1.0	0.9	1.1
		n	10	10							10	9	9	9	9
P. troglodytes	M	Mean	25.1	25.5	19.5	19.9	8.7	6.3	8.4	9.4	43.8	13.1	13.9	15.1	14.6
		St. dev.	1.3	1.3	1.5	1.5	1.4	1.4	0.9	1.4	3.8	1.1	1.3	1.4	1.0
		n	22	22	5	5	5	5	5	5	23	22	22	22	22
	F	Mean	23.1	23.8	17.5	18.1	7.6	6.3	7.9	8.9	40.1	12.1	13.2	14.1	13.9
		St. dev.	1.7	1.7	1.5	0.9	0.5	0.6	0.8	0.7	3.8	1.2	1.6	1.2	1.3
		n	32	32	11	11	11	11	11	11	33	33	33	31	31
G. g. gorilla (western)	M	Mean	33.9	34.3	25.3	25.3	13.7	10.6	14.6	16.4	63.6	18.4	18.6	19.7	20.7
		St. dev.	2.8	2.5	2.3	3.1	1.8	2.1	1.8	2.6	4.8	1.8	2.0	1.7	1.8
		n	29	30	13	14	14	13	14	14	10	29	29	31	31
	F	Mean	26.4	26.9	19.6	19.7	10.4	8.3	10.2	13.1	49.4	14.4	15.0	16.0	16.3
		St. dev.	1.7	1.6	1.0	1.2	1.3	1.3	1.5	1.0	4.2	1.4	1.3	1.3	1.6
		n	27	27	7	7	7	7	7	7	8	27	27	26	26
G. g. berengei G. g. graueri (eastern)	M	Mean	34.8	35.4	25.3	25.9	14.9	11.4	14.6	16.1	62.0	18.4	17.5	18.3	20.8
		St. dev.	2.2	2.0	1.8	1.9	1.5	1.1	1.6	1.1	5.2	2.2	2.0	1.6	1.7
		n	10	10	5	5	5	5	5	5	31	10	10	10	10
	F		27.1	28.2	18.7	20.1	10.7	7.2	11.0	11.4	48.8	13.2	13.7	14.4	15.7
			1.6	1.7	1.0	0.3	0.8	0.2	1.6	1.6	3.7	1.5	1.7	0.6	0.8
			8	7	3	3	3	3	3	3	27	9	9	9	9
P. pygmaeus	M		25.0	25.7	20.4	20.9	9.9	7.9	9.0	10.6	45.1	14.1	15.2	15.1	15.1
			2.3	2.2	2.0	2.0	1.9	1.4	1.5	2.1	3.6	1.9	1.6	1.5	1.9
			15	15	12	12	12	11	11	12	15	15	15	15	15
	F		20.7	21.3	17.0	17.2	8.6	7.1	7.6	9.0	37.9	10.8	12.2	11.9	12.2
			0.9	1.0	1.0	0.8	1.3	1.4	1.4	1.2	3.7	1.1	1.4	1.1	1.4
			20	20	17	17	17	17	17	17	20	20	20	20	20
Hylobates			12.9	13.3	9.2	9.5	5.9	5.8	5.9	6.0	31.5	6.6	7.1	7.4	7.6
			1.3	1.4	1.1	0.9	1.3	1.4	0.8	1.2	9.2	0.6	0.7	0.7	0.7
			30	30	9	9	9	9	9	9	40	15	15	9	9
A. africanus															
Sts 68			21.1		15.9	16.1					34.2		14.8		
Sts 2198b					13.8	12.3*									
StW 105 (juvenile)			11.4[b]	13.5[b]							14.1	8.6	9.5	8.1	8.3
StW 139				23.2	15.4	16.3	8.1			8.3		13.3			
StW 431a			22.2	22.6	14.4	14.4	8.2	8.4	7.9		38.7	12.7	14.1	12.7	13.7
StW 516			22.6	22.8		17.7							13.3		
StW 528														10.2	11.4
A. afarensis															
A.L. 288-1			14.34	15.18	10.2	11.75	5.15	5.02	5.72	5.76	30.2	9.89	10.3	10.3	11.0
A.L. 333x-14			23.4*		16.6										

continued

Table 7.2 Continued

Taxon/specimen	Sex	Stat.	RHML	RHAP	RFML	RFAP	RHSA	RHSL	RHSP	RHSM	RNL	RBML	RBAP	RTML	RTAP
A.L. 333x-16			22.9	23.3	17.5	16.6									
A.L. 333w-33														12.5	15.3
A. anamensis															
KNM-ER 20419[c]			22.2		16.5	16.9					39.7	12.2	13.3		
Homo cf. *erectus*															
SK18[d]				19.6	15.1	15.5	8.6		8.7		29.7	11.4	12.2	11.1	
SKX 2045[d]			21.0	21.2	16.6	17.6					34.5	13.1	13.6	14.9	13.8
P. boisei															
KNM-ER 1500e[e]			20.0	19.5	14.7	15.7					34.3	9.3	10.1	9.1*	12.5
OH 80[f]			*25.3*	*26.3*							*43.8***	*14.8*	*19.1*	*14.0***	
P. robustus															
SK 24601[g]			17.3		12.3	12.8					29.1	9.9	10.9	12.4	9.7
SKX 3699[h]			18.9	19.6	13.3	15.0						11.6			

[a] Sex undetermined.
[b] Measured at the metaphyseal plate.
[c] Attributed to *A. anamensis* by Leakey et al. (1995).
[d] Attributed to *Homo* cf. *erectus* by Susman et al. (2001).
[e] Attributed to *P. boisei* by Grausz et al. (1988).
[f] Attribution and data from Dominguez-Rodrigo et al. (2013).
[g] Attributed to *P. robustus* by Susman et al. (2001).
[h] Attributed to *P. robustus* by Susman (1989).
RHML: Mediolateral diameter of head
RHAP: Anteroposterior diameter of head
RFML: Mediolateral diameter of fovea
RFAP: Anteroposterior diameter of fovea
RHSA: Proximodistal height of head anteriorly
RHSL: Proximodistal height of head laterally
RHSP: Proximodistal height of head posteriorly
RHSM: Proximodistal height of head medially
RNL: Proximodistal neck length
RBML: Mediolateral diameter of the neck
RBAP: Anteroposterior diameter of the neck
RTML: Mediolateral diameter of diaphysis immediately distal to the tuberosity
RTAP: Anteroposterior diameter of diaphysis immediately distal to the tuberosity
* Estimated.
** Estimated from publication.

specimen, the heavily worn postcanine teeth also support that interpretation; Drapeau et al., 2005).

The Sterkfontein specimens' trochlear notches are more anteriorly oriented than any extant hominoid taxa including humans, but are comparable in orientation to those in the Hadar sample (Figure 7.10; Table 7.1). Anteriorly oriented trochlear notches are common in Plio-Pleistocene hominins, from *Ardipithecus* to Neandertals (Churchill et al., 1996; Hambücken, 1998; Drapeau et al., 2005; Lovejoy et al., 2009), even in specimens with pronounced anteroposterior diaphyseal curvature such as OH 36 and L40-19 (Drapeau et al., 2005). This orientation may be functionally related to the proximally oriented olecranon also found in the fossils (Drapeau, 2004; Drapeau et al., 2005) and to general upper limb robusticity combined with non-locomotor use of the upper limb. Lovejoy et al. (2009) have proposed that the anterior orientation represents the primitive morphology for hominins from which *Pan* and recent humans

would have been derived independently. The anteriorly oriented notch in *Ardipithecus* is interpreted as reflecting above-branch quadrupedalism with a flexed forearm (Lovejoy et al., 2009), and it could reflect a similar arboreal locomotor behavior in australopithecines. Alternatively, if we accept Lovejoy et al.'s evolutionary scenario, another interpretation would be that the morphology is maintained in australopithecines as an exaptation to non-locomotor use of the upper limb with a forearm in a flexed position, likely for manipulation.

Keeling of the trochlear notch is modest in the Sterkfontein sample, comparable to the Hadar sample and more similar to humans than apes (Figure 7.11; Table 7.1; Drapeau, 2008). The human trochlear notch is less keeled than that of all apes except western male *Gorilla* (Figure 7.11; Table 7.1). Since the greatest keeling is found in the most arboreal taxa, keeling is likely an adaptation to resist lateral displacement at the elbow during powerful finger

Table 7.3 Variables used in the discriminant function analysis of the proximal ulna. All measurements in mm.

Taxon/Specimen	Stat.	UOLM	UOAP	UOML	UTD	USND	UTW	USML	UBAP	UMH	ULH	URAP	URPD
H. sapiens	Mean	29.8	15.2	15.2	22.0	15.2	15.2	15.2	23.4	15.2	15.2	15.2	15.2
n=29	St. dev.	2.8	2.4	2.4	2.4	2.4	2.4	2.4	3.0	2.4	2.4	2.4	2.4
P. troglodytes	Mean	32.9	16.8	22.5	20.5	21.4	15.7	18.2	24.4	16.7	14.4	15.9	11.4
n = 22	St. dev.	3.1	2.5	3.2	3.0	2.4	3.1	2.2	3.2	2.2	2.8	2.2	1.6
G. gorilla	Mean	40.2	20.4	28.3	27.3	26.7	32.2	28.5	28.4	20.0	17.8	24.8	18.1
n =37	St. dev.	6.9	4.9	5.2	5.5	4.5	5.9	4.1	4.5	4.1	4.5	3.5	2.4
P. pygmaeus	Mean	28.4	17.1	23.0	20.3	18.9	17.0	19.5	20.5	14.0	12.0	20.1	14.2
n =18	St. dev.	4.5	3.4	3.2	3.8	2.9	3.3	3.2	3.1	2.5	3.0	2.7	1.8
A. africanus													
StW 113		28.7	22.0	24.1	19.5	17.6	14.6	16.9	23.3	13.0	11.2	10.0	8.0
StW 380		23.8	19.5	22.1	18.2	16.2	15.4	16.7	19.4	12.2	8.6	9.2	10.2
StW 431b		29.7	23.4	25.2	21.0	18.3	13.7	16.3	18.8	13.3	12.5	14.4	11.7
A. afarensis													
A.L. 438-1		30.7	19.2	27.8	20.4	16.2	11.4	19.6	20.0	12.6	10.2	16.4	10.2
A.L. 288-1		22.1	12.6	13.4	15.9	12.4	12.3	15.2	14.7	10.0	9.3	9.8	6.8

UOLM: Mechanical olecranon length
UOAP: Anteroposterior length of olecranon
UOML: Mediolateral width of olecranon
UTD: Diameter of a circle fitted to the profile of the trochlear notch in lateral view
USND: Maximum depth of trochlear notch measured from UNH
UTW: Mediolateral width of the central trochlear notch
USML: Maximum mediolateral width of diaphysis at the level of the brachialis attachment
UBAP: Anteroposterior depth of diaphysis immediately distal to the radial notch
UMH: Anteroposterior depth from posterior margin of trochlear notch articular surface to the posterior margin of the bone, measured medially
ULH: Anteroposterior depth from posterior margin of trochlear notch articular surface to the posterior margin of the bone, measured laterally
URAP: Anteroposterior width of radial notch
URPD: Proximodistal height of radial notch

Table 7.4 Variables used in the PCA of the distal ulna. All measurements in mm.

Taxa/Specimen	Stat	UDAP	UDML	USAP	UDNS	UHT
Homo	Mean	20.5	17.2	5.0	1.4	10.7
n =30	St. dev.	2.2	2.5	0.7	0.7	1.3
Pan	Mean	20.5	18.1	6.5	3.3	7.8
n=56	St. dev.	1.7	1.6	1.2	1.0	1.1
Gorilla	Mean	28.2	26.1	9.7	5.8	12.3
n =59	St. dev.	4.1	4.2	2.1	1.8	2.2
Pongo	Mean	22.8	21.5	6.5	4.1	9.6
n=23	St. dev.	2.7	3.2	1.0	1.1	1.6
A. africanus						
StW 326		15.1	14.0	5.4	2.6	6.9
StW 398b		16.7	14.3	4.3	2.8	7.2
A. afarensis						
A.L. 438-1		20.1	17.5	5.5	3.5	8.9
A.L. 288-1		13.3	12.2	3.5	1.7	6.0
A.L. 333-12		18.0	18.3	5.4	3.2	7.7

UDAP: Maximum anteroposterior diameter of the distal extremity (head)
UDML: Maximum mediolateral diameter of the distal extremity (head)
USAP: Anteroposterior diameter of styloid
UDNS: Depth of notch between styloid and head measured from the distal-most articular surface of the head
UHT: Anteroposterior diameter of the articular surface of the head

and wrist flexors contraction causing shear loads going through the elbow joint (Preuschoft, 1973; Andersson, 2004; Drapeau, 2008). The reduced keeling in the fossils, relative to arboreal apes, likely reflects reduced use of the forearm in suspensory or above head positions and/or possibly less powerful forearm muscles than in apes (Drapeau, 2008).

The waisting of the trochlear notch in the Sterkfontein sample is a characteristic shared with *A. afarensis*, humans, *Pan* and *Pongo* (Figure 7.12). In contrast, western and eastern *Gorilla* have a notch that is less constricted, but the latter morphology may reflect the larger size of gorillas combined with locomotor-related loading of the elbow (Figure 7.12; Drapeau, 2008). Not surprisingly, all fossils lack the expanded distomedial quadrant of the trochlear notch associated with knuckle-walking (Figures 7.1, 7.3, and 7.4; Drapeau, 2008).

The majority of the Sterkfontein ulnae have trochlear notches that tilt laterally in anterior view (Figures 7.1, 7.3, and 7.4), while in *A. afarensis* and *P. robustus* (SKX 8761) this tilt is less marked. The lateral tilt of the notch should accentuate the carrying angle of the forearm. It is slightly greater in humans than in other hominoids, but large overlap between taxa (Knussmann, 1967) precludes the inference of any clear functional significance of this trait.

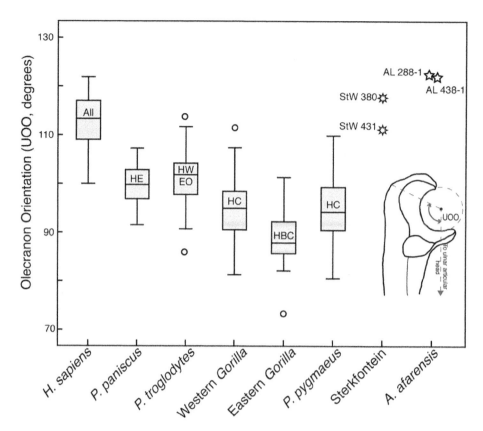

Figure 7.9 Box plots of olecranon orientation relative to the dorsum of the proximal diaphysis. The box represents the 25–75% percentiles, the whiskers the range, circles represent outliers, and small asterisks represent extreme outliers. The horizontal bar within the box represents the median, and the letters within the box signal a difference at the $p \leq 0.05$ confidence level from *H. sapiens* (H); *P. paniscus* (B); *P. troglodytes* (C); western *Gorilla* (W); eastern *Gorilla* (E); *P. pygmaeus* (O); or all other groups (All). Sterkfontein specimens are represented by eight-point stars and *A. afarensis* specimens by five-point stars.

The radial notch is oriented mostly laterally in the Sterkfontein sample and its size is variable, with the notch appearing generally large in StW 431b, intermediate in StW 390 and StW 398, and small in StW 113 and 380 (Table 7.1). Its size appears to be independent of the general size of the specimen (i.e., larger specimens do not necessarily have larger notches). The size of the notch relative to size does not appear to be related to locomotion and does not have a clear functional signal since it is large in gorillas and orangutans, but small in chimpanzees, bonobos, humans, the Hadar and Swartkrans fossil specimens, and even smaller in the Sterkfontein specimens (data not presented).

In posterior view, the Sterkfontein specimens have a ridge delineating the *m. anconeus* attachment laterally that is always more marked than that, when present, for the *m. flexor carpi ulnaris* medially. The degree to which the attachment of these muscles is delineated with a crest could be related to size in the Hadar sample, with the large form (A.L. 438-1) having well-delineated attachments and the small form (A.L. 288-1) having less well-marked attachments. In contrast, in the Sterkfontein sample, the same

variation is observed but does not seem to relate to size. It is marked in StW 398 (small) and StW 113 (large) while in StW 380 (medium) and StW 431 (large) it is much less so. The development of these crests is difficult to interpret functionally and is as likely to reflect activity as chronological age of the specimens or a combination of both (Villotte et al., 2010). Medial to the coronoid process, there is a depression variable in depth for *m. flexor digitorum profundus* (not deep in StW 113 to fairly deep in StW 431b; Figure 7.13). The very faint or absent ridge for *m. flexor carpi ulnaris* and the moderate depression for *m. flexor digitorum profundus* of the Sterkfontein specimens is reminiscent of humans and of A.L. 288-1. However, in African apes, the highly excavated appearance of the *m. flexor digitorum profundus* attachment and the height of the crest for *m. flexor carpi ulnaris* may be exaggerated due to the extreme medial positioning of the olecranon in those taxa (Aiello et al., 1999). It may not be a coincidence that the Sterkfontein ulna with the most excavated origin for *m. flexor digitorum profundus* (StW 431b) is also the one with an olecranon that protrudes medially past the articular trochlear notch (Figure 7.13). A similar morphology, albeit

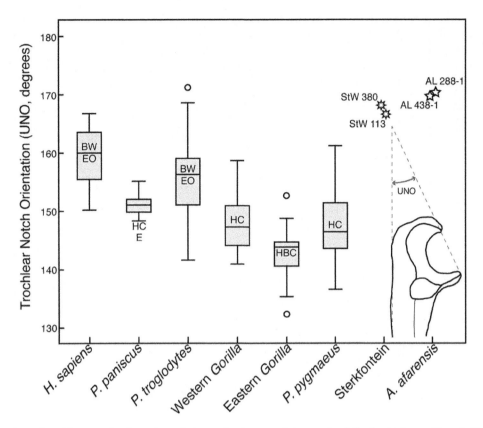

Figure 7.10 Box plots of trochlear notch orientation relative to the dorsum of the proximal diaphysis. Legend is as in Figure 7.9.

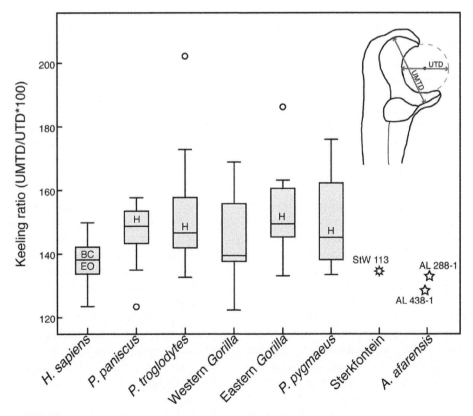

Figure 7.11 Box plots of the trochlear notch keeling (see methods). Legend is as in Figure 7.9.

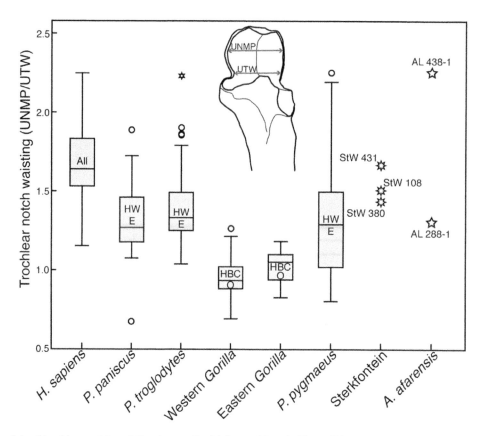

Figure 7.12 Box plots of trochlear notch waisting (see methods). Legend is as in Figure 7.9.

to a lesser degree, is observed on A.L. 438-1 and on the fragmentary TM 1517. The medial projection of the olecranon is interpreted by Lovejoy et al. (2009) to result from enthesis expansion of the deep digital flexor in association with knuckle-walking. If this is the case, the African ape/human difference reflects differences in development of the flexor muscles of the wrist and fingers. Given that none of the fossils are assumed to be knuckle-walkers, the expansive origin *m. flexor digitorum profundus* in some Sterkfontein specimens suggest that they are capable of powerful finger flexion.

The anteromedially facing, proximodistally elongated, excavated ulnar tuberosity seen in the Sterkfontein sample is very similar to that of *A. afarensis* and *P. robustus* specimens. However, *A. afarensis* is more variable in the orientation of the tuberosity (medial in A.L. 438-1 and anterior in A.L. 288-1) than the Sterkfontein sample. In contrast, the posterior margin of the tuberosity is variably crested in the Sterkfontein sample, while in *A. afarensis* it is always well marked. The size, shape, and relief of the tuberosity in the Sterkfontein and Hadar fossil specimens are more like that of *Pan* and unlike the anteriorly oriented, proximodistally short, proximally positioned, often projecting tuberosity of modern humans. However, given that much more recent fossil specimens, such as KNM-BK

66 (dated to about at 600–700 Ka and attributed to *H. erectus*; Aiello et al., 1999, Senut, 1981; Solan and Day, 1992), also have a morphology that is more *Pan*-like, it is difficult to provide a clear functional significance to this variation.

The *m. supinator* crest in the Sterkfontein specimens is well marked proximally at a level just distal to the radial notch (in StW 108, the crest is more proximal, most apparent at the level of the radial notch). It is generally thick and tapers distally relatively quickly, while it is generally a long, thin crest in the Hadar specimens (except for A.L. 438-1, which has a slightly thicker crest). The Sterkfontein specimens are actually comparable to A.L. 438-1 in both respects but differ from the other Hadar specimens (A.L. 288-1 right and left, A.L. 333w-36, and A.L. 333x-5). This crest is more marked in *Gorilla* and *Homo* and less so in *Pan* and *Pongo* (Knussmann, 1967), but large within-taxon variation and among-taxon overlaps obscure clear functional interpretations.

A discriminant function analysis on 12 variables of the proximal ulna of size-adjusted data separates extent taxa fairly well (Figure 7.14, Tables 7.3, 7.5, 7.6, and 7.7; 92.4% correct cross-validated classification). The first function appears to reflect size and, since the data are size-controlled, it suggests that some variables are allometric. Larger taxa have proportionally wider trochlear

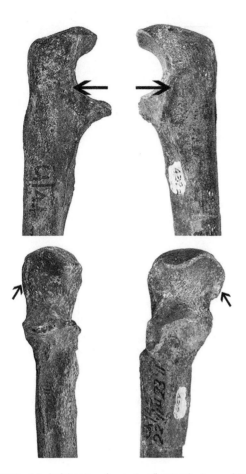

Figure 7.13 Medial (top) and anterior (bottom) views of StW 113 (left) and StW 431b (right). The arrows on the top photos point to the deep depression for the attachment of *m. flexor digitorum profundus* in StW 431b and the shallow depression in StW 113. The arrows in the bottom photos show the olecranon projecting medially in StW 431b and much less so in StW 113.

diaphyses that are anteroposteriorly narrower immediately distal to the radial notch, proportionally shorter olecranons, and proximodistally longer radial notches.

The second function separates African apes from the other taxa, with the former being characterized by deeper trochlear notches in lateral view and mediolaterally narrower diaphyses at the level of the brachialis insertion. The third function separates *Pongo* from all other taxa, the former being characterized by a mediolaterally narrow olecranon and an anteroposteriorly short radial notch. StW 113 is most *Pan*-like (68%), while StW 380 and StW 431b are more *Pongo*-like (64% and 100%, respectively). In contrast, the two *A. afarensis* specimens are clearly more like humans (98% for A.L. 438-1 and 100% for A.L. 288-1). However, when the fossils' distances to the extant taxa centroids are plotted (Figure 7.14B), we can see that StW 113 and StW 380 resemble humans more than A.L. 438-1 does. This is explained by the shape of the Sterkfontein specimens that are intermediate between *Pongo* or *Pan* and

humans, while the shape of A.L. 438-1 is not intermediate between that of humans and apes but instead is on the periphery of only the human distribution (Figure 7.14A).

Ulnar diaphysis

There are no complete specimens; therefore, it is not possible to measure anteroposterior diaphyseal curvature in the Sterkfontein sample. However, the better-preserved specimens suggest that they are very similar to the Hadar sample with some large specimens, such as StW 431b, StW 113, and A.L. 438-1, being moderately curved while others, generally smaller, such as StW 326, StW 349, and A.L. 288-1, are straighter. In this respect, all *Australopithecus* ulnae are somewhat intermediate in diaphyseal curvature between those of apes and humans but tend to more closely resemble humans (Drapeau et al., 2005). Forearm bone curvature is positively allometric in anthropoids (Swartz, 1990), so it is not too surprising to find that the larger fossil specimens tend to be more curved than the smaller specimens.

Development and location of the interosseous crest relative to other features of the ulna are similar in the Sterkfontein and Hadar samples. In two specimens (StW 380 and 568), the interosseous crest starts proximally with a tubercle, something that is only hinted at in A.L. 438-1 and 288-1. In that respect, StW 431 is similar to the Hadar fossils, while StW 113 is intermediate between the Hadar specimens and StW 380/StW 568. Despite this variation, the morphology of the crest is strikingly different from that of modern humans and is more reminiscent of African apes. However, there is no obvious functional signal to variation in morphology in the crest given its variability among extant hominoids (relatively crested in *Pongo*, relatively inconspicuous in *Pan*, and slightly less so in *Gorilla*; Knussmann, 1967). Around midshaft, the Sterkfontein specimens all have a flattening of the posterolateral surface and a triangular cross-sectional shape with posterior, anterolateral (interosseous), and medial margins, while Hadar specimens are more oval. The triangular shape of the Sterkfontein sample is reminiscent of the human shape but with generally a more modest development of the posterior margin and particularly of the interosseous crest in the fossils, while the more oval shape of the Hadar sample appears more *Pan*-like. The etiology of a sharp interosseous crest and posterior margin on the forearm bones remains unknown. If diaphyseal shape variation were simply related to muscular robusticity, apes would have sharper margins than humans, but it is generally the opposite. Alternatively, diaphyseal form may relate to the pattern of use of the different muscles, which can be speculated to be very different between apes and humans. Apes use these muscles predominantly for locomotion, recruiting them at a high frequency and generating large forces, while humans use forearm muscles almost exclusively for manipulation, likely recruiting

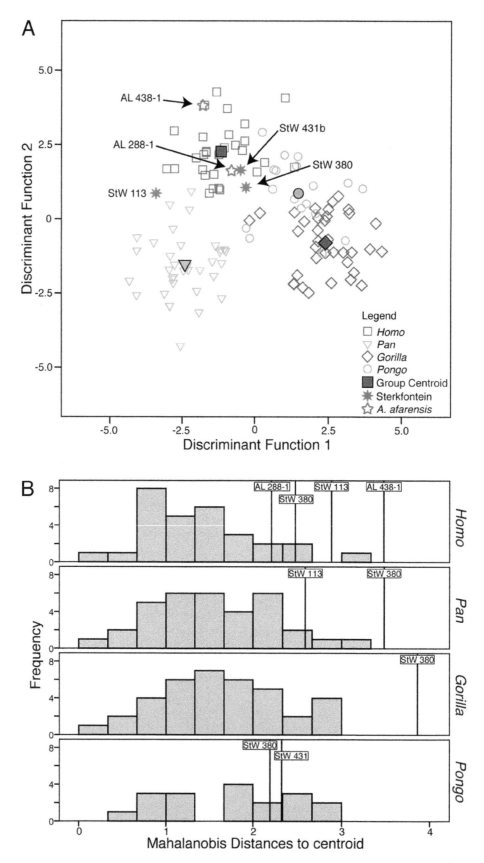

Figure 7.14 A: Scatter plot of the first two functions of the discriminant analysis of the proximal ulna. B: Mahalanobis distances of each specimen to his taxon's centroid and distance of the fossils to each taxon (omitted fossils have distances greater than the presented scale).

Table 7.5 Eigenvalues and % variance of the discriminant functions of the proximal ulna.

Function	Eigenvalue	% variance	Cumulative % variance
1	4.281	53.8	53.8
2	2.322	29.2	83.0
3	1.356	17.0	100

Table 7.6 Structure matrix of the discriminant functions of the proximal ulna (size-controlled data).

Variables	Function		
	1	2	3
UBAP	**−0.572**	−0.001	−0.158
UTW	**0.560**	−0.325	−0.377
UOLM	**−0.448**	−0.238	−0.143
URPD	**0.383**	−0.028	0.346
UOAP	**−0.244**	0.073	0.222
ULH	**−0.116**	−0.032	−0.112
USND	−0.205	**−0.588**	0.132
USML	0.188	**0.585**	−0.470
UMH	−0.244	**−0.296**	0.015
UTD	−0.155	**−0.243**	−0.083
URAP	0.329	0.291	**0.363**
UOML	−0.038	0.150	**0.302**

Bold font indicates a variable's largest absolute correlation with any discriminant function.

Table 7.7 Cross-validated classification of the discriminant function analysis of the proximal ulna. Values for the fossils are the probability of belonging to the comparative extant taxa.

Taxon/specimen	Predicted group			
	Homo	*Pan*	*Gorilla*	*Pongo*
Homo	28	0	1	0
	96.6%	0%	3.4%	0%
Pan	1	32	0	1
	2.9%	94.1%	0%	2.9%
Gorilla	2	0	32	3
	5.4%	0%	86.5%	8.1%
Pongo	0	0	1	17
	0%	0%	5.6%	94.4%
A. africanus				
StW 113	0.32	**0.68**	0	0
StW 380	0.34	0.02	0	**0.64**
StW 431b	0.00	0	0	**1.0**
A. afarensis				
A.L. 438-1	**0.98**	0	0	0.02
A.L. 288-1	**1.0**	0	0	0

Bold font indicates a fossil's largest probability of belonging to each extant taxon.

them at a lower frequency and generating smaller forces overall than if they were also used for locomotion. If this is so, the variation between the Sterkfontein and Hadar samples may reflect modest differences in the use of the hand in the two samples, with the Sterkfontein sample emphasizing more human-like finger use and possibly less use in locomotion, while the *A. afarensis* sample emphasizes more *Pan*-like patterns of use of the forearm muscles, possibly being used in higher frequency for locomotor activities. This hypothesis is extremely speculative but would be interesting to investigate in the future.

Distal ulna

The attachment for *m. pronator quadratus* is well marked in both Sterkfontein specimens that preserve the distal part of the ulna. In contrast, the *A. afarensis* specimens have a crest but it is much less marked (A.L. 288-1o, A.L. 137-48b, A.L. 333-12, and A.L. 438-1). A rugose crest is often interpreted as evidence of use of the muscle (e.g., Aiello and Dean, 1990), although experimental evidence does not support this assumption, at least regarding variation within species (Zumwalt, 2006; Alves Cardoso and Henderson, 2010; Rabey et al., 2015). This crest is well developed in Neandertals (Aiello and Dean, 1990) and, relative to size, it is larger in modern humans than in apes (Aiello et al., 1999), suggesting that crest prominence is not a function of interspecific differences in muscle sizes or in degrees of arboreal locomotion.

Very few studies have focused on the distal ulna of fossil hominins despite the fact that a number of specimens have been discovered. The relatively modest variation of the distal ulna among large-bodied hominoids has been investigated directly (Tallman, 2015), but little is understood of the functional significance of this variation.

The groove for *m. extensor carpi ulnaris* on the postero-lateral aspect of the styloid is variable in the Sterkfontein sample, but erosion of the head of StW 326 (which may have reduced the appearance of the groove) may exaggerate the variation in that trait. In contrast, all the Hadar ulnae have a well-marked groove even on the diminutive A.L. 288-1. Extant apes are characterized by a well-defined groove, while in humans it tends to be inconspicuous. The *m. extensor carpi ulnaris* (with the *mm. extensor carpi radialis longus* and *brevis*) functions to keep the wrist in extension during finger flexion, and hence allows the finger flexor muscles to maintain powerful contractions. The presence of a marked groove may indicate a large tendon and more powerful muscle in taxa that require powerful grips during arboreal locomotion. If so, the conspicuous presence of the groove in the fossils may reflect powerful grips in these taxa, possibly related to overall greater muscular robusticity.

The Sterkfontein specimens have a distal ulna with a fairly well-defined fovea between the styloid and articular facet combined with a relatively gracile styloid (Table 7.4; Figures 7.15 and 7.17). Large-bodied extant hominoids are derived relative to other primates in having no bony ulnar-carpal articulation but instead typically have a triangular articular disk interposed between the ulna and triquetral. The fovea between the styloid and the articular head and the anterior surface of the styloid is the attachment

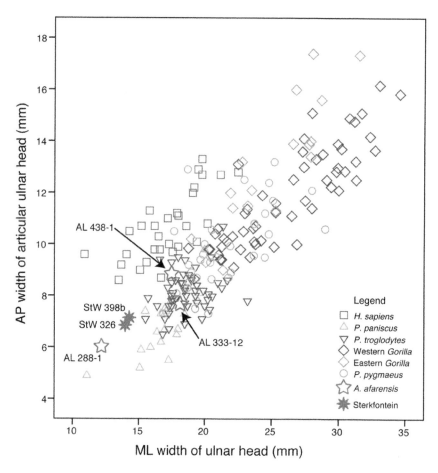

Figure 7.15 Scatter plot of the mediolateral width to the anteroposterior width of the articular ulnar head.

site for the disc and the bordering distal radioulnar ligaments that confer stability to the distal radioulnar articulation (Standring, 2005). This fovea is deeper in apes than in humans (Figure 7.16; Hlusko et al., 2015). Apes are also characterized by a well-defined articular surfaces with the distal radius and triangular disc that does not extend as distally as in humans (Figure 7.18), emphasizing the gap between the styloid and articular surfaces. In addition, apes have a styloid process that is buttressed posteromedially (Figure 7.17) that serves to provide a strong anchor for the palmar and dorsal radioulnar ligaments, the ulnar collateral ligament, and the subsheet of the *m. extensor carpi ulnaris* tendon, all of which stabilizes the distal radioulnar articulation during the wide range of pronation and supination (Standring, 2005). The ligaments and their anchor (i.e., the styloid) might need to be particularly strong in contexts where pronation and supination occur when the hand is loaded in compression or tension, as occurs during locomotion in apes. In general, *A. africanus* and *A. afarensis* present relatively deep foveas (Figure 7.16) while having relatively gracile styloids (except for gracile specimen StW 326, which has a relatively robust styloid; Figure 7.17). The intermediate morphology of the hominin fossils for

these traits suggests that stability was important in these taxa, although possibly proportionally less than in extant apes. Given that little research has been done on the distal ulna, this interpretation remains speculative. As stated above, humans are characterized by an anteroposteriorly expanded, distally facing articular head that extends toward the styloid (Figure 7.15) giving the articular surface of the head a half-moon shape when viewed distally (Hlusko et al., 2015), while that of apes is more mediolaterally wide with a deep fovea, making the articular surface C-shaped (Figure 7.15). For that feature, *A. africanus* and *A. afarensis* have an ape-like morphology that tends toward the human form (Figure 7.15).

A PCA of the distal ulna on size-controlled variables (Tables 7.4, 7.8, and 7.9; Figure 7.18A) does not separate apes from one another, but the first function separates humans from apes. Factor loading of that function supports the observations made above: humans are characterized by anteroposteriorly expanded articular ulnar heads, shallow foveas, and, to a lesser degree, gracile (anteroposteriorly short) styloids (Table 7.9). All the fossil hominins included in the analysis (StW 326, StW 398b, A.L. 288-1, A.L. 333-12, A.L. 438-1) are ape-like and different from

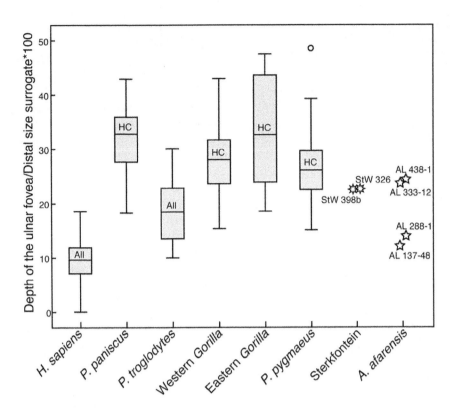

Figure 7.16 Box plots of the depth of the fovea of the ulnar head relative to the size surrogate. Legend is as in Figure 7.9.

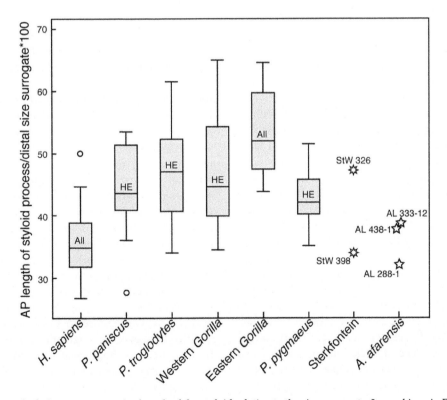

Figure 7.17 Box plots of relative anteroposterior length of the styloid relative to the size surrogate. Legend is as in Figure 7.9.

Table 7.8 Eigenvalues and % variance of the PCA of the distal ulna (size-controlled data).

Function	Eigenvalue	% variance	Cumulative % variance
1	2.09	41.8	41.8
2	1.52	30.3	72.1

Table 7.9 Varimax rotated component matrix of PCA of the distal ulna (size-controlled data).

Variables	Functions	
	1	2
UHT	**−0.915**	0.089
UDNS	**0.746**	0.127
USAP	0.662	**−0.704**
UDAP	−0.107	**0.670**
UDML	0.344	**0.823**

Bold font indicates a variable's largest absolute correlation with any discriminant function.

humans (Figure 7.18B), although their positions tend to be on the edge of the ape distribution that is closest to the human distribution (Figure 7.18A).

Radius

In general, the Sterkfontein radii are of comparable size to those of the Hadar *Australopithecus afarensis* radii and the *A. anamensis* KNM-ER 20419 specimen (Table 7.2).

Proximal radius shape does not provide a clear locomotor signal within extant hominoids (Patel, 2005); however, some differences are apparent among taxa. Apes have a head that is similar in breadth relative to diaphyseal breadth as that of humans, but they tend to have a larger head relative to their neck (Figures 7.19 and 7.20). This indicates that apes may not have a large radial head but instead have a somewhat constricted neck. The Sterkfontein specimens can be described as having a morphology intermediate between humans and apes, with StW 516 being somewhat more ape-like and StW 431 being somewhat more human-like. On the other hand, A.L. 288-1 has very little constriction like humans, so does, surprisingly, the large *P. boisei* specimen OH 80. The human value of the latter is likely due to its antero-posteriorly robust neck (Dominguez-Rodrigo et al., 2013). The two specimens from Swartkrans (both attributed to *H.* cf. *erectus*; Susman et al., 2001) are more human-like, while KNM-ER 1500e (*P. boisei*) is ape-like. Unfortunately, the functional significance of neck constriction in fossil hominins is unclear, but it does not appear to be related to bone size since the largest and smallest fossil specimens

(A.L. 288-1 and OH 80) have similar low, human-like values (Figure 7.19).

African apes and hylobatids are characterized by a fovea that occupies a relatively small portion of the articular surface of the proximal head (Patel, 2005). This results in beveled profiles of the walls of the head, particularly anteriorly and medially (Grine and Susman, 1991; Patel, 2005). Although there is a wide variation among australopithine and paranthropine fossils, they all have a morphology that is more ape-like with relatively small foveas (Figures 7.5 and 7.21). The more human-like morphology of SKX 2045 supports the *H.* cf. *erectus* attribution of this specimen based on its general morphology (Susman et al., 2001). The beveling of the head is usually interpreted as a sign of increase stability at the elbow joint (Patel, 2005) and is associated with the development of the lateral lip of the trochlea of the humerus and articulation at the zona conoidea (Rose, 1988). Its function in extant taxa is assumed to resist proximoulnar displacement of the radius by providing resistance to loads that would be incurred from disto-radial to medioulnar, the direction taken by the muscles originating at the medial epicondyle that cross the elbow diagonally, and as well as from substrate reaction forces during terrestrial locomotion (Schmitt, 2003). The beveling observed in the Sterkfontein specimens (Figures 7.5 and 7.21; also observed on Sts 68, A.L. 333x-14, and the *A. anamensis* specimen KNM-ER 20419 in the mediolateral axis and reported for the very large and robust *P. boisei* OH 80; Dominguez-Rodrigo et al., 2013) hints at similar functions requiring stability, usually assumed to be suspensory behaviors and more powerful wrist and finger flexors muscle than in humans. However, reduced keeling observed on the ulna, also hypothesized to resist obliquely directed loads generated by muscle crossing the elbow diagonally (Preuschoft, 1973; Andersson, 2004; Drapeau, 2008), points toward a decrease in the resistance to the forces directed obliquely, and hence the force of contraction of the elbow and wrist flexors. This suggests that the fossil hominins may not have been using arboreal locomotion to the degree observed in extant apes, or, at the very least, had less powerful muscles. The morphology of the Sterkfontein, *A. afarensis* and *P. boisei* specimens contrasts with specimens attributed to *H.* cf. *erectus*, such as SK18 and particularly SKX 2045, which are clearly human-like in lacking the beveling of the head and reduced neck constriction (Susman et al., 2001).

Radial neck length affects the lever length of the *m. biceps brachii* and can be interpreted functionally only when in reference to the load arm length (radial length). However, in fragmentary fossils, if neck length is compared to diaphyseal size immediately distal to the tuberosity, humans are characterized by a short neck while apes have a relatively longer neck (Figure 7.22).

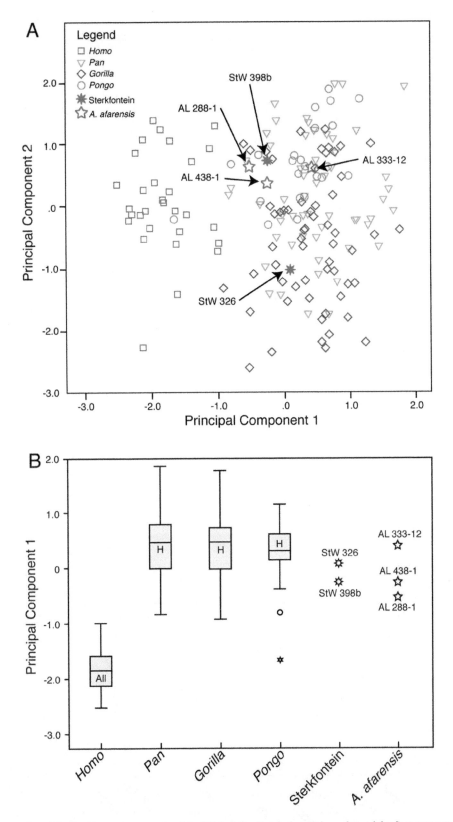

Figure 7.18 A: Scatter plot of the first two components of the PCA of the distal ulna. B: box plots of the first component values for each taxon. Legend is as in Figure 7.9.

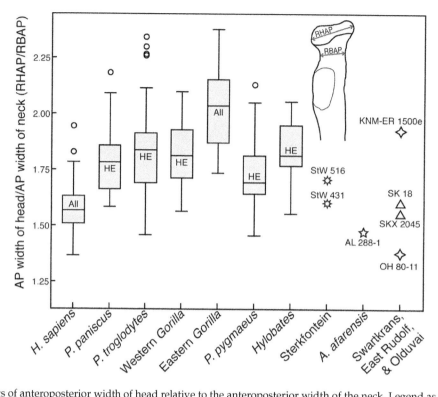

Figure 7.19 Box plots of anteroposterior width of head relative to the anteroposterior width of the neck. Legend as in Figure 7.9 and Swartkrans specimens (both attributed to *H.* cf. *erectus*) are triangles and *P. boisei* specimens are four-point stars.

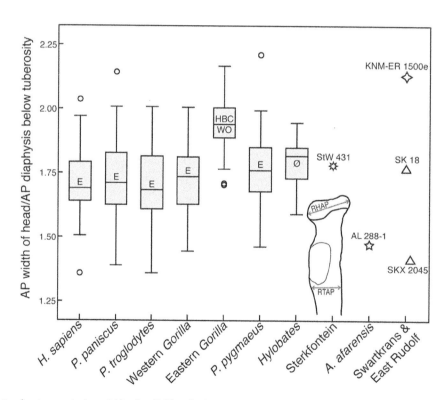

Figure 7.20 Box plots of anteroposterior width of radial head relative to the anteroposterior width of the diaphysis distal of the tuberosity. Legend as in Figure 7.19.

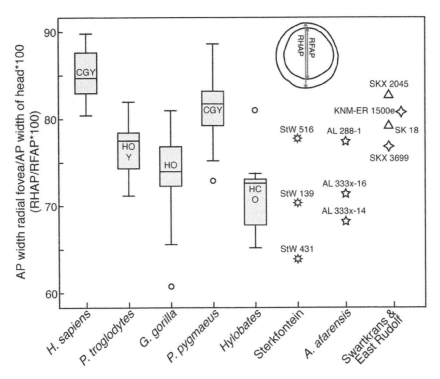

Figure 7.21 Box plots of anteroposterior width of the fovea relative to the anteroposterior width of the head. Legend as in Figure 7.19; *Hylobates* (Y). The Swartkrans specimen SKX 3699 is attributed to *P. robustus* and is represented by a four-point star (as the other *Paranthropus* specimen).

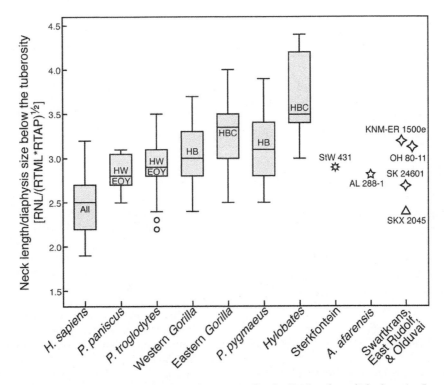

Figure 7.22 Box plots of neck length relative to the diaphyseal size immediately distal to the radial tuberosity. Legend as in Figure 7.19. The Swartkrans specimen SK 24601 is attributed to *P. robustus* and is represented by a four-point star (as the other *Paranthropus* specimen). For the Olduvai specimen (OH 81-11 attributed to *P. boisei*), relative neck length is calculated as RNL/RTML.

The Sterkfontein (StW 431) and *A. afarensis* (A.L. 288-1) specimens tend to have necks that are longer than in humans, falling closest to the *Pan* median (Figure 7.22). The probable *P. boisei* specimens (KNM-ER 1500e and OH80-11) are ape-like, while the Swartkrans *P. robustus* specimen SK 24601 has a relative neck length that is not too different from the Sterkfontein specimens but more human-like. On the other hand, the other Swartkrans specimen SKX 2045, attributed *to H.* cf. *erectus*, is clearly human-like. The interosseous crest is generally well marked in the fossils, although it tends to get rounder distally. This morphology is neither ape-like nor human-like, but somewhat intermediate, mirroring what is observed on the ulna.

In general, the Sterkfontein radii, as well as those attributed to *A. anamensis*, *A. afarensis*, *P. boisei*, and *P. robustus* (SK 24601, SKX 3699) with their very beveled heads and their relatively long and constricted necks, can be described as being more ape-like than human-like. On the other hand, specimens from Swartkrans attributed to *H.* cf. *erectus*, without being completely human-like, tend to resemble human radii more than do the other fossils, particularly in relative neck length and constriction and in their reduced beveling of the head.

Discussion and conclusions

The relatively small size of ulna and radius specimens, comparable to the small end of the size range of *Homo*, *Pan*, and *Pongo* ulnae and radii, agree with Sterkfontein fossils representing relatively small-bodied hominins. The proximal orientation of the olecranon process and anterior orientation of the trochlear notch suggest that the forelimb was used in flexed position and that use of the *m. triceps brachii* had the greatest mechanical advantage in that position. The relatively straight diaphysis, the reduced trochlear notch keeling, and the less marked attachments for *mm. flexor digitorum profundus* and flexor *carpi ulnaris* relative to extant apes, but similar to *A. afarensis* specimens, suggest that the Sterkfontein and Hadar individuals had reduced forearm muscularity relative to apes. Cross-sectional shape of the diaphysis on the radius and ulna and the inconspicuous interosseous crests are comparable to other australopithecines, and suggest a pattern of forearm muscle use that may be neither completely human-like nor ape-like but somewhat intermediate, perhaps reflecting a reduced use in locomotion and an increased use in manipulation compared with extant apes.

In contrast to the diaphysis, the proximal radius is more clearly ape-like in the Sterkfontein sample, with beveled heads and constricted and relatively long necks comparable to the Hadar specimens. This underscores

unique morphology of the elbow (neither completely human-like nor ape-like) in these hominins. The distal ulna, with the well-marked groove for the *m. extensor carpi ulnaris* tendon, a relatively deep fovea, and an articular head shape intermediate between apes and humans, again comparable to *A. afarensis* specimens, suggest that stability was required during loading and use of the hand and wrist, although possibly less so than in extant apes. The overall picture of the Sterkfontein forelimb is that of a taxon that had powerful arms but less so than apes, and that they were using them for tasks such as manipulation and probably less so for locomotion.

Acknowledgments

We would like to thank the editors for the invitation to participate to this volume and for their constructive comments on earlier version of this chapter. We would also like to thank the following individuals and institutions for granting access and facilitating work on the osteological collections under their care: Jerome Cybulski and Janet Young, Canadian Museum of Civilization; the Nunavut Inuit Heritage Trust; Isabelle Ribot, Université de Montréal; François Bélanger, Ville de Montréal; Yohannes Haile-Salessie and Lyman Jellema, Cleveland Museum of Natural History; Richard W. Thorington and Linda Gordon, National Museum of Natural History (Washington, D.C.); Wim Van Neer, Musée Royal de l'Afrique Centrale; Marcia Ponce de León and Christoph Zollikofer, Anthropological Institute of the University of Zurich; Stephany Potze, Ditsong Museum of Natural History, Getachew Senishaw, Tomas Getachew and Yonas Desta, National Museum of Ethiopia; and the Authority for Research and Conservation of the Cultural Heritage of Ethiopia.

References

Aiello, L., Dean, C., 1990. An introduction to human evolutionary anatomy. Academic Press: London.

Aiello, L.C., Wood, B., Key, C., Lewis, M., 1999. Morphological and taxonomic affinities of the Olduvai ulna (OH 36). American Journal of Physical Anthropology 109, 89–110.

Alves Cardoso, F., Henderson, C.Y., 2010. Enthesopathy formation in the humerus: data from known age-at-death and known occupation skeletal collections. American Journal of Physical Anthropology 141, 550–560.

Andersson, K., 2004. Elbow-joint morphology as a guide to forearm function and foraging behaviour in mammalian carnivores. Zoological Journal of the Linnean Society 142, 91–104.

Brain, C.K., 1981. The hunters or the hunted? An introduction to African cave taphonomy. University of Chicago Press: Chicago.

Churchill, S.E., Pearson, O.M., Grine, F.E., Trinkaus, E., Holliday, T.W., 1996. Morphological affinities of the proximla ulna from Klasies River main site: archaic or modern? Journal of Human Evolution 31, 213–237.

Dominguez-Rodrigo, M., Pickering, T.R., Baquenado, E., Mabulla, A., Mark, D.F., Musiba, C., Bunn, H.T., Uribelarrea, D., Smith, V., Diez-Barboni, F., Pérez-Gonzaléz, A., Sánchez, P., Santonja, M., Barboni, D., Gidna, A., Ashley, G., Yravedra, J., Heaton, J.L., Arriaza, M.C., 2013. First partial skeleton of a 1.34-million-year-old *Paranthropus boisei* from Bed II, Olduvai Gorge, Tanzania. Plos One 8, e80347.

Drapeau, M.S.M., 2004. Functional anatomy of the olecranon process in hominoids and Plio-Pleistocene hominins. American Journal of Physical Anthropology 124, 297–314.

Drapeau, M.S.M., 2008. Articular morphology of the proximal ulna in extant and fossil hominoids and hominins. Journal of Human Evolution 55, 86–102.

Drapeau, M.S.M., 2012. Forelimb adaptations in *Australopithecus afarensis*. In: Reynolds, S.C., Gallagher, A. (Eds.), African Genesis Perspectives on Hominin Evolution. Cambridge University Press: Cambridge, pp. 223–247.

Drapeau, M.S.M., Ward, C.V., Kimbel, W.H., Johanson, D.C., Rak, Y., 2005. Associated cranial and forelimb remains attributed to *Australopithecus afarensis* from Hadar, Ethiopia. Journal of Human Evolution 48, 593–642.

Grabowski, M., Hatala, K.G., Jungers, W.L., Richmond, B.G., 2015. Body mass estimates of hominin fossils and the evolution of human body size. Journal of Human Evolution 85, 75–93.

Grausz, H.M., Leakey, R.E., Walker, A., Ward, C.V., 1988. Associated cranial and post-cranial bones of *Australopithecus boisei*. In: Grine, F.E. (Ed.), Evolutionary History of the Robust Australopithecines. Aldine: Chicago, pp. 127–132.

Green, D.J., Gordon, A.D., 2008. Metacarpal proportions in *Australopithecus africanus*. Journal of Human Evolution 54, 705–719.

Grine, F.E., Susman, R.L., 1991. Radius of *Paranthropus robustus* from Member 1, Swartkrans formation, South Africa. American Journal of Physical Anthropology 84, 229–248.

Hambücken, A., 1998. Morphologie et fonction du coude et de l'avant-bras des néandertaliens. Bulletin et Mémoire de la Sociéte d'Anthropologie de Paris, 10, 213–236.

Hlusko, L.J., Reiner, W.B., Njau, J.K., 2015. A one-million-year-old hominid distal ulna from Olduvai Gorge, Tanzania. American Journal of Physical Anthropology 158, 36–42.

Knussmann, V.R., 1967. Humerus, Ulna, und Radius der Simiae. S. Karger: Basel.

Leakey, M.G., Feibel, C.S., McDougall, I., Walker, A., 1995. New four-million-year-old hominid species from Kanapoi and Allia Bay, Kenya. Nature 376, 565–571.

Lovejoy, C.O., Simpson, S.W., White, T.D., Asfaw, B., Suwa, G., 2009. Careful climbing in the Miocene: the forelimbs of *Ardipithecus ramidus* and humans are primitive. Science 326, 70e71–70e78.

McHenry, H.M., Brown, C.C., McHenry, L.J., 2007. Fossil hominin ulnae and the forelimb of *Paranthropus*. American Journal of Physical Anthropology 134, 209–218.

Menter, C.G., 2002. Hominid distal humeri, proximal radii and proximal ulnae from Sterkfontein, in comparison with the elbow bones of other Plio-Pleistocene fossil hominids. Ph.D. Dissertation, University of the Witwatersrand.

Patel, B.A., 2005. The hominoid proximal radius: re-interpreting locomotor behaviors in early hominins. Journal of Human Evolution 48, 415–432.

Preuschoft, H., 1973. Functional anatomy of the upper extremity. In: Bourne, G.H., (Ed.), The Chimpanzee. Karger: Basel, pp. 34–120.

Rabey, K.N., Green, D.J., Taylor, A.B., Begun, D.R., Richmond, B.G., McFarlin, S.C., 2015. Locomotor activity influences muscle architecture and bone growth but not muscle attachment site morphology. Journal of Human Evolution 78, 91–102.

Rose, M.D., 1988. Another look at the anthropoid elbow. Journal of Human Evolution 17, 193–224.

Schmitt, D., 2003. Mediolateral reaction forces and forelimb anatomy in quadrupedal primates: implications for interpreting locomotor behavior in fossil primates. Journal of Human Evolution 44, 47–58.

Senut, B., 1981. L'humérus et ses Articulations chez les Hominidés Plio-Pleistocène. Éditions du Centre National de la Recherche Scientifique: Paris.

Senut, B., Tobias, P.V., 1989. A preliminary examination of some new hominid upper limb remains from Sterkfontein (1974-1984). Comptes Rendus de l'Académie des Sciences Série II 308, 565–571.

Skinner, M.M., Stephens, N.B., Tsegai, Z.J., Foote, A.C., Nguyen, N.H., Gross, T., Pahr, D.H., Hublin, J-J., Kivell, T.L., 2015. Human-like hand use in *Australopithecus africanus*. Science 347, 395–399.

Solan, M., Day, M.H., 1992. The Baringo (Kapthurin) ulna. Journal of Human Evolution 22, 307–313.

Standring, S. (Ed.), 2005. Gray's Anatomy, 39th Edition: the Anatomical Basis of Clinical Practice. Churchill Livingstone: London.

Susman, R.L., 1989. New hominid fossils from the Swartkrans Formation (1979-1986): postcranial specimens. American Journal of Physical Anthropology 79, 451–474.

Susman R.L., de Ruiter, D., Brain, C.K., 2001. Recently identified postcranial remains of *Paranthropus* and early *Homo* from Swartkrans Cave, South Africa. Journal of Human Evolution 41, 607–629.

Susman, R.L., Grine, F.E., 1989. New *Paranthropus robustus* radius from Member 1, Swartkrans Formation. American Journal of Physical Anthropology Suppl. 78, 311–312.

Swartz, S.M., 1990. Curvature of the forelimb bones of anthropoid primates: overall allometric patterns and specializations in suspensory species. American Journal of Physical Anthropology 83, 477–498.

Tallman, M., 2015. Phenetic and functional analyses of the distal ulna of *Australopithecus afarensis* and *Australopithecus africanus*. The Anatomical Record 298, 195–211.

Tobias, P.V., 1985. 19th annual report of PARG and its precursors: September 1985—September 1985. Palaeo-Anthropology Research Group, University of the Witwatersrand: Johannesburg, pp. 1–27.

Tobias, P.V., 1986. 20th annual report of PARG and its precursors: September 1985-September 1986. Palaeo-Anthropology Research Group, University of the Witwatersrand: Johannesburg, pp. 1–23.

Tobias, P.V., 1992. New researches at Sterkfontein and Taung with a note on Piltdown and its relevance to the history of palaeo-anthropology. Transactions of the Royal Society of South Africa 48, 1–14.

Tobias, P.V., Copley, K., Brain, C.K., 1977. South Africa. In: Oakley, B.G., Campbell, B.G., Molleson, T.I., (Eds.). Catalogue of Fossil Hominids Part I: Africa, 2nd edition, British Museum of Natural History: London, pp. 95–151.

Toussaint, M., Macho, G.A., Tobias, P.V., Partridge, T.C., Hughes, A.R., 2003. The third partial skeleton of a late Pliocene hominin (StW 431) from Sterkfontein, South Africa. South African Journal of Science 99, 215–223.

Villotte, S., 2008. Enthésopathies et activités des hommes préhistoriques. Recherche méthodologique et application aux fossiles européens du Paléolithique supérieur et du Mésolithique. Ph.D. Dissertation, Université de Bordeaux.

Villotte, S., Castex, D., Couallier, V., Dutour, O., Knüsel, C.J., Henry-Gambier, D., 2010. Enthesopathies as occupational stress markers: evidence from the upper limb. American Journal of Physical Anthropology 142, 224–234.

Zumwalt, A., 2006. The effect of endurance exercise on the morphology of muscle attachment sites. Journal of Experimental Biology 209, 444–454.

8

Carpals

Matthew W. Tocheri and Job Kibii

The hominin wrist is among the least well-represented anatomical areas recovered from Sterkfontein. Thus far, only two carpal bones have been recovered and both derive from Member 4: a right capitate (TM 1526) and a left scaphoid (StW 618). Both of these fossils have been described previously (Broom and Schepers, 1946; Le Gros Clark, 1947; Kibii et al., 2011) and discussed and/or analyzed in various studies (e.g., Le Gros Clark, 1967; Robinson, 1972; Lewis, 1973; McHenry, 1983; Marzke, 1983; Tocheri et al., 2007; Orr et al., 2013; Kivell et al., 2015). Interested readers are encouraged to consult these previously published descriptions and analyses for additional details. The anatomical descriptions presented here are not intended to supersede the previously published descriptions. Here we use the terms "proximal," "distal," "dorsal," "palmar," "ulnar," and "radial" in reference to the hand placed in anatomical position. Appendix I presents member, locus with grid reference, and date of discovery where known of each specimen.

Descriptions

TM 1526—Right capitate

Preservation. This specimen is in excellent condition (Figure 8.1). It is complete and well preserved, and all of its articular facets are well defined. Its maximum dimensions are approximately 18.5 mm proximodistally, 15.9 mm dorsopalmarly, and the head 9.8 mm radioulnarly. It was recovered by Robert Broom in the late 1930s within a block of matrix (Broom and Schepers, 1946). Initially, because of its small size, Broom did not remove it from the surrounding matrix for two years as he thought it was most likely a carpal from a baboon. However, when he finally did remove it, he realized that it was a hominin capitate. Some light preparation scratches are present from its removal from the surrounding matrix, but these do not detract from the preserved morphology.

Morphology. The head is round and globular, and almost as wide mediolaterally as the dorsodistal part of the bone. The proximal surface of the bone is encompassed entirely by the lunate facet, which extends distally up the proximal third of the dorsal side and only marginally up the palmar side. This facet is considerably more convex dorsopalmarly than radioulnarly. On the radial side, the neck is slightly constricted (i.e., "waisted") in its central portion. The proximal half is occupied by the scaphoid facet, which is moderately convex proximodistally and slightly convex dorsopalmarly. Its distal half displays a small, flat, roughly triangular-shaped facet for the trapezoid dorsally. The trapezoid facet is confluent with the facet for the second metacarpal, which extends along almost the entire distal margin of the radial side, is slightly concave dorsopalmarly, and is set at an oblique angle to the distal surface of the bone. The third metacarpal facet occupies almost all of the distal side, except for the dorso-ulnar corner which likely articulated at least in part with the base of the fourth metacarpal. The facet for the third metacarpal declines radioulnarly, particularly in its palmar half. On the ulnar side, the hamate facet begins narrowly along the dorsodistal half of the bone and then expands to encompass the entire proximal half. This facet is slightly concave proximodistally and flat dorsopalmarly.

Matthew W. Tocheri and Job Kibii, *Carpals.* In: *Hominin Postcranial Remains from Sterkfontein, South Africa, 1936–1995.* Edited by: Bernhard Zipfel, Brian G. Richmond, and Carol V. Ward, Oxford University Press (2020). © Oxford University Press.
DOI: 10.1093/oso/9780197507667.003.0008

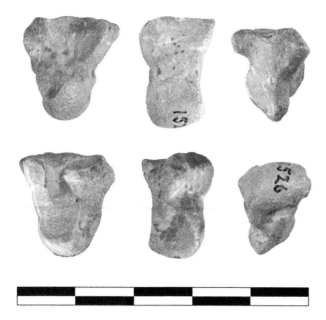

Figure 8.1 TM 1526 right capitate. Views (clockwise from top left): radial, dorsal, distal, ulnar, palmar, proximal.

Figure 8.2 StW 618 left scaphoid. Views: ulnar (bottom left), radial (bottom right), and distal (top right).

StW 618—Left scaphoid

Preservation. This specimen is in excellent condition (Figure 8.2). It is complete and well preserved, and all of its surfaces are in good condition except for a small eroded area on the radial side of its tubercle. Its maximum dimensions are approximately 26 mm dorsopalmarly and 16.6 mm proximodistally.

Morphology. The radial facet is roughly equally convex proximodistally and radioulnarly, and dominates the radial portion of the body. Ulnarly, the flat, half-moon-shaped lunate facet is positioned proximally to the larger, oval-shaped capitate facet, which is concave in all directions. Distally, the trapezium and trapezoid facets are continuous and slightly convex dorsoventrally and radioulnarly. A deep sulcus is present immediately radial to the

trapezium and trapezoid facets. The tubercle is conically shaped with a wide base and narrow tip.

Comparative morphology and analysis

Materials and methods

To explore how these two carpals from Sterkfontein compare with those of other hominins and African apes, we use the total surface areas of each carpal as a measure of absolute bone size and canonical variates analysis of three-dimensional shape variables (i.e., relative articular areas and angles between articular surfaces) (e.g., Tocheri et al., 2007; Orr et al., 2013). The comparative sample includes specimens of both extant species of *Gorilla*, both extant species of *Pan*, *A. afarensis* (A.L. 288-1w and A.L. 333-40), *A.* cf. *afarensis* (KNM-WT 22994H), *A. sediba* (MH2), *H. habilis* (OH7), *H. floresiensis* (LB1 and LB20), *H. naledi* (U.W. 101-1730 and U.W. 101-1726), Neandertals (Amud 1, Kebara 2, Krapina, La Chapelle-aux-Saints, La Ferrassie 1, Regourdou 1, Regourdou 2, Shanidar 3, Shanidar 4, Shanidar 8, and Tabun 1), and modern humans (Table 8.1). Tables 8.2 and 8.3 provide descriptive statistics for the comparative samples and TM 1526 and StW 618, respectively.

Results

Capitate

TM 1526 is larger in absolute size than capitates attributed to *A. afarensis* (A.L. 288-1w), *H. floresiensis* (LB1 and LB20), *Homo naledi* (UW-101-1730), and *Australopithecus sediba* (MH2), but smaller than others attributed to *Australopithecus afarensis* (A.L. 333-40), *Australopithecus* cf. *afarensis* (KNM-WT 22994), and Neandertals (Figure 8.3).

Table 8.1 Samples used in the comparative analyses.

Taxon	Capitates	Scaphoids
Gorilla beringei	40	41
Gorilla gorilla	20	44
Pan paniscus	21	22
Pan troglodytes	22	48
Australopithecus afarensis	2	–
Australopithecus cf. *afarensis*	1	–
TM 1526	1	–
StW 618	–	1
Australopithecus sediba	1	1
Homo habilis	–	1
Homo floresiensis	2	1
Homo naledi	1	1
Homo neanderthalensis	6	8
Homo sapiens	76	125

Table 8.2 Capitate metrics1 of TM 1526 and other early hominin specimens along with means and standard deviations (SD) of extant taxa and Neandertals.

	Articular angles[a]						Relative surface areas				
	ham-mc2	ham-sca/lun	ham-mc3	mc2-mc3	mc2-sca/lun	mc3-sca/lun	ham	mc2	mc3	sca/lun	nart
TM 1526	48	98	56	128	76	27	8.6	4.3	11.6	24.7	48.0
A.L. 333-40	47	92	60	133	74	30	8.8	6.3	11.5	25.0	46.6
A.L. 288-1w	25	86	61	118	94	41	8.9	4.0	10.9	28.2	45.9
KNM-WT 22994H	51	119	55	109	76	30	7.4	5.7	5.2	32.5	48.0
MH2	36	99	50	112	94	35	8.9	5.1	11.6	25.3	48.1
LB1	17	80	58	117	105	43	11.1	3.3	14.1	20.8	49.3
LB20	31	84	54	125	97	42	8.9	3.4	12.9	22.5	51.4
UW 101	55	106	64	122	65	11	7.2	4.2	7.7	29.9	49.8
Neandertal	41	63	86	132	79	32	8.9	3.8	10.2	21.3	53.0
SD	9	7	3	10	10	7	1.1	1.7	1.9	3.2	3.8
Homo sapiens	39	65	77	139	77	40	9.5	4.7	10.3	24.4	46.3
SD	9	5	5	9	8	6	1.2	0.9	1.4	2.6	3.8
Pan	20	56	92	102	112	38	10.2	3.6	12.9	28.4	43.3
SD	9	8	8	9	12	7	1.1	0.9	1.0	2.6	3.3
Gorilla	18	62	87	96	119	36	9.1	2.0	14.9	28.5	45.4
SD	8	6	4	9	10	7	0.8	0.7	1.1	2.3	2.4

[a]ham, hamate; mc2, second metacarpal; sca/lun, scaphoid-lunate; mc3, third metacarpal; nart, nonarticular.

Table 8.3 Scaphoid metrics[1] of StW 618 and other early hominin specimens along with means and standard deviations (SD) of extant taxa and Neandertals.

	Articular angles[a]			Relative surface areas			
	cap-rad	cap-tpm/tzd	rad-tpm/tzd	cap	rad	tpm/tzd	nart
STW 618	41	59	86	14.6	20.4	14.8	47.7
MH2	27	80	87	13.3	23.0	14.7	47.5
OH7	36	58	87	19.5	15.8	10.4	49.3
LB1	40	75	76	14.5	22.3	9.2	50.9
UW101	34	69	93	16.7	23.3	18.9	37.0
Neandertals	38	77	77	17.5	23.1	15.3	43.1
SD	6	5	5	1.5	2.1	1.6	2.6
Homo sapiens	32	82	79	15.1	23.0	14.7	43.8
SD	4	8	7	1.8	2.6	1.8	4.2
Pan	32	68	84	15.5	22.6	12.5	45.7
SD	4	6	6	1.9	1.9	1.6	3.4
Gorilla	31	69	83	11.6	21.0	11.5	52.0
SD	5	7	6	1.7	2.0	1.9	4.7

[a]cap, capitate; rad, radius; tpm/tzd, trapezium-trapezoid; nart, nonarticular.

It falls within the low end of size variation for adult modern human capitates, near the lower limit of size variation for adult *Pan* capitates, and beyond the lower limit for adult *Gorilla* capitates.

In terms of capitate shape, TM 1526 falls in between the clusters for extant African apes and humans along the first canonical axis (CAN1) and plots nearest to capitates attributed to *Australopithecus* and some Neandertals (Figure 8.4). CAN1 explains 95% of the variance and is mostly driven by differences in the orientation of the second metacarpal facet relative to the third metacarpal and scapholunate facets (Table 8.2).

Scaphoid

StW 618 is larger in absolute size than scaphoids attributed to *H. floresiensis* (LB1), *H. naledi* (UW-101-1726), *Australopithecus sediba* (MH2), and *H. habilis* (OH7, although this is a minimum estimate as it is incomplete) (Figure 8.5). It falls within the low range of scaphoid size variation for adult modern humans, Neandertals, and *Pan*, and beyond the lower limit of size variation in *Gorilla*.

In terms of scaphoid shape, StW 618 falls within an area along CAN1 where the clusters for modern humans and African apes overlap slightly, similar to the position of LB1's scaphoid (Figure 8.6). CAN1 explains 81% of the variance and is mostly driven by differences in the angle between the capitate and trapezium-trapezoid facets, and the relative areas of the capitate, trapezium-trapezoid, and nonarticular surfaces (Table 8.3).

Comparative morphology of carpals

Similar to the results of previous studies (McHenry, 1983; Marzke, 1983; Tocheri et al., 2007; Kibii et al., 2011; Orr et al., 2013; Kivell et al., 2015), both TM 1526 and StW 618

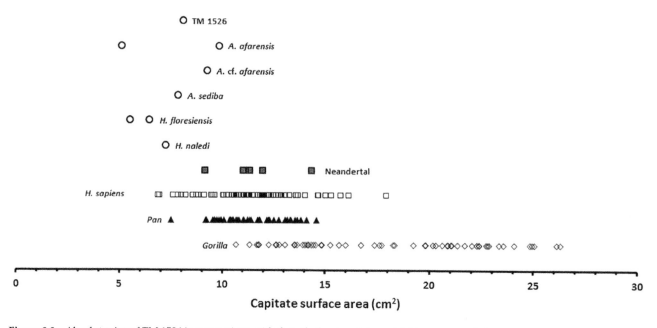

Figure 8.3 Absolute size of TM 1526 in comparison with that of other hominin and African ape capitates.

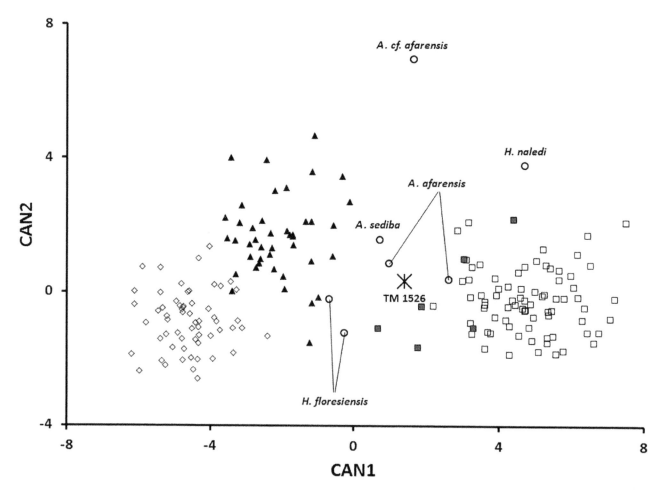

Figure 8.4 Plot of the canonical variables (CAN1 and CAN2) generated from the analysis of capitate shape. The resulting shape space is based on variation in extant genera only. CAN1 and CAN2 explain 95% and 5% of the observed variance, respectively. Modern humans, open squares; Neandertals, closed squares; *Pan*, closed triangles; *Gorilla*, open diamonds; fossil hominins, as labeled.

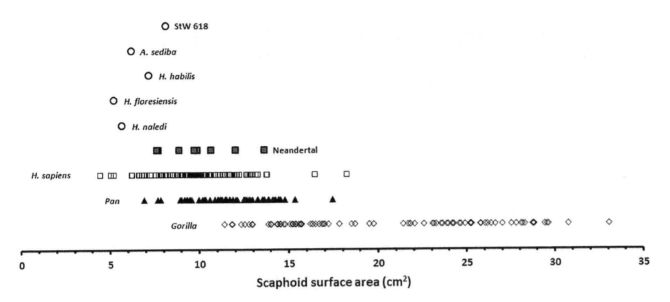

Figure 8.5 Absolute size of StW 618 in comparison with that of other hominin and African ape scaphoids.

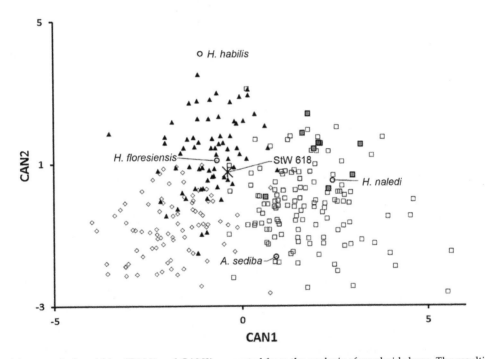

Figure 8.6 Plot of the canonical variables (CAN1 and CAN2) generated from the analysis of scaphoid shape. The resulting shape space is based on variation in extant genera only. CAN1 and CAN2 explain 81% and 19% of the observed variance, respectively. Modern humans, open squares; Neandertals, closed squares; *Pan*, closed triangles; *Gorilla*, open diamonds; fossil hominins, as labeled.

are broadly similar to the same elements from other fossil hominin taxa, modern humans, and African apes. However, both carpals are noticeably smaller than that of any *Gorilla* included here and are near the lower limit of sizes for the *Pan* sample, but they fall comfortably within the small size ranges for the modern human sample and appear reasonably typical in comparison with the carpal sizes of other early hominins (Figures 8.3 and 8.5). Despite

the absolute carpal size similarities with other early hominins, their shape affinities reveal a more complex picture.

Both are similar in shape overall to other fossil hominin carpals, falling somewhat intermediate between the shapes seen in modern humans on one hand and African apes on the other. This is especially the case for the capitate, which shows no overlap between African apes and modern humans and only slight overlap between *Gorilla*

Table 8.4 Correlations between the capitate and scaphoid variables and their respective canonical axes (abbreviations as in Tables 8.2 and 8.3).

	Capitate			Scaphoid		
	Variable	CAN1	CAN2	Variable	CAN1	CAN2
Angles	ham-mc2	0.27	–0.17	cap-rad	0.05	0.05
	ham-mc3	–0.22	**0.58**	cap-tpm/tzd	**0.58**	**–0.35**
	ham-sca/lun	0.08	**–0.39**	rad-tpm/tzd	–0.21	0.15
	mc2-mc3	**0.54**	–0.28			
	mc2-sca/lun	**–0.47**	0.17			
	mc3-sca/lun	0.05	0.04			
Relative areas	ham	0.02	**0.40**	cap	**0.42**	**0.87**
	mc2	**0.31**	**0.43**	rad	0.21	0.24
	mc3	**–0.38**	–0.27	tpm/tzd	**0.50**	0.04
	sca/lun	–0.19	0.17	nart	**–0.47**	**–0.50**
	nart	0.05	–0.31			

and *Pan* (Figure 8.4). In this respect, TM 1526 is more similar to other australopith and *H. floresiensis* capitates. Shape differences in the scaphoid are subtler, at least based on the metrics employed here, as StW 618 falls at the respective edges of both African ape and modern human variation, and near to the scaphoid of *H. floresiensis* (Figure 8.6).

Discussion and conclusions

In total, these quantitative comparative data suggest that TM 1526 and StW 618 belong to one or more hominin taxa that lacked the derived carpal features that characterize the wrist of modern humans and Neandertals (Tocheri, 2007; Tocheri et al., 2008). Although both of these fossils are reasonably attributed to *A. africanus*, this attribution is based solely on the geological and paleontological context of where these specimens were recovered. The more recent discoveries of *A. sediba* and *H. naledi* from the same geographical region (Berger et al., 2010, 2015) underscore the complexity and diversity of hominin evolution in southern Africa throughout the Pleistocene and suggest caution is needed in attributing isolated hominin elements to particular taxa. Indeed, the lack of clear autapomorphies in the carpals of many early hominin taxa makes confident taxonomic attribution almost impossible without direct association with more diagnostic elements. Similarly, functional interpretations based on single fossil carpals also must be treated cautiously until more reasonably complete fossil hominin hand skeletons are recovered. That said, however, the comparative morphology of TM 1526 and StW

618 suggests that they belonged to hominins in which hand functional morphology was likely broadly similar to that of other early hominins, such as australopiths and *H. floresiensis*.

Acknowledgments

We thank the Wenner-Gren Foundation for funding the workshop that led to this volume. Aspects of this research were also supported by a Smithsonian Institution Scholarly Studies Grant Program Award and a Wenner-Gren Foundation post-Ph.D. grant (Grant Number 7822). For curatorial assistance and access to specimens in their care, we thank Eileen Westwig (American Museum of Natural History), William Kimbel and Donald Johanson (Arizona State University's Institute of Human Origins), Emma Mbua (National Museum of Nairobi), Richard Thorington, Richard Potts, David Hunt, and Linda Gordon (National Museum of Natural History), Judith Chupasko (Museum of Comparative Zoology), Emmanuel Gilissen and Wim Wendelen (Royal Museum for Central Africa), Patrick Semal and Georges Lenglet (Royal Belgian Institute of Natural Sciences), Erik Trinkaus (Washington University in St. Louis), Tony Djubiantono (National Research and Development Center for Archaeology in Indonesia), and Ned Gilmore (Academy of Natural Sciences in Philadelphia).

References

Berger, L.R., Hawks, J., de Ruiter, D.J., Churchill, S.E. Schmid, P., Williams, S.A., DeSilva, J.M., Kivell, T., Skinner, M., Musiba, C.M., Cameron, N., Holliday, T.W., Harcourt Smith, W., Ackermann, R.R., Bastir, M., Barry Bogin, R.R., Bolter, D., Brophy, J., Cofran, Z.D., Congdon, K.A., Deane, A.S., Delezene, L., Dembo, M., Drapeau, M., Elliott, M., Feuerriegel, E.M., Garcia-Martinez, D., Garvin, H.M., Green, D.J., Gurtov, A., Irish, J.D., Kruger, A., Laird, M.F., Marchi, D., Meyer, M. R., Nalla, S., Negash, E.W., Orr, C.M., Radovcic, D., Scott, J.E., Schroeder, L., Throckmorton, Z., Tocheri, M.W., VanSickle, C., Walker, C.S., Wei, P., Zipfel, B., 2015. A new species of the genus *Homo* from the Dinaledi Chamber, South Africa. eLife, e09560.

Broom, R., Schepers, G.W.H., 1946. The South African fossil ape men: the Australopithecinae. Transvaal Museum Memoire No. 2.

Kibii, J.M., Clarke, R.J., Tocheri, M.W., 2011. A hominin scaphoid from Sterkfontein, Member 4: morphological description and first comparative phenetic 3D analyses. Journal of Human Evolution 61, 510–517.

Kivell, T.L., Deane, A.S., Tocheri, M.W., Orr, C.M., Schmid, P., Hawks, J., Berger, L.R., Churchill, S.E., 2015. The hand of *Homo naledi*. Nature Communications 6, 8431.

Le Gros Clark, W.E., 1947. Observations on the anatomy of the fossil Australopithecinae. Journal of Anatomy 81, 300–333.

Le Gros Clark, W.E., 1967. Man-apes or ape-man? Holt, Rinehart and Winston: New York.

Lewis, O.J., 1973. The hominid os capitatum, with special reference to the fossil bones from Sterkfontein and Olduvai Gorge. Journal Human Evolution 2, 1–12.

Marzke, M.W., 1983. Joint function and grips of the *Australopithecus afarensis* hand, with special reference to the region of the capitate. Journal of Human Evolution 12, 197–211.

McHenry, H.M., 1983. The capitate of *Australopithecus afarensis* and *A. africanus*. American Journal of Physical Anthropology 62, 187–198.

Orr, C.M., Tocheri, M.W., Burnett, S.E., Awe, R.D., Saptomo, E.W., Sutikna, T., Wasisto, S., Morwood, M.J., Jungers, W.L., 2013. New wrist bones from *Homo floresiensis*. Journal of Human Evolution 64, 109–129.

Robinson, J.T., 1972. Early hominid posture and locomotion. University of Chicago Press: Chicago.

Tocheri, M.W., Orr, C.M., Jacofsky, M.C., Marzke, M.W., 2008. The evolutionary history of the hominin hand since the last common ancestor of *Pan* and *Homo*. Journal of Anatomy 212, 544–562.

Tocheri, M.W., Orr, C.M., Larson, S.G., Sutikna, T., Jatmiko, W., Saptomo, E., Awe, R.D., Djubiantono, T., Morwood, M.J., Jungers, W.L., 2007. The primitive wrist of *Homo floresiensis* and its implications for hominin evolution. Science 317, 1743–1745.

9

Metacarpals and manual phalanges

Tracy L. Kivell, Kelly R. Ostrofsky, Brian G. Richmond, and Michelle S.M. Drapeau

There are several hominin metacarpal and phalangeal specimens from Sterkfontein that represent elements from all five rays, although none can be clearly associated with the same individual. Most of these metacarpal and phalangeal fossils derive from Member 4, while a few may come from Member 5 (see Appendix 1; also Kuman and Clarke, 2000). None of these bones has yet to be formally described, although many have been included in morphological analyses. For example, Ricklan (1987, 1990) provided the first functional interpretations, focusing on the metacarpal morphology and suggesting that *A. africanus* had strong power grip abilities. Others have assessed the potential intrinsic hand proportions (e.g., thumb length relative to the fingers) (Green and Gordon, 2008; Ostrofsky and Richmond, 2015), while Skinner et al. (2015) have investigated the internal trabecular morphology of the metacarpals. Although the phalanges have been included as comparative data in several fossil hominin hand publications (e.g., Tocheri et al., 2008; Kivell et al., 2011, 2015, 2018a, 2018b; Kivell, 2015; Richmond et al., 2016; Pickering et al., 2018), only morphological aspects of the distal pollical phalanx StW 294 have been published in detail (Ricklan, 1987, 1990; Susman, 1998; Almécija et al., 2010).

In addition to the metacarpal and phalangeal fossils described below, other hand fossils have been uncovered from Sterkfontein. Perhaps the most intriguing is the articulated, complete left hand and isolated right hand bones associated with the StW 573 "Little Foot" skeleton, which will offer the first glimpse into the hand functional morphology of a single individual at Sterkfontein (Clarke, 1999, 2008, 2013). Only superficial descriptions of some of its morphology have been published thus far (Clarke, 1999, 2013). In addition, Partridge et al. (2003: 611) show one image of a partial manual phalanx (StW 605) from Jacovec Cavern, but describe it as "heavily eroded, and

[that] little can be said about it." Stratford et al. (2016) describe a manual proximal phalanx (StW 668) from Milner Hall, while Pickering et al. (2018) provide comparative morphological descriptions of several metacarpal and phalangeal specimens discovered between 1998 and 2003 in the Jacovec Cavern and Member 4. When possible, these specimens are included within the comparative sample below.

The state of preservation is described for each individual specimen. Morphology of each bony element is described in aggregate (e.g. MC1, MC2, etc.).

Descriptions

First metacarpal

StW 418—Left first metacarpal

Preservation. This specimen is a complete left first metacarpal of an adult (Figure 9.1). The surface preservation is excellent, with the exception of a minor abrasion on the palmar-proximoulnar margin of the distal facet, exposing trabeculae, and a small fragment missing from the dorsal margin of the proximal facet at roughly the sagittal midline.

StW 583—Left distal first metacarpal

Preservation. This distal end of an adult left first metacarpal preserves the shaft distal to midshaft (Figure 9.1). The maximum preserved length is 20.7 mm from the head to the proximodorsal end of an oblique break that continues 15.0 mm distally to the palmar surface, exposing the cortical thickness and medullary cavity. A fragment of bone is missing from the radiopalmar surface of the distal articular surface. Otherwise, the surface preservation is excellent.

Tracy L. Kivell, Kelly R. Ostrofsky, Brian G. Richmond, and Michelle S.M. Drapeau, *Metacarpals and manual phalanges*. In: *Hominin Postcranial Remains from Sterkfontein, South Africa, 1936–1995*. Edited by: Bernhard Zipfel, Brian G. Richmond, and Carol V. Ward, Oxford University Press (2020). © Oxford University Press.
DOI: 10.1093/oso/9780197507667.003.0009

StW 418

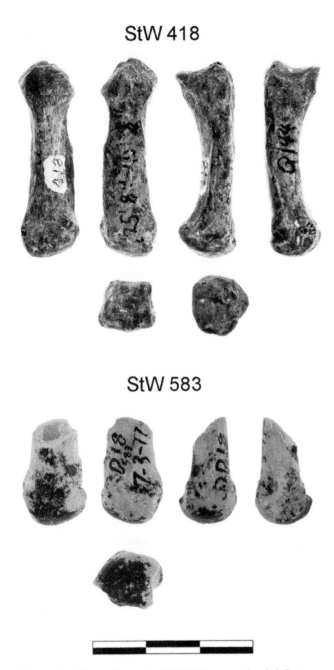

StW 583

Figure 9.1 First metacarpals. StW 418 is a complete left first metacarpal and StW 583 is the distal portion of a left first metacarpal. Both metacarpals are shown in, from left to right, palmar, dorsal, radial, and ulnar views above (proximal is toward the top) and distal and proximal (for StW 418 only) views below. All specimens are to scale.

Morphology. StW 418 is a gracile metacarpal with slight longitudinal curvature. The head of StW 583 is absolutely larger than that of StW 418, but the preserved morphology is very similar (Table 9.1). Both specimens have small heads that are broader palmarly and narrower dorsally. The heads' ulnar margins are parasagittal, whereas the radial margins angle from the broader palmar to the

narrower dorsal surface. The dorsoradial aspect of the heads is slightly flattened, particularly on StW 583. The articular surfaces are slightly convex radioulnarly, and only slightly more convex dorsopalmarly. The articular surfaces extend as far (StW 418) or farther (StW 583) proximally on the dorsal surface as it does on the palmar surface, with limited articular area on the palmar surface. The palmar aspect of the articular surfaces each have a midline ridge flanked by slight depressions. In each, a crest runs from the radiopalmar margin of the head toward the palmar shaft midline; this crest is pronounced in StW 583 and faint in StW 418.

The proximal end of StW 418 is approximately the same radioulnar breadth as its head. The saddle-shaped trapezium facet is more radioulnarly convex than it is dorsopalmarly concave. The ulnar border of the articular surface extends further distally than its radial border. The shaft is generally gracile and roughly triangular-shaped in cross-section, such that the palmar surface of the shaft is somewhat "pinched." There is a small but distinct flange projecting from the radial side of the shaft for the attachment of the *m. opponens pollicis* tendon, starting 15.7 mm from the proximal end and running 14.3 mm distally. Along the dorsoulnar border there is a slight ridge extending from the base to a roughened area at roughly midshaft, together extending approximately 11.0 mm distally, that is the likely attachment of the *m. first dorsal interosseous* tendon. At the break of the StW 583 distal shaft, the cortical thickness is 2.0 mm dorsally, 2.3 mm ulnarly, 3.2 mm palmarly, 3.8 mm radially, with a total radioulnar breadth of 9.0 mm and dorsopalmar height of 6.8 mm.

Second metacarpal

StW 382—Left second metacarpal

Preservation. This specimen is a complete left second metacarpal of an adult (Figure 9.2; Table 9.1). Aside from a small hole in the dorsal aspect of the head at the sagittal midline, the surface preservation is excellent.

Morphology. The shaft of StW 382 has moderate longitudinal curvature that is slightly more strongly curved at the distal end. The head is aligned with the shaft, without appreciably projecting palmarly, and is noticeably asymmetrical. Palmarly, the radioulnar curvature peaks ulnarly and slopes to the radial side, although the articular surface extends proximally to an equal extent on both the radial and ulnar margins. In lateral view, the curvature of the head is more convex distally. The articular surface is radioulnarly broad palmarly and narrows rapidly to a point at the ulnar epicondyle. The shaft proximal to the head is flat, with interosseous ridges originating from the epicondyles that converge toward the base and become increasingly faint. There is also a short crest that extends from the head proximally approximately 13.0 mm along the palmar surface of the shaft.

Table 9.1 Linear measurements of Sterkfontein metacarpals. All measurements in mm.

Specimen	Element	Side	TL	IA	BB	BH	PAB	PAH	MSB	MSH	HB	HH
StW 418	MC1	L	37.4	35.8	11.1	11.1	9.6	9.4	8.5	6.8	11.3	9.7
StW 583	MC1	L	20.3 pres.	–	–	–	–	–	–	–	11.7	10.6
StW 382	MC2	L	66.4	63.8	13.2	18.3	11.0	15.5	8.3	9.2	13.0	14.9
StW 27	MC3	R	23.8 pres.	–	–	–	–	–	–	–	[10.2]	12.5
StW 64	MC3	L	55.8	53.9	12.3	14.0	12.1	12.5	7.4	[7.2]	11.2	12.0
StW 68	MC3	R	54.2	52.7	[13.2]	[15.6]	–	[13.5]	7.6	8.4	11.3	11.5
StW 394	MC3	L	44.64 pres.	–	–	–	–	–	6.8	8.4	12.5	14.0
StW 26	MC4	R?	43.3 pres.	–	–	–	–	–	[6.9]	[7.4]	10.7	11.2
StW 65	MC4	R	44.6 pres.	–	10.5	12.3	8.2	10.6	[7.7]	[8.2]	–	–
StW 292	MC4	R?	31.2 pres.	–	–	–	–	–	–	–	11.0	11.0
StW 330	MC4	L	43.7 pres.	–	11.2	11.7	10.1	10.7	6.4	7.4	–	–
StW 552	MC4	R?	40.0 pres.	–	–	–	–	–	[6.1]	[7.8]	10.5	11.6
StW 63	MC5	L	[43.3]	42.7	11.3	10.0	8.5	8.6	7.8	6.2	11.1	–

Abbreviations: '[x]', estimated value; 'MC', metacarpal; 'L', left; 'R', right; 'pres.', maximum preserved length; 'TL', maximum total length; 'IA', interarticular length; 'BB', proximal base radioulnar breadth; 'BH', proximal base dorsopalmar height; 'PAB', proximal articular radioulnar breadth; 'PAH', proximal articular dorsopalmar height; 'MSB', midshaft radioulnar breadth; 'MSH', midshaft dorsopalmar height; 'HB', distal head radioulnar breadth; 'HH', distal head dorsopalmar height; 'pres.', preserved.

StW 382

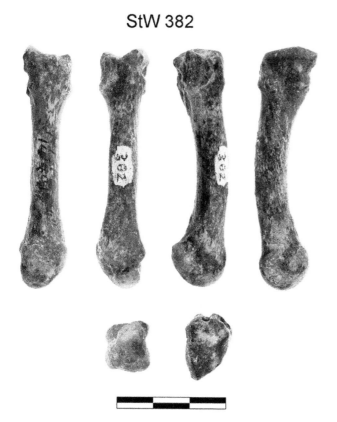

Figure 9.2 Second metacarpal. StW 382 is a complete left second metacarpal shown in, from left to right, palmar, dorsal, radial, and ulnar views above (proximal is toward the top), and distal and proximal views below.

In StW 382 the articulation for the trapezoid is V-shaped in the coronal plane. The angle of the two sides is greater than 90°. The ulnar side of the articulation is larger and slightly convex dorsopalmarly. The articulation for the third metacarpal has rounded dorsal and palmar facets with a groove that almost separates them but stops just distal to the capitate facet. The dorsal facet is 2.6 mm proximodistally and about 5.2 mm dorsopalmarly, and the palmar facet is 4.0 mm proximodistally and 8.2 mm dorsopalmarly. The capitate facet is gently convex, narrow proximodistally (4.6 mm), and elongated dorsopalmarly (15.5 mm). The articulation for the trapezium faces radially, more or less parallel to the articulation for the third metacarpal on the other side. The articulation is triangular in shape and is 6.2 mm palmarly and 9.7 mm dorsopalmarly.

The StW 382 base has a moderately rugose prominence on its palmar surface about 10.0 mm from the proximal end in the area of *m. flexor carpi radialis* insertion. The dorsal surface preserves a distinctly demarcated region (9.0 mm proximodistally and 9.5 mm radioulnarly) on the radial side in the area of *m. extensor carpi radialis longus* tendon insertion. On the ulnar side of the dorsal base is a distinct rugosity (8.6 mm proximodistally and 4.8 mm radioulnarly) where the *m. extensor carpi radialis brevis* tendon inserts. The raised rugosities produce a 1.5 mm-wide groove that is continuous distally with the shaft.

Third metacarpal

StW 27—Probably a right distal third metacarpal

Preservation. This specimen is the distal end of what is probably a right third metacarpal of an adult (Figure 9.3). Approximately 25% of the shaft is preserved extending to an oblique fracture. The greatest preserved length is 23.7 mm to the proximoulnar margin of the broken shaft. The proximal end of the break extends about 7.0 mm dorsodistally. Surface preservation is very good.

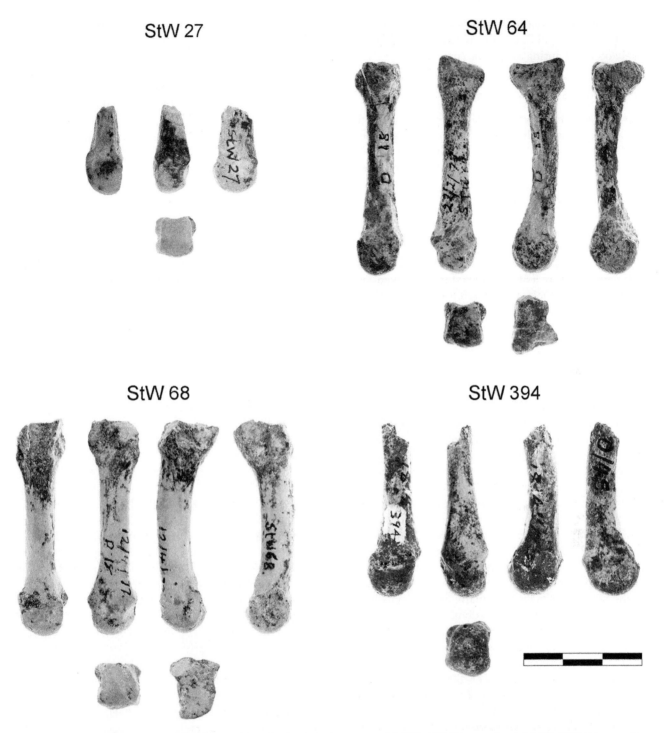

Figure 9.3 Third metacarpals. StW 27 is a probable right distal third metacarpal (MC3), StW 64 is complete left MC3, StW 68 is complete right MC3, and StW 394 is a left distal MC3. All metacarpals are shown in, from left to right, palmar, dorsal, radial, and ulnar views above (proximal is toward the top) and distal and proximal (when preserved) views below. Radial view is not shown for StW 27. All specimens are to scale.

StW 64—Left third metacarpal

Preservation. This specimen is a complete left third metacarpal of an adult (Figure 9.3). The head is undistorted and is taphonomically twisted in a supinated direction relative to the shaft. The palmar and radial shaft surface is missing a large fragment of cortex extending approximately 35.0 mm from the edge of the proximal shaft up to edge of the radial articular surface of the head and radial epicondyle. The base is intact and well preserved, although there is scratch along the coronal midline of the distal articular surface that likely occurred during excavation/preparation.

StW 68—Right third metacarpal

Preservation. This specimen is a complete right third metacarpal of an adult (Figure 9.3). The surface preservation is excellent. Small chips are missing from the dorso-radial aspect of the head just distal to the epicondyle, the radial margin of the proximal articular surface, and the radial and ulnar aspect of the dorsal surface of the base.

StW 394—Left distal third metacarpal

Preservation. This specimen is a distal left third metacarpal of an adult (Figure 9.3). The maximum preserved length is 44.7 mm. The shaft is preserved just beyond midshaft. The break is diagonal with the lateral shaft margin broken about 7.2 mm more distal than the ulnar side. The surface preservation is excellent. There are only minor abrasions exposing trabecular bone at the ulnar margin of the distal articular surface and the palmar proximal margin of the articular facet, exposing trabeculae.

Morphology. StW 64 and StW 68 are very similar in absolute length, although StW 68 is more robust, while StW 394 represents a larger individual and StW 27 is a smaller individual (Table 9.1). Specimens StW 64, StW 68, and StW 394 share similar morphology. The shafts have slight longitudinal curvature. The shafts of StW 64 and StW 68 are robust, with the latter being particularly robust, whereas the shaft of StW 394 shows more pronounced tapering moving proximally. The heads are asymmetric. The distal articular surfaces are radioulnarly broader palmarly than they are dorsally. In the dorsal view, the articular surface is flattened dorsoradially and in the palmar view, the radial portion of the head projects more palmarly and proximally than the ulnar side. The proximal extension of the radial side is particularly prominent in StW 68. In StW 64, the articular surface extends dorsally onto the epicondyles, whereas in StW 68 and StW 394 the articular surface extends only onto the ulnar epicondyle. StW 68 differs slightly in having narrower, spike-like extensions of articular surface on the ulnar and, especially, radial margins of the head.

StW 27 shares similar morphology to the specimens described above but differs in being smaller in overall size, and in the distal view, it is slightly more asymmetric such that the convexity of the articular surface slopes more radially than in StW 64 and StW 394 but is similar to StW 68. The articular surface extends dorsally to both the radial and the ulnar epicondyles as in StW 64. There is a strong ridge starting from the ulnar base of the palmar articular surface and continuing proximally, which is more prominent than that of StW 394.

Along the shaft, dorsal interosseous ridges take origin from the epicondyles and converge toward the base. In StW 64 and, especially, StW 394 these ridges lose definition, whereas in StW 68 they remain distinct and converge just proximal to midshaft. On the palmar surface, the palmar ridges of StW 68 and StW 394 take origin from the ulnar margin of the head and run proximally along the base (this region is obscured by damage and taphonomic preservation in StW 64). The shafts narrow proximally in StW 64 and StW 394, but in StW 68 there is a more consistent radioulnar breadth throughout the shaft. The shaft break in StW 394 exposes the cortical bone measuring 2.8 mm dorsally, 2.3 mm ulnarly, 2.7 mm palmarly, and 2.3 mm radially in thickness (total dorsopalmar diameter is 8.1 mm and radioulnar diameter is 6.8 mm). At the break of StW 27, the cortical thickness is 1.7 mm dorsally, 1.7 mm ulnarly, 1.8 mm palmarly, and 1.6 mm radially.

The bases of StW 64 and StW 68 preserve articular surfaces for the capitate and MC2 and MC4. In each, the articulation for the MC4 has two separate, flat facets separated by a groove. The dorsal facets are larger than the palmar facets and ovoid with greater proximodistal (StW 64: 5.2 mm; StW 68: 6.3 mm) than dorsopalmar (StW 64: 3.6 mm; StW 68: 4.9 mm) lengths. In StW 64, the dorsal facet is oriented equally ulnarly and palmarly, while in StW 68 the facet is primarily ulnarly oriented. The palmar facet of StW 64 is circular, well defined, and oriented ulnarly while in StW 68 it is ovoid and oriented ulnarly and slightly distally (proximodistal StW 64: 3.4 mm; StW 68: 5.6 mm; dorsopalmar StW 64: 4.0 mm; StW 68: 5.4 mm). The MC2 facet of StW 64 is not well defined but runs most of the dorsopalmar length of the base, measuring 11.8 mm proximodistally, and appears dorsopalmarly taller at the palmar portion, measuring 3.1 mm in height. It is shallowly dorsopalmarly concave and is oriented laterally. The MC2 facet is not preserved in StW 68.

In both StW 64 and StW 68 the capitate articulation is radioulnarly broader dorsally (StW 64: 12.0 mm; StW 68: 12.7 mm) than it is palmarly (StW 64: 8.4 mm; StW 68: 9.1 mm), particularly the ulnar portion, and the entire surface is shallowly dorsopalmarly concave. The base of StW 64 lacks a styloid process, but the radial side projects proximally approximately 2.0 mm relative to the remainder of the base. This region is not well preserved in StW 68. Each has a distinct tubercle just distal to the dorsodistoradial border in the area of the *m. extensor carpi radialis brevis* tendon attachment.

Fourth metacarpal

StW 26—Distal fourth metacarpal

Preservation. This specimen is a metacarpal of an adult preserving the head and most of the shaft, with an oblique fracture (Figure 9.4). It is likely a left fourth metacarpal based on the relative symmetry of the head and shaft shape. The surface preservation is excellent. There is small chip of bone missing just proximal to the radial epicondyle (when viewing the bone dorsally).

StW 65—Right fourth metacarpal

Preservation. This right fourth metacarpal preserves the base and entire shaft distally to the region where the shaft expands at the base of the head (Figure 9.4). The maximum preserved length is 44.9 mm, and 44.2 mm parallel to the shaft's axis. It likely represents an adult individual based on the size and well-developed carpometacarpal facets. The shaft surface contains many longitudinal cracks, but they have not visibly expanded the cortex. The proximal joint surfaces are well preserved.

StW 292—Distal fourth metacarpal

Preservation. This specimen is a metacarpal of an adult preserving the head and about half of the distal shaft, with a section of the ulnopalmar cortex missing (Figure 9.4). It is likely a right fourth metacarpal based on the relative symmetry of the head and shaft shape. The surface preservation is excellent with the exception of an area of the dorsal portion of the distal articular surface exposing the underlying trabecular bone.

StW 330—Left fourth metacarpal

Preservation. This left fourth metacarpal preserves the base and entire shaft distally to the base of the head, exposing the medullary cavity and relatively thin cortical bone (Figure 9.4). The maximum preserved length is 43.5 mm, and this length occurs parallel to the shaft's axis. Without the head, it is not possible to determine whether it is an

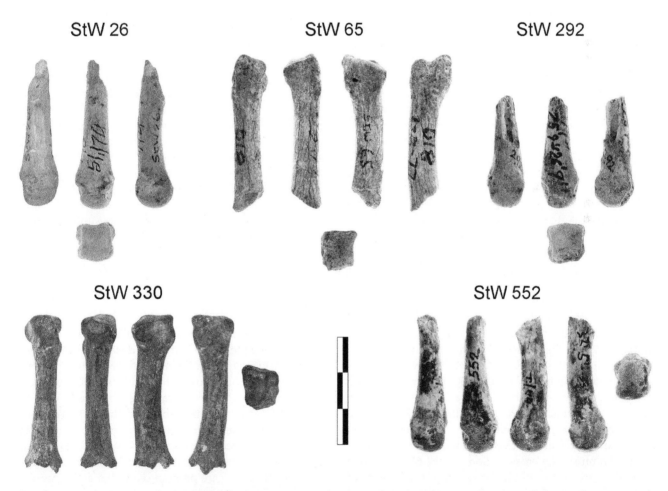

Figure 9.4 Fourth metacarpals. StW 26 is likely left distal fourth metacarpal (MC4), StW 65 is a right proximal MC4, StW 292 and 552 are likely right distal MC4s, and StW 330 is left proximal MC4. From left to right, StW 65, 292, and 552 are shown in, palmar, dorsal, ulnar, and radial views, as well as a proximal (StW 65) and distal views (StW 292 and 552); StW 26 and 330 are shown in palmar, dorsal, radial, and ulnar (StW 330 only) views, as well as distal (StW 26) and proximal (StW 330) views. Proximal is toward the top and all specimens are to scale.

adult, but the size and well-developed carpometacarpal facets indicate that it represents an individual near or at adulthood. The surface preservation is excellent. There is a small crack in the cortex just distal to the dorsal MC3 facet that obscures somewhat the full extent of this articular surface.

StW 552—Distal fourth metacarpal

Preservation. This specimen is a metacarpal of an adult preserving the head and most of the shaft (Figure 9.4). It is likely a right fourth metacarpal based on the relative symmetry and shape of the head and shaft shape. The surface preservation is excellent with the exception of small areas at the palmar margin and dorsal margin at the midline of the distal articular surface that expose the underlying trabecular bone.

Morphology. These five specimens are fairly similar in size, but they show some variation in robusticity and curvature (Table 9.1). StW 65 is the most robust among them, StW 330 is notably gracile, and the others vary between these two specimens. All of them show slight to moderate longitudinal curvature (dorsally convex), with a "bend" in the shaft just distal to midshaft that is most pronounced in StW 330 and 552.

In lateral view, the heads of StW 26, 292 and 552 extend only very slightly beyond the palmar border of the shafts. The heads are only slightly asymmetric, with slightly flattened dorsoradial surfaces. The sides of the heads are subparallel, with slightly radioulnarly wider palmar surfaces than dorsal articular surfaces. The articular surfaces do not extend onto the epicondyles. On the palmar surface, there are narrow proximal extensions of articular surface on the radial and ulnar margins of the head that are relatively equal in how far they extend. Deep fossae are present dorsally on the radial and ulnar sides of the head that pinch the head just distal to the epicondyles, making them very pronounced.

The shafts have faint bilateral interosseous dorsal ridges that take origin from the epicondyles. Moving proximally, the shafts taper; this tapering is most pronounced in what is preserved of StW 552 while there is minimal tapering in StW 65 and 330. The breaks expose the cortical bone at various levels along the shaft. StW 26 has a particularly narrow medullary cavity (1.3 mm diameter), exposed at about 25–30% of its length from the proximal end, and thick cortical bone (dorsal 3.7 mm, radial 2.8 mm). The cross-section of StW 65 is exposed near the base of the head, with cortical thicknesses of 1.6 mm dorsally, 1.7 mm ulnarly, 2.1 mm palmarly, and 1.8 mm radially. In StW 292 the cross-sectional thickness of the palmar (2.1 mm) and radial (2.0 mm estimated) is exposed at about 75% of length (from the proximal end), while it is exposed near midshaft for the dorsal (2.1 mm)

and ulnar (2.5 mm) sides. The cross-section of StW 330 is exposed just proximal to the epicondyles and head, and the internal trabecular bone is missing; the cortical thicknesses are 1.2 mm dorsally, 1.8 mm ulnarly, 1.5 mm palmarly, and 1.9 mm radially. The cross-section of StW 552 is exposed at roughly the same level as in StW 26; the cortical thicknesses are 2.3 mm dorsally, 1.8 mm (estimate) ulnarly, 2.4 mm palmarly, and 1.9 mm radially.

The morphology of the base in StW 65 and 330 is similar. The MC5 articulations are oriented ulnarly and are dorsopalmarly concave, very shallowly in StW 330 and more deeply in StW 65 (depth = 1.2 mm). The facet is dorsopalmarly tall (StW 65: 10.0 mm, StW 330: 10.5 mm) and proximodistally short (StW 65: 6.6 mm, StW 330: 4.5 mm). In both specimens the dorsal portion of the MC5 facet is proximodistally longer than the palmar portion (StW 65: 5.4 mm; StW 330: 3.1 mm). Both specimens have a small, round raised dorsal facet for the MC3 (dorsopalmar StW 65: 5.2 mm, StW 330 4.3 mm; proximodistal StW 65: 5.6 mm, StW 330: estimate 4.0 mm). A palmar MC3 articulation is not present in StW 330; while this facet in StW 65 is small, round, and raised, measuring an estimated 2.1 mm dorsopalmarly (slight abrasion means the facet is not well-defined) and 3.7 mm proximodistally, and is separated from its dorsal counterpart by a deep sulcus (1.5 mm deep). In both specimens, the hamate articulation is roughly rectangular in shape and is sloped dorsoulnarly. In StW 65, the ulnar half of the facet is dorsopalmarly convex while the radial portion is dorsopalmarly concave. In StW 330, the hamate facet is generally flat throughout, although the most radiodorsal portion is oriented more proximally than the remainder of the facet, and the articular surface of this portion is continuous with the MC3 facet on the radial side.

Fifth metacarpal

StW 63—Left fifth metacarpal

Preservation. This left fifth metacarpal of an adult individual is missing the dorsal half of the distal end but is otherwise intact (Figure 9.5). It was recovered in two pieces, a distal one-third and proximal two-thirds, which are glued together. The surface preservation is very good.

Morphology. The shaft is gracile and the preserved morphology suggests mild dorsally convex longitudinal curvature. The head is radioulnarly wide palmarly with the preserved radial margin of the articular surface tapering dorsally. There are pronounced radial and ulnar palmar articular surface projections; the ulnar one projects slightly further proximally while the radial one projects more palmarly. The shaft tapers proximally such that the narrowest point (6.8 mm) on the shaft is just distal to the base. The ulnar side of the shaft at roughly midshaft

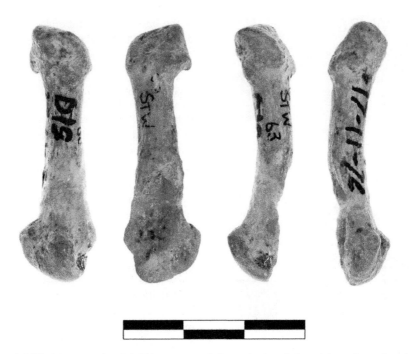

Figure 9.5 Fifth metacarpal. StW 63 is a complete left fifth metacarpal shown in, from left to right, palmar, dorsal, radial, and ulnar views.

preserves a roughened projection in the region of the *m. opponens digiti minimi* attachment, starting approximately 22.0 mm from the proximal end.

The articulation for the fourth metacarpal is not well defined, but it is clearly obliquely oriented, is mildly convex dorsopalmarly and proximodistally, and wraps slightly onto the dorsal surface. It measures an estimated 9.0 mm dorsopalmarly by 5.0 mm proximodistally. The hamate articulation faces dorsoproximally and is flat radioulnarly and convex in the sagittal plane with the greatest convexity at its proximal-most extent. The surface extends approximately 3.0 mm onto the palmar surface and extends further dorsally on the radial side compared to the ulnar side. The ulnar side of the base has a prominent projection in the region of the attachment of several ligaments, including the pisometacarpal, palmar carpometacarpal, and ulnar collateral ligaments. On the radial side of the base is a rugosity in the area of the attachment of the palmar interosseous metacarpal ligament. On the dorsal surface, distal to the hamate facet, the nonarticular surface is raised and rugose. The ulnar side of the base has a prominent tubercle projecting 3.5 mm from the shaft in the region of the attachment of the *m. extensor carpi ulnaris*.

Pollical proximal phalanx

StW 478—Left pollical proximal phalanx

Preservation. This right pollical proximal phalanx of an adult preserves a little over half of the proximal end to

an irregular break (Figure 9.6). The surface preservation is excellent. In dorsal view it angles to the left relative to the base.

StW 575—Right pollical proximal phalanx

Preservation. This specimen is a complete left pollical proximal phalanx of a subadult individual, based on almost complete fusion of the proximal epiphysis (Figure 9.6). The surface preservation is very good with no visible damage. In the dorsal view it angles to the right relative to the base.

Morphology. The proximal half of StW 478 is absolutely larger than StW 575 (Table 9.2). In both specimens, the shaft is robust, is oval-shaped in cross-section, and shows mild longitudinal curvature (dorsally convex), especially the distal portion in StW 575. The palmar surface of the shaft is flat proximodistally and, in StW 575, slightly radioulnarly convex. The shaft is waisted relative to the ends, and in StW 575 the narrowest breadth (radioulnar = 8.2 mm) occurs close to the head, at 17.0 mm from the proximal end. On StW 575, there is a slight but distinct crest along the radial border of the shaft (this region is not well-preserved in StW 478).

The metacarpal articulation is oval, being radioulnarly broader than it is dorsopalmarly tall, although it is more circular (i.e., dorsopalmarly taller) and deeper in StW 478 (Table 9.2). The two palmar tubercles are fairly symmetrical and not strongly distinct from the palmar shaft. On the ulnar side, ridges extend dorsally from the

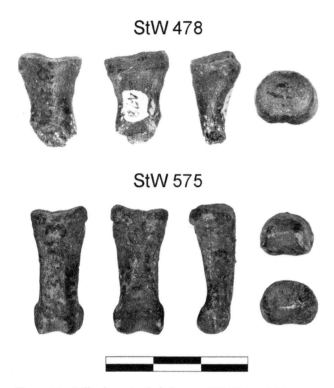

StW 478

StW 575

Figure 9.6 Pollical proximal phalanges. StW 478 is a right proximal pollical proximal phalanx (PP1) and StW 575 is a complete subadult left PP1. Both phalanges are shown in, from left to right, dorsal, palmar, radial, proximal, and, for StW 575 only, distal (above) views. Both specimens to scale.

tubercle. The head of StW 575 sits squarely on the end of the shaft. The trochlea is much wider palmarly than dorsally, and the condyles are evenly curved sagittally. In the distal view, the radial condyle is radioulnarly broader and projects more palmarly than its ulnar counterpart.

Non-pollical proximal phalanges

StW 28—Proximal phalanx

Preservation. This complete proximal phalanx of an adult individual is likely from the right fifth digit based on its small size, expanded ulnar tubercle, and asymmetry of the trochlea (Figure 9.7). It is missing fragments of bone on the palmar surface of the ulnar trochlea and the dorsal surface proximal to the head. An oblique crack runs through the shaft at about one-third of the length from the base, and small fragments of bone are missing from the radial and ulnar margins of the crack. Other than this damage, the surface preservation is very good, with no abrasion or weathering.

StW 29—Proximal phalanx

Preservation. This proximal phalanx of an adult individual preserves the base and proximal half of the shaft up to a radioulnarly oblique break (Figure 9.7). The dorsal margin

of the base and one tubercle are abraded and expose the underlying trabecular bone. Its size and symmetry of the base suggest that it is from a third or fourth digit.

StW 122—Proximal phalanx

Preservation. This proximal phalanx preserves a portion of the head and entire shaft proximally to the edge of a palmar tubercle (Figure 9.7). Without the base, it is not possible to determine whether it represents an adult or subadult, but the size and anatomy of the head indicate it was not a young immature individual. The surface is slightly abraded in areas. Based on size alone, it likely derives from digits 2, 3, or 4.

StW 293—Proximal phalanx

Preservation. This specimen is a complete proximal phalanx of an adult (Figure 9.7). Aside from some longitudinal micro-cracks and a larger crack along the dorsal midline surface of the distal shaft, it is in good condition. Its size and slight asymmetry of the basal tubercles suggest that it may represent a fourth digit from the right side.

StW 400—Proximal phalanx

Preservation. This proximal phalanx of an adult individual preserves the base and about half of the proximal shaft (Figure 9.7). Other than abrasion to the dorsal margin of the base, its preservation is good. Based on size and slight asymmetry of the basal tubercles, it likely derives from digits 3 or 4.

Morphology. The proximal phalanges show moderate longitudinal curvature (dorsally convex) with slightly greater curvature at the distal end. They are robust, with the radioulnar breadth of the shaft that is approximately as broad as that of the head (Table 9.2). The trochleae do not project far from the palmar surface of the shaft. StW 293 is the only specimen to preserve a complete trochlea and it slightly asymmetric with the ulnar side flaring laterally slightly more than the radial side.

In StW 293, the margins of the shaft taper slightly distally such that the narrowest radioulnar breadth is just proximal to the trochlea, while in StW 28 and 122 the margins run parallel throughout the proximodistal length of the shaft. The palmar surfaces of the shafts are slightly proximodistally concave in StW 29 and 400, while generally flat in StW 122. In StW 293 and, especially StW 28, the palmar surface is proximodistally concave but also slightly radioulnarly convex, particularly at the proximal end where there this convexity radiates from the ulnar (StW 28) or radial (StW 293) tubercles. The flexor sheath ridges are generally distinct and run most of the proximodistal length of the shafts but do not project far beyond the

Table 9.2 Linear measurements Sterkfontein phalanges. All measurements in mm.

Specimen	Element	Side	TL	IA	BB	BH	PAB	PAH	MSB	MSH	TB	TH
StW 478	PP1	L	19.5 pres.	–	13.6	11.3	12.0	9.5	–	–	–	–
StW 575	PP1	R	27.0	–	12.3	9.3	10.6	8.3	8.3	7.1	10.9	6.8
StW 28	PP5	R	32.5	30.8	12.6	12.6	10.5	9.3	8.5	5.7	[10.4]	[6.9]
StW 29	PP3 or 4	?	26.6 pres. [28.3]	–	12.7	9.2	10.4	7.6	[9.9]	[6.0]	–	–
StW 122	PP2-4	?	36.4 pres.	–	–	–	–	–	9.6	6.2	[10.2]	–
StW 293	PP4	R?	36.3	34.7	12.9	10.2	9.2	8.1	10.2	6.3	9.7	7.1
StW 400	PP3 or 4	?	30.4 pres.	–	12.9	10.9	[10.9]	8.6	10.6	6.4	–	–
StW 331	IP2-4	?	30.3	27.7	13.1	9.9	10.5	6.8	9.3	5.6	8.8	6.2
StW 294	DP1	?	15.9	–	10.5	[7.0]	8.9	4.9	6.0	4.5	8.6	3.9

Abbreviations: 'PP', proximal phalanx; 'IP', intermediate phalanx; 'DP', distal phalanx; 'pres.', preserved. For additional abbreviations, see Table 9.1.

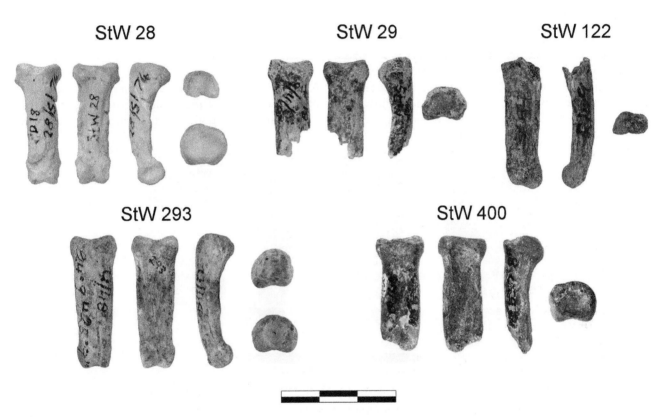

Figure 9.7 Non-pollical proximal phalanges. StW 28 is a likely right fifth proximal phalanx (PP5), StW 29 and StW 400 are likely either a PP3 or PP4, StW 122 likely derives from digits 2, 3, or 4, and StW 293 is possibly a right PP4. Most phalanges are shown in, from left to right, dorsal, palmar, sagittal (proximal toward the top), distal (above), and proximal (below) views. StW 122 shown in palmar, sagittal, and distal views only. All phalanges are to scale.

palmar surface of the remaining shaft. They are only clearly defined and preserved in three specimens: StW 28, 29, and 293. In StW 28, only the radial flexor sheath ridge is distinct, starting 10.7 mm from the proximal end and running 8.8 mm distally, but does not project beyond the palmar surface because of the radioulnar convexity of the shaft. In StW 29, the left one (when viewing palmarly) is preserved and starts 12.0 mm from the proximal end and runs at least 15.0 mm distally to where the specimen is broken. In StW 293, the ulnar flexor sheath ridge starts 14.0 mm from the proximal end and extends 10.4 mm distally, while the radial ridge, which is more well defined, starts 10.4 mm from the proximal end and extends 12.4 mm distally.

The bases have oval, radioulnarly wide, concave articulations for the metacarpal heads. The palmar tubercles are well developed and less symmetrical in some specimens (i.e., StW 28 and 400) than others (i.e., StW 293). StW 28 has a particularly prominent ulnar tubercle, projecting about 3.0 mm from the shaft, for the attachment of the *mm. abductor digiti minimi* and *flexor digiti minimi brevis*.

Manual intermediate phalanx

StW 331—Intermediate phalanx

Preservation. This specimen is a complete intermediate phalanx of an adult individual (Figure 9.8; Table 9.2). There is no visible damage and the surface preservation is good. Based on size and relative symmetry, it is likely from digits 2, 3, or 4.

Morphology. The trochlear condyles are more or less circular in the sagittal view, and the proximal extent of the articular surface is almost as great dorsally and palmarly. The condyles are slightly asymmetric with, when viewing palmarly, the right condyle being slightly larger and extending slightly more distally than the left condyle.

The shaft has slight longitudinal curvature (dorsally convex), but most of this curvature is accentuated distally. In the lateral view, the proximodorsal border is concave, flaring dorsally and extending further distally then the palmar half of the bone. In the palmar view, the shaft margins taper distally, with the narrowest radioulnar breadth (8.2 mm) occurring at about one-third of the length from the distal end, and is marginally narrower than the head (8.8 mm). The possible attachment sites of the *m. flexor digitorum superficialis* tendons (Marzke et al., 2007) are projecting, semicircular-shaped and radioulnarly broad (2.1 mm in maximum thickness). The raised insertion sites occupy the proximal half of the shaft. The distal half of the palmar shaft has a median bar that is radioulnarly convex, which makes the proximal surface appear depressed relative to the flexor attachments.

The proximal articular surface has paired oval-shaped depressions separated by a shallow ridge and is slightly asymmetric. The entire proximal margin is bounded by a proximodistally thick (2.0–3.0 mm) rim that projects from the shaft.

Pollical distal phalanx

StW 294—Pollical distal phalanx

Preservation. This specimen is a complete pollical distal phalanx of an adult individual (Figure 9.9). It is likely to represent a left digit based on the asymmetry of the

Figure 9.8 Intermediate manual phalanx. StW 331 is a complete intermediate phalanx likely from digits 2, 3, or 4, shown in, from left to right, dorsal, palmar (proximal is toward the top), sagittal, distal (above), and proximal (below) views.

Figure 9.9 Distal pollical proximal phalanx. StW 294 is a complete, likely left, distal pollical phalanx shown in, from left to right, dorsal, palmar, radial (proximal is toward the top) and proximal views.

palmar fossa. There is a small fragment missing from the distoulnar region of the apical tuft and a small fragment from the palmar-ulnar border of the base, which obscures the full extent of the proximal articular facet in that region. Otherwise, the surface preservation is good.

Morphology. The ungual tuberosity or apical tuft is almost as wide as the base, although it is dorsopalmarly flat (Table 9.2). The palmar surface of the tuft is continuous with the shaft and lacks an ungual fossa and ungual spines. The ridge for the attachment of the *m. flexor pollicis longus* lies just proximal to the tuft at about two-thirds of the length (7.7 mm) from the proximal end. Proximal to that is a large palmar fossa, measuring 6.6 mm radioulnarly and 6.5 mm proximodistally, occupying the proximal half of the palmar shaft. The margins of the fossa are slightly asymmetric with a more prominent radial margin, although there is some abrasion to the ulnar side. A peripheral rim surrounds the base and is slightly accentuated dorsally in the region of the attachment of the *m. extensor digitorum communis* tendon. The shallow articular depressions of the proximal surface are moderately concave dorsopalmarly.

Comparative anatomy

Materials and methods

Qualitative and quantitative comparisons of the Sterkfontein metacarpals and manual phalanges were made to all available, original fossil specimens of *Ardipithecus ramidus*, *Australopithecus afarensis*, *Australopithecus sediba*, the Swartkrans sample attributed to either *Australopithecus* (*Paranthropus*) *robustus* or early *Homo*, *Homo naledi*, *Homo habilis*, *Homo neanderthalensis*, and early *Homo sapiens*, as well as high-quality 3D prints of *Homo floresiensis* and a cast of KNM-WT 51260 (*Homo* sp.; Ward et al., 2013).

Comparative quantitative analyses were conducted using two comparative extant samples. First, the morphometric analyses (i.e., box-and-whisker plots) were conducted using the comparative metacarpal and phalangeal sample described in Tables 9.3 and 9.4, respectively. Most morphometric measurements were taken directly on original fossils. Data for *A. ramidus* specimens are taken from published values in Lovejoy et al. (2009), as well as additional measurements taken on original fossils by one of the authors (TLK) and G. Suwa. Some of the values published in Lovejoy et al. (2009) have been adjusted after remeasurement of the original fossil (see Kivell et al., 2018b, for details). Additional comparative data was taken from the literature for *Homo* sp. (Ward et al., 2013; Domínguez-Rodrigo et al., 2015; Lorenzo et al., 2015), *H. antecessor* (Lorenzo et al., 1999), *H. neanderthalensis* (Trinkaus, 1983, 2016a; Niewoehner et al., 1997; Trinkaus and Walker, 2017), and early *H. sapiens* (Sladek

et al., 2000; Trinkaus et al., 2001, 2010, 2014; Groucutt et al., 2018). Quantitative comparisons are made via shape ratios to account for differences in absolute size across the Sterkfontein and comparative samples. For most comparisons, this ratio is relative to interarticular proximodistal length of the metacarpal or to total proximodistal length of the phalanx. The extant comparative sample derived from adult, non-pathological, associated hand skeletons from a diverse sample of recent humans, including robust Tierra del Fuegians and smaller-bodied Khoisan individuals, *Pan troglodytes*, *Pan pansicus*, and *Gorilla gorilla* (Tables 9.3 and 9.4).

Second, quantitative analyses of the intrinsic hand proportions were conducted using a different comparative sample, which included metacarpals of modern humans (n = 86), *Pan* (n = 24), *Gorilla* (n = 26), and *Pongo* (n = 42), as well as phalanges from a subsample of these groups (humans, n = 10; *Pan*, n = 15; *Gorilla*, n = 11; *Pongo*, n = 20) (Table 9.5). All specimens were adults of known sex, with no pathology in the appendicular skeleton. Morphometric measurements for fossil taxa were taken on original fossils or from published values (Kivell et al., 2011; Ward et al., 2013; Kivell et al., 2015). All Sterkfontein hand specimens that were complete enough to confidently measure interarticular length were included in randomization analysis (see following paragraph). This included one first metacarpal (StW 418), two third metacarpals (StW 64 and 68), one pollical proximal phalanx (StW 575), two complete non-pollical proximal phalanges (StW 28 and 293), and one intermediate phalanx (StW 331). The extant hominoid sample included the same elements (Table 9.5) and the assignment of the phalanges to a particular ray was determined by size and morphological differences (e.g., Susman, 1979; Christensen, 2009).

To assess intrinsic hand proportions among the Sterkfontein sample, we used a randomization approach that accounts for the inability to securely identify isolated phalangeal elements by individual, side, or ray, following a simplified but similar protocol used by Rolian and Gordon (2013). This protocol generated three morphological indices of intrinsic hand proportions based on all possible combinations of hand elements using both the fossil and extant samples. Randomization generates a more conservative range of metacarpal, phalangeal, and digital ray proportions for extant hominoids to which we could compare the Sterkfontein hand assemblage and determine which resampled intrinsic hand proportions overlap.

The resampling strategy followed three steps. First, from the extant hominoid samples, we randomly drew a subsample matching the size and composition of the fossil hand sample. That is, we randomly selected one first metacarpal (MC1), two MC3s, one pollical proximal phalanx (PP1), two non-pollical proximal phalanges (PPs),

Table 9.3 Comparative metacarpal sample for quantitative analyses.

	Element				
	MC1	MC2	MC3	MC4	MC5
Site/taxon					
Sterkfontein	StW 418 StW 583	StW 382 StW 664	StW 64 StW 68	StW 26 StW 65 StW 292 StW 330 StW 552	StW 63
Swartkrans	SK 84 SKX 5020			SKX 2954 SK 85	SK(W) 14147
Ar. ramidus	ARA-VP-6/500-015 ARA-VP-6/1638			ARA-VP-7/2G ARA-VP-6/500-010	ARA-VP-6/500-036
A. afarensis	A.L. 333w-39 A.L. 333-58	A.L 333-48 A.L. 438-1f A.L. 438-1e	A.L. 333-16 A.L. 438-1d	A.L. 333-18 A.L. 333-56 A.L. 333-122	A.L. 333-14 A.L. 333-89 A.L. 333-141
A. sediba	U.W. 88-119	U.W. 88-115	U.W. 88-116 U.W. 88-112	U.W. 88-117	U.W. 88-118
H. erectus			KNM-WT 51260		
H. naledi	U.W. 101-007 U.W. 101-270 U.W. 101-917 U.W. 101-1321 U.W. 101-1641 U.W. 101-1282	U.W. 101-1320	U.W. 101-1319	U.W. 101-1318 U.W. 102a-028	U.W. 101-1309
H. neanderthalensis	Kebara 2 Tabun C1 Amud 1-57 Palomas 96T Shanidar 4	La Chapelle La Ferrassie 1 & 2 Regourdou 1 Shanidar 4 & 5 Spy 2 Spy 21A Tabun C1-160 Kebara 2 Moula-Guercy M-G2-648	La Chapelle La Ferrassie 1 & 2 Regourdou 1 Shanidar 4 & 6 Kebara 2 Amud 1 Tabun C1-151 Spy 22A Moula-Guercy M-D3-78 Palomas 96RR	Shanidar 4 &5 Spy 21C Spy 22C Tabun 1-166 Palomas 96X	Tabun C1-164 Shanidar 4 &5
early *H. sapiens*	Qafzeh 9 Ohalo II H2 Barma Grande 2 Arene Candide 2 Dolni Vestonice 3 Dolni Vestonice 16 Pavlov 31	Qafzeh 9 Ohalo II H2 Arene Candide 2 Dolni Vestonice 13 Dolni Vestonice 15 Dolni Vestonice 16 Dolni Vestonice 59 Pavlov 31 Palomas 96U	Qafzeh 8 Qafzeh 9 Ohalo II H2 Barma Grande 2 Arene Candide 2 Dolni Vestonice 13 Dolni Vestonice 16 Dolni Vestonice 58	Qafzeh 8 Qafzeh 9 Ohalo II H2 Barma Grande 2 Arene Candide 2 Dolni Vestonice 3 Dolni Vestonice 15 Dolni Vestonice 16	Qafzeh 9 Ohalo II H2 Barma Grande 2 Dolni Vestonice 16 Arene Candide 2
recent *H. sapiens*	n=43	n=45	n=42	n=40	n=37
small-bodied *H. sapiens*	n=25	n=25	n=25	n=25	n=25
P. troglodytes	n=10	n=11	n=12	n=12	n=11
P. paniscus	n=11	n=11	n=11	n=11	n=11
Gorilla	n=9	n=10	n=11	n=11	n=9

Table 9.4 Comparative manual phalangeal sample for quantitative analyses.

		Element		
	PP1	PP2-5	IP2-5	DP1
Site/taxon				
Sterkfontein	StW 478	StW 28	StW 331	StW 294
	StW 575	StW 29	StW 620	StW 617
		StW 122	StW 635	
		StW 293	StW 657	
		StW 400		
		StW 597		
		StW 668		
Swartkrans/Kromdraai		SKX 5018	SKX 5019	SKX 5016
		SKX 15468	SKX 5021	TM 1517k
		SKX 22741	SKX 9449	
		SKX 27431	SKX 13476	
			SKX 35439	
			SKX 36712	
Ar. ramidus	ARA-VP-7/21	ARA-VP-6/500-030	ARA-VP-6/500-002	ARA-VP-6/500-049
		ARA-VP-6/500-022	ARA-VP-6/500-059	
		ARA-VP-6/500-069	ARA-VP-6/500-078	
		ARA-VP-7/2H	ARA-VP-6/500-092	
		ARA-VP6/507		
A. afarensis	A.L. 333-69	A.L. 288-1x	A.L. 333x-18	A.L. 333-159
	A.L. 438-4	A.L. 333w-4	A.L. 333-32	
		A.L. 333-19	A.L. 333-46	
		A.L. 333-57	A.L. 333-64	
		A.L. 333-62	A.L. 333-88	
		A.L. 333-63	A.L. 333-49	
		A.L. 333-93	A.L. 333-50	
		A.L. 1044-1		
		A.L. 444-4		
A. sediba	U.W. 88-91	U.W. 88-108	U.W. 88-122	U.W. 88-124
	U.W. 88-160	U.W. 88-120	U.W. 88-161	
		U.W. 88-121	U.W. 88-162	
		U.W. 88-164		
H. habilis			OH 7 FLK NN-D	OH 7 FLK NN-A
			OH 7 FLK NN-E	
			OH 7 FLK NN-F	
			OH 7 FLK NN-G	
H. erectus?		OH 86		
Homo sp.		ATE9-2		
H. antecessor			ATD6-28	
			ATD6-53	
H. naledi	U.W. 101-428	U.W. 101-558	U.W. 101-381	U.W. 101-1351
	U.W. 101-1055	U.W. 101-754	U.W. 101-924	U.W. 101-1453
	U.W. 101-1721	U.W. 101-923	U.W. 101-1027	
		U.W. 101-1025	U.W. 101-1308	
		U.W. 101-1326	U.W. 101-1310	
		U.W. 101-1327	U.W. 101-1311	
		U.W. 101-1328	U.W. 101-1325	
		U.W. 101-1454	U.W. 101-1646	
		U.W. 101-1460	U.W. 101-1647	
		U.W. 101-1644	U.W. 101-1648	
		U.W. 101-1645		
		U.W. 101-1725		
H. floresiensis		LB6/8	LB1/48	LB1/55
			LB6/9	

(continued)

Table 9.4   Continued

	Element			
	PP1	PP2-5	IP2-5	DP1
H. neanderthalensis	Kebara 2 Shanidar 4, 5, 6 Tabun 1 Moula Guersy M-E1-123 Spy 25H	Kebara 2 Shanidar 4, 5, 6 Tabun 1 Spy 24A, B, C Spy 426a Spy 748a Spy 766a Palomas 92T, U Palomas 96 SP96W Palomas 96 SP96AA Palomas 96 SP96YY	Kebara 2 Shanidar 3, 4, 5, 6 Amud 1 Tabun 1 Moula Guersy M-G1-154 Spy 222b Spy 390a Spy 430a Spy 484a Krapina 205.1-205.10 Krapina 205.12-205.19 Krapina 205.22-205.24 Palomas 65 Palomas 92W Palomas 96CC, DD, ZZ	Kebara 2 Shanidar 3, 4, 5, 6
early *H. sapiens*	Dolni Vestonice 14 Dolni Vestonice 16 Ohalo II H2 Qafzeh 9 Barma Grande 2 Arene Candide 2	Dolni Vestonice 3 Dolni Vestonice 13, 14, 15, 16 Ohalo II H2 Qafzeh 8 Qafzeh 9 Barma Grande 2 Arene Candide 2	Dolni Vestonice 3 Dolni Vestonice 13, 14, 15, 16 Dolni Vestonice 34 Dolni Vestonice 53 Ohalo II H2 Qafzeh 8 Qafzeh 9 Al Wusta-1 Barma Grande 2 Arene Candide 2 Skhul IV Tianyuan 1 Sunghir 1 Pavlov 33 Caldeirao 9	Dolni Vestonice 16 Ohalo II H2 Qafzeh 9 Palomas 96DDD Barma Grande 2 Arene Candide 2
recent *H. sapiens*	n =38	n = 19	n = 15	n = 34
small-bodied *H. sapiens*	n = 24	n = 6	n = 6	n = 22
P. troglodytes	n = 10	n = 6	n = 4	n = 1
P. paniscus	n = 10	n = 5	n = 4	n = 6
Gorilla	n = 8	n = 11	n = 9	n = 2

Table 9.5   Comparative sample of extant hominoids for analysis of intrinsic hand proportions.

Taxon	Analysis	MC1	MC3	PP1	PP2-5	IP2-5
Homo sapiens	Intrinsic ratios	n = 86	n = 86	n = 10	n = 10	n = 10
	resampling	n = 29	n = 10	n = 10	n = 12	n = 12
Pan[a]	Intrinsic ratios	n = 24	n = 24	n = 15	n = 15	n = 15
	resampling	n = 22	n = 17	n =15	n = 21	n = 22
Gorilla[b]	Intrinsic ratios	n = 26	n = 26	n = 11	n = 11	n = 11
	resampling	n = 23	n = 12	n = 11	n = 12	n = 11
Pongo[c]	Intrinsic ratios	n = 42	n = 42	n = 20	n = 20	n = 20
	resampling	n = 21	n = 21	n = 19	n = 20	n = 20

[a]*Pan* sample includes *Pan troglodytes*.
[b]*Gorilla* sample includes *Gorilla beringei* and *Gorilla gorilla*.
[c]*Pongo* sample include *Pongo pygmaeus* and *Pongo abelii*.

and one intermediate phalanx (IP). Second, we calculated the metrics of manual proportions from the extant subsamples. These included three length ratios: (1) a metacarpal ratio, which is the ratio of MC1 to MC3 length; (2) a phalangeal ratio, which is the ratio of PP1 to ulnar PP length; and (3) a digital ratio, which is the ratio of first ray (MC1 + PP1) to ulnar ray (MC3 + PP + IP) length. For elements that have duplicates in the Sterkfontein sample (i.e., MC3 and ulnar PPs), we used a geometric mean of the interarticular lengths to calculate the metacarpal and digital ratios. For each metric, we took the natural logarithm of the ratio, to normalize data derived from the ratio of two positive numbers (see Smith, 1999; Green et al., 2007). Third, we repeated the first two steps, resampling the extant subsamples to derive distributions of manual proportions and their confidence intervals for the extant hominoids. We used all possible combinations (i.e., exact resampling) to calculate the metacarpal and phalangeal ratios, and resampled 5,000 times for the digital ratio (as 5,000 is less than 10% of the total possible combinations for this ratio). Each resulting distribution, excluding the upper and lower 2.5% of ratios, provides 95% confidence intervals, which can be used to assess the morphological similarity of the Sterkfontein intrinsic hand proportions. Ratios for the Sterkfontein sample that fell outside the 95% confidence interval of the extant hominoid distribution were considered significantly different.

Results

First metacarpal

In absolute length, the StW 418 first metacarpal (MC1) is slightly longer than *Ar. ramidus* ARA-VP-6/500-014, *A. sediba* MH2, and SK 84, but shorter than ARA-VP-6/1638, *A. afarensis* A.L. 333w-39, and later *Homo*, including most *H. naledi* specimens. The overall morphology of the StW 418 is notably more gracile than that of SK 84, SKX 5020, and later *Homo*, including *H. naledi*. A comparison of radioulnar breadth at midshaft relative to interarticular length shows that StW 418 is only more robust than A.L. 333w-39 and MH2 but falls within the lower range of variation of recent humans and *H. naledi* (Figure 9.10).

The distinct *m. opponens pollicis* flange of StW 418 distinguishes it from A.L. 333w-39 and MH2, neither of which presents well-developed entheses (Figure 9.1). This difference in entheseal morphology may suggest more habitual and/or powerful abduction and flexion of the thumb in the StW 418 individual than that of MH2 or the A.L. 333w-39 individual, as demonstrated by a recent study that found a correlation between modern human entheseal morphology and the intensity of manual labor (Karakostis et al., 2017). However, it is important

to note that no relationship between entheseal area and *m. opponens pollicis* muscle architecture was found in elderly, cadaveric human samples (Williams-Hatala et al., 2016) and that several functional and non-functional factors can influence the size and shape of entheses (Milella et al., 2012; Rabey et al., 2015).

The distally extended *m. first dorsal interosseous* insertion is similar to MH2, SK 84, and SKX 5020 and later *Homo* (Musgrave, 1971; Tocheri et al., 2008; Jacofsky, 2009). This extended tendon insertion is typical of the condition found in humans and unlike the more proximally limited insertion found in *A. afarensis* (*contra* Bush et al., 1982) and African apes, in which the *m. first dorsal interosseous* is much smaller (Tocheri et al., 2008; Jacofsky, 2009). The entheseal position of StW 418, as in humans and most other hominins, would provide a longer moment arm for adduction of the thumb (Tocheri et al., 2008).

In relative radioulnar breadth of the MC1 base, StW 418 is intermediate, being most similar to *A. afarensis* and falling within the upper range of variation in *P. paniscus* and lower range of variation in modern humans and other African apes (Figure 9.10). It is relatively broader than MH2 and *H. naledi* but narrower than SK 84 and SKX 5020, Neandertals, and early *H. sapiens*. The trapezium facet overall is strongly curved, being more radioulnarly convex than it is dorsopalmarly concave, as in African apes, *A. afarensis*, and SK 84, and unlike the radioulnarly flatter and broader (inferred) morphology of OH 7, Neandertals, and *H. sapiens*. Previous studies that have quantified this curvature using 3D methods confirm these comparative differences, showing similarly strong articular curvature in StW 418, *A. afarensis*, and SK 84 (Marzke et al., 2010; Marchi et al., 2017). In StW 418, the ulnar border of saddle-shaped trapezium facet extends farther distally than its radial border, but this asymmetry is not as accentuated as it is in *Ar. ramidus* and A.L. 333w-39. The base does not have a "palmar beak" as in A.L. 333w-39 or MH2, and instead looks more similar to *H. naledi*, although its articular surface is absolutely larger than that of U.W. 101-1321, 101-1641, and 101-1282. Altogether, the strongly curved articular morphology suggests a greater degree of stability but more limited range of motion and proportionally less joint surface area to cope with axial loading of the thumb at the MC1-trapezium articulation compared to that of humans (Hamrick, 1996; Marzke et al., 2010; Marchi et al., 2017). The absence of a "palmar beak" suggests less resistance to dorsal sliding of the MC1 during pinch grips than that of *A. afarensis*, MH2, or SK 84 (Marzke et al., 2010). However, it is important to emphasize that relative stability of the MC1-trapezium joint is also largely dependent on the ligamentous structure (Bettinger et al., 1999; Bettinger and Berger, 2001), which does not preserve in the fossil record.

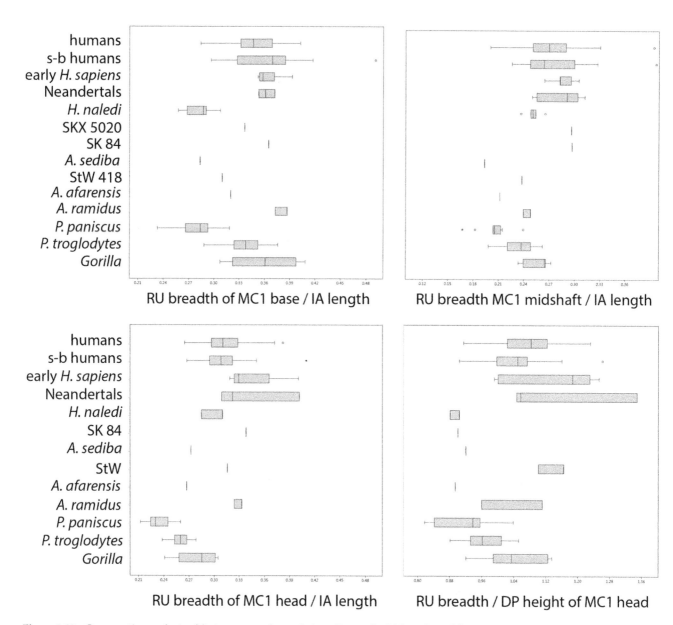

Figure 9.10 Comparative analysis of first metacarpal morphology. Box-and-whisker plots of first metacarpal (MC1) shape, in which each variable is shown as a ratio of interarticular (IA) length: radioulnar (RU) breadth of the MC1 base (top, left), midshaft (top, right), and head (bottom left). To include StW 583, which is only a distal MC1, RU breadth relative to dorsopalmar (DP) height of the MC1 head is also compared (bottom, right). See Table 9.3 for details of the comparative extant and fossil sample. Here and in all box-and-whisker plots that follow, the box represents the 25th and 75th percentiles, the center line represents the median, the whiskers represent the non-outlier range, open circles represent outliers, and asterisks represent extreme outliers.

The MC1 heads of StW 418 and StW 583 have a palmar midline ridge with shallow grooves on either side for radial and ulnar sesamoid bones (Figure 9.1). These midline ridges are slightly more pronounced than that of A.L. 333w-39 but far less so than the prominent "beaks" of MH2 and SK 84. In the radial view, the dorsopalmar curvature of the distal articular surface is intermediate between the more rounded facet of African apes and A.L. 333w-39 and the flatter facet of MH2, SK 84, and

particularly *H. naledi*. The articular surfaces of StW 418 and 583 are not extended as far palmarly or dorsally as in *Ar. ramidus* or A.L. 333w-39, but they extend farther onto the dorsal surface than in MH2, *H. naledi*, and later *Homo*.

Radioulnar articular head breadth relative to interarticular length shows that StW 418 is most similar to the median values of Neandertal and *H. sapiens*, being relatively radioulnarly broader than African apes, *A. afarensis*, MH2, and *H. naledi* (although the maximum radioulnar breadth of the *H. naledi* MC1 head is notably broader, the breadth of the articular surface is absolutely and relatively smaller than StW 418) but narrower than *Ar. ramidus* and SK 84 (Figure 9.10). StW 583 preserves only the distal portion of the MC1 and thus a comparative analysis of head articular breadth relative to dorsopalmar height reveals that StW 583 and StW 418 are broadly similar in their morphology, with StW 418 having a relatively broader head (Figure 9.10). Both specimens are most similar to Neandertals and *H. sapiens*, but also overlap with *Gorilla*, in having MC1 heads that are radioulnarly broader than they are tall. In contrast, A.L. 333w-19 and all other South African specimens, particularly MH2 and SK 84 with a palmar "beak," have dorsopalmarly taller heads. Overall, the curved, broad distal articular morphology suggests that StW 418 may have had a larger range of motion and a less stable metacarpophalangeal joint than that of Neandertals and humans, but one that potentially functioned slightly differently from that of *A. sediba* and SK 84, although the biomechanical implications, if any, of the distal "beak" are unclear (Galletta et al., 2019).

Second metacarpal

StW 382 is large, robust specimen that is almost identical in absolute length to the large *A. afarensis* A.L. 438-1e/f MC2 specimens but is absolutely longer than StW 664 (Pickering et al., 2018), A.L. 333-48, MH2, and *H. naledi* Hand 1. In comparison to *Pan*, StW 382 is shorter, more robust, and relatively straight, which is most similar humans and other hominins. Among hominins, StW 382 has a slight proximodistal curvature of the shaft, particularly at the distal end (Figure 9.2), which is similar to *H. naledi* (this region is not well preserved in StW 664), and it is more curved than *A. afarensis*, MH2, Neandertals, and *H. sapiens*. The interosseous crest morphology in StW 382 is less prominent than that of most *H. naledi* specimens but more well defined than that of *A. afarensis* and converges closer to the proximal end than in MH2 (where they converge just proximal to the midshaft). In relative radioulnar breadth at the midshaft, StW 382 is among the broadest in the sample, with only *A. afarensis* A.L. 438-1e/f, *H. naledi*, and *Gorilla* being broader, although this specimen does fall within the high range of variation in Neandertals and *H. sapiens* (Figure 9.11).

StW 382 has slightly more well-developed muscle insertions along the dorsal base than StW 664, A.L. 438-1e (this region is not well preserved in A.L. 438-1f), and MH2, and much more developed than the more gracile morphology of A.L. 333-48. StW 382 is almost identical in relative radioulnar breadth of the base to that of StW 664, both of which are narrower than the median values of all other fossil hominins but fall within the low range of

variation of recent *H. sapiens* and *Gorilla* (Figure 9.11). The angle of the V-shaped trapezoid articulation is similar to (i.e., greater than 90°) StW 664, A.L. 438-1e/f, and MH2. It is more obtuse than A.L. 333-48 and *H. naledi* U.W. 101-1474 but more strongly V-shaped than Neandertals and humans. Like most other hominins, StW 382 has a dorsopalmarly continuous articulation for the capitate suggesting the absence of the carpometacarpal ligament that is found in African apes, and thus permitting greater rotational capability at the MC2-capitate joint. The degree of dorsopalmar curvature of the capitate facet is similar to StW 664 and MH2 but less accentuated than that of *A. afarensis*, *H. naledi*, and later *Homo*. In the dorsal view, the StW 382 trapezium facet is oriented primarily laterally (~22°), which is similar to *A. afarensis* and African apes and differs from the more proximally oriented facets of MH2, U.W.101-1474, and humans (Drapeau et al., 2005; Drapeau, 2012; Kivell et al., 2011, 2015). In proximal view, the StW 382 trapezium facet is less palmarly oriented (~17°) than all other hominins, including *A. afarensis*, and is most similar to *Pan* (Drapeau et al., 2005). Thus, the trapezium articular morphology of the StW 382 base is most similar to *Pan*, and generally primitive compared to australopiths and *Homo*. Many have drawn inferences about the shape and function of the radial carpometacarpus from the articular morphology of the MC2 base (e.g., Marzke et al., 1992; Niewoehner et al., 1997; Drapeau et al., 2005; Niewoehner, 2006; Tocheri, 2007; Tocheri et al., 2008). The StW 382 morphology suggests it may have been associated with a palmarly narrow, wedge-shaped trapezoid (rather than the palmarly broad, boot-shaped trapezoid of humans, Neandertals, and *H. naledi*), with more limited pronation of the index finger and a more limited ability to cope with axial loads from the thumb than later *Homo* and possibly *A. afarensis* (Marzke et al., 1992; Niewoehner et al., 1997; Drapeau et al., 2005;). However, the radial carpometacarpal region is complex; there is substantial variation in different aspects of its morphology across hominins (e.g., Niewoehner et al., 1997; Niewoehner, 2006; Kivell et al., 2015; Trinkaus, 2016b), and different aspects of its morphology may compensate for others to produce similar overall function. Thus, any inferences about the shape and/or orientation of the carpus and thumb in the absence of fossil evidence should be treated with caution.

In palmar view, the curvature of the distal articulation in StW 382 is strongly radioulnarly asymmetric, which is similar to the morphology of A.L. 438-1e/f (Drapeau, 2012), MH2, and *H. naledi* but unlike the more symmetrical head of A.L. 333-48. In the sagittal view, the distal articulation is more strongly convex dorsally and flatter palmarly, such that the head projects palmarly more than that of *A. afarensis*, and is similar in this way to MH2 and

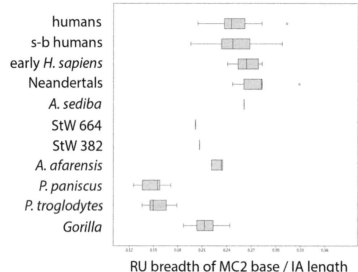

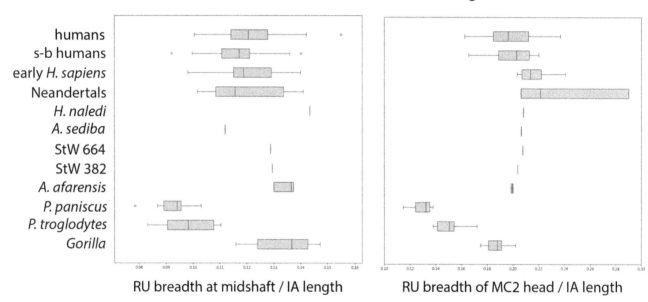

Figure 9.11 Comparative analysis of second metacarpal morphology. Box-and-whisker plots of second metacarpal (MC2) shape, in which each variable is shown as a ratio of interarticular (IA) length: radioulnar (RU) breadth of the MC2 base (top), midshaft (bottom, left), and head (bottom right). See Table 9.3 for details of the comparative extant and fossil sample. Note that the IA value for StW 664 is measured using ImageJ from the published image in Pickering et al. (2018), as they only provide total length.

H. naledi. This asymmetrical shape to the articular surface creates a proximodistally longer articular surface than the more rounded and shorter articulation of A.L. 333-48, MH2, and *H. naledi* Hand 1. In relative radioulnar head breadth, StW 382, StW 664, australopiths, and *H. naledi* are all virtually identical, falling out as more broad than African apes but less so than Neandertals, and most similar to the median values of early and recent *H. sapiens* (Figure 9.11). Altogether, the distal articulation of StW 382 suggests ulnar rotation (i.e., pronation) of the proximal phalanx during flexion that would facilitate opposition to the thumb and other fingers as in other hominins (Lewis, 1989; Marzke, 1997).

Third metacarpal

StW 64 and StW 68 are absolutely proximodistally longer than MH2 and *H. naledi* Hand 1 but shorter than the complete *A. afarensis* MC3 specimens, KNM-WT 51260 (Ward et al., 2013), and later *Homo*. The slight proximodistal curvature of shaft in StW 64 and 68 (Figure 9.3) is similar across the comparative fossil sample, apart from *H. naledi* Hand 1, which is more curved. Interosseous crests along the dorsum are more well defined than those of the *A. afarensis* specimens, but not as prominent as in MH2, *H. naledi*, or KNM-WT 51260. Relative to interarticular length, StW 68 is slightly radioulnarly broader at

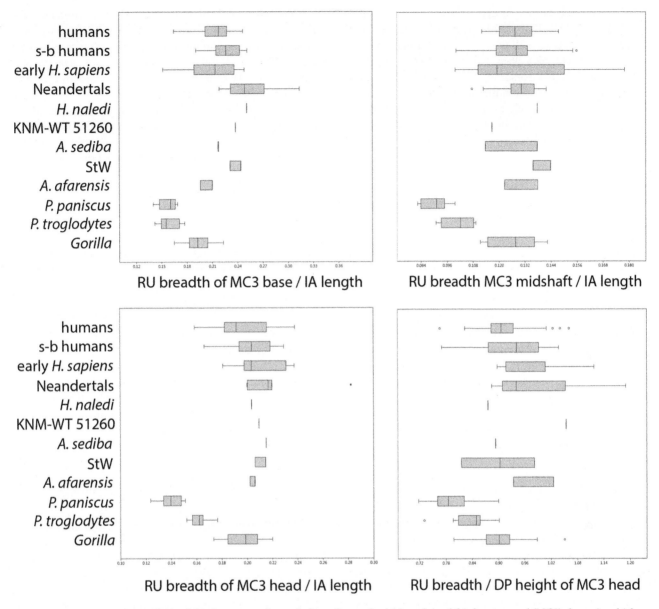

Figure 9.12 Comparative analysis of third metacarpal morphology. Box-and-whisker plots of third metacarpal (MC3) shape, in which each variable is shown as a ratio of interarticular (IA) length: radioulnar (RU) breadth of the Mc3 base (top, left), midshaft (top, right), and head (bottom left). To include StW 27 and 394, which only preserve the distal MC3, RU breadth relative to dorsopalmar (DP) height of the MC3 head is also compared (bottom, right). See Table 9.3 for details of the comparative extant and fossil sample. Note that the *A. sediba* MH1 subadult U.W. 88-112 specimen is missing its distal epiphysis and thus IA length is only an estimation. All variables were also analyzed using total length that includes the MC3 styloid process in KNM-WT 51260, *H. sapiens* and Neandertals and the relative relationships among taxa did not change.

midshaft than StW 64 (Figure 9.12). Both specimens are broader than MH2 (but similar to the estimated value of the juvenile MH1 MC3), A.L. 333-16, and KNM-WT 51260. StW 64 and 68 fall out as most similar to *H. naledi* Hand 1 and within the upper range of variation found in Neandertals and *H. sapiens*, as well as *Gorilla*.

Proximally, neither StW 64 nor 68 show the strongly concavoconvex capitate articulation typical of African

apes and instead are flat or shallowly concave like that of other hominins and humans (McHenry, 1983; Begun, 2004). The palmar half of the StW 64 and 68 capitate articulations are more radioulnarly expanded than the palmarly narrow and triangular-shaped articulations in *Ar. ramidus*, *A. afarensis*, and possibly some specimens of *H. naledi* (the palmar region appears narrow but is not well preserved in either U.W. 101-1319 or 101-1749, and is more radioulnarly expanded in juvenile MC3 specimen U.W. 101-559) (Rein and Havarti, 2012). In this way, the capitate facets of StW 64 and 68 are similar in overall shape to MH2, SKX 3646, KNM-WT 51260, and later *Homo*. In StW 64, the dorsoradial border extends slightly more proximally than the remainder of the base and has been described previously as a styloid process (Ricklan, 1987) (Figure 9.3). This morphology is similar to SKX 3646, which has also been described as a "faint styloid process" (Susman, 1988: 162), and *H. naledi* Hand 1, but this morphology is distinct from the more projecting styloid process of Neandertals and, particularly KNM-WT 51260 (Ward et al., 2013) and *H. sapiens*. The relative radioulnar breadth of the StW 64 and 68 bases is similar to KNM-ER 51260, *H. naledi*, and Neandertals in being broader than *A. afarensis*, MH2, and African apes and falling out in the upper range of variation found in *H. sapiens* (Figure 9.12). The continuous, dorsopalmarly concave and radially oriented MC2 facet in StW 64 suggests the absence of the carpometacarpal ligament that is found in apes. The MC2 facet morphology is similar to that of SKX 3646 and MH2 and unlike the more radiopalmarly oriented MC2 facets of *A. afarensis* and *H. naledi* Hand 1. Altogether, the proximal MC3 morphology of StW 64 and 68 is consistent with that of StW 382, suggesting the absence of the carpometacarpal ligamentous connection and pronation of the MC2 that facilitates opposition to the thumb and precision gripping as in other hominins (Marzke, 1997; Niewoehner et al., 1997). The larger articulation with the capitate, particularly palmarly, suggests potentially greater stability (Rein and Havarti, 2012) and load transfer across the MC3-capitate joint than that of *A. afarensis*. The styloid process in humans and Neandertals is thought to stablize the MC3 carpometacarpal joints during forceful precision and power gripping (Marzke, 1983; Markze and Marzke, 1987; Niewoehner et al., 1997). The absence of a projecting styloid process in the Sterkfontein specimens suggests less stabilization of the MC3 carpometacarpal joints than is found in humans, Neandertals, or KNM-WT 51260 (Ward et al., 2013), although it cannot be ruled out that the slight proximal projection of the dorsoradial border in StW 64 provides some additional stability to this region than does *A. afarensis* or MH2.

The distal articular morphology of the Sterkfontein MC3s is generally similar to that of *A. afarensis*, MH2, and *H. naledi*. StW 64 and 68 both demonstrate a radial orientation of the head relative to the shaft, like that humans and other fossil hominins, and unlike African apes, which facilitates opposition of the third ray with the thumb (Susman, 1979; Marzke, 1997; Drapeau, 2015). In palmar view, StW 27 shows slightly more radioulnar asymmetry in the shape of the articular surface than that of other Sterkfontein specimens, but there is variation in asymmetry within the *A. afarensis* sample as well (e.g., A.L. 438-1d

is more asymmetrical than A.L. 333-16). Similarly, the radiopalmar margin of the distal articular facet extends notably more proximally than the ulnar-palmar margin in StW 27, 68 and 394, which is similar to *A. afarensis* and MH2, while StW 64 is more symmetrical and, in this way, more similar to *H. naledi* Hand 1, KNM-WT 51260, and later *Homo* (Figure 9.3). The articular surface extends to approximately the same extent both palmarly and dorsally as in *A. afarensis*, MH2, *H. naledi*, and KNM-WT 51260 (although the dorsal region of the latter two taxa is poorly preserved).

There is limited variation in radioulnar breadth of the MC3 head relative to interarticular length across fossil hominins, with all being broader than that of *Pan* and similar to *Gorilla* and *H. sapiens* (Figure 9.12). However, radioulnar breadth relative to dorsopalmar height of the head shows a large range of variation among the Sterkfontein sample, with StW 68 having a head that is almost as equally broad as it tall, which is most similar to *A. afarensis*, Neandertals, and *H. sapiens*. In contrast, the head of StW 27 is radioulnarly narrower than it is tall, and is most similar to *Pan* and the lower range of variation found in recent humans. The intermediate head breadths of StW 64 and 394 are similar to MH2 and *H. naledi*. This variation also does not support Ricklan's (1988, cited in Green and Gordon, 2008) suggestion that StW 64 and 68 derive from the same individual.

Fourth metacarpal

In absolute size, StW 330 is smaller than StW 65 (Figure 9.4). The StW 330 proximal end is similar in size to that of A.L. 333-56/81, SKX 2954, and MH2, although what is preserved of the shaft indicates that it was shorter in proximodistal length than A.L. 333-56/81 but longer than that of SKX 2954 and, especially, MH2. A ratio of radioulnar breadth of the base relative to the midshaft shows substantial variation among the Sterkfontein sample; StW 330 has a relatively broad base, similar to that of *H. naledi* Hand 1, Neandertals, and *H. sapiens*, while StW 65 has a comparatively narrower base, being more similar to *Ar. ramidus* and SKX 2954 and overlapping with African apes (Figure 9.13). Radioulnar breadth versus dorsopalmar height at midshaft again reveals variation among the Sterkfontein metacarpals, with StW 552 having a relatively taller shaft while StW 65 and 26 have a shaft that is almost as equally broad as it is tall. This range of morphological variation overlaps with all other fossil hominins and extant taxa, with StW 552 being most similar to MH2, and StW 65 and 26 most similar to *A. afarensis* A.L. 333-56/81 (Figure 9.13). The "bend" of the dorsal surface of the distal MC4 shaft is most pronounced in StW 330 and 552, which is more prominent than that of A.L. 333-56/81 and MH2 but less so than *H. naledi* Hand 1 and, especially, SKX 2954.

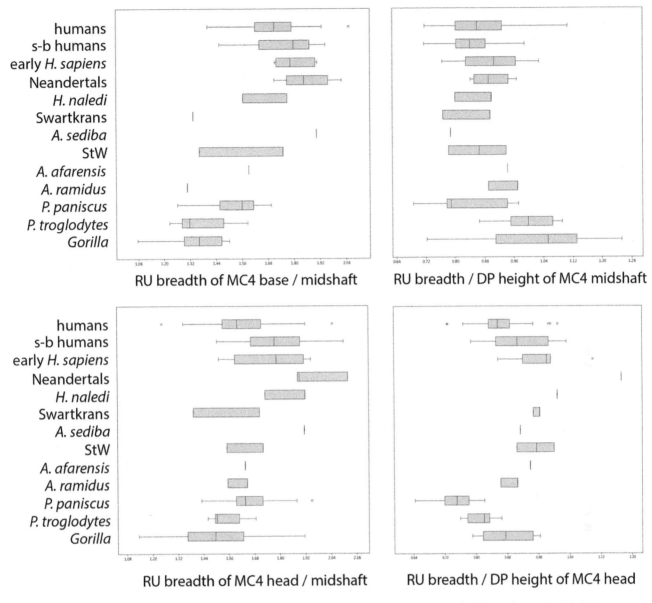

Figure 9.13 Comparative analysis of fourth metacarpal morphology. Box-and-whisker plots of fourth metacarpal (MC4) shape ratios that maximize the inclusion of the Sterkfontein MC4 sample, none of which is complete: radioulnar (RU) breadth of the MC4 base relative to midshaft breadth (top, left), including StW 65 and 330; RU breadth relative to dorsopalmar (DP) height at midshaft (top, right), including StW 26, 65, 330, and 552; RU breadth of the MC4 head relative to midshaft breadth (bottom left), including StW 26 and 552; and RU breadth relative to DP height of the MC4 head, including StW 26, 292, and 552. See Table 9.3 for details of the comparative extant and fossil sample.

The StW 65 and 330 proximal bases differ somewhat in their articular morphology. The more dorsopalmarly curved MC5 facet of StW 65 is most similar to *Ar. ramidus* (ARA-VP 7/26), *A. afarensis* (A.L. 333-56/81 and -122), MH2, and *H. naledi* (U.W. 101-1318 and U.W. 102a-028), while the flatter facet of StW 330 is more similar to SKX 2954. StW 65 also differs from StW 330 in having a palmar (and dorsal) MC3 facet, although there is variation in this articulation within and across other hominin taxa as well; both dorsal and palmar facets are present in ARA-VP 7/

26, A.L. 333-56/81, and MH2, while only a dorsal MC3 facet is found in A.L. 333-122, SKX 2954, and *H. naledi* (U.W. 101-1318 and U.W. 102a-028). The concavoconvex hamate articulation of StW 65 is similar to that of A.L. 333-122, while the generally flat facet of StW 330 is similar to A.L. 333-56/81, MH2, SKX 2954, and *H. naledi*. This subtle variation in articular morphology both within and across taxa suggests that there may be limited functional consequences for this varying morphology, especially across joints with limited mobility and that are bound by carpometacarpal ligaments. Indeed, previous studies have documented substantial variation in MC4 carpometacarpal articulations across humans and other primates (Viegas et al., 1991; Marzke et al., 1994).

Distally, there is greater "pinching" of the dorsal articular surface in StW 26, 292, and 552 compared to that of A.L. 333-56/81, MH2, SK 85, SKX 2954, and *H. naledi* U.W. 102a-028 (this region is not preserved in Hand 1). The heads of the Sterkfontein specimens do not project far palmarly from the shaft and in this way are most similar to A.L. 333-56/81, SK 85, and SKX 2954 but differ from the more palmarly projecting heads of all other hominins in our sample. In radioulnar breadth of the MC4 head relative to the midshaft, StW 552 has a broader head than StW 26, although one or both specimens overlap with the range of variation found in African apes, *A. ramidus*, *A. afarensis*, the Swartkrans specimens, and *H. sapiens* (Figure 9.13). Both Sterkfontein MC4s have relatively narrower heads than MH2, *H. naledi*, and Neandertals. In radioulnar breadth compared with dorsopalmar height of the MC4 head, StW 552 is slightly narrower than that of StW 26 and 292, but all are similar to *A. afarensis*, MH2, the Swartkrans specimens, and *H. sapiens*. The Sterkfontein MC4 heads are radioulnarly broader than *Ar. ramidus* but narrower than *H. naledi* and Neandertals (Figure 9.13).

Fifth metacarpal

In absolute proximodistal length, StW 63 is just slightly longer than MH2 and *H. naledi* Hand 1 and is shorter than *Ar. ramidus* ARA-VP-6/500-036, *A. afarensis* A.L. 333-14 and 333-89, and SK(W) 14147. In relative radioulnar breadth at midshaft, StW 63 is among the broadest specimen within our comparative sample, falling within the extreme upper range of variation found in small-bodied recent humans and only SK(W) 14147 is broader (Figure 9.14). In radioulnar breadth versus dorsopalmar height of the MC5 midshaft, StW 63 is broader than it is tall and is most similar to StW 639, MH2, SK(W) 14147, and some *A. afarensis* specimens (A.L. 333-89 and -141), all of which fall within the upper range of variation found in Neandertals, *H. sapiens*, and African apes (Figure 9.14). Although preservation is poor, the rugosity along the ulnar shaft within the region of *m. opponens digiti minimi* insertion is more pronounced than that of *Ar. ramidus* and *A. afarensis*, and is constrained to the distal end of the shaft as in *H. naledi* Hand 1 (Figure 9.5). The StW 63 morphology is unlike StW 639, which has the large, projecting enthesis just proximal to midshaft (Pickering et al., 2018), or the longer, proximally extended rugosity found in MH2 or SK(W) 14147, which suggests a more human-like pattern in which the *m. opponens digiti minimi* inserts along the full length of the MC5 shaft. This variation in

entheseal morphology and position may suggest a slight mechanical advantage of the *m. opponens digiti minimi* among the South African hominins, a muscle that is recruited by both the non-dominant and, especially, dominant hands during tool behaviors in humans (Marzke et al., 1998). However, a previous study of human cadavers documented substantial variation in the length of *m. opponens digiti minimi* enthesis and no correlation between entheseal morphology and muscle size (Williams-Hatala et al., 2016).

In relative radioulnar breadth of the MC5 base, StW 63 is among the broader specimens in our comparative sample, falling out as most similar *H. naledi* Hand 1 and *H. sapiens* (Figure 9.14). It is broader than the other Sterkfontein specimens MC5, StW 689, but narrower than MH2 and SK(W) 14147. At the base of StW 63, the ulnar projection for attachment of the metacarpal and collateral ligaments and the dorsoulnar tubercle in the region of the *m. extensor carpi ulnaris* insertion are both more well developed than that of *Ar. ramidus*, *A. afarensis*, *H. naledi* Hand 1, and SK(W) 14147 but less prominent than those of MH2. Ricklan (1987) suggested that the *m. extensor carpi ulnaris* in *A. africanus* had a greater mechanical advantage for ulnar deviation at the wrist than that of humans. This, in combination with the prominence of the *m. extensor carpi ulnaris* tubercle and the robusticity of the StW 63 base and shaft, have been interpreted as evidence that this muscle was powerful and used consistently during power gripping (Ricklan, 1987; Marzke et al., 1992). The StW 63 hamate facet is similar in its strong dorsopalmar convexity and slight radioulnar convexity to *A. afarensis*, MH2, and SK(W) 14147, and unlike the saddle-shaped articulation of the *H. naledi* Hand 1 and later *Homo*. The extent to which the hamate facet extends onto the palmar surface is less than that of MH2 and SK(W) 14147 and similar to that of *Ar. ramidus*, *A. afarensis*, and *H. naledi* Hand 1.

In relative radioulnar breadth of the head, StW 63 is the broadest specimen in our extant and fossil comparative sample, falling even outside the range of variation found in *H. sapiens* (Figure 9.14). Qualitatively, the broad head of StW 63 is most similar to MH2 and *H. naledi* Hand 1, but their broad heads are accentuated by radioulnarly narrow shafts, particularly in MH2. StW 63 contrasts that of SK(W) 14147, in which the radioulnar breadth of the head is very similar to that of its broad shaft. The StW 63 distal articular surface extends quite far proximally onto the palmar surface and in this way is most similar to MH2 and *H. naledi* Hand 1, and unlike the less-extended palmar articulation found in *A. afarensis* and particularly SK(W) 14147. In palmar view, the StW 63 head is oriented ulnarly relative to the long axis of the shaft and its articular surface is more asymmetrical than that of *Ar. ramidus*

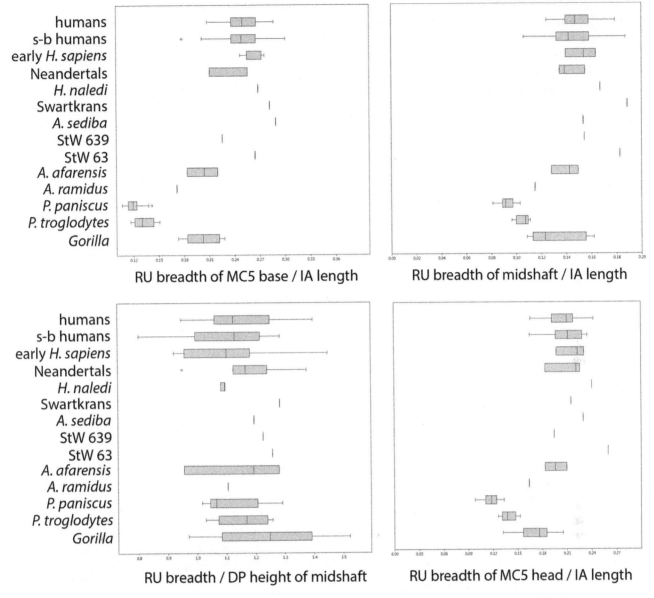

Figure 9.14 Comparative analysis of fifth metacarpal morphology. Box-and-whisker plots of fifth metacarpal (MC5) shape ratios: radioulnar (RU) breadth of the MC5 base (top, left) and midshaft (top, right) relative to interarticular (IA) length, RU breadth relative to dorsopalmar (DP) height at midshaft (bottom, left), and RU breadth of the MC5 head relative to IA length (bottom, right). See Table 9.3 for details of the comparative extant and fossil sample.

and *A. afarensis*. Overall, the StW 63 distal morphology is similar to that of other South African hominins and *Homo* and would facilitate opposition of the fifth finger toward the index finger and thumb, while the absence of a saddle-shaped hamate facet suggests this ability was more limited than that of *H. naledi* and other later *Homo*.

Proximal pollical phalanx

StW 575 proximal pollical phalanx (PP1) is similar in absolute proximodistal length to A.L. 333-69 and *H. naledi*

Hand 1 (U.W. 101-1721) but longer than MH2. In overall robusticity, StW 575 is similar to *A. afarensis* A.L. 438-4 and *H. naledi*, and more robust than *Ar. ramidus* ARA-VP-7/2I, *A. afarensis* A.L. 333-69, and MH2. What is preserved of StW 478 demonstrates that it derives from a larger hand, with the absolute size of the base being larger than all of our comparative sample, apart from Neandertals and *H. sapiens* (Figure 9.6). In relative radioulnar breadth versus dorsopalmar height of the PP1 base, StW 575 is relatively broader than StW 478 (Figure 9.15), although both specimens are similar to other South and East African

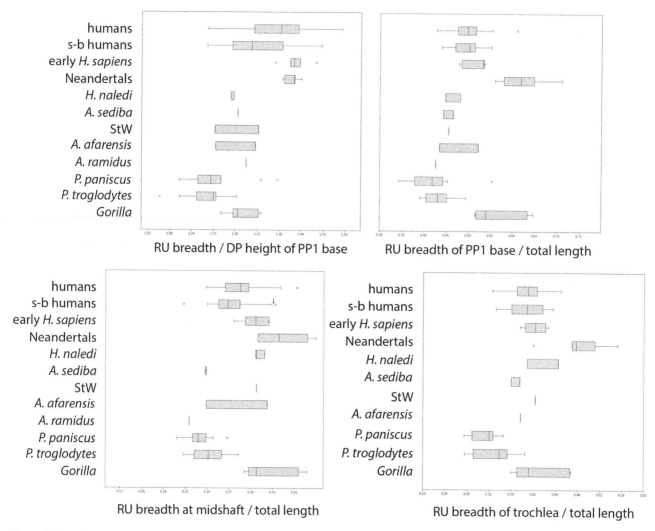

Figure 9.15 Comparative analysis of proximal pollical phalanx morphology. Box-and-whisker plots of proximal pollical phalanx (PP1) shape ratios: radioulnar (RU) breadth vs. dorsopalmar (DP) height of the base, including both StW 478 and StW 575 (top, left), and RU breadth of the base (top, right), midshaft (bottom, left), and trochlea (bottom, right) relative to total length of the PP1, including just StW 575. See Table 9.4 for details of the comparative extant and fossil sample. Note that the total length of *A. afarensis* A.L. 438-4 is estimated, as this specimen does preserve the distal end, and should be considered with caution. In the plots below, it has a higher ratio in RU breadth of the base and midshaft relative to total length compared with A.L. 333-69.

fossil hominins, overlapping with the range of variation found in African apes and recent humans but narrower than Neandertals and early *H. sapiens*. Both specimens show less asymmetry in the base than is found in MH2 and that is typical of Neandertals and *H. sapiens*. StW 478 has more palmarly projecting basal tubercles than StW 575, but the overall morphology of both specimens is most similar to A.L. 438-4 and *H. naledi*.

In radioulnar breadth of the base relative to total PP1 length, StW 575 is similar to all other South African hominins and *A. afarensis* in overlapping with the range of variation found in both *Pan* and *H. sapiens* (Figure 9.15). StW 575 is slightly broader than *Ar. ramidus* but narrower than Neandertals and *Gorilla*. The sides of the PP1 shaft in StW 575 (and what is preserved of StW 478) run

relatively parallel to each other, tapering slightly distally, such that the narrowest point is just proximal to the trochlea. This morphology is similar to some *H. naledi* specimens (U.W. 101-428 and -1055) but contrasts the shafts of A.L. 333-69, MH2, and *H. naledi* Hand 1 that are more hour-glass-shaped, such that the narrowest point is closer to the midshaft. The palmar surface of the shaft in both Sterkfontein specimens is flat or radioulnarly convex throughout the proximodistal length of the shaft, which is similar to the morphology of *H. naledi* and later *Homo* but unlike the radioulnarly concave distopalmar shaft of A.L. 333-69 and, especially, MH2, which makes the trochlea project farther palmarly from the shaft. In relative radioulnar breadth at midshaft, StW 575 is most similar to *H. naledi* (Figure 9.15). It is broader than ARA-VP-7/2I,

A.L. 333-69, and MH2, is narrower than most Neandertals, and falls within the range of variation found in *Gorilla* and *H. sapiens*. The Sterkfontein PP1 base and shaft morphology is consistent with more well-developed adduction and flexion of the thumb (via the *mm. adductor pollicis oblique* and *flexor pollicis brevis*) than that of African apes and the more gracile specimens of *Ar. ramidus*, *A. afarensis*, and MH2.

In relative radioulnar breadth of the trochlea, StW 575 is, again, most similar to *H. naledi*, as well as *H. sapiens* and *Gorilla* (Figure 9.15). It is broader than *Pan*, A.L. 333-69, and MH2, but narrower than all Neandertals except Moula Guersy M-E1-123. The distal articular surface extends further proximally onto the dorsal surface than in A.L. 333-69, MH2, and U.W. 101-1721. In the distal view, the trochlea is more asymmetrical than A.L. 333-69 (this region is not well-preserved in MH2 or any of the *H. naledi* specimens) and dorsopalmarly taller than MH2 and U.W. 101-1721. The trochlear breadth and asymmetry of StW 575 suggest greater loading and conjunct rotation of distal pollical phalanx during flexion to oppose to the fingers than that of *A. afarensis* and MH2, but less so than that of later *Homo* (Shrewsbury et al., 2003).

Non-pollical proximal phalanx

One of the most discussed functional aspects of non-pollical proximal phalanx morphology is the longitudinal curvature of the dorsal surface, which has been shown to positively correlate with the degree of arboreality among extant primates (Susman, 1979; Hunt, 1991; Jungers et al., 1997; Deane and Begun, 2008) and to change throughout ontogeny, with more arboreal juveniles having more curved phalanges than their adult counterparts (Richmond, 1998). Phalangeal curvature has also been demonstrated to biomechanically reduce overall strain experienced by the phalanx during flexed-fingers grasping (Richmond, 2007; Nguyen et al. 2014). Several studies have quantified the dorsal longitudinal curvature of the StW non-pollical proximal phalanges. Using the first polynomial coefficient to quantify phalangeal curvature (Deane and Begun, 2008), the Sterkfontein specimens are less curved than the median values of *A. afarensis*, MH2, OH 7, and *H. naledi* but more curved than the Swartkrans specimens and *H. sapiens*. Using the traditional included angle method, StW 28 is less curved than other, more recently discovered Sterkfontein specimens StW 355, 597 and 668 (Appendix I) and most *A. afarensis* specimens, but more curved than the Swartkrans phalanges and *Homo* sp. specimens OH 86 and AT-9-2 (Domínguez-Rodrigo et al., 2015; Stratford et al., 2016; Pickering et al., 2018). However, it is important to highlight that there is a substantial degree of intraspecific variation and overlap across fossil and extant taxa in phalangeal curvature.

The overall morphology of the Sterkfontein non-pollical proximal phalanges described here (Figure 9.7) is most similar to that of more recently discovered Sterkfontein specimens (Stratford et al., 2016; Pickering et al., 2018), the Swartkrans sample, and to *H. naledi* in being relatively robust but with limited palmar-projection of the flexor sheath ridges. None of the Sterkfontein proximal phalanges is as proximodistally long as those of *Ar. ramidus* or A.L. 333w-4, the longest proximal phalanx in the *A. afarensis* sample. They differ from those of *A. afarensis* in having relatively radioulnarly broader and dorsopalmarly taller shafts, especially when comparing specimens of similar absolute length (e.g., StW 293 vs. A.L. 333-57 and A.L. 1044-1), and differ from those of MH2 in having more robust shafts and less projecting flexor sheath ridges. The Sterkfontein proximal phalanges have straight shaft margins that differ from those of OH 7, some *A. afarensis* specimens (A.L. 333-57 and -63), and possibly *A. anamensis* KNM-KP 30503, which are radioulnarly broader around midshaft and taper at the ends. Finally, they are unlike the *H. floresiensis* LB1 proximal phalanges, which are dorsopalmarly tall for their size, and distinct in having a radioulnarly convex palmar surface and almost circular cross-section. Quantitatively, radioulnar breadth at midshaft relative to total length shows that StW 28 and 293 are relatively broad and most similar to StW 597 (Pickering et al., 2018), StW 668 (Stratford et al., 2016), and other South African hominins (Figure 9.16). Both StW 28 and 293 are relatively broader than *Pan*, *Ar. ramidus*, *A. afarensis*, and recent humans.

StW 28 is most similar to SKX 5018; both specimens are similar in size and robusticity, have projecting and asymmetric palmar tubercles, and a ridge extending distally from the ulnar tubercle in the region of the *mm. flexor digiti minimi brevis* and *abductor digiti minimi* insertions. StW 28, like SKX 5018 (*contra* Susman, 1989), is considered to be a fifth proximal phalanx (PP5) and, if so, is absolutely longer and more robust than the PP5 of MH2 and *H. naledi* Hand 1, but shorter than the purported PP5 specimens StW 597 (Pickering et al., 2018), OH 86 (Domínguez-Rodrigo et al., 2015), and the *Homo* sp. ATE9-2 from Spain (Lorenzo et al., 2015), and lacks the palmarly projecting flexor sheath ridges of ATE9-2 (Lorenzo et al., 2015). The PP5 of MH2 also has a well-developed ridge extending from the ulnar tubercle like StW 28, but this is not found in other (purported) PP5 specimens, including A.L. 333-62, StW 597, OH 86, ATE9-2, and *H. naledi* Hand 1 (Domínguez-Rodrigo et al., 2015; Lorenzo et al., 2015; Pickering et al., 2018).

In radioulnar breadth versus dorsopalmar height of the PP2-PP5 base, the Sterkfontein proximal phalanges demonstrate a high degree of variation with, at one extreme, StW 28 having a base that is almost as equally radioulnarly broad as it is dorsopalmarly tall, while at the other extreme, StW 29 is much broader (Figure 9.16).

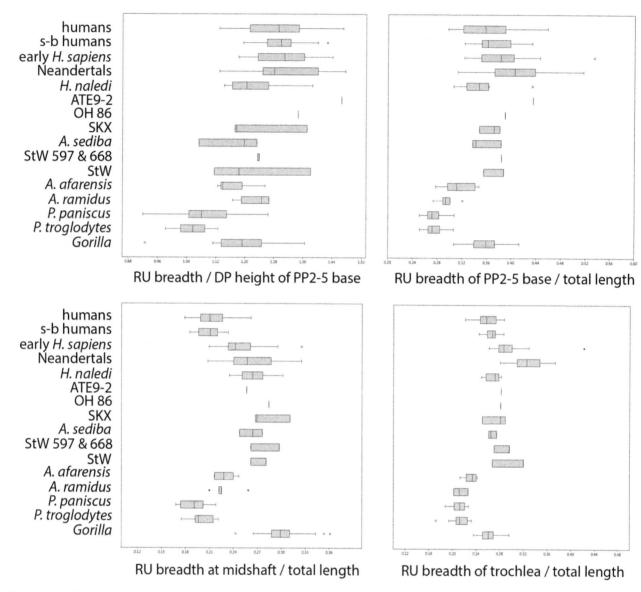

Figure 9.16 Comparative analysis of proximal phalangeal morphology, including phalanges from rays 2–5. Box-and-whisker plots of non-pollical proximal phalanx (PP2-5) shape ratios: radioulnar (RU) breadth vs dorsopalmar (DP) height of the base, including StW 28, 29 and 293 (top, left), and RU breadth of the base (top, right), midshaft (bottom, left), and trochlea (bottom, right) relative to total length, including complete specimens StW 28 and 293. See Table 9.4 for details of the comparative extant and fossil sample.

In this way, the Sterkfontein proximal phalanges overlap with the range of variation found in all other extant and fossil taxa, apart from ATE9-2, which is extremely broad (Lorenzo et al., 2015). Relative to total length, the bases of StW 28 and 293 (the only complete proximal phalanges) are radioulnarly broader than *Pan*, *Ar. ramidus*, and *A. afarensis* and similar to other South African hominins and *H. sapiens* (Figure 9.16).

The trochlea projects palmarly from the surface of the shaft to a similar degree as that of the *A. afarensis*, MH2, the Swartkrans specimens, but less so than OH 7 and *H. naledi*.

The distal articulation is similar to other South African hominins and differs most notably from OH 7, which has a deeper and broader depression between the trochlear condyles and the articular surface extends farther proximally on both the palmar and dorsal surfaces. In radioulnar breadth of the trochlea relative to total length, StW 28 and 293 are similar to all other South African hominins, OH 86, ATE9-2, Neandertals, and *H. sapiens*, and are broader than most African apes, *A. ramidus* and *A. afarensis* (Figure 9.16).

Overall, the intermediate curvature of the Sterkfontein proximal phalanges, robusticity of the shaft but limited development of the flexor sheath ridges suggest some degree of flexed-finger grasping may have still been a significant component of their locomotor repertoire, but that this was less significant than that of *A. afarensis*, MH2, OH 7, and *H. naledi*. However, it is important to highlight that there is substantial degree of intraspecific variation, including across different rays, and overlap across fossil and extant taxa in phalangeal curvature and flexor sheath ridge development, and thus direct functional inferences about potential locomotor behaviors are unclear.

Intermediate phalanx

StW 331 is a large intermediate phalanx (IP). It is absolutely proximodistally longer than any of the more recently discovered Sterkfontein IPs (StW 620, 635, and 657; Pickering et al., 2018), as well as all IPs in the *A. afarensis*, Swartkrans, *A. sediba*, OH 7, and *H. naledi* samples. It is only shorter than the IP3 and IP4 of the *Ar. ramidus* ARA-VP-6/500 hand, and is most similar to the absolute lengths of the (purported) IP3s in some *H. sapiens* and Neandertals. Its dorsal curvature is most similar to that of *A. sediba* and African apes, and is more curved than most *A. afarensis* specimens, OH 7, and the Swartkrans specimens (Kivell et al., 2015).

StW 331 is most similar in overall morphology to the more recently discovered StW 657 (Pickering et al., 2018) and *H. naledi*. All share straight shaft margins that taper distally, a trochlea that is equally or just slightly more radioulnarly broad than the narrowest breadth of the shaft, and raised, semicircular-shaped projections, possibly for the (partial) attachment of the *m. flexor digitorum superficialis* (FDS) (Marzke et al., 2007). StW 331 shares a similar shaft shape with the Swartkrans sample, although in the latter most specimens have semicircular fossae (e.g., SKX 5021, -13476, and 36712), rather than projections, an area around which the FDS tendons may attach (Marzke et al., 2007). StW 331 differs from *H. naledi* is having a less curved dorsal surface (Kivell et al., 2015), a more prominent median bar, and a pronounced peripheral rim around the proximal base, which is most similar to *A. afarensis*. StW 331 differs from the "bottle-shaped" IPs of OH 7, in which the rounded shaft margins bulge laterally, or the more distally apering shaft margins of LB1. The IPs of *Ar. ramidus*, *A. afarensis*, and OH 7 all have a well-developed median bar and semicircular fossae, which is distinct from StW 331. Finally, StW 331 does not have the distinct "proximal phalanx" morphology of the MH2 IPs (Kivell et al., 2011), which have straight shaft margins, thin flexor sheath ridges, and lack both a median bar and semicircular fossae.

The base of StW 331 has dorsopalmarly tall, oval-shaped articular facets that are most similar to *A. afarensis*, MH2, LB1-48, and some specimens at Swartkrans (e.g.,

SKX 5021). Quantitatively, there is substantial overlap in the IP-shape ratios across the comparative sample, due in part to the inclusion of IPs from all four rays in the analyses (Figure 9.17). Relative to total length, the radioulnar breadth of the StW 331 base is most similar to that of other Sterkfontein IPs (StW 620, 635, and 657; Pickering et al., 2018) and LB1. StW 331 has a broader base than *Pan*, *Ar. ramidus* and most *A. afarensis* specimens, but narrower than MH2, the Swartkrans sample, *H. antecessor*, and most Neandertal and *H. sapiens* specimens.

In relative radioulnar breadth at midshaft, StW 331 is most similar to *A. afarensis*, MH2, and *H. sapiens* in being relatively narrow compared to most other fossil hominins, including the other Sterkfontein IPs; only the IPs of *A. ramidus* and *H. floresiensis* are narrower (Figure 9.17). In relative radioulnar breadth of the trochlea, StW 331 is among the narrowest in the sample, overlapping only with the other Sterkfontein IPs, while only *Ar. ramidus* has a narrower trochlea (Figure 9.17).

Altogether, the large StW 331 IP morphology may have had similar development of the FDS to that of *H. naledi*, suggesting powerful flexion of the fingers. However, Marzke et al. (2007) have shown there is little correlation between FDS size and insertion morphology across extant primates. A comparison of relative curvature in StW 331 together with that of the Sterkfontein proximal phalanges, suggests the Sterkfontein hominins had finger curvature that was less than that of *H. naledi* and most similar to *A. sediba*. However, without associated phalanges with the same individual and given intraspecific variation in morphology and curvature across rays and ontogeny, drawing functional inferences from StW 331 alone is challenging.

Distal pollical phalanx

The morphology of the StW 294 distal pollical phalanx (DP1) is most similar to the preserved morphology of MH2 in being small and gracile, with a radioulnarly narrow apical tuft (although still broad relative to African apes and earlier hominins) and a proximodistally long and "waisted" shaft. Both StW 294 and MH2 have a moderately deep proximopalmar fossa and a clear gabled ridge for attachment of *m. flexor pollicis longus* (FPL) that projects palmarly above the rest of the surface, although in MH2 this ridge is more pronounced and it also has an ungual fossa. StW 294 differs from *Orrorin* (Almécija et al., 2010), *Ar. ramidus*, *A. afarensis*, and *H. floresiensis* in having a more radioulnarly expanded base and apical tuft. It shares a FPL insertion ridge with *Orrorin* (Almécija et al., 2010), *A. afarensis*, and *H. floresiensis* that is not found in *Ar. ramidus* (*contra* Lovejoy et al., 2009) or OH 7. StW 294 differs from the extremely radioulnarly broad DP1s of SKX 5016, TM 1517k (but see Day, 1978), OH 7, and

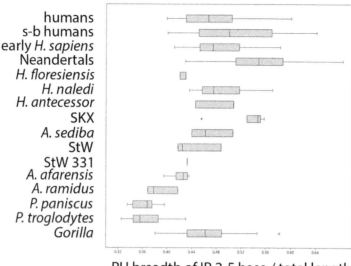

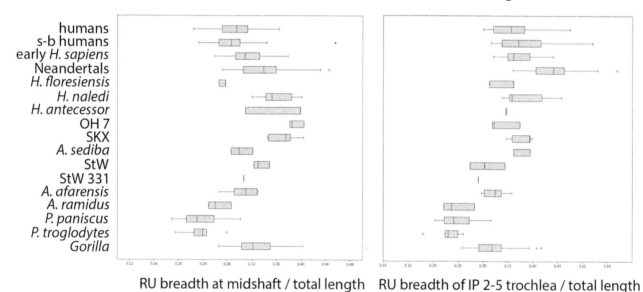

RU breadth at midshaft / total length   RU breadth of IP 2-5 trochlea / total length

Figure 9.17 Comparative morphology of the intermediate phalanges, including phalanges from rays 2–5. Box-and-whisker plots of intermediate phalanx (IP2-5) shape ratios: radioulnar (RU) breadth of the base (top), midshaft (bottom, left), and trochlea (bottom, right). See Table 9.4 for details of the comparative extant and fossil sample. Note that the total length of the OH 7 juvenile phalanges, which are missing their proximal epiphyses, has been estimated and results should be interpreted with caution.

H. naledi, which have relatively equally broad bases and apical tufts, with proximodistally short and broad shafts.

StW 294 is also distinct from the other (potential) DP1 from Sterkfontein, StW 617. Ruff et al. (Chapter 16, this volume) describe this specimen as a hallucal DP1, although we consider it to possibly be a pollical phalanx instead. Aspects of StW 617 (and TM 1517k; see Day, 1978) suggest it may be a hallucal distal phalanx (see Ruff et al., Chapter 16, this volume, Figure 16.8). However, its morphology is similar to the DP1 associated with the articulated hand of StW 573 (Clarke, 2013). StW 617 has a dorsopalmarly thick apical tuft, a proximodistally short

and radioulnarly broad shaft, a deep palmar fossa, and an extremely radioulnarly broad base that is much broader than its apical tuft. If it is a pollical phalanx, the differences between StW 294 and StW 617 would suggest two DP1 morphs among the Sterkfontein sample.

In relative radioulnar breadth of the DP1 base, StW 294 is similar to most other earlier hominins, as well as to Neandertals and early *H. sapiens* in being broader than *Pan, Ar. ramidus* and *H. floresiensis* but narrower than OH7, *H. naledi*, and, in particular, the extremely broad base of StW 617 (Figure 9.18). In radioulnar breadth of the apical tuft, StW 294 is similar to most other hominins in

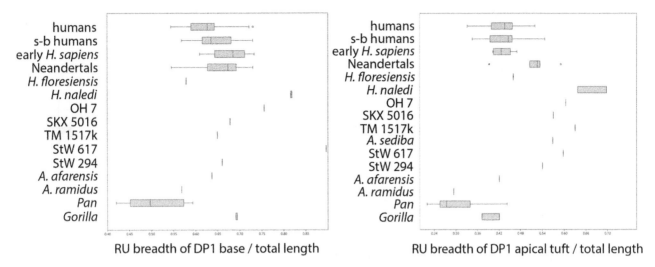

Figure 9.18 Comparative morphology of the distal pollical phalanx. Box-and-whisker plots of distal pollical phalanx (DP1) shape ratios: radioulnar (RU) breadth of the base (left) and apical tuft (right) relative to total length of the phalanx. See Table 9.4 for details of the comparative extant and fossil sample. Note that, due to preservation, all measurements in *H. floresiensis* LB1/55 are estimated from the original fossil (Larson et al., 2009) or from the 3D print and apical tuft breadth is estimated in *A. sediba* MH2.

being broader than African apes, *Ar. ramidus, A. afarensis*, and *H. sapiens* (Figure 9.18).

Together, the morphology of StW 294 suggests that it had an independent FPL, like that of most other hominins, that would provide greater flexion of the distal thumb than that of African apes (but see van Leeuwen et al., 2018). Although the apical tuft is radioulnarly broad, the absence of an ungual fossa or ungual spines (Susman, 1998; but see Ricklan 1987, 1990) suggests that the StW 294 individual may have lacked a fully compartmentalized palmar pulp that helps to facilitate object manipulation in humans.

Intrinsic hand proportions

We assessed the intrinsic hand proportions among the Sterkfontein sample via three ratios (metacarpal, phalangeal and digit ratios) using total length of skeletal elements from the first ray (thumb) and third ray (middle finger; Table 9.5). Figure 9.19 shows a box-and-whisker plot of the intrinsic metacarpal ratio (MC1 total length/MC3 total length). The Sterkfontein metacarpal ratio is distinct from that of great apes, which have a relatively short MC1, and is more similar to that of recent humans and the estimated ratio of *A. afarensis* (also a composite hand from multiple individuals; also see Rolian and Gordon, 2013). These results are consistent with the study of metacarpal proportions by Green and Gordon (2008), which showed that while the relative breadth of MC1 was more narrow and ape-like, the relative lengths of the metacarpals were more human-like. The Sterkfontein sample is also distinct from the relatively long MC1 of

H. naledi and, particularly, *A. sediba*, both of which fall outside the range of variation found in humans (Kivell et al., 2011).

Figure 9.19 shows the intrinsic phalangeal ratio. For humans and great apes, typically the phalanges of the third ray are longest compared to the other rays (e.g., Susman, 1979; Christensen, 2009). Since not all non-pollical proximal phalanges (PPs) can always be reliably assigned to a ray without a complete hand skeleton, the comparisons here used the proximodistally longest, complete non-pollical PP within the preserved fossil samples to calculate the phalangeal ratio. A comparison of the phalangeal ratios shows that great apes have a relatively short PP1, recent humans, as well as *A. sediba* and *H. naledi*, have a relatively long PP1, and *A. afarensis* is intermediate between these two groups. The Sterkfontein phalangeal ratio is distinct from the rest of the extant and fossil comparative sample in having the relatively longest PP1. The digital ratio, which compares the length of the first (MC1 + PP1) to third (MC3 + PP3 + IP3) rays, is shown in Figure 9.19 shows the digital ratio, which compares the length of the first (MC1 + PP1) to third (MC3 + PP3 + IP3) rays. As with the non-pollical PPs, the intermediate phalanges (IPs) could not be reliably assigned to a ray in *A. afarensis* and the Sterkfontein sample only includes one IP, StW 331. The results here are similar to the metacarpal ratio in that the Sterkfontein estimate is distinct from apes, and closer to *A. afarensis* and modern humans. The estimated digit ratio for Sterkfontein also differs from that of *A. sediba* and *H. naledi*, which have a relatively long thumb, outside the range of our modern human sample (Kivell et al., 2011, 2015).

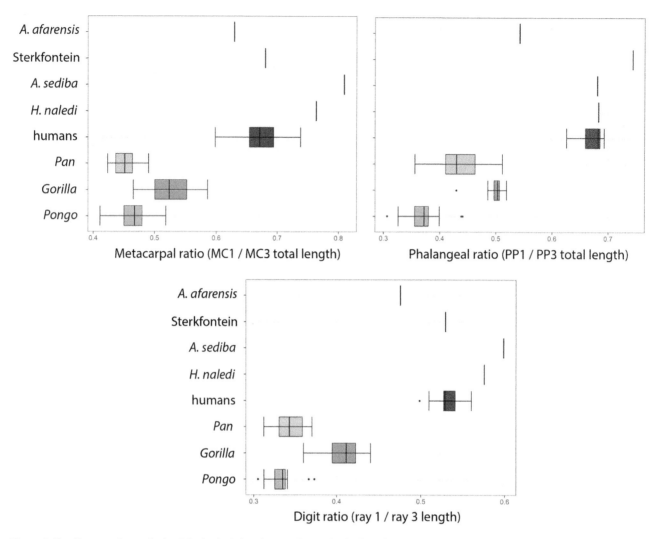

Figure 9.19 Comparative analysis of the intrinsic hand ratios. Box-and-whisker plots comparing the ratio of first metacarpal (MC1) to third metacarpal (MC3) total lengths (top left), first proximal phalanx (PP1) total length to the proximodistally longest non-pollical proximal phalanx, assumed here to be from the third ray (PP3; top, right), and the summed proximodistal length of the first ray (MC1 + PP1) to third ray (MC3 + PP3 + third intermediate phalanx; bottom). Hand elements used to estimate intrinsic ratios for Sterkfontein sample and *A. afarensis* are unlikely to be associated to the same individuals. Sterkfontein specimens include StW 418 MC1, the geometric mean of StW 64 and 68 complete MC3s, StW 575 PP1, StW 293, the longest complete non-pollical phalanx, and StW 331, the only preserved intermediate phalanx (IP). *A. afarensis* specimens include A.L. 333w-39 MC1, the geometric mean of A.L. 438-1d and A.L. 333-16 complete MC3s, A.L. 33-69 PP1, A.L.333w-4, the longest complete non-pollical phalanx, and A.L. 333-46, the longest available intermediate phalanx. All other extant and fossil intrinsic ratios are calculated from associated hand skeletons. See Table 9.5 for details on extant comparative sample. Note that for the *A. sediba* MH2 hand, IP4 is slightly longer than IP3, and thus the length of IP4 was used in the digit ratio.

It is important to emphasize that these intrinsic hand ratios are biased by the limitations of the fossil record. For example, while the PP1s used here from the Sterkfontein sample and for *A. afarensis* are relatively similar in proximodistal length (27.0 mm vs. 25.9 mm, respectively), the longest non-pollical PP available from Sterkfontein (StW 293, 36.3) is much shorter than the longest non-pollical PP preserved for *A. afarensis* (AL 333w-4, 47.8; Ward et al., 2012). Similarly, the only IP preserved in the Sterkfontein sample (StW 331) is absolutely one of the longest specimens (30.3 mm) in our comparative hominin sample

while the longest *A. afarensis* IP (A.L. 333-46) is comparatively short (26.8 mm), especially relative to the long length of the A.L. 333w-4 PP (Rolian and Gordon, 2013). Thus, caution is needed when drawing functional inferences from intrinsic proportions based on hand elements that derive from different individuals.

To mitigate this bias, a resampling approach can take into account that elements may not belong to the same individual and that phalanges may not be assigned to a specific ray, by drawing subsamples of elements that match the fossil assemblage and generating

Table 9.6 Intrinsic hand ratio 95% confidence intervals (CI) for resampling distributions.

Proportion	Sterkfontein	*Homo*	*Pan*	*Gorilla*	*Pongo*
Metacarpal	−0.354	−0.519 to −0.113	**−0.931 to −0.567**	**−0.786 to −0.378**	**−0.962 to −0.486**
Phalangeal	−0.272	−0.609 to −0.012	**−1.017 to −0.376**	**−0.823 to −0.419**	**−1.198 to −0.704**
Digital ray	−0.602	−0.692 to −0.356	**−1.134 to −0.783**	**−0.968 to −0.672**	**−1.214 to −0.860**

Values in bold indicate no overlap between the Sterkfontein values and extant hominoid 95% CI ranges.

a distribution of possible ratios. In its metacarpal proportions, the Sterkfontein composite ratio overlaps with the 95% confidence interval of humans is just outside the higher end of the confidence interval of *Gorilla*, and is outside the range of *Pan* and *Pongo* (Table 9.6). This is similar to the results reported in Green and Gordon (2008) regarding the metacarpal length proportions for *A. africanus*; however, in their analysis, the Sterkfontein estimate was not significantly different from *Gorilla*. For the phalangeal ratio, the Sterkfontein sample falls outside the 95% confidence intervals of all the apes, and only within the interval for humans. In a resampling analysis by Rolian and Gordon (2013) for *A. afarensis*, this fossil estimate fell within the confidence interval of *Gorilla* and humans for both metacarpal and phalangeal ratios (for ratios calculated using the geometric mean method). The estimated digital ratio for the Sterkfontein sample also falls outside the confidence intervals of all the apes, and within the range of humans; these results are similar to that of *A. afarensis*, which had digit proportions that were inside the lower confidence interval of humans, and slightly outside the confidence interval of *Gorilla* (Rolian and Gordon, 2013).

While resampling provides a more conservative approach to compare intrinsic manual proportions given the unassociated fossil sample, it is important to note, as also shown by Rolian and Gordon (2013) for *A. afarensis*, it is possible, if rare, to generate identical manual proportions to those found in the fossil taxa from the distributions within *Gorilla*, *Pan*, and modern humans, when the same sampling constraints are imposed (e.g., for the phalangeal ratio, although outside the 95% confidence interval, the Sterkfontein estimate falls within the full distribution ranges of *Pan*). Overall, however, the results suggest that based on the available Sterkfontein manual elements, this taxon had manual proportions that were likely significantly different from *Pan*, *Pongo*, as well as *Gorilla*. Although there are limitations in sample size, here, there was no overlap between the 95% confidence intervals of *Gorilla* and the manual ratio estimates for Sterkfontein, unlike for *A. afarensis*, where a similar approach showed that in some cases the manual proportions do fall within the range of overlap in the *Gorilla* and human distributions (Rolian and Gordon, 2013; but see Watkins et al., 1993; Alba et al., 2003; Almécija and Alba, 2014).

Discussion and conclusions

Although most elements of the hand skeleton are represented at Sterkfontein, none can be confidently attributed to the same individual or potentially even to the same species (Tobias, 1978; Susman, 1998; Lockwood and Tobias, 2002). The discovery of the StW 573 associated skeleton may also add to the taxonomic diversity at Sterkfontein (Clarke, 1999, 2013). Therefore, it is challenging to draw conclusions about overall hand function in *A. africanus*—the taxon to which most of these specimens are traditionally attributed—and how it might differ from that of other fossil hominins. Although the morphology is generally similar across the sample where there are duplicates of the same element, there are differences in size that are quite remarkable within the context of all South African hominins. In particular, StW 382 MC2, StW 478 PP1, and the StW 331 intermediate phalanx are among some of the largest specimens in our South African fossil hominin sample. These specimens suggest the presence of individuals at Sterkfontein with much larger hands, and presumably larger body size, than those of *A. sediba* MH2, *H. naledi*, and the Swartkrans hominins. Indeed, Ricklan (1987: 658) proposed that StW 382 possibly represented a large male individual and "may not be entirely typical of *A. africanus*." If we assume that all of the Sterkfontein hand bones described here are attributed to *A. africanus* (e.g., Ricklan, 1987, 1990; Green and Gordon, 2008), then there is also a potentially large degree of sexual dimorphism in this taxon (as well as potential temporal variation in size). StW 382 is approximately the same size as the MC2 from the A.L. 438-1 *A. afarensis*, presumed male, partial skeleton, one of the largest known Hadar individuals (Kimbel et al., 1994; Drapeau et al., 2005). The StW 331 intermediate phalanx is longer than any IPs in our comparative sample, apart from *Ar. ramidus*, which has much longer fingers than later hominins (Lovejoy et al., 2009; Almécija et al., 2015a), and is only similar to the IP3 (the longest ray) of *H. sapiens* and Neandertals. This specimen may belong to a larger individual (e.g., similar in size to or the same individual as the StW 382 specimen) or indicate longer fingers in the Sterkfontein hominins, although the latter is not supported by resampling analyses here.

Our analyses of the intrinsic hand proportions suggest that, if the Sterkfontein hand elements all belong to the same species, then *A. africanus* had human-like

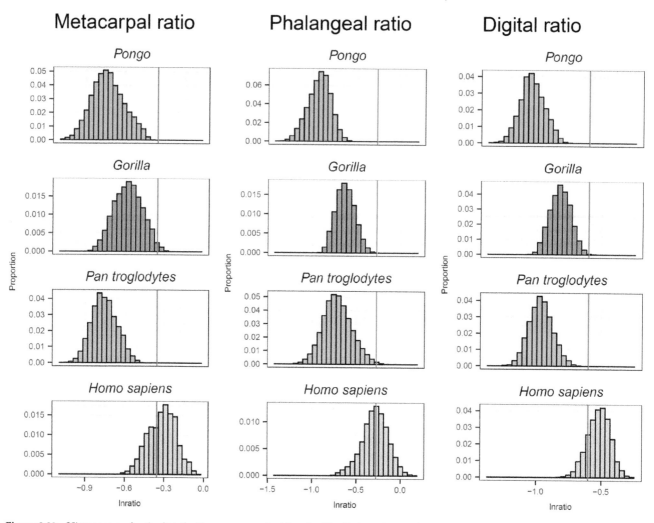

Figure 9.20 Histograms of ratio distributions among extant hominoids. Ratio estimates for the Sterkfontein sample are indicated by the red vertical line. Extant hominoid distributions whose 95% confidence limits exclude the ratio estimate from Sterkfontein are indicated with an asterisk (see also Table 9.6).

proportions in the length of the thumb and third ray, as well as in the length of the metacarpals and proximal phalanges within these rays (Ostrofsky and Richmond, 2015). Green and Gordon's (2008) resampling analysis of Sterkfontein metacarpals across rays 1–4 (and based on a larger sample size) found a similar result with regard to length proportions, but that *A. africanus* was more ape-like in the relative breadth of the metacarpals. The long thumb relative to the fingers that characterizes the human hand has long be accepted as critical for forceful, pad-to-pad precision grip abilities (e.g., Napier, 1956; Marzke, 1997; Susman, 1998) and similar intrinsic hand proportions in *A. africanus* suggest that it, like *A. sediba* (Kivell et al., 2011), *H. naledi* (Kivell et al., 2015), and possibly *A. afarensis* (Alba et al., 2003; Rolian and Gordon, 2013; Almécija and Alba, 2014), may have been capable of similar precision grips and frequent tool use (Ricklan, 1987, 1990; Green and

Gordon, 2008). However, such hand proportions may be plesiomorphic within the hominoid clade and later exapted in (some) hominins for tool-related behaviors (Alba et al., 2003; Almécija et al., 2015a). Furthermore, the relative gracility of *A. africanus* metacarpals (Green and Gordon, 2008) suggests that their hands did not undergo the same magnitude or frequency of stress from manipulative activities as is reflected in the external morphology of later *Homo*, including *H. naledi*, hands. The potentially less mobile trapezium-MC1 joint, absence of a fully developed palmar pulp on the distal thumb, more limited pronation of the index finger, and potentially more wedge-shaped trapezoid inferred from the preserved external morphology, is consistent with lower manipulative loading of the thumb than is typical of later *Homo* (Tocheri, 2007; Tocheri et al., 2008; Marzke et al., 2010). That being said, analyses of the internal bone structure, which can respond more quickly to loads incurred during life than external morphology, show a human-like distribution of trabecular bone within *A. africanus* metacarpals that has been interpreted as evidence of frequent use of forceful, human-like precision grips (Skinner et al., 2015; but see Almécija et al., 2015b). Stone tools have yet to be found in direct association with *A. africanus* fossils but were present within the South African archaeological record during its time period (Kuman and Clarke, 2000), with much earlier evidence of stone tool use and tool-making in East Africa (McPherron et al., 2010; Harmand et al., 2015). Thus, frequent and adept tool use in Sterkfontein hominins, be it of stone or organic tools, is not necessarily surprising (see also Alba et al., 2003; Almécija et al., 2015b).

The hand, of course, does not exist in isolation from the remainder of the skeleton. For example, limb proportions within *A. africanus* (e.g., StW 431 partial skeleton) have been shown to be more ape-like, with a relatively longer upper limb and smaller lower limb joint sizes than those of *A. afarensis* and humans (McHenry and Berger, 1998a, 1998b; Dobson, 2005; Green et al., 2007; see also Gordon et al., Chapter 17). Humeral diaphyseal strength in *A. africanus* StW 431 is strong (relative to distal humeral articular breadth), being more similar to that of chimpanzees than to humans (Ruff et al., 2016; Chapter 16, this volume). This broader comparative context, in addition to moderately curved manual phalanges (Kivell et al., 2015; Pickering et al., 2018), suggests a greater reliance on forelimb-dominated locomotor behaviors and perhaps selection for more frequent use of an arboreal environment in *A. africanus* than is found in *A. afarensis* (McHenry and Berger, 1998a, 1998b; Green et al., 2007) and most later hominins (e.g., Niewoehner, 2006; Domínguez-Rodrigo et al., 2015; but see Moyà-Solà et al., 2007; Kivell et al., 2015). This behavioral reconstruction is also consistent with a more ape-like inner ear morphology (semicircular canals) in *A. africanus*, suggesting a locomotor repertoire with arboreal postures (Spoor et al., 1994) and the palaeoenvironmental reconstruction of Member 4, from which most of the hand elements derive (Appendix I), as a tropical to subtropical gallery forest and woodland (Kuman and Clarke, 2000; but see Pickering et al., 2018). Mosaic morphologies of ape-like, human-like and unique features characterize South African hominins, including the skeletons of *A. sediba* (e.g., Berger et al., 2010; Churchill et al., 2013, 2018; DeSilva et al., 2018), *H. naledi* (e.g., Berger et al., 2015; Harcourt-Smith et al., 2015; Feuerriegel et al., 2017), and what little postcrania are confidently attributed of *A. robustus* (e.g., Lague and Jungers, 1996; Gebo and Schwartz, 2006), suggesting locomotor repertories that included both distinct bipedal mechanics (see also Gordon et al., Chapter 17, this volume) as well as some component of arboreal climbing.

Thus, within this broader context, the Sterkfontein fossil hominin remains are not unusual. The Sterkfontein hand fossils suggest an overall manipulative and locomotor loading regime that was more similar to that of other South African australopiths and distinct from that of later *Homo*, but more refined functional interpretations require additional fossil evidence, particularly from associated hand skeletons.

Acknowledgments

We are grateful to the Editors for the opportunity to contribute to this valuable volume on the Sterkfontein postcrania. We thank all of the curators and researchers who have generously shared the specimens in their care, including F. Mayer and S. Jancke (Berlin Natural History Museum), C. Boesch and J-J Hublin (Max Planck Institute for Evolutionary Anthropology), M. Teschler-Nicola and R. Muehl (Vienna Natural History Museum), E. Gilissen, M. Louette, and W. Wendelin (Royal Museum for Central Africa), J. Moggi Cecchi and S. Bortoluzzi (University of Florence), B. Zipfel and L. Berger (University of the Witwatersrand), S. Potze and L. Kgasi (Ditsong National Museum of Natural History), A. Kweka and A. Gidna (National Museum of Tanzania), B. Kimbel, T. White, B. Asfaw, G. Suwa, and G. Shimelies (National Museum of Ethiopia), and Y. Rak and I. Hershkovitz (Tel Aviv University). We thank Nick Stephens for sharing comparative data of Barma Grande 2 and Arene Candide 2 and Job Kibii for collecting some of the modern human comparative data. This research was funded by the European Research Council Starting Grant 336301 and the Max Planck Society (TLK).

References

Alba, D.M., Moya-Sola, S., Kohler, M., 2003. Morphological affinities of the *Australopithecus afarensis* hand on the basis of manual proportions and relative thumb length. Journal of Human Evolution 44, 225–254.

Almécija, S., Alba, D.M., 2014. On manual proportions and pad-to-pad precision grasping in *Australopithecus afarensis*. Journal of Human Evolution 73, 88–92.

Almécija, S., Moyà-Solà, S., Alba, D.M., 2010. Early origin for human-like precision grasping: a comparative study of pollical distal phalanges in fossil hominins. PLoS ONE 5, e11727.

Almécija, S., Smaers, J.B., Jungers, W.L., 2015a. The evolution of human and ape hand proportions. Nature Communications 6, 7717.

Almécija, S., Wallace, I.J., Judex, S., Alba, D.M., Moyà-Solà, S., 2015b. Comment on "Human-like hand use in *Australopithecus africanus*." Science 348, 1101-a.

Begun, D.R., 2004. Knuckle-walking and the origin of human bipedalism. In: From Biped to Strider. Springer: Boston, pp. 9–33.

Berger, L.R., de Ruiter, D.J., Churchill, S.E., Schmid, P., Carlson, K.J., Dirks, P.H.G.M., Kibii, J.M., 2010. *Australopithecus sediba*: a new species of *Homo*-like australopith from South Africa. Science 328, 195–204.

Berger, L.R., Hawks, J., de Ruiter, D.J., Churchill, S.E., Schmid, P., Delezene, L.K., Kivell, T.L., Garvin, H.M., Williams, S.A., DeSilva, J.M., Skinner, M.M., Musiba, C.M., Cameron, N., Holliday, T.W., Harcourt-Smith, W., Ackermann, R.R., Bastir, M., Bogin, B., Bolter, D., Brophy, J., Cofran, Z.D., Congdon, K.A., Deane, A.S., Dembo, M., Drapeau, M., Elliot, M.C., Feuerriegel, E.M., Garcia-Martinez, D., Green, D.J., Gurtov, A., Irish, J.D., Kruger, A., Laird, M.F., Marchi, D., Meyer, M.R., Nalla, S., Negash, E.W., Orr, C.M., Radovcic, D., Schroeder, L., Scott, J.E., Throckmorton, Z., Tocheri, M.W., VanSickle, C., Walker, C.S., Wei, P., Zipfel, B., 2015. *Homo naledi*, a new species of the genus *Homo* from the Dinaledi Chamber, South Africa. eLife 205, e09560.

Bettinger, P.C., Berger, R.A., 2001. Functional ligamentous anatomy of the trapezium and trapeziometacarpal joint (gross and arthroscopic). Hand Clinic 17, 151–168.

Bettinger, P.C., Linscheid, R.L., Berger, R.A., Cooney, W.P. III, An, K.-N., 1999. An anatomic study of the stabilizing ligaments of the trapezium and trapeziometacarpal joint. Journal of Hand Surgery A24, 786–798.

Bush, M.E., Lovejoy, C.O., Johanson, D.C., Coppens, Y., 1982. Hominid carpal, metacarpal, and phalangeal bones recovered from the Hadar Formation—1974-1977 collections. American Journal of Physical Anthropology 57, 651–677.

Christensen, A.M., 2009. Techniques for siding manual phalanges. Forensic Science International 193, 84–87.

Churchill, S.E., Green, D.J., Feuerriegel, E.M., Macias, M.E., Mathews, S., Carlson, K.J., Schmid, P., Berger, L.R., 2018. The shoulder, arm, and forearm of *Australopithecus sediba*. PaleoAnthropology 2018, 234–281.

Churchill, S.E., Holliday, T.W., Carlson, K.J., Jashashvili, T., Macias, M.E., Matthews, S., Sparling, T.L., Schmid, P., de Ruiter, D.J., Berger, L.R. 2013. The upper limb of *Australopithecus sediba*. Science 340, 1233477.

Clarke, R.L., 1999. Discovery of complete arm and hand of the 3.3 million-year-old *Australopithecus* skeleton from Sterkfontein. South African Journal of Science 95, 477–480.

Clarke, R.J., 2008. Latest information on Sterkfontein's *Australopithecus* skeleton and a new look at *Australopithecus*. South African Journal of Science 104, 443–449.

Clarke, R.J., 2013. *Australopithecus* from Sterkfontein Caves, South Africa. In: Reed, K.E., Fleagle, J.G., Leakey, R.E. (Eds.), The Paleobiology of *Australopithecus*. Springer: Dordrecht, pp. 105–123.

Day, M.H., 1978. Functional interpretations of the morphology of postcranial remains of early African hominids. In: Jolly, C.J. (Ed.), Early Hominids of Africa. St. Martin's Press: New York, pp. 311–345.

Deane, A.S., Begun, D.R., 2008. Broken fingers: retesting locomotor hypotheses for fossil hominoids using fragmentary proximal phalanges and high-resolution polynomial curve fitting (HR-PCF). Journal of Human Evolution 55, 691–701.

DeSilva, J.M., Carlson, K.J., Claxton, A.G., Harcourt-Smith, W.E., MCNutt, E.J., Sylvester, A.D., Walker, C.S., Zipfel, B., Churchill, S.E., Berger, L.R., 2018. The anatomy of the lower limb skeleton of *Australopithecus sediba*. PaleoAnthropology 2018, 357–405.

Dobson, S.D., 2005. Are the differences between StW 431 (*Australopithecus africanus*) and A.L. 288-1 (*A. afarensis*) significant? Journal of Human Evolution 49, 143–154.

Domínguez-Rodrigo, M., Pickering, T.R., Almécija, S., Heaton, J.L., Baquedano, E., Mabulla, A., Uribelarrea, D., 2015. Earliest modern human-like hand bone from a new >1.84-million-year-old site at Olduvai in Tanzania. Nature Communications 6, 7987.

Drapeau, M.S.M., 2012. Forelimb adaptations in *Australopithecus afarensis*. In: Reynolds, S.C., and Gallagher, A. (Eds.), African Genesis: Perspectives on Hominin Evolution. Cambridge University Press: Cambridge, pp. 223–247.

Drapeau, M.S.M., 2015. Metacarpal torsion in apes, humans, and early *Australopithecus*: implications for manipulatory abilities. PeerJ 3, e1311.

Drapeau, M.S.M., Ward, C.V., Kimbel, W.H., Johanson, D.C., Rak, Y., 2005. Associated cranial and forelimb remains attributed to *Australopithecus afarensis* from Hadar, Ethiopia. Journal of Human Evolution 48, 593–642.

Feuerriegel, E.M., Green, D.J., Walker, C.S., Schmid, P., Hawks, J., Berger, L.R., Churchill, S.E., 2017. The upper limb of *Homo naledi*. Journal of Human Evolution 104, 155–173.

Galletta, L., Stephens, N.B., Kivell, T.L., Marchi, D., 2019. Three-dimensional geometric morphometric analysis of the first metacarpal distal articular surface in humans, great apes and fossil hominins. Journal of Human Evolution 132, 119–136.

Gebo, D.L., Schwartz, G.T., 2006. Foot bones from Omo: implications for hominid evolution. American Journal of Physical Anthropology 19, 499–511.

Green, D.J., Gordon, A.D., 2008. Metacarpal proportions in *Australopithecus africanus*. Journal of Human Evolution 55, 705–719.

Green, D.J., Gordon, A.D., Richmond, B.G., 2007. Limb-size proportions in *Australopithecus afarensis* and *Australopithecus africanus*. Journal of Human Evolution 52, 187–200.

Groucutt, H.S., Grün, R., Zalmout, I.A., Drake, N. A., Armitage, S.J., Candy, I., Clark-Wilson, R., Louys, J., Breeze, P.S., Duval, M., Buck, L.T., Kivell, T.L., Pomeroy, E., Stephens, N.B., Stock, J.T., Stewart, M., Price, G.J., Kinsley, L., Sung, W.W., Alsharekh, A., Al-Omari, A., Zahir, M., Memesh, A.M., Abdulshakoor, A.J., Al-Masari, A.M., Bahameem, A.A., Al Murayyi, K.M.S., Zahrani, B., Scerri, E.L.M., Petraglia, M.D., 2018. *Homo sapiens* in Arabia by 85,000 years ago. Nature Ecology and Evolution 2, 800–809.

Hamrick, M.W., 1996. Locomotor adaptations in the wrist joints of early Tertiary primates (Adaptiformes). American Journal of Physical Anthropology 100, 585–604.

Harcourt-Smith, W.E.H., Throckmorton, Z., Congdon, K.A., Zipfel, B., Deane, A.S., Drapeau, M.S.M., Churchill, S.E., Berger, L.R., DeSilva, J.M., 2015. The foot of *Homo naledi*. Nature Communications 6, 8432.

Harmand, S., Lewis, J.E., Feibel, C.S., Lepre, C.J., Prat, S., Lenoble, A., Boës, X., Quinn, R.L., Brenet, M., Arroyo, A., Taylor, N., Clément, S., Daver, G., Brugal, J.-P.,, Leakey, L., Mortlock, R.A., Wright, J.D., Lokorodi, S., Kirwa, C., Kent, D.V., Roche, H., 2015. 3.3-million-year-old stone tools from Lomekwi 3, West Turkana, Kenya. Nature 521, 310–315.

Hunt, K.D., 1991. Mechanical implications of chimpanzee positional behavior. American Journal of Physical Anthropology 86, 521–536.

Jacofsky, M.C., 2009. Comparative muscle moment arms of the primate thumb: *Homo, Pan, Pongo*, and *Papio*. Ph.D. Dissertation, Arizona State University.

Jungers, W.L., Godfrey, L.R., Simons, E.L., Chatrath, P.S., 1997. Phalangeal curvature and positional behavior in extinct sloth lemurs (Primates, Palaeopropithecidae). Proceedings of the National Academy of Sciences USA 94, 11998–12001.

Karakostis, F.A., Hotz, G., Scherf, H., Wahl, J., Harvati, K., 2017. Occupational manual activity is reflected on the patterns among hand entheses. American Journal of Physical Anthropology 164, 30–40.

Kimbel, W.H., Johanson, D.C., Rak, Y., 1994. The first skull and other new discoveries of *Australopithecus afarensis* at Hadar, Ethiopia. Nature 368, 449–451.

Kivell, T. L., 2015. Evidence in hand: recent discoveries and the early evolution of human manual manipulation. Philosophical Transactions of the Royal Society B: Biological Sciences 370, 20150105.

Kivell, T.L., Churchill, S.E., Kibii, J.M., Schmid, P., Berger, L.R., 2018b. The hand of *Australopithecus sediba*. PaleoAnthropology 2018, 282–333.

Kivell, T.L., Deane, A.S., Tocheri, M.W., Orr, C.M., Schmid, P., Hawks, J., Berger, L.R., Churchill, S.E., 2015. The hand of *Homo naledi*. Nature Communications 6, 8431.

Kivell, T.L., Kibii, J.M., Churchill, S.E., Schmid, P., Berger, L.R., 2011. *Australopithecus sediba* hand demonstrates

mosaic evolution of locomotor and manipulative abilities. Science 333, 1411–1417.

Kivell, T.L., Rosas, A., Estralrrich, A., Huguet, R., García-Tabernero, A., Ríos, L., de la Rasilla, M., 2018a. New Neanderthal wrist bones from El Sidrón, Spain (1994–2009). Journal of Human Evolution 114, 445–75.

Kivell, T. L., Churchill, S. E., Kibii, J. M., Schmid, P., & Berger, L. R., 2018). The hand of *Australopithecus sediba*. PaleoAnthropology, 282–333.

Kuman, K., Clarke, R.J., 2000. Stratigraphy, artefact industries and hominid associations for Sterkfontein Member 5. Journal of Human Evolution 38, 827–847.

Lague, M.R., Jungers, W.L., 1996. Morphometric variation in Plio-Pleistocene hominid distal humeri. American Journal of Physical Anthropology 101, 401–427.

Lewis, O. J., 1989. Functional morphology of the evolving hand and foot. Clarendon Press: Oxford.

Lockwood, C.A., Tobias, P.V., 2002. Morphology and affinities of new hominin cranial remains from Member 4 of the Sterkfontein Formation, Gautang Province, South Africa. Journal of Human Evolution 42, 389–450.

Lorenzo, C., Arsuaga, J.L. Carretero, J.M., 1999. Hand and foot remains from the Gran Dolina Early Pleistocene site (Sierra de Atapuerca, Spain). Journal of Human Evolution 37, 501–522.

Lorenzo, C., Pablos, A., Carretero, J.M., Huguet, R., Valverdú, J., Martinón-Torres, M., Arsuaga, J.L., Carbonell, E., Bermúdez de Castro, J.M., 2015. Early Pleistocene human hand phalanx from the Sima del Elefante (TE) cave site in Sierra de Atapuerca (Spain). Journal of Human Evolution 78, 114–121.

Lovejoy, C.O., Simpson, S.W., White, T.D., Asfaw, B., Suwa, G., 2009. Careful climbing in the Miocene: the forelimbs of *Ardipithecus ramidus* and humans are primitive. Science 326, 70e1–70e8.

Marchi, D., Proctor, D.J., Huston, E., Nicholas, C.L., Fischer, F., 2017. Morphological correlates of the first metacarpal proximal articular surface with manipulative capabilities in apes, humans and South African early hominins. Comptes Rendus Palevol 16, 645–654.

Marzke, M.W., 1983. Joint functions and grips of the *Australopithecus afarensis* hand, with special reference to the region of the capitate. Journal of Human Evolution 12, 197–211.

Marzke, M.W., 1997. Precision grips, hand morphology, and tools. American Journal of Physical Anthropology 102, 91–110.

Marzke, M.W., Marzke, R.F., 1987. The third metacarpal styloid process in humans: origin and functions. American Journal of Physical Anthropology 73, 415–431.

Marzke, M.W., Shrewsbury, M.M., Horner, K.E., 2007. Middle phalanx skeletal morphology in the hand: can it predict flexor tendon size and attachments?

American Journal of Physical Anthropology 134, 141–151.

Marzke, M.W., Tocheri, M.W., Steinberg, B., Femiani, J.D., Reece, S.P., Linscheid, R.L., Orr, C.M., Marzke, R.F., 2010. Comparative 3D quantitative analyses of trapeziometacarpal joint surface curvatures among living catarrhines and fossil hominins. American Journal of Physical Anthropology 141, 38–51.

Marzke, M.W., Toth, N., Schick, K., Reece, S., Steinberg, B., Hunt, K., Linscheid, R.L., An, K.-N., 1998. EMG study of hand muscle recruitment during hard hammer percussion manufacture of Oldowan tools. American Journal of Physical Anthropology 105, 315–332.

Marzke, M.W., Wullstein, K.L., Viegas, S.F., 1992. Evolution of the power ("squeeze") grip and its morphological correlates in hominids. American Journal of Physical Anthropology 89, 283–298.

Marzke, M.W., Wullstein, K.L., Viegas, S.F., 1994. Variability at the carpometacarpal and midcarpal joints involving the fourth metacarpal, hamate and lunate in Catarrhini. American Journal of Physical Anthropology 93, 229–240.

McHenry, H. M., 1983. The capitate of *Australopithecus afarensis* and *Australopithecus africanus*. American Journal of Physical Anthropology 62, 187–198.

McHenry, H.M., Berger, L.R., 1998a. Body proportions in *Australopithecus afarensis* and *A. africanus* and the origin of the genus *Homo*. Journal of Human Evolution 35, 1–22.

McHenry, H.M., Berger, L.R., 1998b. Limb lengths in *Australopithecus* and the origin of the genus *Homo*. South African Journal of Science 94, 447–450.

McPherron, S.P., Zeresenay, A., Marean, C.W., Wynn, J.G., Reed, D., Geraads, D., Bobe, R., Béarat, H.A., 2010. Evidence for stone-tool-assisted consumption of animal tissues before 3.39 million years ago at Dikika, Ethiopia. Nature 466, 857–860.

Milella, M., Belcastro, M.G., Zollikofer, C.P.E., Mariotti, V., 2012. The effect of age, sex, and physical activity on entheseal morphology in a contemporary Italian skeletal collection. American Journal of Physical Anthropology 148, 379–388.

Moyà-Solà, S., Köhler, M., Alba, D.M., Almécija, S., 2007. Taxonomic attribution of the Olduvai Hominid 7 manual remains and the functional interpretation of hand morphology in robust australopithecines. Folia Primatologica 79, 215–250.

Musgrave, J.H., 1971. How dexterous was Neanderthal man? Nature 233, 538–541.

Napier, J. R., 1956. The prehensile movements of the human hand. Journal of Bone and Joint Surgery 38(B), 902–913.

Nguyen, N.H., Pahr, D.H., Gross, T., Skinner, M.M., Kivell, T.L., 2014. Micro-finite element (µFE) modeling of the siamang (*Symphalangus syndactylus*) third proximal phalanx: the functional role of curvature and the flexor sheath ridge. Journal of Human Evolution 67, 60–75.

Niewoehner, W.A., 2006. Neandertal hands in their proper perspective. In: Harvati, K., Harrison, T. (Eds.), Neanderthals Revisited: New Approaches and Perspectives. Springer: Dordrecht, pp. 157–190.

Niewoehner, W.A., Weaver, A.H., Trinkaus, E., 1997. Neandertal capitate-metacarpal articular morphology. American Journal of Physical Anthropology 103, 219–223.

Ostrofsky, K.R., Richmond, B.G., 2015. Manual proportions in *Australopithecus*: a comparative analysis including new material from Sterkfontein. American Journal of Physical Anthropology 156, 243.

Partridge, T.C., Granger, D.E., Caffee, M.W., Clarke, R.J., 2003. Lower Pliocene hominid remains from Sterkfontein. Science 300, 607–612.

Pickering, T.R., Heaton, J.L., Clarke, R.J., Stratford, D., 2018. Hominin hand bone fossils from Sterkfontein Caves, South Africa (1998-2003 excavations). Journal of Human Evolution 118, 89–102.

Pickering, R., Herries, A.I.R., Woodhead, J.D., Hellstrom, J.C., Green, H.E., Paul, B., Ritzman, T., Strait, D.S., Schoville, B.J., Hancox, P.J., 2018. U-Pb-dated flowstones restrict South African early hominin record to dry climate phases. Nature 565, 226–229.

Rabey, K.N., Green, D.J., Taylor, A.B., Begun, D.R., Richmond, B.G., McFarlin, S.C., 2015. Locomotor activity influences muscle architecture and bone growth but not muscle attachment site morphology. Journal of Human Evolution 78, 91–102.

Rein, T.R., Harvati, K., 2012. Exploring third metacarpal capitate facet shape in early hominins. Anatomical Record 296, 240–249

Richmond, B.G., 1998. Ontogeny and biomechanics of phalangeal form in primates. Ph.D. Dissertation, State University of New York at Stony Brook.

Richmond, B.G., 2007. Biomechanics of phalangeal curvature. Journal of Human Evolution 53, 678–690.

Richmond, B.G., Roach, N.T., Ostrofsky, K.R., 2016. Evolution of the early hominin hand. In: Kivell, T.L., Lemelin, P., Richmond, B.G., Schmitt, D. (Eds.), The evolution of the primate hand: anatomical, developmental, functional and paleontological evidence. Springer: New York, pp. 515–543.

Ricklan, D.E. 1987. Functional anatomy of the hand of *Australopithecus africanus*. Journal of Human Evolution 16, 643–664.

Ricklan, D. E., 1990. The precision grip in *Australopithecus africanus*: anatomical and behavioural correlates. In: Sperber, G.E. (Ed.), From Apes to Angels: Essays in Anthropology in Honor of Phillip V. Tobias. Wiley-Liss: New York, pp. 171–183.

Rolian, C., Gordon, A.D., 2013. Reassessing manual proportions in *Australopithecus afarensis*. American Journal of Physical Anthropology 152, 393–406.

Ruff, C.B., Burgess, M.L., Ketcham, R.A., Kappelman, J., 2016. Limb bone structural proportions and locomotor behavior in A.L. 288–1 ("Lucy"). Plos ONE 11, e01666095.

Shrewsbury, M.M., Marzke, M.W., Linscheid, R.L., Reece, S.P., 2003. Comparative morphology of the pollical distal phalanx. American Journal of Physical Anthropology 121, 30–47.

Skinner, M.M., Stephens, N.B., Tsegai, Z.J., Foote, A.C., Nguyen, N.H., Gross, T., Pahr, D.H., Hublin, J.-J., Kivell, T.L., 2015. Human-like hand use in *Australopithecus africanus*. Science 347, 395–399.

Sládek, V., Trinkaus, E., Hillson, S.W., Holliday, T.W., 2000. The People of the Pavlovian: Skeletal Catalogue and Osteometrics of the Gravettian Fossil Hominids from Dolní Věstonice and Pavlov, Dolní Věstonice Studies 5. Archeologicky ústav AV ČR: Brno.

Smith, R.J., 1999. Statistics of sexual size dimorphism. Journal of Human Evolution 36, 423–459.

Spoor, F., Wood, B., Zonneveld, F., 1994. Implications of early hominid labyrinthine morphology for evolution of human bipedal locomotion. Nature 369, 645–648.

Stratford, D., Heaton, J.L., Pickering, T.R., Caruana, M.V., Shadrach, K., 2016. First hominin fossils from Milner Hall, Sterkfontein, South Africa. Journal of Human Evolution 91, 167–73.

Susman, R.L., 1979. Comparative and functional morphology of hominoid fingers. American Journal of Physical Anthropology 60, 215–236.

Susman, R.L., 1988. New postcranial remains from Swartkrans and their bearing on the functional morphology and behavior of *Paranthropus robustus*. In: Grine, F.E. (Ed.), Evolutionary History of the "Robust" Australopithecines. Aldine de Gruyter: New York, pp. 149–172.

Susman, R.L., 1989. New hominid fossils from the Swartkrans formation (1979-1986 excavations): postcranial specimens. American Journal of Physical Anthropology 79, 451–474.

Susman, R.L., 1998. Hand function and tool behavior in early hominids. Journal of Human Evolution 35, 23–46.

Tobias, P.V., 1978. The earliest Transvaal members of the genus *Homo* with another look at some problems of hominid taxonomy and systematics. Zeitschrift für Morphologie und Anthropologie 69, 225–265.

Tocheri, M.W., 2007. Three-dimensional riddles of the radial wrist: derived carpal and carpometacarpal joint morphology in the genus *Homo* and the implications for understanding the evolution of stone tool-related behaviors in hominins. Ph.D. Dissertation, Arizona State University.

Tocheri, M. W., Orr, C. M., Jacofsky, M. C., Marzke, M. W., 2008. The evolutionary history of the hominin hand since the last common ancestor of *Pan* and *Homo*. Journal of Anatomy 212, 544–562.

Trinkaus, E., 1983. The Shanidar Neandertals. Academic Press: New York.

Trinkaus, E., 2016a. The Krapina human postcranial remains: morphology, morphometrics and paleopathology. University of Zagreb: Zagreb.

Trinkaus, E., 2016b. The evolution of the hand in Pleistocene *Homo*. In: Kivell, T.L., Lemelin, P., Richmond, B.G., Schmitt, D., (Eds.), The evolution of the primate hand: anatomical, developmental, functional and paleontological evidence. Springer: New York, pp. 545–571.

Trinkaus, E., Bailey, S.E., Zilhão, J., 2001. Upper Paleolithic human remains from the Gruta do Caldeirão, Tomar, Portugal. Revista Portuguesa de Arqueologia 4, 5–17.

Trinkaus, E., Buzhilova, A.P., Mednikova, M.B., Dobrovolskaya, M.V., 2014. The People of Sunghir: Burials, Bodies, and Behavior in the Earlier upper paleolithic. Oxford University Press: Oxford.

Trinkaus, E., Svoboda, J.A., Wojtal, P., Nývltová Fišáková, M., Wilczynski, J., 2010. Human remains from the Moravian Gravettian: morphology and taphonomy of additional elements from Dolní Věstonice II and Pavlov I. International Journal of Osteoarchaeology 20, 645–669.

Trinkaus, E., Walker, M.J., 2017. The people of Palomas: Neandertals from the Sima de las Palomas del Cabezo Gordo, Southeastern Spain. Texas A&M University Press: College Station.

Van Leeuwen, T., Vanhoof, M.J., Kerkhof, F.D., Stevens, J.M., Vereecke, E.E., 2018. Insights into the musculature of the bonobo hand. Journal of Anatomy 233, 328–340.

Viegas, S.F., Crossly, M., Marzke, M.W., Wullstein, K., 1991. The fourth metacarpal joint. Journal of Hand Surgery (Am) 16, 525–533.

Ward, C.V., Tocheri, M.W., Plavcan, J.M., Brown, F.H., Manthi, F.K., 2013. Early Pleistocene third metacarpal from Kenya and the evolution of modern human-like hand morphology. Proceedings of the National Academy of Sciences 111, 121–124.

Ward, C.V., Kimbel, W.H., Harmon, E.H., Johanson, D.C., 2012. New postcranial fossils of *Australopithecus afarensis* from Hadar, Ethiopia (1990–2007). Journal of Human Evolution 63, 1–51.

Watkins, B., Parkinson, D., Mensforth, R., 1993. Morphological evidence for the abandonment of forelimb dominance in *Australopithecus afarensis*. American Journal of Physical Anthropology 16 (Suppl.), 205.

Williams-Hatala, E.M., Hatala, K.G., Hiles, S., Rabey, K. N., 2016. Morphology of muscle attachment sites in the modern human hand does not reflect muscle architecture. Scientific Reports 6, 28353.

10

Thoracolumbar vertebrae and ribs

Carol V. Ward, Burt Rosenman, Bruce Latimer, and Shahed Nalla

Thoracic and lumbar vertebrae from Sterkfontein include those attributed to the two partial skeletons Sts 14 and StW 431 (see Chapter 4, this volume as well as the series of four vertebrae preserved in articulation StW 8/41). Most of these fossils have been discussed and figured in previous publications to varying degrees, and many have been analyzed by multiple researchers. Here, these elements are described briefly and figured, and their comparative morphology summarized. Isolated lumbar vertebrae Sts 73 and StW 572 also are described here for the first time, despite Sts 73 being among the earliest fossil discoveries at Sterkfontein (see Appendix 1). In addition, we provide new analyses of the thoracolumbar transition in *Australopithecus africanus*.

Descriptions

Vertebrae

Sts 14—Associated thoracic and lumbar vertebrae

The Sts 14 partial skeleton preserves the sacrum and inferiormost 15 presacral elements (Ward et al., Chapter 4, this volume). Haeusler and Ruff (Chapter 11, this volume) describe the sacrum with the pelvis and presacral elements. These vertebrae were treated briefly by Robinson (1972) and Benade (1980). However, the latter researcher described only the two caudalmost elements (Sts 14a and b). Because the original number of thoracic elements is unknown for the Sts 14 and the StW 431 specimens, vertebral positions are described relative to the sacrum, and numbered superiorly as presacral vertebrae (PSV) 1, 2, and so on, continuing superiorly. As Benade (1980) notes, the catalogue numbers of the Sts 14 vertebrae do not always correspond to their anatomical order from inferior

to superior, so they are described in anatomical sequence rather than by catalogue number. The following discussion of vertebrae are oriented anatomically, rather than in alphanumeric order.

The lumbar vertebrae of Sts 14 were reconstructed with a plaster mix compounded by John Robinson in the 1960s. The exact materials used in the reconstructions are unknown, but it is clear that this material cannot be safely removed, especially given the fragile nature of the fossils themselves. In light of these reconstructed areas, the following descriptions focus primarily on the preserved portions of the fossils. Because the plaster cannot be safely removed mechanically, the fossils were CT-scanned and the plaster elements were, when possible, removed digitally. This effort was unexpectedly difficult as many of the fossils are partially demineralized as a consequence of the acid preparation used to remove the adhering matrix. Digital polygonal models of the original, unreconstructed bones are presented alongside models with plaster removed, accompanying the description of each specimen. These models should represent what the elements looked like prior to reconstruction. These digitally dissected models are approximately accurate, given the standard medical CT data available. It is possible that with higher contrast imaging, and imaging at finer resolution, we would be able to refine these models.

Sts 14p - PSV 15

Preservation. This specimen (Figure 10.1) is preserved in two fragments. The first fragment consists of the vertebral body with seven of the left pedicle and almost none of the right one. The second fragment includes the left transverse process and lamina, the inferior zygapophyseal facets, and part of the spinous process. The fragments do not join. There is some mild abrasion along the posterior half

Carol V. Ward, Burt Rosenman, Bruce Latimer, and Shahed Nalla, *Thoracolumbar vertebrae and ribs*. In: *Hominin Postcranial Remains from Sterkfontein, South Africa, 1936–1995*. Edited by: Bernhard Zipfel, Brian G. Richmond, and Carol V. Ward, Oxford University Press (2020). © Oxford University Press.
DOI: 10.1093/oso/9780197507667.003.0010

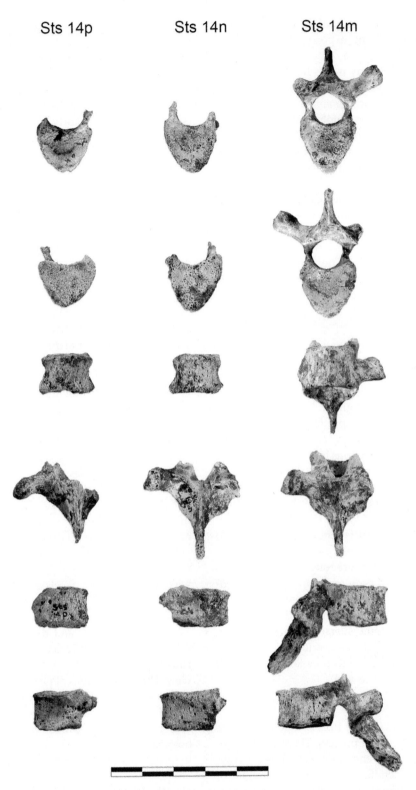

Figure 10.1   Sts 14p, Sts 14n, and Sts 14m in (from top to bottom) superior, inferior anterior, posterior, right lateral, and left lateral views.

of the superior endplate, and most of the original surface of inferior endplate is missing exposing the underlying cancellous bone.

Morphology. The vertebral body is less heart-shaped than in the others preserved. No anterior extension of bone is evident on either surface, and the anterior margin of the bone is rounded. The superior apophyseal ring is 4.2 mm wide where preserved anterolaterally. The body is not wedged. Dimensions of the costovertebral demifacets are difficult to quantify, as the margins grade gently into the surrounding bone, but they are slightly smaller than in Sts 14n relative. The transverse process was angled posteriorly, but its inclination cannot be quantified because of postmortem damage. The costotransverse foramen is 4.1 mm superoinferiorly by 5.4 mm transversely. It is angled somewhat inferiorly. The spinous process is inclined inferiorly, but this observation cannot be accurately measured due to damage.

Sts 14n - PSV 14

Preservation. This vertebra (Figure 10.1) is preserved in two parts. The body and broken pedicles are one fragment, and the laminae, left transverse process, and most of the spinous process form the other. The two pieces do not join. The vertebral body fragment has a mostly intact superior endplate with abraded portions especially anteriorly. The surface of the inferior endplate is missing, exposing the underlying cancellous bone. The left pedicle is preserved for about 4 mm, and the right pedicle for about 7 mm. The posterior fragment is missing the tip of the spinous process, and the margins of the zygapophyseal facets are abraded except for the right inferior one.

Morphology. The vertebral body is concave posteriorly and the anterior median keel is barely present. The superior costovertebral demifacets measure roughly 3 mm superoinferiorly by 4 mm transversely on both sides. The pedicles are thin mediolaterally. The transverse process was angled posteriorly, but its inclination is difficult to quantify as preserved. The costotransverse foramen is 4.7 mm in diameter and inclined slightly superiorly. The spinous process is inferiorly inclined, but its orientation is difficult to quantify accurately as preserved.

Sts 14m - PSV 13

Preservation. This vertebra (Figure 10.1) is largely complete except for the right transverse process. There are areas of abrasion along the vertebral endplates, including the right inferior costovertebral demifacet. The superior margins of the articular facets are broken, but very little bone appears to be missing. The costotransverse facet is also abraded, obscuring its dimensions but not its general morphology.

Morphology. The vertebral body is heart-shaped in the superior and inferior views. Anterior extensions of about 3 mm of bone are present along both surfaces, with

a rounded keel between them. The vertebral body is wedged anteriorly. The superior apophyseal ring is about 3 mm wide laterally where it is preserved, and the inferior one is up to 5 mm thick. The inferior endplate has a shallow depression anterolaterally. The zygapophyseal facets are set at about a 5° angle to a coronal plane both superiorly and inferiorly, with the superior facets facing slightly laterally and the inferior ones medially. The superior demifacets are just under 4 mm superoinferiorly and just over 6 mm transversely. The preserved left inferior one is about 5 mm wide transversely. The left transverse process is inclined posteriorly 35° as measured using a line through costovertebral and costotransverse facets (following Jellema et al., 1993). The costovertebral facet is 5.5 mm in diameter and angled slightly superiorly. The spinous process declines 55° to the inferior endplate.

Sts 14i - PSV12

Preservation. This specimen (Figure 10.2) is complete, except for the right transverse process which is broken at a point 13 mm from the internal surface of the neural foramen. The endplates have some abraded areas but these do not distort the original contours. The left transverse process has a small area of abraded bone along the posterior surface near its distal extent.

Morphology. The superior and inferior vertebral endplates are roughly heart-shaped. Both have 3 mm anterior extensions of bone beyond the apophyseal rings. The anterior keel of the body is present but rounded. In the lateral view the body is wedged anteriorly. The superior and inferior apophyseal rings are approximately 3.8 mm wide. The inferior one is slightly wider on the right anterolateral corner. The inferior endplate has a shallow depression anterolaterally. The superior costovertebral demifacets both measure 4.0 mm superoinferiorly by 6.6 mm transversely. The inferior facets are smaller but distinct. The pedicles are transversely narrow. The preserved left transverse process has a costal facet measuring 6.6 mm tall by 7.2 mm transversely, and this facet is angled superiorly. A line between costovertebral and costotransverse facets runs at an angle of 35° to the median plane in superior view. The zygapophyseal facets are set at about a 5° angle to a coronal plane both superiorly and inferiorly, with the superior facets facing slightly laterally and the inferior ones medially. The spinous process is long and declines posteriorly about 55° to the inferior endplate and tapers to its distalmost extent with no expansion of bone here.

Sts 14k - PSV11

Preservation. This vertebra (Figure 10.2) is complete except for much of the spinous process distal to a point 20 mm from the internal margin of the neural foramen, and about half of the right transverse process. There is a thin wedge of bone missing from the superior endplate and

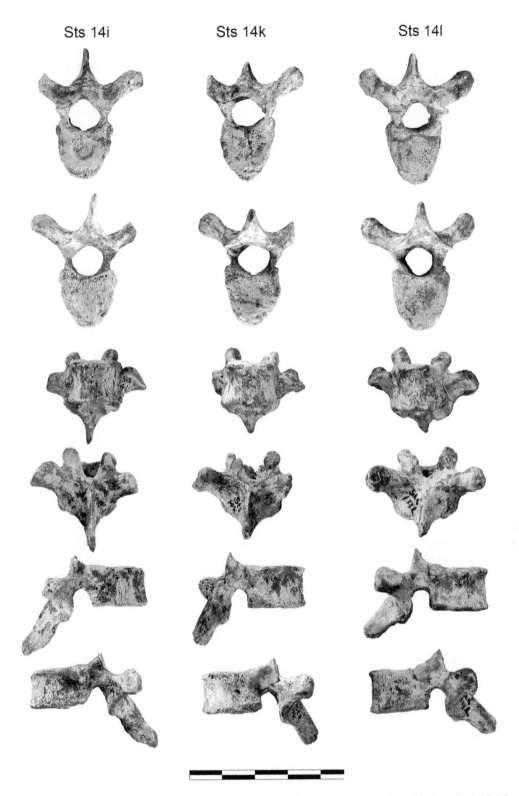

Figure 10.2 Sts 14i, Sts 14k, and Sts 14l in (from top to bottom) superior, inferior anterior, posterior, right lateral, and left lateral views.

the endplate margins are slightly abraded in a few places. A crack extends along the superior margin of the right superior zygapophyseal facet.

Morphology. The vertebral body appears unusually elongate anteroposteriorly due to an anterior extension of bone about 3 mm along the superior margin and 2 mm long inferiorly from the original body margin. The body is not appreciably wedged. The apophyseal rings are about 5 mm wide anteriorly and laterally and thin to about 2 mm posteriorly on both endplates. Nutrient foramina are visible on the anterolateral margins of the body. No anterior keel is present. Costovertebral demifacets are present superiorly and inferiorly. The left superior facets are 3.7 mm tall and 6.3 mm wide, and the right 3.5 mm by 6.2 mm. Dimensions of the inferior demifacets are difficult to measure as they round gently to the intervertebral surfaces, but both are about 4 mm wide anteroposteriorly. The transverse process has a strong, superiorly inclined costotransverse foramen 5.5 mm in diameter. This vertebra also displays asymmetries of the posterior elements, with the spinous process slightly skewed to the right. The spinous process sharply declines 60° relative to the inferior endplate.

Sts 14l - PSV10

Preservation. This vertebra (Figure 10.2) is complete except for the tip of the spinous process and surface abrasion over most of the superior endplate. There are fragments of bone missing along the left side and in the left superior costovertebral facet so that measurements cannot be taken on that side. The right inferior demifacet is weathered and cannot be measured.

Morphology. The endplates of this vertebra are heart-shaped. There is no anterior median keel. Superiorly, there is a small anterior extension of bone about 1.8 mm long. Inferiorly, there is a 2.5 mm extension of bone anterior to the apophyseal ring. The ring itself is 3.7 mm wide anteriorly, tapering to just over 2 mm wide posteriorly. Vertebral body wedging is difficult to determine due to damage. There is a slight depression internal to the ring along the left anterolateral surface that matches the elevation in the subjacent vertebral body Sts 14o. Nutrient foramina are apparent near the lateral midpoints of the body. There are costovertebral demifacets present superiorly and interiorly on both sides, at the body-pedicle junction superiorly and posterolateral margin of the body inferiorly. On the right, the superior facet is 4.5 mm superoinferiorly by 4.0 mm transversely, and the right one is 4.2 mm by 3.2 mm. The inferior demifacets have indistinct margins but measure about 2 mm by 4 mm. The pedicles are transversely thin. Strong and thick transverse processes are present and the costotransverse facets incline superiorly. The left one measures 4.4 mm superoinferiorly by 6.0 mm transversely, and the right one 4.4 mm by

6.1 mm. The angle of the axis through the costovertebral and costotransverse facets is 30° relative to midline. The superior zygapophyseal facets angle slightly superiorly and laterally about 5° from a coronal plane. The inferior facets are angled about 5° medially and slightly inferiorly. There is asymmetry to the posterior elements, such that the spinous process is deflected somewhat to the right. The right transverse process projects more directly laterally in superior and anterior views than does the right transverse process. The spinous process is inclined 45° inferiorly to the plane of the inferior endplate.

Sts 14o - PSV9

Preservation. This specimen (Figure 10.3) preserves only the vertebral body. Neither of the pedicles remain. The superior endplate is mostly intact, with the posterior margin slightly abraded and a piece of bone missing from the anterolateral edge of the apophyseal plate. The anterior portion of the inferior endplate is also missing, as is surface bone along the vertebral body's right side.

Morphology. The vertebral endplates are roughly heart-shaped with a rounded anterior margin. The body does not appear to have been anteriorly wedged, but this is difficult to discern with certainty due to abrasion of the superior surface. The internal margins of the superior apophyseal ring are obscure, making measurement unreliable. The inferior ring is about 2 mm wide where preserved laterally and posteriorly. There is a superoinferior thickening along the left anterolateral portion of the ring that matches a corresponding depression on the suprajacent vertebral body of Sts 14l. Remains of costal facets can be seen along the superior margin of the body-pedicle junctions. These both are facing superolaterally, but their margins are damaged and cannot be measured. No inferior costal demifacet is present.

Sts 14h - PSV8

Preservation. This bone (Figure 10.3) is mostly complete but missing its left transverse process and a piece of bone from the posterior superior portion of the body. The zygapophyseal facet margins are chipped somewhat but the original contours are not obscured. Anteroposterior body diameters cannot be reliably measured.

Morphology. Morphology of the endplates is unclear owing to postmortem damage. Any evidence of a median keel has been abraded. The body is anteriorly wedged. Single costal facets lie along the superior endplate on both sides at the body-pedicle junction and are angled slightly superiorly. They measure 7.0 mm superoinferiorly and 6.1 mm transversely on the right, and 6.6 mm by 5.5 mm on the left. The preserved right transverse process is thick and has a strong costotransverse facet. The facet is flat and faces slightly superiorly. The transverse process extends 13 mm from the internal margin of the

Sts 14o   Sts 14h   Sts 14g

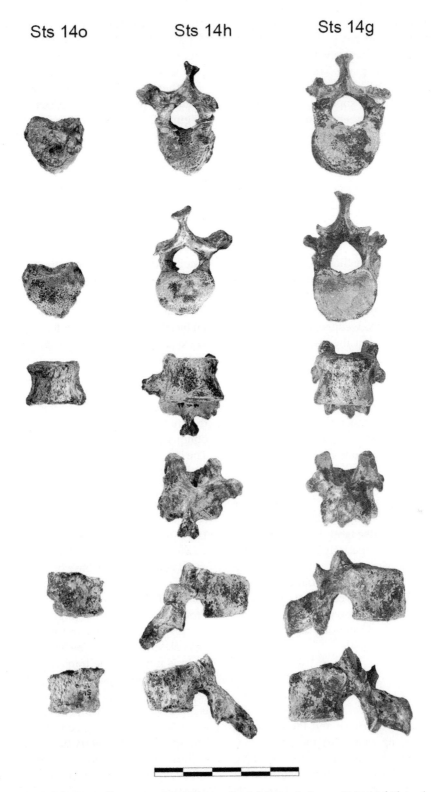

Figure 10.3 Sts 14o, Sts 14h, and Sts 14g in (from top to bottom) superior, inferior anterior, posterior, right lateral, and left lateral views.

neural foramen and a line through the centers of the costal facets is inclined 35° to the median plane. The superior zygapophyseal facets are transversely flat and angled only 8° laterally from a coronal plane. They project about 4 mm superior to the endplate. The inferior zygapophyses are angled 8° medially, and their superior extent reaches roughly to the level of the inferior endplate. The spinous process is strongly inclined inferiorly at an angle of about 45° to the inferior endplate. The tip is bilobate along its inferior margin, with the right side projecting slightly further posteriorly than the left.

Sts 14g - PSV7

Preservation. This bone is complete except for minor surface abrasion (Figure 10.3).

Morphology. This vertebra is the thoracolumbar transitional vertebra, with flat, thoracic-like paracoronal zygapophyseal facets superiorly and transversely convex, anterolaterally facing facets inferiorly. The endplates are reniform and the body mildly wedged anteriorly. The superior endplate is abraded, but the preserved inferior apophyseal ring is 3.6 mm wide laterally thinning to 1.6 mm posteriorly. There is too much abrasion to assess the possible presence of an anterior median keel. Large single costal facets are located on the body-pedicle junctions on each side and occupy most of the height of the pedicle. The left one measures 7.0 mm superoinferiorly by 8.2 mm transversely, and the left one 7.8 mm by 8.7 mm. Small transverse processes flank each posteriorly, measuring 8.5 mm long on the left and 10.1 mm on the right. They are anteroposteriorly compressed in section. The end of the left one is rounded, the right one is bilobate, and neither presents a costotransverse facet. From their bases, strong bony processes project superiorly and would have extended posterior to the suprajacent articulating zygapophysis. The left one is 5.3 mm superoinferiorly and the right one only 3.8 mm. These are confluent posteriorly with the posterior laminar crest that runs toward midline. The spinous process is inclined inferiorly at about a 20° angle relative to the inferior endplate.

Sts 14f - PSV6

Preservation. This vertebra (Figure 10.4) is nearly complete, except for the right anterolateral portion of the body. The superior endplate surface and the adjoining apophyseal ring are missing as a consequence of abrasion. Most of the inferior endplate is preserved, missing only a strip about 13 mm long from the right anterolateral margin. The apophyseal ring is preserved along its posterior and lateral portions but missing anteriorly. A small, true costotransverse foramen traverses the transverse process on the left side. On the right side, however, only a costovertebral articular facet remains that would have articulated with a rib. Except for some

missing chips of surface bone, the superior zygapophyses are intact. The right superior laminar margin and a segment about 6 long along the top of the spinous process were reconstructed, but not much bone is missing in either area (Figure 10.5). The central portion of the distal end of the spinous process was also reconstructed (Figure 10.5), but the top and bottom corners remain and the original contours are preserved. The inferior half of the left inferior zygapophysis is broken and was reconstructed. On the right side, posterior to the costal facet is a low anapophysis, but its surface is abraded exposing the underlying trabecular bone.

Morphology. In the lateral view, the vertebral body is wedged anteriorly. The endplates are reniform in outline. The inferior apophyseal ring is uniformly narrow, about 3 mm wide laterally and 2 mm posteriorly.

The right inferior pedicle margin is sharp, the left one rounded. A transverse process is present on the left side, whereas a costal facet is present on the right side. The costal facet is 5.4 mm superoinferiorly by 6.4 mm wide transversely, and it is situated at the superoinferior midpoint of the body-pedicle junction. It faces inferolaterally and projects to a level about 5 mm above the surrounding surface. It is flanked posteriorly by a tubercle that represents the anapophysis. The superior zygapophyses as well as the inferior ones are transversely curved and oriented obliquely, resembling joints typical of lumbar vertebrae. The left and right superior facets are oriented 50° to a coronal plane, and the inferior ones 55°. The superoinferior middle of the inferior zygapophyseal joint is at the level of the inferior endplate of the body. The transverse process is inclined posteriorly about 65° and inferiorly about 15°. It is slightly curved superiorly. There is a large costotransverse foramen on its inferior margin near the base, and a small pit at the base of the process.

Sts 14e - PSV5

Preservation. This bone (Figures 10.4 and 10.5) is nearly complete. The superior endplate is abraded along its right anterolateral margin. The left posterolateral corner of the body is missing. The transverse processes are preserved except for their distalmost tips. The base of the left one has been reconstructed. The right superior zygapophysis is preserved, as are the lateral portions of the left one. The articular surface was reconstructed on its posterior side. The apparent asymmetry between the coronal orientation of these facets may be due to the posterior portion of the left zygapophysis having been plastically distorted. There is some minor abrasion along the superior margin of the lamina. Both inferior zygapophyses have been reconstructed. There is a straight, thin tunnel about 1 mm in diameter that traverses the entire left neural arch vertically through a spot where the costotransverse foramen should be. It is not a natural feature.

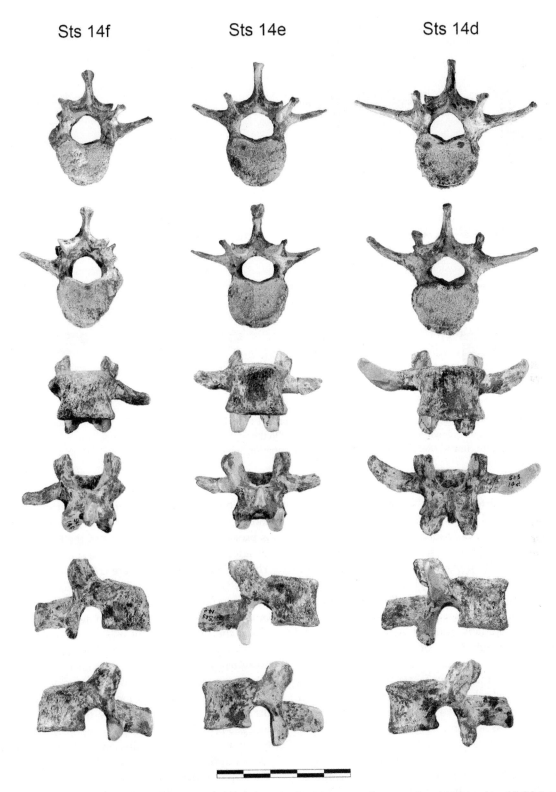

Figure 10.4 Sts 14f, Sts 14e, and Sts 14d in (from top to bottom) superior, inferior anterior, posterior, right lateral, and left lateral views.

Original   Reconstructed   Original   Reconstructed

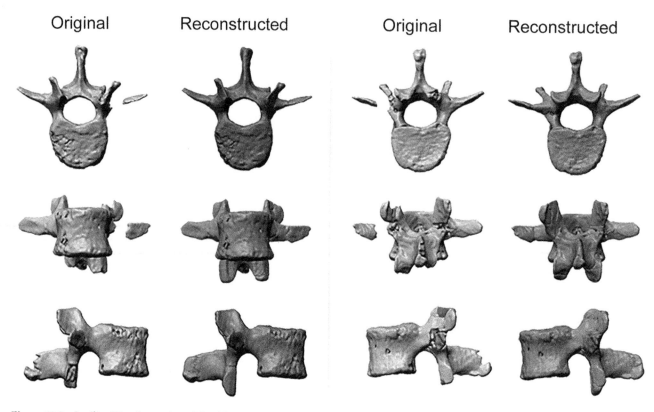

Figure 10.5 *In silico* 3D polygonal models of Sts 14e constructed from CT scan data. On the left in light color, the specimen is shown as originally preserved (Original), made by digitally removing plaster. On the right in dark color, the specimen is show as reconstructed with plaster and currently curated (Reconstructed). Each shown in superior, anterior, and right lateral (left pane) and inferior, posterior, and left lateral (right pane).

Morphology. The vertebral body endplates are slightly reniform, and the body is wedged anteriorly. The body is blocky, similar in superoinferior and mediolateral dimensions. The inferior zygapophyseal ring is about 2–3 mm wide throughout. The transverse processes angle posteriorly about 20° from a coronal plane but project almost directly laterally in anterior view. Both superior and inferior zygapophyseal facets are transversely curved and almost parasagittally oriented. Relative to a coronal plane, the left and right superior facets are oriented 60° medially. The inferior ones have been reconstructed. The articular surfaces of the inferior facets are situated mostly inferior to the vertebral endplate, a location that is probably reasonably accurate based on preserved morphology. The spinous process is inclined about 10° inferiorly.

Sts 14d - PSV4

Preservation. This bone (Figures 10.4 and 10.6) is complete except for the broken right transverse process 13.7 mm distal to the internal neural arch surface and some areas of abrasion and chipping along the body and zygapophyseal margins. The superior endplate preserves parts of the ring except along the right side, where it is largely missing. A flake of bone measuring about 6 mm wide and

3 mm high is missing along the posterior surface of the inferior endplate at midline, and most of the right side is broken away. The right side of the body is rough and abraded and appears to still have some adhering breccia. The zygapophyseal joints are cracked and chipped along their edges, particularly the inferior ones, but these are minor and do not disturb the original contours. The end of the spinous process is broken away, but the basal flaring at the inferior end of it is mostly preserved and its dimensions can be quantified fairly accurately.

Morphology. The vertebral endplates are reniform, but otherwise the vertebral body is blocky in appearance. The inferior endplate is mediolaterally broader than is the superior one. In lateral view, the body is only slightly wedged posteriorly. The superior apophyseal ring is about 6 mm wide anterolaterally, and the inferior one 5 mm wide anteriorly narrowing to only 2 mm wide posteriorly. The superior zygapophyses are transversely concave and face almost sagittally, angled 60° to a coronal plane. Thick, strong mammillary processes extend posteriorly from the facet surface. The superior and inferior zygapophyses project about equally far superiorly and inferiorly, respectively, from the vertebral body. The superior margins of the laminae are marked by grooves that reach to the

Original   Reconstructed   Original   Reconstructed

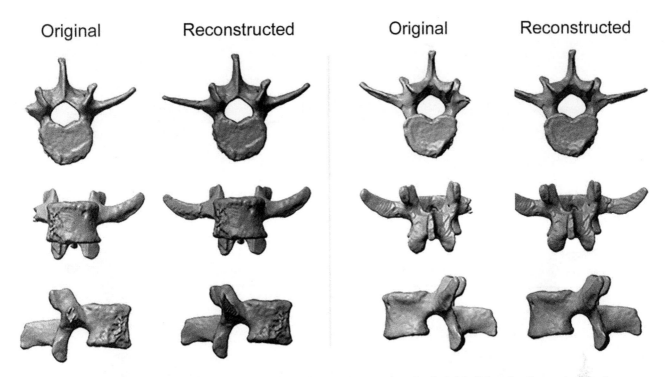

Figure 10.6 *In silico* 3D polygonal models of Sts 14d constructed from CT scan data. On the left in light color, the specimen is shown as originally preserved (Original), made by digitally removing plaster. On the right in dark color, the specimen is show as reconstructed with plaster and currently curated (Reconstructed). Each shown in superior, anterior, and right lateral (left pane) and inferior, posterior, and left lateral (right pane).

midline crest. The preserved left transverse process angles about 60° posteriorly and is curved posteriorly toward its tip. It also arcs superiorly from its midpoint. The spinous process projects almost directly posteriorly, angling only slightly inferiorly. The inferior zygapophyses project almost straight inferiorly. They are transversely convex and oriented 45° to a coronal plane.

Sts 14c - PSV3

Preservation. This vertebra (Figures 10.7 and 10.8) is nearly complete, with most of the neural arch and posterior structures except for the right transverse process and the distal end of the spinous process. The medial margin of the left zygapophysis was broken and has been reconstructed. The inferior and lateral margins of the left inferior zygapophysis have also been reconstructed. The superior surface and the adjoining apophyseal ring are largely preserved except for some pitting through the anterior portion of the ring and some minor abrasion. Inferiorly, the ring is broken and much of it is missing, but its dimensions can still be estimated reasonably.

Morphology. The vertebral body is mediolaterally wide and reniform. It is roughly symmetrical in anterior view and barely wedged posteriorly in lateral view. The superior ring is 6 mm wide anteriorly, 5 mm laterally, and 2 mm posteriorly on both endplates. The transverse processes project almost directly laterally in superior view

and about 20° superiorly from their midpoint in anterior view. There is a shallow groove along the inferior margin of the bases of both transverse processes. The superior zygapophyses have mammillary processes evident along their posterior margins but these do not extend superiorly from the facets. The superior zygapophyses are transversely concave and angled 30° to the coronal plane, and the inferior facets are convex and oriented 45° to a coronal plane. The superior margins of the laminae are marked by grooves that reach to the midline crest. The inferior facets lie almost completely inferior to the vertebral body, and face almost anteriorly when viewed from a lateral perspective. The original spinous process length as reconstructed is probably reasonable, comparing it to that of Sts 14d. However, the reconstruction of the Sts 14c spinous process appears to be too inferiorly inclined based on the preserved contours of the base of the spine.

Sts 14b - PSV 2

Preservation. This specimen (Figure 10.7 and 10.9) consists of the left half of a vertebral body, the left pedicle and transverse process, much of the lamina, and the base of the spinous process. Posteriorly, the body is preserved, but anteriorly only the left half remains. The superior endplate preserves almost no morphology, and trabecular bone is visible across the surface. Inferiorly, the apophyseal ring remains along with the adjacent portion

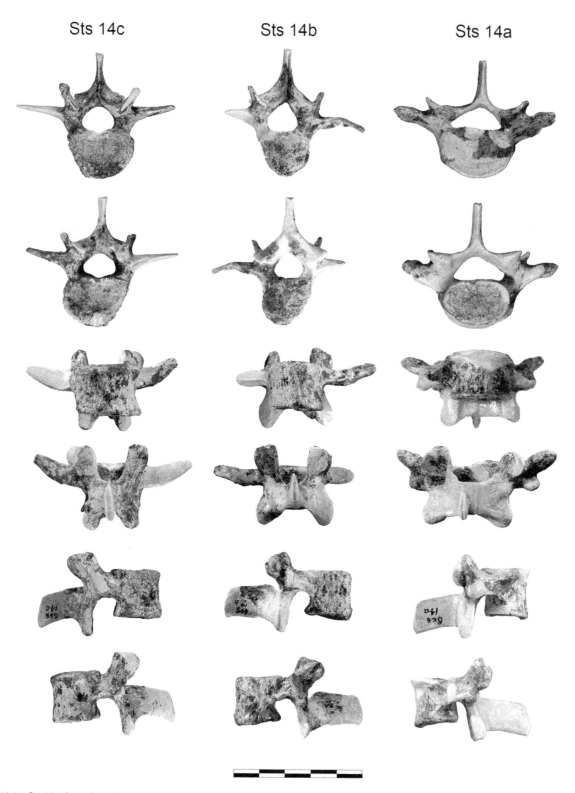

Sts 14c **Sts 14b** **Sts 14a**

Figure 10.7 Sts 14c, Sts 14b and Sts 14a in (from top to bottom) superior, inferior anterior, posterior, right lateral, and left lateral views.

Original Reconstructed Original Reconstructed

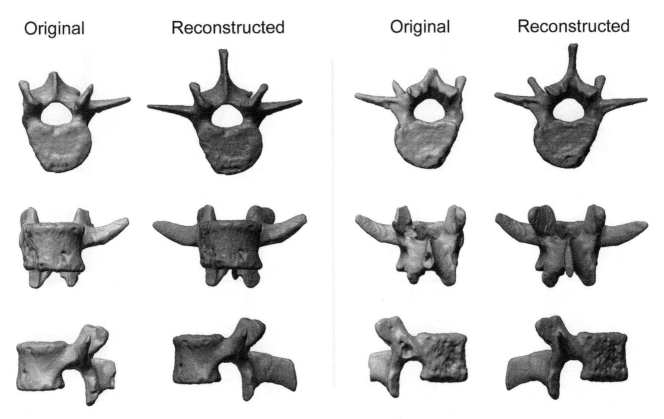

Figure 10.8 *In silico* 3D polygonal models of Sts 14c constructed from CT scan data. On the left in light color, the specimen is shown as originally preserved (Original), made by digitally removing plaster. On the right in dark color, the specimen is show as reconstructed with plaster and currently curated (Reconstructed). Each shown in superior, anterior and right lateral (left pane) and inferior, posterior, and left lateral (right pane).

of intervertebral surface which is preserved to the right pedicle posteriorly.

The right pedicle is missing and is entirely reconstructed, but the left one is preserved. The left transverse process is broken immediately distal to the tubercle at a point about 16.5 mm from the internal surface of the neural arch. Lateral to the glue joint adhering the two fracture pieces, the transverse process is deflected superiorly and anteriorly more than it would have been originally. The right superior zygapophysis is abraded along its medial margin.

The only preserved part of the left side of the lamina is the posterior margin which connects the spinous process portion and the zygapophysis; the rest has been reconstructed. The medial part of the left inferior zygapophysis is sheared away so that only 7 mm of the width of the original facet is preserved. In life this facet would have been closer to the width of the subjacent superior zygapophyseal facet (on Sts 14a) which measures about 11 mm across mediolaterally. The superior margin of the spinous process is preserved to a point 16 from the internal neural arch surface, but none of the posterior or inferior portions remain and have been reconstructed.

Importantly, the inferior zygapophyses and pedicle have been reconstructed incorrectly; the right facet is about as wide as only the preserved portion of the left one, the inferior zygapophyses are not set widely enough apart mediolaterally, and the pedicle meets the vertebral body too far medially. This misreconstruction results in the vertebral body appearing to be mediolaterally narrower than it would have been originally. It is tempting to assume that the junction of preserved and missing portions of the anterior body represent a median keel, but this is not where the midline would have been. The inferior endplate preserves its surface morphology, and the original body can be reconstructed as being roughly double what is currently preserved. Thus, the original anteroposterior axis of the bone would have been oblique to that which has been reconstructed. As a result of this erroneous reconstruction, the pedicle is too narrow mediolaterally, and probably reconstructed as too thin. Also, the cross-sectional shape of the neural canal is too narrow and rounded as compared to the vertebra above and below it. The inaccurate reconstruction is obvious in that the inferior zygapophyses of Sts 14b do not articulate with those of Sts 14a but instead are set too narrowly. The right transverse process is much shorter than the other

Original　　　Reconstructed　　　Original　　　Reconstructed

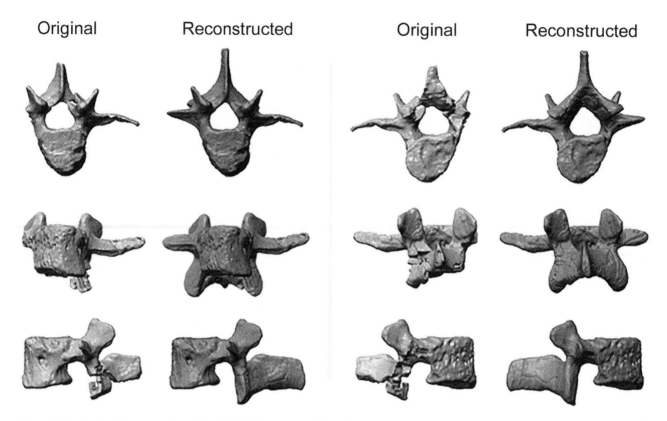

Figure 10.9 *In silico* 3D polygonal models of Sts 14b constructed from CT scan data. On the left in light color, the specimen is shown as originally preserved (Original), made by digitally removing plaster. On the right in dark color, the specimen is show as reconstructed with plaster and currently curated (Reconstructed). Each shown in superior, anterior, and right lateral (left pane) and inferior, posterior, and left lateral (right pane).

side, suggesting that the distal fragment (lateral to what is now a glue join) was identified after Robinson completed the reconstruction but before Benade (1980) wrote her descriptions.

Morphology. Despite the inaccurate reconstruction, it is apparent that the vertebral body would have been much wider mediolaterally than it was deep anteroposteriorly. The vertebral body appears asymmetric in the anterior view, about 2 mm shorter superoinferiorly on the right than on the left. In the lateral view, the vertebral body is strongly wedged posteriorly. The apophyseal ring on the inferior endplate is quite wide, measuring 6.5 mm along the lateral margin and narrowing to 3.7 mm posteriorly. A large nutrient foramen is present on the right lateral side of the body.

The pedicles were thick mediolaterally and angled laterally from where they arose from the body.

Enough of the left superior zygapophysis remains to see a crest extending from the lateral margin of its facet toward the spinous process, and this is seen on the right side as well. The facets are mildly concave transversely, and angled medially 40° to a coronal plane. Very small mammillary processes flank the posterior margin of the facets. Benade (1980) describes an accessory process but

none is evident. The superior margin of the lamina is marked by a groove that extends to the midline crest. The spinous process projects almost straight posteriorly, with its superior margin at roughly the level of the superoinferior midpoint of the body, and slightly superior to the laminar margin. Exactly how much is uncertain due to damage and subsequent reconstruction, but in the superior or inferior view the spinous process is angled slightly to the right of a midsagittal plane.

Sts 14a - PSV 1

Preservation. This fossil (Figures 10.7 and 10.10) preserves almost all of the inferior half of the vertebral body, portions of the superior endplate, most of the transverse processes, and both superior zygapophyses. The inferior endplate is nearly entirely intact with the apophyseal ring preserved, except for a small area along the right posterolateral corner. The inferior and internal surfaces of the left pedicle are completely reconstructed, as are the entirety of the laminae, inferior zygapophyses, and the spinous process. The right accessory process is missing and has been reconstructed. The left superior zygapophysis and facet are intact, except for the inferomedial portion of the right facet, which is missing and has been reconstructed.

Original Reconstructed Original Reconstructed

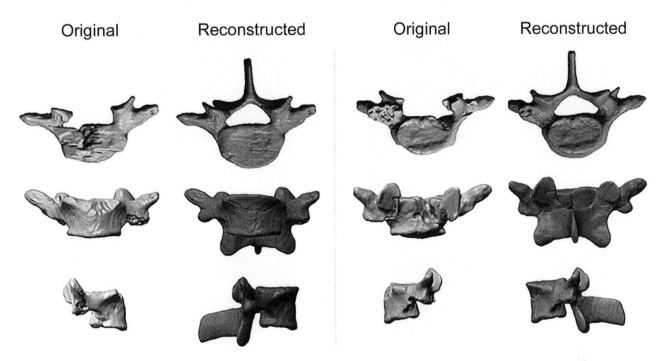

Figure 10.10 *In silico* 3D polygonal models of Sts 14a constructed from CT scan data. On the left in light color, the specimen is shown as originally preserved (Original), made by digitally removing plaster. On the right in dark color, the specimen is show as reconstructed with plaster and currently curated (Reconstructed). Each shown in superior, anterior, and right lateral (left pane) and inferior, posterior, and left lateral (right pane).

The right superior zygapophyseal facet preserves a small patch of original surface that is only 5.5 mm wide mediolaterally and 7.2 mm superoinferiorly.

Morphology. There appear to be significant asymmetries in this vertebra. The body, as far as can be discerned given the reconstruction, is less than 16.5 mm superoinferiorly on the left but 18.0 mm tall on the right. The inferior surface is eroded just along its left margin, and the left portion is also more abraded on the left than the right. However, it would be difficult to reconstruct this surface to be sufficiently thicker as to make the body symmetrical. The inferior surface is also much smaller than the superior one. The vertebral body is posteriorly wedged. It is oval in section, with no posterior concavity typically seen in lumbar vertebrae and found in the other Sts 14 vertebrae, but also sometimes found in the lowest presacral vertebrae of hominoids. The apophyseal ring is about 3.5 mm thick anteroposteriorly along its anterior portion, and only 2 mm posteriorly. A single large foramen for the basivertebral vein is present on the posterior surface of the body.

The superior portion of the right side of the body appears to flare out more laterally than does the left, but this seems to be an artifact of the reconstruction. The transverse processes are slightly asymmetric, with the right one being slightly thicker anteroposteriorly and taller superoinferiorly by about 0.5 mm on either side. It is also slightly longer and more superiorly inclined than

the left one. At its base, the laminae measure 11.5 mm tall superoinferiorly on the left but only 9.2 mm on the right. On the better-preserved left side, there is a prominent accessory process present which forms a large blunt projection flanking the inferior margin of the base of the process.

The articular facets appear to be asymmetrical, but this could be a product of the reconstruction. The left facet is flat for most of its length, but 4.5 mm of the superior facet arcs anteriorly in a gentle convexity, as if it occasionally contacted the inferior laminar margin in extension. This is not present on the right as preserved, and the reconstructed portion of the facet does not include this. However, the superior zygapophysis is broken off, as is the medial margin of the articular surface. It is difficult to determine how much further superiorly the process would have extended, but it seems plausible that it would match the left side. The facets are oriented only 20° medially from a coronal plane, and the preserved right one is only mildly concave transversely.

Sts 73—Lumbar Vertebra

Preservation. This is a nearly complete lower thoracic vertebral body with pedicles (Figure 10.11). About 9 mm of the right pedicle is preserved and about 6 mm of the left. The body is broken obliquely through the area just inferior to the superior endplate on the left, through to about

157

Figure 10.11 Sts 73 in (top row) superior and inferior and (bottom row) right and left lateral views.

midline inferiorly. A groove of missing bone about 6 mm wide runs across the entire posterior vertebral surface. Most of the margins of the inferior endplate are abraded as well. The superior endplate appears preserved but no ring is visible.

Morphology. This is most likely one of the lowest 2 rib-bearing vertebrae, as a single large costal facet is present at the superoinferior center of the body-pedicle junction. The body is heart-shaped in superior and inferior views. It is wedged anteriorly. The anterior keel is very mild and rounded. The costovertebral facet measures 9.4 mm superoinferiorly and 6.7 mm transversely. Its margins are raised above the surrounding bone, but its middle sits below the surface in general. The pedicles are tall and thin.

StW 8/41—Associated vertebrae

These vertebrae are preserved as a set of two elements fused (StW 41) and another of four fused (StW 8), fossilized with breccia connecting them (Figure 10.12). They are preserved in a flexed position, but this relative position is almost certainly post-depositional. The StW 41 vertebral bodies are touching across their entire endplates, whereas most of the StW 8 vertebral elements are partially separated from each other by gaps of up to 1–3 mm, but this may not have been the spacing during life. Based on size and preservation, they are almost certainly

adjacent vertebrae from the same adult individual, as discussed in previous publications that describe and figure these specimens (Tobias, 1973; Benade, 1980; Sanders, 1998). The cortical shells of all the specimens are slightly weathered with localized areas of surface bone abraded, but this damage is minor and does not significantly obscure the original contours. These vertebrae appear to be from an adult, because even though preparation scratches obscure the surface morphology of the exposed endplates, the margins of the vertebral bodies show no unfused physeal lines.

StW 41—Two articulated thoracic vertebral bodies

Preservation. This specimen is comprised of two vertebral bodies articulated with one another (Figure 10.12). The vertebral bodies are complete, although there is some abrasion around their margins. They are almost certainly from the same individual as Stw 8 based on a match of preservation, size, and contours.

Morphology. These vertebrae are from an adult individual, as the apophyseal rings are fully fused to the vertebral bodies. The bodies are reniform in endplate shape. Neither shows appreciable wedging of the vertebral body in lateral view. Both vertebrae have a single distinct costal facet adjacent to its superior margin that are angled slightly superiorly and elevated above the adjacent bone surface. No inferior demifacet is evident on either bone.

StW 8

StW 41

StW 41

Figure 10.12 StW 8/41. From left to right, (top row) StW 41 in anterior, right lateral, and left lateral views, then top to bottom in superior and inferior views, and StW 41 in superior view; (bottom row) StW 41 in anterior, left lateral, right lateral, and posterior views.

StW 8—Four articulated lumbar vertebrae
Preservation. These vertebrae (Figure 10.12) are almost certainly from the same individual as StW 41, based on a match of preservation, size, and contours. The bodies are articulated by the intervening matrix as an anteriorly concave unit. This curvature is taphonomic, not anatomical. Each vertebra is described here individually, although they are not separated from one another as preserved.

StW 8a
Preservation. This specimen preserves most of the body with portions of the posterior elements (Figure 10.12). The superior intervertebral surface is missing all surface features. This vertebra is fused to the one subjacent to it by matrix, although matrix has been removed along the margins between them, except at the center of the intervertebral space. The right inferior zygapophyseal joint adheres to the subjacent superior zygapophysis with about 3 mm of its surface exposed. The left side of the vertebral body has much of its surface bone abraded away. The right posterior corner of the superior endplate has been sheared off, and this continues through the superoinferior center of the pedicle so that its superior margin,

the transverse process, and right superior zygapophysis are missing. On the left side, the pedicle, transverse process, and zygapophyses are missing. The right lamina and only the base of the spinous process remain.
Morphology. The presence of costal facets is uncertain, as the bone is damaged on both sides in the region where they would have been. It is unlikely that it had any, because this vertebra sits inferior to StW 41 that has single costovertebral facets, so it is almost certain that if it had a facet, that facet would be situated near the superoinferior center of the vertebral body. Thus, it is most likely that this vertebra was not rib bearing. Furthermore, this suggests that the specimen identified as StW 41 represents two non-rib-bearing vertebrae.

Although the right inferior zygapophyseal facet is present, it is glued to the subjacent vertebra so its curvature cannot be fully appreciated. The joint is inclined about 10–15° from a median plane, but the articular surfaces do not appear to be very curved transversely.

StW 8b
Preservation. This vertebra is intact except for the left transverse process and inferior zygapophysis, and the

lateral portions of the left pedicle, superior zygapophysis, and lamina (Figure 10.12). There is some surface abrasion on the sides of the vertebral body and tip of the transverse process. There is also a sliver of bone about 2 mm wide missing along the posteromedial aspect of the right inferior zygapophysis. Endplate morphology and zygapophyseal facet morphology are not observable as this vertebra is fused to the ones above and below it with matrix between the vertebral bodies and glue at the zygapophyseal joints. There are small areas on the superior margin of the transverse process, right pedicle, and right inferior zygapophysis that have been reconstructed with plaster.

Morphology. This vertebral body is wedged anteriorly in the lateral view. The transverse process is about 23 mm long and projects almost straight laterally in the superior and anterior view. No accessory process is present. On the inferior side of the base of the transverse process, a large costotransverse foramen is present. The dimensions of this foramen appear to have been slightly enlarged during preparation. As with the superior zygapophyseal joint, the inferior one is oriented about 15–20° relative to a sagittal plane. The spinous process is about 13 mm long and projects almost straight posteriorly with no inferior inclination.

StW 8c

Preservation. This vertebral body preserves none of the pedicles and is abraded over much of its surface (Figure 10.12). Much of the right neural arch and superior zygapophysis are glued into articulation with StW 8b, and connected to the body with plaster.

Morphology. The anterior margin of the endplate is visible and shows a fully fused apophyseal ring. Wedging is difficult to evaluate because of damage along the inferior margin of the vertebral body, but there was likely little to none. As noted for StW 8b, the zygapophyseal joint is oriented about 15–20° out of the sagittal plane.

StW 8d

Preservation. This vertebral body (Figure 10.12) preserves only the superior endplate, fused to StW 8c, and small portions of the anterior surface. A fragment of the posterior surface of the vertebral body is also present adjacent to the superior endplate.

Morphology. This vertebra also seems to have had a fully fused apophyseal ring. It appears to have had a superoinferiorly short vertebral body with a sloping anterior face that makes an obtuse angle with the superior endplate, as is often seen in ultimate lumbar vertebrae. This observation may, however, be misleading and a consequence of taphonomic damage to this specimen. It is unlikely to have been the ultimate lumbar given the narrow set

to the zygapophyses of StW 8b, which is only two levels superior to StW 8d, and the likelihood that StW 41, which are inferior thoracic vertebrae, would have articulated with StW 8.

StW 431—Associated vertebrae

This specimen was originally described with different accession numbers for each vertebra noted in Appendix I. These are associated elements from the adult partial skeleton StW 431 (Ward et al., Chapter 4, this volume). As with the other StW 431 bones, the surface bone of all elements is generally well preserved with only occasional abrasions.

StW 431o—PSV 10 (StW 455 originally)

Preservation. This is most of a vertebral body and the base of the left pedicle (Figure 10.13). The left posterior corner of the superior endplate is sheared off, and there are some areas of abraded surface bone along the right anterior region of the inferior apophyseal ring. The right pedicle is broken unevenly along the posterior margin of the superior costovertebral facet.

Morphology. The apophyseal rings are up to 6 mm wide along their anterior margins. The body is anteriorly wedged. The anterior face of the body is very mildly keeled. A superior costotransverse facet measures 7.0 mm in diameter. An inferior demifacet is also present measuring about 3.0 mm superoinferiorly and 5.0 mm anteroposteriorly.

StW 431na and nb—PSV 9 (StW 454 originally)

Preservation. This vertebra (Figure 10.13) is nearly complete, missing only the left transverse process, lamina, and inferior zygapophysis. It is preserved in two pieces broken through the pars interarticularis of the right lamina. About half of the body along its right inferior corner is broken away and there is surface bone missing along much of the superior and inferior endplate margins as well as along the caudolateral margin of the right inferior zygapophysis.

Morphology. Both the superior and inferior physeal rings are up to 6 mm wide. The body is not appreciably wedged in lateral view. No anterior extension of bone is evident. A costovertebral demifacet roughly 8 mm in diameter is present at the superior body-pedicle junction. Along the inferior endplate, a small demifacet measuring 4 mm superoinferiorly by 6 mm anteroposteriorly is evident. The superior zygapophyses are angled just a few degrees lateral relative to a coronal plane. Even taking some abrasion along the left articular margin into account, the right superior zygapophyseal facet is larger (9.8 mm) mediolaterally than the left (8 mm).

StW 431o   StW 431na   StW 431nb

Figure 10.13 StW 431o, na, and nb in (from top to bottom) superior, inferior anterior, posterior, right lateral, and left lateral views.

The transverse process projects directly laterally in superior and anterior views. It is 14.0 mm long. A flat costotransverse facet measuring 8.0 mm superoinferiorly by 9.7 mm mediolaterally is present. Posterior to the facet the transverse process is 8.5 mm thick transversely. Only part of the inferior zygapophyseal facet remains, but it is flat and faces about 20° lateral to a coronal plane. The spinous process is about 36 mm long, and angled inferiorly, though this cannot be quantified accurately as preserved. Its posterior end is 10.0 mm superoinferiorly and up to 7.9 mm wide mediolaterally. The end is asymmetrical with the left side thicker and more flaring laterally and inferiorly.

StW 431ma and mb—PSV8 (StW 453 originally)
Preservation. This vertebra (Figure 10.14) is preserved in two pieces that do not articulate. StW 431mb is the left

portion of the vertebral body. It is broken obliquely so that roughly the left third of the superior endplate but almost the entire left half of the inferior endplate are preserved. The pedicle is broken immediately posterior to the costal facet. The preserved surfaces of bone show no abrasion. StW 431ma includes the lamina, right transverse process, spinous process, inferior zygapophyses, and base of the left transverse process. The inferior third of the right inferior zygapophyseal facet is broken away, the right side of the posterior end of the spinous process and the distal end of the transverse process also have most of their surface bone abraded away.

Morphology. The superior apophyseal ring cannot be measured at midline anteriorly, but the inferior one is 5 mm wide. The body does not appear to be wedged in the lateral view. There is a large costovertebral facet that measures about 8 mm superoinferiorly by 10 mm

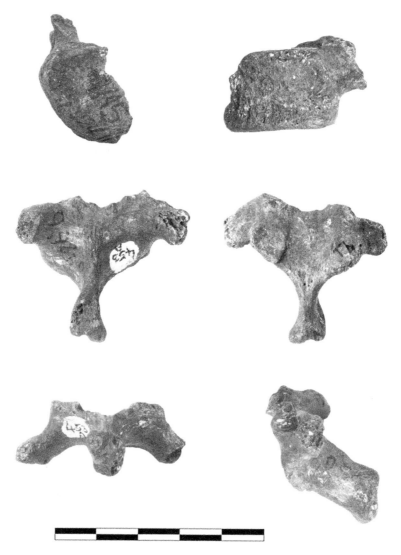

Figure 10.14 StW 431m. Top row, StW 431ma, vertebral body, in superior and left lateral views. Middle row, StW 431mb in posterior and anterior views, and bottom row StW 431mb in superior and left lateral views.

anteroposteriorly situated immediately adjacent to the superior endpoint and pedicle. The transverse processes are just over 15 mm long and angled about 45° posteriorly from a coronal plane. Although the more complete right one is abraded at the tip, it is clear that no costotransverse foramen was present, nor is there evidence of a strong ligamentous attachment of the rib. The posterior inclination of the transverse process also argues against substantive attachment between the rib and transverse process. The inferior zygapophyses are transversely flat and face only a few degrees medially from a coronal plane. The spinous process is 26 mm long and angled about 60° inferiorly relative to the zygapophyseal facets. At its tip, the spinous process is 13.3 mm superoinferiorly and up to 8.4 mm wide mediolaterally. The right side is larger and projects 5.7 mm further inferiorly than the smaller left portion.

StW 431l—PSV7 (StW 452 originally)

Preservation. This is the lamina with the right transverse process, spinous process, and inferior zygapophyses (Figure 10.15). Some surface bone is missing from the posterior and anterior sides of the transverse process.

Morphology. The inferior zygapophyseal facets are transversely flat and are posterolaterally oriented, as is typical for most lumbar vertebrae. Although the superior zygapophyses are not preserved, preserved contours of the bone suggest that they would have been in a paracoronal plane, typical for thoracic vertebrae, so this bone appears to the transitional vertebra of StW 431. The transverse process is only about 5 long and projects posteriorly almost parallel with the spinous process in the superior view. It is angled slightly inferiorly in anterior view. No costotransverse facet is present. The spinous process is 21.6 mm long, up to 15.3 mm tall superoinferiorly, and inclined inferiorly. Its posterior end is 15.6 mm thick near its inferior margin. There is a slight vertical groove along

the midline, and the right side of the posterior extent of the spinous process extends just over 3 mm further inferiorly than does the left side.

StW 431qa and qb—PSV6 (sixth presacral vertebra; StW 457 originally)

Preservation. This vertebra (Figure 10.16) is preserved in two pieces. StW 431qa is a partial vertebral body and StW 431qb is a neural arch fragment that preserves the spinous process and inferior zygapophyses. Originally, StW 431qb was described as articulating with StW 431r (Benade, 1980; Toussaint et al., 2003). However, upon closer observation it is clear that StW 431qa and StW 431qb are parts of the same bone, which is how they are accessioned.

StW 431qa is a partial vertebral body broken obliquely so that it preserves less than the right third of superior endplate but almost the entire inferior one. The inferior endplate is intact except for some bone missing along its left margin. Surface bone on the preserved right side of the body is intact. StW 431qb preserves the area immediately inferior to the left superior zygapophyseal facet and the base of the pedicle.

Morphology. The inferior apophyseal ring is less than 4 mm thick anteriorly. The vertebral body is anteriorly wedged. On the posteriormost edge of the lateral surface of the body, there is a small (~1 mm) tubercle of bone. This is likely the margin of a costal facet, suggesting that this is the inferiormost rib-bearing vertebra.

StW 431r—PSV5 (StW 458 originally)

Preservation. This is a complete vertebral body and parts of both pedicles, with the rest of the posterior elements missing (Figure 10.16). The left pedicle is preserved for probably half its length, and the right one a bit further posteriorly, flaring toward where the transverse process would have originated.

Figure 10.15 StW 431l in posterior and right lateral views.

StW 431qa & qb   StW 431r   StW 431s

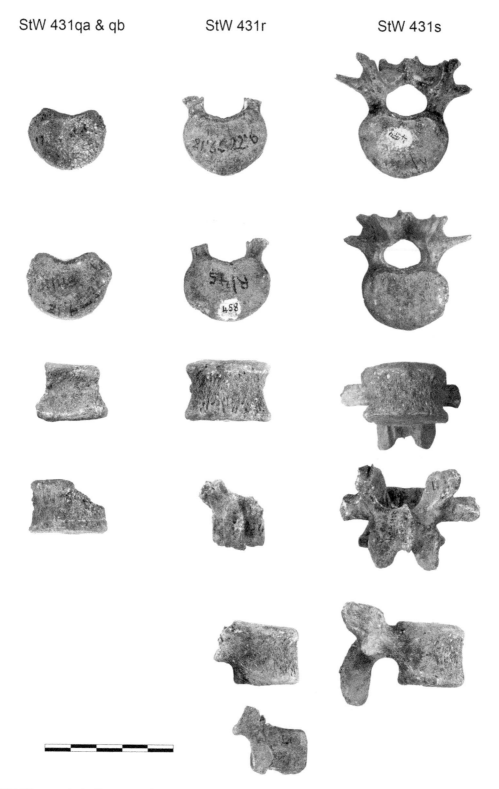

Figure 10.16   StW 431q, r, and s in (from top to bottom) superior, inferior anterior, posterior, right lateral, and left lateral views.

Morphology. The apophyseal rings are about 4 wide along the anterior margins of the endplates. In lateral view, the body is anteriorly wedged. The body is slightly asymmetrical, with the left anterior corner superoinferiorly shorter (21.7 mm) than the right (22.6 mm).

The inferior zygapophyseal joints are nearly flat transversely and oriented about 45° to the median plane. The spinous process is 21.8 mm long and 18.8 mm in maximum superoinferior height with a maximum mediolateral breadth of 8.5 mm along at its posterior end. The spine projects almost directly posteriorly, typical of lumbar vertebrae.

StW 431s—PSV4 (StW 459 originally)
Preservation. This specimen (Figure 10.16) is complete except for most of the left transverse process, the spinous process, and the tip of the right transverse process. There are small areas of abraded surface bone along the right side of the inferior endplate, left lamina, left mammillary process, and posterior margin of the right inferior zygapophysis.
Morphology. The superior endplate is less ovoid than the inferior one. The superior apophyseal ring is about 6 mm at its thickest anteriorly, and the inferior one only 5 mm. The endplates are nearly parallel in lateral view. The transverse processes angle slightly posteriorly in the superior view and project straight laterally in anterior view. The superior zygapophyseal joints are transversely curved to a depth of just over 3 mm and angle 30° to a coronal plane. The mammillary processes project up to 6.5 mm posteriorly and are 3.7 mm thick, measured on the better-preserved right side. There are no imbrication pockets. The inferior zygapophyses extend directly inferiorly from the neural arches. They are slightly asymmetrical with the right one more tightly convex than the left. The right one faces 35° to the median plane but the left one 40° from it.

StW 431t—PSV3 (StW 460 originally)
Preservation. This vertebra (Figure 10.17) is missing its left pedicle and part of the left lamina, left superior zygapophysis, and both transverse processes. Most of the spinous process is also broken away. There are just a few other minor spots where surface bone is missing, but this does not obscure any major contours. The right pedicle is broken through its midpoint and the vertebra remains in two separate pieces.
Morphology. The superior endplate is visibly less oval than the inferior one. The apophyseal rings are roughly 7 mm thick anteriorly on both endplates. The endplates appear roughly parallel in lateral view. The superior zygapophyses are about 2 concave transversely and face about 50° to the median plane. A mammillary process projects about 5 mm posteriorly from the subchondral surface and is

up to about 5 mm thick. No imbrication pocket is present flanking the facet. The inferior zygapophyseal facet faces laterally and is only mildly convex transversely and its posterior margin hooks sharply making a "J" shape transversely. The spinous process is broken but appears to have projected almost straight posteriorly.

StW 431ua and ub—PSV2 (StW 461 & 462 originally)
Preservation. This vertebra (Figure 10.17) is preserved in two separate, non-conjoining pieces. StW 431ua consists of the body, right pedicle, superior zygapophysis, and base of the transverse process. There are minor areas of abraded surface bone, but these do not obscure important contours. A crack completely traverses the specimen sagittally at about the right third of the vertebral body, and there are missing pieces of bone but no major contours are affected appreciably. There are some areas of missing surface bone at the center of the inferior endplate surface. StW 431ub is the spinous process and left inferior zygapophysis. It is undamaged in the parts preserved.
Morphology. StW 431ua has osteophytic lipping along the entire anterior half of the body along the inferior margin. This reactive bone is inclined inferiorly toward the comparable pathological bone on the subjacent vertebra (StW 431v) and projects inferior to the endplate on both the right and left sides. Adjacent to midline, it extends up to the margin of the superior endplate. The margin of the apophyseal ring on the superior endplate is difficult to discern, but the inferior ring is up to 7 mm wide. In lateral view, there is a slight posterior wedging of the vertebral body.

The preserved portion of the base of the transverse process angles slightly posteriorly in the superior view but appears to be directly lateral in anterior view. The superior zygapophyseal facet is transversely concave with a depth of just over 1 mm, and faces about 45° posteromedially relative to a coronal plane. Just inferior to the superior zygapophyseal facet is a depression for contact with the suprajacent inferior zygapophysis in extension. The inferior articular facet is transversely convex but unevenly so, with the anterior two-thirds only mildly convex but the posterior portion more so making a "J" shape transversely. Inferior to the facet, there is a flattened and smooth area of bone that would have contacted the subjacent lamina in its imbrication pocket during peak extension. The spinous process is 30.2 mm long and 20.5 mm superoinferiorly near its posterior end. Its superior margin is 5 mm thick, and the inferior edge is 9.7 mm thick. It is very slightly asymmetric, with the posteroinferior margin projecting further to the right than the left as seen in posterior view. The spinous process is inclined inferiorly about 20° relative to the anterior margin of the inferior zygapophyseal facet.

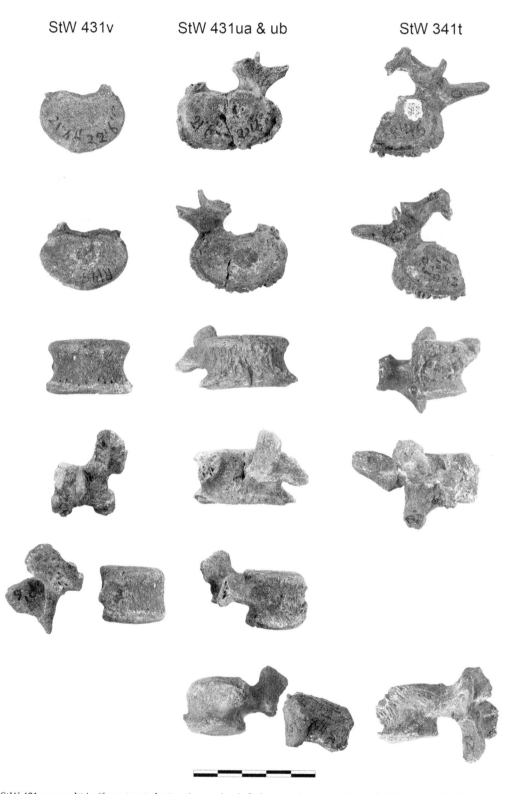

Figure 10.17 StW 431v, u, and t in (from top to bottom) superior, inferior anterior, posterior, right lateral, and left lateral views.

StW 431v—PSV1 (First presacral vertebra; StW 463 originally)
Preservation. This vertebra (Figure 10.17) preserves roughly half of the body and neural arch on the left side. It is broken at the root of the pedicle (StW 431va) and an additional piece preserves a portion of the neural arch (StW 431vb). These pieces have been glued together. Only the base of the spinous process is preserved. Superiorly, about two-thirds of the endplate surface remains and is well preserved with the apophyseal ring intact except for abrasion along the anterior margin. This abraded region is flanked inferiorly by osteophytic growths that cover most of the anterior midline surface and fade out laterally along the superior margin of the vertebral body. Only half of the inferior surface of the vertebral body remains but is also well preserved with the apophyseal ring intact.

Morphology. The vertebral body is wedged posteriorly in lateral view. The apophyseal rings are wide. The superior one cannot be measured accurately due to the pathology, but the inferior one is up to 8 mm wide. Extra pathological bone forms a ridge along the anterior margin of the inferior endplate. Beginning about 3 mm superior to this, there is mass of roughened pathological exostotic bone, or brucellosis (D'Anastasio et al., 2009), covering about 10 of the anterior surface, superior to which the bone is hollowed out so that the anterior margin of the superior endplate is missing. Theses exostoses continue to the left side of the vertebral body angling superiorly to end adjacent to the lateral-most extent of the endplate. The growths are inclined superiorly toward the comparable pathological bone on the suprajacent vertebra (StW 431ua).

The pedicle is 13 wide, and its lateral surface is confluent with the anterior margin of the transverse process. The transverse process projects laterally and slightly posteriorly in superior view and angles slightly superiorly in anterior view. Along its inferior margin, there is a small accessory process. The superior zygapophysis is mildly convex transversely with a depth of about 1. It is angled about 50° medially relative to a coronal plane. A depression adjacent to the inferior margin of the superior zygapophyseal facet likely contacted the suprajacent inferior zygapophysis at peak extension. The inferior zygapophysis projects directly inferiorly, and is also angled slightly laterally. It is uniformly convex transversely.

StW 572—Lumbar vertebra

Preservation. This is a nearly complete lumbar vertebra (Figure 10.18) from an adult individual. The right side of this specimen has been sheared off in a nearly straight plane from the pedicle superiorly through the specimen, leaving roughly two-thirds of the inferior surface intact. The lateral side of the pedicle and right transverse process are missing. The left transverse process and superior zygapophysis are also broken away, as is part of the right superior zygapophysis, about half of the left inferior zygapophysis, and the posterior extent of the spinous process. There is a large crack through the right inferior zygapophysis, and its posteroinferior margin is missing along this crack. The superior endplate surface is marked with numerous small grooves from preparation.

Morphology. Based on the narrow set of its zygapophyses, the very slight anterior wedging of the vertebral body, and the almost sagittally oriented zygapophyseal joints, this specimen is likely to have been a first or second lumbar. The superior endplate preserves no morphology. The inferior endplate is well preserved, with a clear apophyseal ring up to 4.5 mm wide anteriorly. The rest of the endplate surface is roughened and porous but no cancellous bone is exposed. The superior zygapophyseal facet is transversely concave with a depth of over 2 mm. It faces about 130° from a coronal plane. The inferior zygapophyses project directly inferiorly. The facets are transversely convex, but slightly more so along their posterior margins. The preserved part of the spinous process projects directly posterior and is 16.4 mm tall superoinferiorly at the break.

Descriptions: Ribs

Sts 14—Associated ribs

The ribs attributed to Sts 14 have only briefly been described or mentioned in the literature. They only recently have been attributed to sides and specific level, and given accession

Figure 10.18 StW 572 in superior, inferior, anterior, posterior, and left lateral views.

numbers. Originally, the only rib remains with numbers were numbered Sts 14w and Sts 14x. A box of rib fragments was labeled Sts 14w–y. In this box, three fragments have Sts 14y written on them, and one has another number that is difficult to discern. Now, these rib fragments are attributed to specific levels and sides and have been given numbers Sts 14y through Sts 14ad. Thus, Sts 14 now includes parts of the inferior eight pairs of ribs as well as the single right rib (Sts 14aa) that would articulate with vertebra Sts 14g (Figures 10.19 and 10.20). Although none is complete, these ribs provide new information about *Australopithecus africanus*.

In addition to the ribs that can be attributed to particular levels, there is the head of a left rib that could be from right rib 3 to 6. There are also eight pieces of rib diaphysis that derive from somewhere anterolateral to the costal angle but little more can be reliably said other than attribution to side. There is also a box of smaller rib fragments.

The exact number of thoracic lumbar vertebrae and rib pairs in this skeleton is indeterminate. So, as with the vertebrae, traditional numbering schemes that run from superior to inferior cannot be employed for this skeleton (e.g., rib 1, 2, 3). It is also not possible to serially number the ribs from caudalmost to cranially (e.g., last, second-to-last, and third-to-last), given that there is one more on the right side of the thorax than on the left (see Figure 10.20 and Ward et al., Chapter 4, this volume). Table 10.1 shows which ribs would have articulated with which vertebrae. Although somewhat unsatisfactory, simply to provide a means of discussing the relative position of the ribs, we employ an arbitrary scheme here. The cranialmost rib preserved is Sts 14y, a right rib that would have articulated with Sts 14n at its caudal costovertebral demifacet and with the vertebra suprajacent to that at its cranial costovertebral demifacet. The ribs are numbered sequentially from Sts 14y caudally, so Sts 14x is referred to as level y + 1, Sts 14W as y + 2, y + 3, y + 4, Sts 14ad as y + 5, and so on (see Table 10.1).

Sts 14y—Cranial-most rib preserved for Sts 14
Preservation. This rib measures 78.4 mm long as preserved (Figure 10.19). It is intact from the tubercular facet to a point about 73 mm from its lateral margin.
Morphology. This appears to be left rib 4. It is smaller and more curved than Sts 14x. The posterior orientation of its costotransverse facet demonstrates that it does not derive from the inferior portion of the rib cage. Because the head and neck are missing, these areas cannot be used to compare to vertebra Sts 14p. A shard of bone is secondarily fused across the articular surface, so its contours also cannot be compared. However, Sts 14y does not differ dramatically from Sts 14x in size and curvature, and is appropriate to the size and shape to represent the rib immediately superior to it.

The facet is 5 mm in diameter anteroposteriorly, but its mediolateral dimension cannot be measured. The angle is 18° from the lateral margin of the facet. The neck is transversely concave. The anteroinferior border of the bone is sharp at the medial end of the fragment but becomes rounder laterally. Immediately lateral to the facet the rib is 6.4 mm in maximum and 5.4 mm in minimum diameter. At the angle, it is 7.9 mm by 5.2 mm and near the lateral break is 7.2 mm by 4.0. mm In posterior view, the inferior margin is concave medially and gently convex laterally. There is only a faint costal groove more easily appreciated by palpation rather than observation. In superior view, the rib is more tightly concave (smaller radius of curvature) toward its medial end than its lateral end.

Sts 14x—Left rib level y+1
Preservation. This is a partial rib (Figure 10.19) measuring 83 mm long as preserved. It extends from a point 3.8 mm proximal to the tubercle to about 68 mm lateral to the angle. It has small spots of adhering breccia but otherwise is undamaged. The superior margin of the bone is missing to a point just lateral to the tubercular facet.
Morphology. Sts 14x appears to be left rib 5. It is slightly narrower and more curved than any of the Sts 14w ribs, and it is thicker than Sts 14y. The contour of its costotransverse facet conforms to the opposing surface on Sts 14m as well.

Based on its slightly smaller diameter and slightly tighter curvature, it appears to have belonged immediately superior to Sts 14w. The tubercular facet measures 6.0 mm in diameter along the axis of the rib and 5.7 mm in diameter transverse to that. Immediately lateral to the facet, the rib measures 7.0 mm by 5.7 mm in maximum and minimum diameter. At the angle, it is 8.9 mm by 5.9 mm. The angle is 22 mm from the lateral margin of the facet. In the posterior view, the inferior margin of the bone is concave medial to the angle and mildly convex lateral to it. At the lateral-most point the bone is 9.0 mm by 7.8 mm in maximum and minimum diameters. In posterolateral view, the inferior margin of the rib is linear medial to the angle and mildly convex lateral to it, indicating a slight declination of the corpus.

Sts 14w—Left ribs level Y+2, Y+3, and Y+4
Preservation. These specimens consist of the proximal ends of three middle ribs (Figure 10.19). They each preserve the head, neck, tubercle, angle, and part of the shaft. They all have various cracks about 65 mm from the proximal end and are distorted to curve further inferomedially than they would have originally been. Proximal to this, the original contours are preserved. There are small cracks in the surface bone even here, but these disturb no meaningful contours. All three ribs show damage to the heads with exposed trabecular bone. The superior one, Sts 14w$_i$,

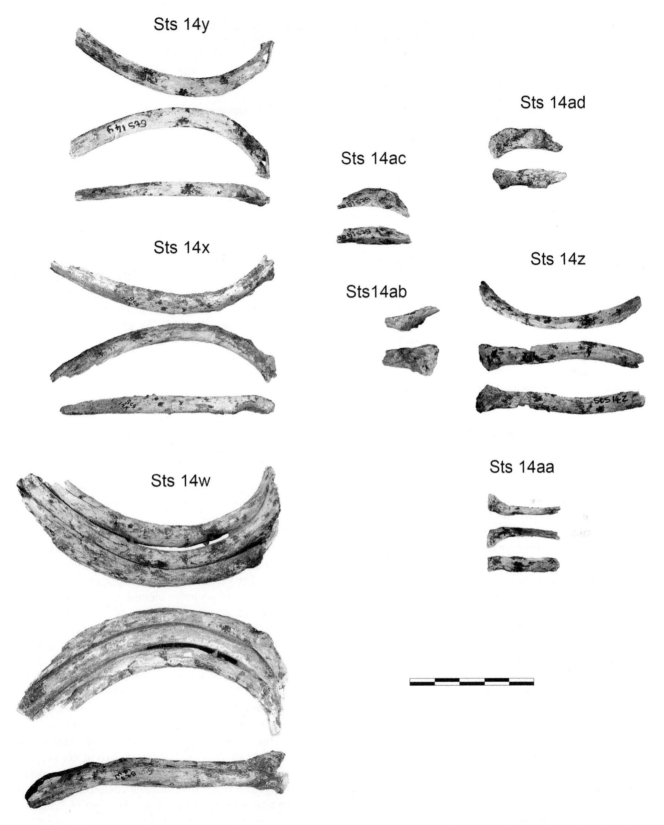

Figure 10.19 Sts 14 ribs. Sts 14y, x, w, z, and aa in superior, inferior, and posterolateral views, Sts14ac in inferior and anterior view, Sts 14ad in inferior and posterior views, and Sts14ab in posterior and medial views.

Figure 10.20 StW 431p rib in inferior and superior views.

Table 10.1 Articulation and segmentation of Sts 14 ribs and vertebrae, along with rib segmentation ordering designation.

Vertebra	Rib-left	Rib-right	Rib level
n	y		y
m	x		y+1
i	w_i		y+2
k	w_{ii}		y+3
l	w_{iii}		y+4
o		ad	y+5
h	ac		y+6
g	ab	z	y+7
f		aa	y+8

is missing most of its costovertebral facet, Sts $14w_{ii}$ is missing most of the head so that only 3 mm remains of the articular surface, and Sts $14w_{iii}$ is missing the inferior corner of the facet and adjacent surface of the neck.

Morphology. Sts $14w_{iii}$ fits well with vertebra Sts 14k in the length of the neck and orientation of the costotransverse facet. Sts $14w_{ii}$ and Sts $14w_i$ cannot be directly compared to any vertebrae as their heads and costovertebral articular surfaces are too damaged. However, given that they were preserved fused to Sts $14w_{iii}$ by breccia, it is clear that they represent the next two superior ribs. Nothing about their morphology contradicts this interpretation.

The vertebral articular facet of Sts $14w_i$ and Sts $14w_{ii}$ are too incomplete to measure. On Sts $14w_{iii}$, the vertebral articular surface is about 7.4 mm tall superoinferiorly, but its mediolateral dimension cannot be measured.

The necks of all three ribs have a crest running along their inferior margins and their anterior surfaces are nearly flat superoinferiorly. The inferior surfaces have a sharp crest running between the articular facets. Another sharp crest runs from the rib heads along the anteroinferior corner, fading away only adjacent to the angle of the ribs. The neck of the superior rib (Sts14 w_i) is 6.3 in diameter superoinferiorly and 5.9 mm anteroposteriorly. The neck of the middle rib (Sts $14w_{ii}$) is 8.1 mm by 4.7 mm and the neck of the inferior rib (Sts $14w_{iii}$) is 8.3 mm by 4 mm. Neck lengths, measured between the centers of the vertebral and transverse facets, of the superior and inferior ribs are both 17.5 mm, but the middle is too damaged to measure.

The angles of the ribs are located 31.4 mm, 30.0 mm, and 27.5 mm from the lateral margin of the transverse facet for the superior, middle, and inferior ribs, respectively. At the angle, the superior rib measures a maximum of 9.0 mm superoinferiorly and 6.4 mm anteroposteriorly, and the inferior one is 8.5 mm by 5.8 mm here. The middle rib cannot be measured. Distal to the angle, cracking and crushing of the diaphyses preclude measurement. The costal grooves become visible at the angle, and become pronounced lateral to this. The grooves are flanked by a blunt ridge superiorly and a sharp margin of the bone inferiorly. The lateral surfaces of the ribs spiral from posterior to anterior.

Sts 14ad- Right rib level y+5
Preservation. This is the proximal end of a rib (Figure 10.19) extending to an oblique break lateral to the costotransverse facet. It is 30.1 mm in length. There are small areas of trabecular bone exposure around the head, but most of the surface is intact and no morphology is obscured by damage.

Morphology. This is the right fourth-to-last rib. It is not the third-to-last rib, as the facet is confined to the posterior margin of the inferior margin of the bone, unlike Sts 14ac. The neck is also too long to articulate with Sts 14h, which would be the corresponding vertebra if this was the third-to-last rib. The vertebra immediately superior to that, Sts 14o, is missing its posterior elements, so it is not possible to assess congruence with this rib. Neck length is appropriate for the next most superior vertebra, Sts 14l, but the conformation of facets is imperfect so this rib does not appear to articulate with this vertebra.

The head is 9.5 mm superoinferiorly by 5.4 mm transversely. The inferior costovertebral facet is 7.0 mm superoinferiorly by 6.6 mm anteroposteriorly, and the smaller superior one is 3.5 mm by 6.2 mm. The neck has a large ridge for ligament attachment along its posterosuperior margin that forms a strong convexity of the superior margin of the bone in anterior or posterior view. Between this ridge and the costotransverse facet, the diameters of the neck are a maximum of 7.5 mm by a minimum of 5.4 mm. The costotransverse facet faces posteromedially and is 7.2 mm long mediolaterally by 5.5 mm transversely. It is raised from the surface of the bone on all sides.

Sts 14ac—Left rib level y+6

Preservation. This is part of the neck and the region flanking the costotransverse facet of a left rib (Figure 10.19). It measures 29 mm in maximum length. The head and much of the neck are missing. The facet and most of the inferior surface of the bone are preserved but the superior margin of the bone is damaged. Cracks traverse the specimen but cause no distortion.

Morphology. This is the left second-to-last rib. This is apparent because the costotransverse facet spans the entire inferior surface of the bone. The facet measures a maximum of 8.0 mm oriented along the rib corpus and 7.3 mm transversely. Immediately lateral to the facet, the diaphysis is 7.5 mm anteroposteriorly and 6.7 mm superoinferiorly. Here, the medial margin forms a sharp crest throughout the length of the specimen. The inferior margin lateral to the facet is sharp, and the corpus is transversely concave. At the level of the facet, the bone is roughly triangular in cross-section with the superior margin of the bone forming the apex of the triangle and the facet forming the base. Neck dimensions cannot be accurately measured.

Sts 14z—Right rib level y+7

Preservation. This is the proximal end of a right rib (Figure 10.19) measuring 66.2 mm in maximum length. There are areas of missing bone along its length, especially along a crack that traverses the specimen transversely 23 mm from the proximal end of the specimen. Trabecular bone is exposed along the superomedial margin of the head obscuring the original inferomedial margin of the facet. The sternal extremity is missing so its original length cannot be determined.

Morphology. This is the second-to-last right rib. It lacks a costotransverse facet, although it is substantially larger than Sts14ab. Given that PLV3 (Sts 14h) has a large and well-developed costotransverse facet, there were only two floating ribs on the right side of this individual. Sts 14z articulates well with Sts14g (PLV2), the seventh presacral and transitional vertebra. When articulated with the Sts14g, the rugose area for ligament attachment on the rib's lateral surface matches the corresponding tubercle on the vertebral transverse process.

The head of this rib is 10.8 mm superoinferiorly and 8.0 mm anteroposteriorly. Facet dimensions cannot be measured accurately due to damage of the surface bone. The superior margin of the bone adjacent to the head is concave in posterior view, and a rugose area for ligament attachment flanks the costotransverse facet inferiorly. A sharp crest delineates the inferior margin of the head. The superior margin of the diaphysis progresses from a rounded ridge near the head to a smooth convexity laterally. The lateral surface of the rib spirals strongly toward its sternal end, and the sternal end of the bone curves inferiorly relative to the vertebral end of the diaphysis.

Immediately lateral to the rib angle, the diaphysis measures 6.7 mm superoinferiorly and 4.8 mm anteroposteriorly. At the sternal-most point that can be measured, it is 7.5 mm by 4.8 mm. No costal groove is apparent.

Sts 14ab—Left rib level y+7, caudal-most left rib

Preservation. This is the head and neck of a right lower rib (Figure 10.19) that is 25 mm in maximum length. Most of the surface bone on the articular portion of the head is abraded, exposing trabecular bone, but the major contours remain unobscured. This specimen articulates with the Sts 14g vertebra.

Morphology. This is the left second-to-last rib, and the antimere of Sts14aa. The rib lacks a costotransverse facet, indicating that it is not anatomically associated with the Sts 14h. Even despite the abrasion, there is clearly only one facet on the head. The head is 11.2 mm superoinferiorly by a maximum of 6.1 mm transversely. The diaphysis narrows almost immediately to 7.8 mm by 4.5 mm in diameter. A rugose set of ridges marking ligamentous insertions flanks the posterior surface of the specimen.

Sts 14aa—Right rib level y+8, caudal-most right rib

Preservation. This is the proximal end of a rib (Figure 10.19) that measures 29.1 mm in maximum length. There is minor surface bone abrasion around the margins of the head, but no contours are obscured.

Morphology. This is the last rib from the right side. It lacks a costotransverse facet and is very small. The costovertebral facet faces almost completely proximally. The rib is only slightly concave along its length and it conforms well to the opposing articular surface on vertebra Sts14g. When articulated, the orientation of this rib matches that of the transverse process of Sts14f, although it is angled further inferiorly than the transverse process.

The costovertebral articular facet comprises the entire vertebral end of the bone, and it measures a maximum of 7.6 mm in diameter by a minimum of 6.0 mm. The neck measures 5.7 mm by 4.2 mm in diameter. Rugose areas of ligamentous attachment flank the posterior and inferior side of the bone in a ridge running from a point 5 mm lateral to the head along the superior margin of the bone. Immediately lateral to the rib head the bone is 5.7 mm by 3.1 mm. The sternal-most point is 5.6 mm by 2.9 mm. In the superior view, the medial portion of the bone angles posteriorly but is nearly straight for the rest of its length. The rib angles about 3–4° inferiorly relative to the costovertebral facet.

StW 431p—Rib (originally StW 456)

Preservation. This is the proximal end of a middle rib (Figure 10.20) that is 34 mm long as preserved. The head, neck, and tubercle are completely preserved, as is up to 12 mm of the shaft.

Morphology. The two costovertebral facets are set at an acute angle separated by a prominent crest along the head. Both facets are oval in shape. The neck is constricted anteroposteriorly and widens as it approaches the tubercle. The posterior margin of the neck is straight and the surface is marked by roughenings from capsular attachments. The beginning of a shallow costal groove is visible on the inferior surface. The tubercle is raised and the costotransverse articular surface faces posteroinferiorly. The beginning of the costal groove is preserved.

Comparative anatomy

Materials and methods

The Sterkfontein vertebrae were compared qualitatively with those of extant hominoids, as well as those of other early hominins attributed to *Australopithecus afarensis* (A.L. 288-1, A.L. 333-152, A.L. 444-7, DIK 1-1) (Johanson et al., 1982; Lovejoy et al., 1982; Alemseged et al., 2006; Ward et al., 2012), *Australopithecus sediba* (MH 1, MH2), and *Homo erectus* (KNM-WT 15000) (Jellema et al., 1993; Latimer and Ward, 1993; Walker and Leakey, 1993), as well as specimens recovered from Swartkrans of uncertain taxonomy (SK 3891, SKX 3342, SKX 41692). Many of these fossils are fragmentary and of uncertain segmental level, and thus could not be included in metric comparisons, but each is included where possible.

For assessing features related to presence or absence of spinal curvatures, the Sterkfontein vertebrae are compared with those of chimpanzee (*Pan troglodytes*), gorilla (*Gorilla gorilla*), and human (*Homo sapiens*) specimens from the Cleveland Museum of Natural History (n=5 males, 5 females of each species). No comparative skeletons show evidence of neoplastic, congenital, traumatic, or degenerative diseases involving the spine. All were fully adult, and the apes wild-shot. Humans were between 30 and 35 years of age, in order to avoid potential problems that can occur in younger specimens due to late fusion of the annular ring and in older individuals to age-progressive deterioration of the vertebral column. No effort was made to assess population variation in humans. This procedure does, of course, hide population variation known to occur in adult spines (Scoles et al., 1988) as well as in vertebral developmental processes (Veldhuizen et al., 1986). However, given the inherent uncertainty about such variations in early hominins, our sample was deemed appropriate.

For each of these specimens, we measured (1) superoinferior height of the anterior margin of the vertebral body (ABH), (2) superoinferior height of the posterior margin of the vertebral body (PBH), (3) anteroposterior diameter of superior endplate at midline (SAP), (4) anteroposterior diameter of inferior endplate at midline (SML), (5) mediolateral distance between centers of superior zygapophyseal facets (IFD), and (6) maximum superoinferior distance from tips of superior to inferior zygapophyses (ZH). We also used calibrated photographs of vertebrae taken in the superior view using a flat macro to avoid issues of parallax and image distortion to quantify (7) orientation of superior zygapophyseal facets to a coronal plane (SFO). Orientation of the facets was made by constructing a line through the medial and lateral margins of the facet. Fossils included in these analyses were those represented by more than one thoracolumbar vertebrae that were adolescent to adult (Sts 14, StW 431, A.L. 288-1, MH1, MH2, and KNM-WT 15000). Table 10.2 provides data for Sterkfontein fossils, with measurements defined in Table 10.3.

To assess thoracolumbar segmentation patterns, we assessed frequencies of morphologies, requiring a larger sample of species and a greater number of species and a larger sample of each. For these analyses, we include *H. sapiens* (N = 93), *Pan troglodytes* (N = 84), *Gorilla gorilla* (N = 117), *Pongo pygmaeus* (N = 45), *Hylobates* spp. (N = 82), and *Macaca fascicularis* (N = 30). Most of the specimens were adult, but we included subadult specimens in the non-metric analysis, for which age was not a factor. All specimens included in the metric analysis were adult, and also free of obvious pathologies. Nonhuman specimens were wild-shot. We measured: (8) last rib length (RL), measured as the maximum length of the last rib, and (9) transverse process span for the first non-rib-bearing lumbar vertebra, measured from tip to tip (LTP). Table 10.4 presents comparative data.

To compare individuals of different body sizes, we expressed each length as a simple ratio by dividing each metric by the geometric mean of five capitate and four talar measurements that are highly correlated with body size (from Lovejoy et al., 2009). The LTP measurement disregards any differences in vertebral body width among taxa. It does, however, reflect the biologically relevant aspect of transverse process span. Since Sts 14 preserves no podials with which to estimate size, and in order to compare lumbar transverse process span for Sts 14, we first calculated the mean difference between Sts 14 and A.L. 288-1 for five vertebral and pelvic measurements (not shown). We then reduced A.L. 288-1's body size estimate by the appropriate percentage. Therefore, the relative transverse process span for Sts 14 should be viewed cautiously, given the compounding effects of several levels of estimation.

To explore the pattern of costovertebral morphology in the lowest rib-bearing vertebrae, we developed a categorical method of assessing the morphology of costovertebral

Table 10.2 Basic metric data. Measurement definitions in Table 10.3. All measurements in mm.

Measurement	Sts 14p	Sts 14l	Sts 14n	Sts 14m	Sts 14i	Sts 14k	Sts 14l	Sts 14o	Sts 14h	Sts 14g	Sts 14f	Sts 14e	Sts 14d	Sts 14c	Sts 14b	Sts 14a
SAP	10.7	…	11.6	12.0	12.6	13.2	13.3	13.4	13.7	14.8	16.6	17.7	17.9	18.1	17.7	18.8
SML	16.7		15.6	16.1	16.9	16.1	17.5	18.6	18.8	22.0	22.8	24.0	25.3	28.0		30.2
IAP	11.0		12.2	13.6	12.3	13.3	14.8		14.9	16.0	17.6	18.2	18.5	18.6	18.7	17.0
IML	18.5		18.5	18.9	18.8	18.7	19.5	21.1	22.1	22.0	23.1	24.4	26.4	8.4	28.9	24.9
ASI	12.3		12.4	13.1	13.0	13.7	14.4	14.5	16.8	17.3	18.6	19.2	19.6	19.5	19.0	19.0
PSI	12.6		12.8	13.3	13.6	13.9	14.9	15.6	17.1	19.3	19.7	20.1	19.9	19.5	18.2	17.6
TDN	11.8		9.8	10.3	10.2	10.9	9.4		11.1	11.0	12.3	13.3	13.4	12.8	12.0	19.3
SDN			13.2	10.6	10.2	9.6	9.5		10.3	11.4	12.1	11.3	11.5	12.5	9.8	8.2
PML	2.0		2.5	2.4	2.6	2.7	3.4		5.3	6.0	5.9	4.3	4.7	6.1	6.0	7.8
PSI	6.3		7.6	7.9	7.0	8.1	8.9		11.2	10.9	11.0	9.8	9.4	9.9	8.0	9.6
XFH			18.3	20.8	20.7	22.2	23.8		24.2	29.5	32.5	34.1	33.3	32.7	31.4	
IFD			12.6	13.3	13.0	13.2	14.5		16.3	17.5	17.7	19.9	20.7	20.6	21.3	29.0
TPL		20.4	19.3	18.8	18.8	18.2	19.6		17.4	15.2	26.8	27.9	34.0	30.5		29.7

Measurement	StW 431o	StW 431n	StW 431m	StW 431l	StW 431q	StW 431r	StW 431s	StW 431t	StW 431u	StW 431v
SAP	21.7	22.6				22.6	23.8	25.6	24.1	23.7
SML		26.0				33.6	34.8	36.3	39.6	34.0
IAP	23.5	22.8	22.8		22.7	22.8	24.2	24.4	21.8	
IML	29.0	28.0		20.7	33.0	33.8	35.7	38.4	39.8	36.0
ASI	16.4	17.0	18.1		22.4	24.3	23.6	22.8	22.5	21.8
PSI	17.7	18.5	19.0		22.7	24.0	24.3	23.6	22.6	19.8
TDN		11.0			15.8	15.7	16.0	14.5	15.4	15.4
SDN		12.0					12.8	10.1		12.8
PML	5.4	5.5	6.5			6.5	7.4	10.0	11.8	14.3
PSI		11.1	12.0			14.0	12.4	12.0	11.9	11.1
XFH	39.7	29.6					38.0	38.5		35.7
IFD		15.0					23.5	21.6	22.8	28.0
TPL		25.8	23.2	20.6			30.0			37.4

Measurement	StW 8a	StW 8b	StW 8c	StW 8d	StW 41	StW 41b	StW 572	Sts 73
SAP	21.8	23.5	24.4	23.0	20.0	20.7	24.5	21.6
SML		33.2	33.0	34.6	28.6	30.9	32.2	31.5
IAP	23.2	24.4	23.6		21.4	21.4	24.6	19.5
IML	29.4	33.9	34.3		29.2	28.6	35.0	
ASI	23.8	22.5	21.7		20.9	23.7	20.4	18.6
PSI	24.7	22.7	20.8		22.3	25.5	22.0	20.4
TDN		12.8					14.5	15.8
SDN	12.3	11.3					10.7	
PML		8.9					8.5	5.4
PSI		13.9					13.8	13.7
XFH	39.7	39.9					36.0	
IFD		22.0					23.6	
TPL								

Table 10.3   Description of measurements used here.

Abbreviation	Definition
SAP	Anteroposterior dimension of superior endplate at midline
SML	Mediolateral maximum diameter of superior endplate
IAP	Anteroposterior dimension of inferior endplate at midline
IML	Mediolateral maximum diameter of inferior endplate
ASI	Superoinferior height of anterior margin of vertebral body
PSI	Superoinferior height of posterior margin of vertebral body
TDN	Transverse diameter of neural foramen
SDN	Sagittal diameter of neural foramen
PML	Mediolateral thickness of pedicle
PSI	Superoinferior height of pedicle
XFH	Maximum distance superior zygapophyses to inferior zygapophyses
IFD	Mediolateral distance between center of superior zygapophyseal facets
TPL	Maximum lateral extent of transverse process from midline
SFA	Angle between superior zygapophyseal facets and a coronal plane

articulations. We chose this method of measurement to compare differences in transitional morphology among species independent of lumbar or thoracic count, and to try to infer morphology of incompletely preserved fossils. Because the original number of thoracic vertebrae in all but one early hominin (*Australopithecus afarensis* DIK 1-1; Ward et al., 2017) is unknown, thoracic vertebrae are described relative to the position of the inferiormost element bearing costal facets, which we refer to as PLV, or prelumbar vertebrae. PLV1, therefore, refers to the inferiormost rib-bearing element, PLV2, as the vertebra immediately superior to it, and so on.

For each vertebra, we coded: (10) costovertebral facet score (CVS) on PLV1-5 as follows: 1 = complete facet located at the superoinferior center of body-pedicle junction; 2 = complete facet adjacent to the superior endplate; 3 = a superior demifacet only; 4 = superior and inferior demifacets. We took the mean of the right/left scores for each level, and then averaged a grand score for PLV1-5. A higher number means a more gradual transition to a single facet from a modal thoracic bivertebral demifacet articulation of the ribs with the vertebral column to a single univertebral articulation at the caudal-most rib-bearing vertebra. We also coded: (11) costotransverse facet score (CTS) defined as the presence/absence of the costotransverse facet on vertebrae PLV1 through PLV5 (0 = absent; 1 = present). We took the mean of the right/left scores for each level and then averaged an overall mean for PLV1-5. A higher number describes more transverse process articulation with the costal tubercle on PLV1. To assess correlations between articular morphology and rib length, we used Kendall's τ, Kruskall-Wallace (with Bonferroni-corrected Mann-Whitney post hoc tests) to compare taxa for each variable.

Results

Taxonomy

All but one of the Sterkfontein hominin vertebrae (StW 572) are plausibly attributable to *Australopithecus africanus*, with otherwise no variation that would exceed that seen within a single species of extant hominoid (see also Ward and Zipfel, Chapter 18, this volume; Grine, 2019). There is no evidence for taxonomic diversity in the ribs, as all are attributed to the two partial skeletons Sts 14 and StW 431.

The morphology of StW 572, though, suggests that this specimen is more likely attributable to *Homo* rather than *Australopithecus*. This lumbar vertebra has a large vertebral endplate and comparatively small neural arch, as well as superoinferiorly short vertebral body relative to endplate size. None of these proportions matches vertebrae known for *Australopithecus* (see Robinson, 1972; Rose, 1975; Cook et al., 1983; McHenry, 1992; Shapiro, 1993; Sanders, 1998; Ward et al., 2012; Williams et al., 2013, 2017). An attribution to *Homo* is consistent with the fact that StW 572 was recovered from a Member 5 square (59, 14'10"–15'10"), a location distant from Member 4 and near where tools were recovered at the site.

Size

The vertebral bodies of Sts 14, StW 431, and StW 8/41 have been shown to be smaller in endplate size relative to presumed body size than are those of later hominins (Robinson, 1972; Rose, 1975; Cook et al., 1983; McHenry, 1992; Shapiro, 1993; Sanders, 1998; Ward et al., 2012; Williams et al., 2013) but also may be relatively taller superoinferiorly (Lovejoy et al., 1982; Shapiro, 1993; Sanders, 1998). Relative

Table 10.4 Comparative data for analysis of morphologies related to spinal curvatures. Wedging values and superior facet angle in degrees, other measurements in mm.

vert	Pan troglodytes mean	sd	Gorilla gorilla mean	sd	Homo sapiens mean	sd	Sts 14	StW 431	A.L. 288-1	MH1	MH2	KNM-WT 15000
BODY WEDGING												
17	8.3	5.1	4.1	3.1	3.2	3.5						3.8
16	7.6	3.2	3.4	1.7	−0.2	2.4						4.6
15	5.2	2.5	3.1	3.5	1.1	2.2	1.6					
14	5.0	3.2	4.1	2.7	2.2	2.5	2.0					5.3
13	4.0	3.1	3.1	1.8	2.4	1.7	5.7		8.0			
12	3.8	2.1	3.0	1.9	4.7	2.0	2.7					5.8
11	3.9	2.5	1.8	1.6	3.9	1.6	0.9					3.1
10	3.5	2.9	1.9	1.4	3.6	1.5	2.2	3.4	2.2			
9	3.7	1.9	1.1	1.3	1.7	2.0	4.7	3.8	4.8			
8	3.6	3.2	0.9	2.1	1.8	1.7			7.4			5.0
7	4.2	2.9	0.8	2.3	2.6	6.7	7.7					
6	6.0	3.2	4.0	2.4	4.6	2.0	3.8					5.4
5	6.1	2.3	4.1	1.4	2.6	2.7	2.9	4.1	5.8	4.1		
4	6.0	3.7	2.7	2.5	0.0	2.1	1.0	1.7				
3	5.3	2.8	2.5	1.8	−2.9	2.1	0.0	1.8		−1.6		
2	2.7	2.6	2.6	2.9	−3.5	2.5	−2.6	−0.2			−1.5	−5.5
1	0.8	2.3	0.0	2.8	−7.7	2.2	−4.3	−4.8			−10.9	−11.2
INTERFACET DISTANCE (IFD)												
17	27.3	3.5			35.9	2.7						27.2
16	22.6	2.6	26.8	3.2	29.4	2.3						24.3
15	20.0	1.8	22.0	2.6	26.0	1.7	12.6					
14	18.8	3.1	19.6	1.5	24.0	1.4	13.3					21.3
13	17.6	1.7	19.0	1.7	22.5	1.3	13.0		16.0			
12	17.6	1.6	19.0	1.4	22.5	1.3	13.2					22.0
11	17.7	1.5	19.2	1.4	22.5	1.3	14.5					20.3
10	18.0	1.3	19.4	1.7	23.0	1.6			17.0			
9	18.5	1.5	19.2	1.3	24.2	1.3	16.3	15.0	20.3			21.4
8	19.4	1.6	20.4	2.0	26.0	1.7	17.5		23.1			23.7
7	19.9	1.8	21.6	2.7	26.5	1.7	17.7					
6	20.9	1.7	23.1	2.0	28.2	3.4	19.9					
5	23.4	2.9	25.4	1.8	26.9	3.0	20.7		25.1	22.0		28.0
4	25.5	2.6	25.5	2.2	27.5	2.1	20.6	23.5				
3	27.9	2.5	24.5	2.2	29.6	2.5	21.3	21.6		19.3		26.0
2	26.6	2.0	23.8	2.5	32.9	3.9	28.4	23.8				31.8
1	25.4	1.8	24.0	2.3	37.9	4.5	29.0	28.0			27.0	33.0
0	23.7	2.3	22.3	1.5	43.9	2.7	33.0	34.0	32.0			34.4
ZYGAPOPHYSEAL HEIGHT (XFH)												
17	23.4	2.1	30.4	2.5	29.5	2.4						
16	25.3	1.3	30.5	2.4	30.4	2.5						
15	26.1	1.2	31.1	2.3	31.2	2.5						
14	25.6	1.4	30.2	2.4	31.8	2.4	18.3					
13	25.7	1.7	29.7	1.7	32.1	1.5	20.8		21.0			
12	25.6	1.6	29.0	1.7	33.0	1.3	20.7					
11	26.1	1.3	28.8	1.7	33.5	3.0	22.2					
10	26.5	1.5	28.0	2.0	34.4	2.1	23.8		21.9			
9	27.8	1.8	28.5	2.0	34.8	1.5		29.6	24.0			
8	28.5	1.7	29.6	1.8	37.4	1.9	24.2		25.3			
7	29.9	1.6	32.5	2.2	40.1	2.6	29.5		28.3			
6	32.2	2.0	37.0	3.0	41.7	8.7	32.5					

(continued)

Table 10.4 Continued

vert	Pan troglodytes mean	sd	Gorilla gorilla mean	sd	Homo sapiens mean	sd	Sts 14	StW 431	A.L. 288-1	MH1	MH2	KNM-WT 15000
5	36.0	2.2	38.6	2.0	46.4	2.6	34.1		35.3	33.0		35.0
4	38.7	1.9	42.5	3.4	47.2	1.9	33.3	39.0				
3	41.1	2.0	43.6	2.1	45.2	2.3	32.7	38.0		38.3		31.0
2	40.1	2.6	43.3	2.3	43.8	1.6	31.4					34.0
1	39.7	2.2	42.0	2.4	41.7	2.3		35.7			36.0	35.0
SUPERIOR FACET ANGLE (SFA)												
6	−14	7	−18	10	25	25	65					
5	6	4	−5	20	62	14	65			66	67	
4	49	6	16	19	60	7	60		60			50
3	45	7	39	13	62	7	45	60		55		40
2	50	8	40	11	56	7	35	50				36
1	51	7	44	9	46	7	20	50			30	25
0	59	9	40	13	36	10	15	45	40			15

vertebral size of Sts 73 and StW 572 cannot be assessed as they are isolated specimens. Both Sts 14 and StW 431 have vertebral endplates that are smaller relative to inferred body mass than typical of humans and Neandertals (Williams et al., 2013, 2017). In addition to apparently being typical of australopithecines, small vertebral bodies are also found in *H. erectus* (KNM-WT 15000; MacLarnon, 1993; MacLarnon and Hewett, 1999) and *H. naledi* (Williams et al., 2017). Sts 14 also appears to have a relatively small neural canal relative to body height, whereas that of SKX 41692 and *Homo* (except perhaps KNM-WT 15000; MacLarnon, 1999) appear to have had larger neural canals (Williams et al., 2017). The vertebrae of Sts 14 is smaller than StW 431 (Shapiro, 1993), supporting the hypothesis that this individual was likely female (see also Wood and Quinney, 1996).

Spinal curvatures and vertebral body wedging

The presence of a human-like lumbar lordosis in the Sterkfontein hominins has been clearly established (Robinson, 1972; Latimer and Ward, 1993; Shapiro, 1993; Sanders, 1998; Whitcome et al., 2007; Been et al., 2012; Williams et al., 2013). Spinal curvatures are a consequence of differential anterior and posterior heights of intervertebral discs as well as wedging of individual vertebral bodies. Vertebral body wedging may be partly genetically determined but appears also to reflect loading history during ontogeny, suggesting that wedging reflects loading environment, and thus likely positional behavior during life, to at least some extent (e.g., Mente et al., 1997; Nakatsukasa et al., 1995; Stokes, 2002; Russo et al., 2018). While vertebral wedging in the lumbar region has been studied extensively, less attention has been paid to the thoracic spine.

Humans have a remarkably consistent pattern of vertebral wedging, reflecting their distinctive spinal curvatures (Figure 10.21; Table 10.4). In the lower lumbar region, human vertebrae are negatively wedged, contributing to the lumbar lordosis (see also Robinson, 1972; Latimer and Ward, 1993; Shapiro, 1993; Sanders, 1998; Whitcome et al., 2007; Been et al., 2012; Williams et al., 2013, 2019). In the thoracic region, though humans differ from other primates. Human thoracic vertebrae are taller along their posterior margins than their anterior margins, producing the kyphosis. The greatest degree of anterior wedging occurs around the first thoracic vertebra (T1; PSV 17), around the midthoracic level (T6-T7; PSV 12), and at the thoracolumbar junction (T12-L1; PSV 6) (see also Digiovanni et al., 1989). This pattern creates three peaks of anterior wedging in the thoracic column with two areas of minimal wedging (at T3-T4 (PSV 16) and T9–T10 (PSV 9)) followed by negative wedging toward the sacrum.

Chimpanzees and gorillas show none of the typical human pattern of variation in vertebral wedging, even in the thoracic region (Figure 10.21; Table 10.4). They exhibit a relatively constant amount of wedging throughout the middle and lower thoracic regions, resulting in their fairly uniform, mild thoracolumbar kyphosis. Chimpanzee and gorilla vertebrae are always anteriorly wedged, except for the variable presence of slight posterior wedging at the ultimate lumbar level in some individuals, especially for gorillas.

Sts 14 and StW 431, as with the other early hominins that are sufficiently preserved to evaluate, clearly demonstrate a fundamentally human pattern of vertebral wedging (Figure 10.21) (Robinson, 1972; Latimer and Ward, 1993; Shapiro, 1993; Sanders, 1998; Whitcome et al., 2007; Been et al., 2012; Williams et al., 2019). Sts 14

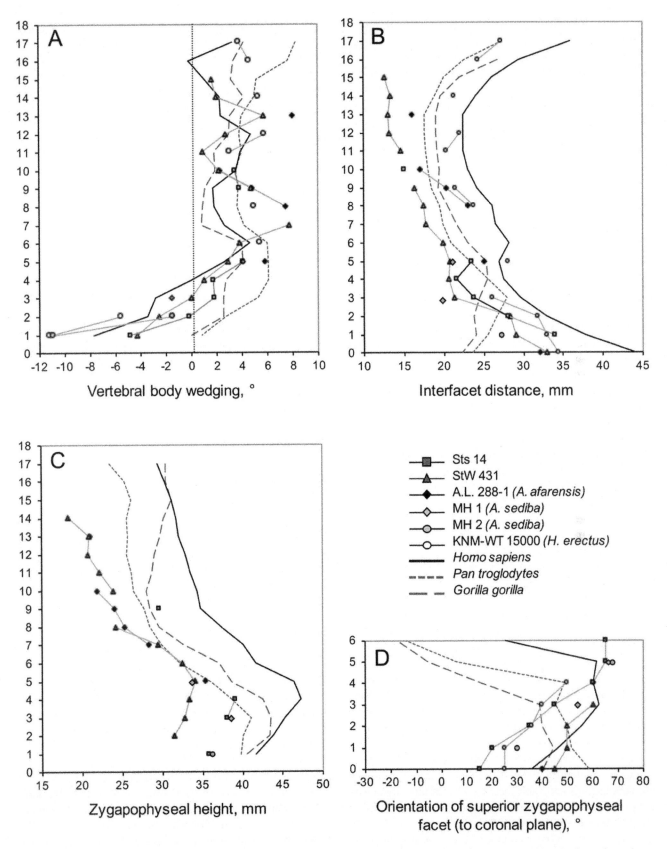

Figure 10.21 Graphs of mean values for (A) vertebral body wedging, (B) interfacet distance, (C) zygapophyseal height throughout the thoracolumbar spine, and (D) zygapophyseal facet orientation throughout lumbar spine for extant hominoids and fossil hominins. Lines connect adjacent values. Data in Table 10.4.

is maximally wedged at PSV 13 and PSV 7 and shows strong negative wedging in the two inferiormost segments. StW 431 does not have enough thoracic vertebrae to assess thoracic wedging, but its lowest lumbar vertebrae are negatively wedged. Other early hominins appear to conform to a similar pattern. *Australopithecus afarensis* (A.L. 288-1) is maximally wedged at PSV 8 and PSV 13. *Australopithecus sediba* (MH1 and MH2; Williams et al., 2013) spines also are anteriorly wedged in the thoracic region and posteriorly in the lumbar. *Homo erectus* (KNM-WT 15000; Ward and Latimer, 1993) also appears to share the thoracic kyphotic peaks and has a negative lower lumbar lordosis.

Although both of the Sterkfontein spines resemble these other hominins in vertebral body shapes, posterior wedging is greater at the lowest portion of the lumbar spine in Sts 14 than in StW 431 (see also Whitcome et al., 2007; Williams et al., 2013). Given that modern human females have a greater lordosis than males this may be another line of evidence for sexual dimorphism in *A. africanus* (Whitcome et al., 2007). It is notable, however, that the presumed juvenile male specimens of *Australopithecus sediba* (MH 2) and *Homo erectus* (KNM-WT 15000) also have especially high wedging in their lowest lumbar vertebrae as well (Latimer and Ward, 1993; Williams et al., 2013, 2019). The high degree of posterior wedging of the lumbar vertebrae is hypothesized to relate to an especially high lumbosacral angle and/or pronounced anterior tilt of the pelvis in early hominins (Abitbol, 1995; Berge and Goularas, 2010; Williams et al., 2013), although these relationships have not been confirmed.

Interfacet distances

The posterior elements of Sts 14 and StW 431 also reflect the presence of a lumbar lordosis, exhibiting morphologies that permit habitually hyperextended posture. As in humans, both of these hominins, as with all others in which it can be assessed, there is a marked incremental increase in the mediolateral separation between zygapophyses, known as interfacet distances, toward the sacrum. This condition stands in clear contrast to their slight convergence in chimpanzees and gorillas (Figure 10.21), even when controlling for the increase in vertebral body breadth inferiorly (Latimer and Ward, 1993; Ward and Latimer, 2005). This increase permits lordotic posture by allowing the zygapophyses to imbricate, thereby protecting the intervening pars interarticularis during hyperlordosis. Modern humans lacking sufficient increase in interfacet distances are prone to the development of spondylolytic defects of the pars (Ward and Latimer, 2005; Ward et al., 2007, 2009; Zehnder et al., 2009). Sufficient facet spacing may be especially important with relative small vertebral bodies,

which also is associated with increased risk of spondylolysis (Wren et al., 2018). All of the hominins display this increase in interfacet distances toward the sacrum, resembling humans and contrasting with the condition seen in apes. This morphology is consistent with the presence of a human-like lumbar lordosis in the Sterkfontein hominins, as with all early hominins. Indeed, it seems likely that the lumbar lordosis and its associated anatomies are among the earliest adaptations to habitual bipedality.

Zygapophyseal height

The Sterkfontein hominins also have zygapophyses that decrease in height progressively throughout the lowest four elements of the lumbar region, as in humans but not extant African apes (Figure 10.21). The same pattern is also evident in the other early hominins for which it can be evaluated, *Australopithecus sediba* (MH 2; Williams et al., 2013) and *Homo erectus* (KNM WT-15000; Latimer and Ward, 1993). This decrease necessarily begins at the same level in which vertebrae become negatively wedged (Figure 10.21), reflecting the necessary functional and spatial geometry associated with the lordosis. The Sterkfontein specimens also show a strong increase in zygapophyseal height throughout the thoracic region, and a progressive decrease in inferior zygapophyseal height inferiorly from LPSV 5 in Sts 14 and at least PSV 4 in StW 431.

Zygapophyseal facet orientation

Sts 14 and StW 431, as with the other early hominin specimens for which it can be evaluated, have an increase in coronal orientation of the lumbar superior zygapophyseal facets toward the sacrum, as in humans (Figure 10.21). In humans, lumbar facet joints become progressively more coronally oriented toward the sacrum (Figure 10.21) to resist anteroinferior slippage of the lowest lumbar vertebrae. The hominin lumbosacral joint is inclined anteriorly and inferiorly, and any vertically aligned compression force will act to displace the lower lumbar vertebrae anteroinferiorly. This tendency is resisted by zygapophyseal facet joint orientation (Latimer and Ward, 1993). The low coefficient of friction in healthy synovial joints necessitates that the primary joint reaction vectors remain normal to the articular surfaces (Latimer et al., 1987). Humans have a progressively coronal orientation of their facet joints inferiorly throughout the lumbar spine, whereas chimpanzees and other primates do not (Latimer and Ward, 1993). Chimpanzees and gorillas do not exhibit this rotation, and instead their facets remain at roughly the same oblique orientation throughout the lower lumbar column. Because this is an adaptation to resist anteroinferiorly-directed force vectors in a lordotic spine, it is further evidence suggesting the presence of a lordosis in early hominins. All of the hominins, show evidence of a coronal shift toward the lower vertebrae.

Pedicle morphology

Commensurate with the relative increase in zygapophyseal facet spacing and reorientation of the lower lumbar vertebrae, Shapiro (1993) noted that humans show a progressive increase in transverse width of the pedicles toward the sacrum, likely related to load transmission in lordotic posture. Sts 14 exhibits this increase (Shapiro, 1993), as does StW 431 (Figures 10.16 and 10.17). However, the Sts 14 pedicles are anteroposteriorly shorter at the ultimate presacral level than typical for humans, which Shapiro (1993) interprets as reflecting a slightly different locomotor adaptation than seen in humans. She suggests that this may reflect decreased robusticity of the iliolumbar ligaments, although the clearly paracoronal orientation of the most inferior zygapophyseal joints would be equally well positioned to accomplish this, and it remains unclear how differences in iliolumbar ligament strength would impact the tendency for anterior dislocation of the lumbosacral joint. Additionally, as the sacrum is situated further inferiorly between the iliac blades of Sts 14 and StW 431 compared with humans, it may be that greater ligamentous support was available between the pelvis and the most inferior region of the vertebral column, resulting in the functionally equivalent ability to resist dislocation of the lumbosacral joint.

Imbrication pockets

The habitual lumbar lordosis in Sts 14 and StW 431 is also apparent in the presence of nonarticular "imbrication pockets" on the laminae of the lumbar vertebrae inferior to the zygapophyseal facets (Lovejoy, 2005; Ward and Latimer 2005; Williams et al., 2013). Imbrication pockets are visible on the three vertebrae from Sts 14 that preserve this region of the laminae, Sts 14b, 14c, and 14d (PSV 2-4) (Figures 10.4, 10.7, 10.9, 10.10), and two vertebrae from StW 431s (PSV4) and StW 431t (PSV1) (Figures 10.16 and 10.17). These pockets are depressions along the superior aspect of the laminae immediately subjacent to the superior zygapophyseal facets, and they are thought to be induced mechanically when vertebrae are placed in extended posture and the suprajacent facet contacts the lamina (Lovejoy, 2005; Ward and Latimer, 2005; Williams et al., 2013). These pockets are typical for humans, and are also present in *Australopithecus afarensis* (A.L. 288-1; Lovejoy et al., 1982), *Australopithecus sediba* (MH 1 and MH 2; Williams et al., 2013, 2019) and *Homo erectus* (KNM-WT 15000; Latimer and Ward, 1993).

Scheuermann's disease

Another line of evidence in support of human-like spinal curvatures is the presence of a pathological anterior deposition of bone on Sts 14n, m, i, k, and l, termed Scheuermann's kyphosis (Scheuermann, 1920; Digiovanni et al., 1989; Scoles et al., 1991; Palazzo et al., 2013). This condition is found in up to 20% of humans, but not in apes (DiGiovanni et al., 1989). These extensions are associated with pronounced anterior wedging of the vertebral bodies (DiGiovanni et al., 1989; Scoles et al., 1991). These anterior extensions are likely a mechanical response to excessive compressive loads along the anterior aspect of the thoracic vertebral column and result from defective anterior endochondral ossification along the apopophyseal rings (Scoles et al., 1991; Palazzo et al., 2013). Scheuermann's lesions are also found in *Australopithecus afarensis* specimens A.L. 288-1 (Johanson et al., 1982; Cook et al., 1983), A.L. 333-152, and perhaps A.L. 333x-12 (Ward et al., 2012). These lesions are further evidence of pronounced thoracic kyphosis in these early hominins.

In one study (Digiovanni et al., 1989), 93% of spines exhibiting anterior extensions of bone associated with Scheuermann's kyphosis also had Schmorl's nodes, which are endplate irregularities that result from disc herniation into the subchondral bone of the vertebral body but were found in only 36% of normal spines (Digiovanni et al., 1989). These lesions were found in only 36% of normal human spines (Digiovanni et al., 1989). Moreover, Schmorl's nodes rarely occur in the African apes, again suggesting an etiological linkage to an sinusoidally curved orthograde spine. The mild depressions in the inferior endplates of Sts 14 m, i, l, and o may represent a defect akin to Schmorl's nodes.

Development

Asymmetry between the right and left sides of the Sts 14g and Sts 14f vertebrae may provide evidence of a developmental disruption and/or instability at the thoracolumbar border. In Sts 14g, the left side displays an unusual development of the mammillary and styloid processes with the mammillary process manifesting as a small rounded protuberance directed superiorly, and the styloid process as a blunted mass directed caudolaterally (Figure 10.3). In contrast, the vertebra's right side has a more typical lumbar phenotype. That is, it exhibits a large, superiorly directed, styliform mammillary process and its styloid process is directed inferiorly, projecting more than its contralateral counterpart. In addition, a rounded external process is directed laterally, located approximately halfway along the superoinferior mammillary-styloid axis. We have not observed the degree of homeotic asymmetry that occurs in the Sts 14g thoracic transverse process in any other catarrhine specimen.

This pattern of asymmetry continues in the contiguous Sts14f vertebra. The most obvious asymmetry in this element is the presence of a lumber transverse process on the left side and a free rib facet on the right Figure 10.22). Closer inspection, however, reveals additional details. On the right side, the mammillary process has fused with the prezygapophyses, which are sagittally oriented. An external process is present, approximately the same size as that found in Sts14g (Figure 10.22). It has been claimed that there is no styloid process (Robinson, 1972); however, closer inspection reveals that the inferior portion of the external process is abraded, thereby obscuring the styloid process-external process relationship.

The superior portion of the left side of Sts14f shows a transitional lumbar transverse process with a flattened and laterally facing external process (Figure 10.4, 10.22). There is also an anterior laminar bridge, or broad pillar of bone proximally, which is fused to the pedicle and which tapers distally to the terminus of a long lumbar transverse process. There is also a deeply excavated costotransverse foramen along the inferior margin of the base of the Sts 14f transverse process (Figure 10.22), an anatomy often found in *Pan* but rarely in *Homo* (Rosenman, 2008). The presence of these combined features indicates the intermediate nature of the Sts 14f vertebra and highlights the observation that lumbarization was as yet incomplete in this individual.

Finally, the change of the thoracic transverse process serial homology between the ipsilateral sides of Sts

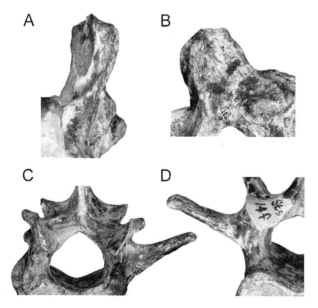

Figure 10.22 Sts 14f showing configuration of external and mammillary processes along right superior zygapophysis (A) posterior and (B) lateral views, (C) posterior elements in superior view, and (D) transverse process in inferior view.

14g and Sts 14f is incongruous. On the left, the change between the penultimate-like thoracic transverse process of Sts 14g, lacking an external process and with only moderately developed mammillary and styloid processes, and the transitional Sts 14f, with a mammillary process fused to the zygapophysis, and flattened external process, appears to have "skipped" an intermediate vertebra. The right side, with Sts 14g's ultimate-like thoracic transverse process configuration, and Sts 14f's fused mammillary process, and rounded external process is also unusual. Both conditions suggest that *Hox* expression may have been in flux at the thoracolumbar boundary.

Vertebral segmentation

Regardless of these developmental perturbations in Sts 14, thoracolumbar segmentation patterns can still be assessed in the Sterkfontein hominins. Segmentation patterns in early hominins has been the subject of a considerable amount of literature, originating with Robinson's (1972) observations about Sts 14. Sts 14 has 5 non-rib-bearing vertebrae, which is the modal pattern for humans, and on PSV6 has a transverse process on the left side and a rib on the right (Figure 10.20). StW 431 also has five non-rib-bearing lumbar vertebrae (Benade, 1980; Toussaint et al., 1993; Haeüsler et al., 2002, 2011; Williams et al., 2013, 2016), as does *Australopithecus afarensis* (DIK 1-1; Ward et al., 2017), *A. sediba* (MH 2; Williams et al., 2013) and *H. erectus* (KNM-WT 15000; Latimer and Ward, 1993; Haeüsler et al., 2002). StW 431 was originally described as having a facet transition at the level of the ultimate rib-bearing vertebra (PLV1) (Benade, 1980). Closer inspection by CVW and BL revealed that the posterior element of PSV5 had incorrectly been glued to the vertebral body of PSV6, however. This glue join has since been removed.

In all of these hominins, the articular facets transition, from a paracoronal, flat morphology typical of thoracic elements to medially oriented, transversely curved facets typical of lumbar vertebrae, is located at the penultimate rib-bearing level rather than the ultimate level which is typical for all extant hominoids (Latimer and Ward, 1993; Toussaint et al., 1993; Haeusler et al., 2002, 2011; Williams et al., 2013, 2016, 2019; Ward et al., 2017). In modern humans, the transition in facet orientation from thoracic to lumbar patterns occurs gradually over 2–3 segments in 43–93% of human individuals (depending on how facet morphology is quantified), but the transition is usually complete no further superior than the last (ultimate) rib-bearing thoracic vertebra (PLV1) (Lanier, 1939; Allbrook, 1955; Davis, 1955; Singer et al., 1989; White and Panjabi, 1990; Panjabi et al., 1993; Shinohara, 1997; Pal and Routal, 1999; Masharawi et al., 2004; Russo, 2010). When an abrupt transition occurs within a single vertebra, it typically occurs at the last rib-bearing element (PLV1) but can be found one segment higher at the penultimate level (PLV2) in up to approximately 10–23% of individuals (Singer et al., 1988, 1989; Shinohara, 1997; Williams et al., 2016). Thus, when a transition is abrupt, most modern humans and extant great apes have a transitional vertebra at the last rib-bearing vertebra.

All of the early hominins (DIK 1-1, MH2, KNM-WT 15000), including Sts 14, StW 431, and StW 8/41, have an abrupt transition in facets, and in all of them this transition occurs at the penultimate rib-bearing level (PLV2). Because an abrupt transitional element at the penultimate rib-bearing vertebra is only found in up to 23% of modern humans (Singer et al., 1988; Pal and Routal, 1999; Williams et al., 2016), the probability of finding all five of these specimens (Sts 14, StW 431, MH 2, KNM-WT 15000, and DIK 1-1) in a modern human population with a similar frequency is less than 0.0007%, and if the same indeed characterizes A.L. 288-1, the probability drops to less than 0.0001% (Ward et al., 2017). These data suggest that the Sterkfontein hominins exhibit a typical early hominin pattern that would have been lost subsequently, perhaps after 1.6 Ma, in human evolution. Given that *Australopithecus* had a deeper capture of the lowest lumbar vertebra between the iliac blades (Lovejoy, 2005) though, this cranial shift may have been compensated for functionally and the resulting spinal mobility may have been equivalent to that seen in humans (Ward et al., 2017).

Rib cage

Rib cage morphology in *Australopithecus africanus* is largely unknown due to the fact that few of these inherently fragile costal remains have been recovered. The identification of partial ribs in Sts 14 and consideration of costotransverse and costovertebral joint morphology provide some insights.

In great apes, the lower ribs are relatively long (Figure 10.23, Table 10.5) (see also Gómez-Olivencia et al., 2018). *Homo sapiens* is striking the shortness of its two inferiormost ribs. Throughout most of the thoracic cage, ribs articulate with the vertebrae at both the costovertebral joints with full synovial contact and costotransverse joints with either synovial joints or strong fibrous attachments (Rosenman, 2008). In humans, the last two diminutive "floating ribs" do not articulate with the vertebral transverse processes (Fischel, 1906). Having smaller ribs that are less tightly bound to the vertebral column may contribute to the "waist" in humans, and allows greater mobility at the thoracolumbar transition as well.

The Sts 14 PSV6 (Sts 14g) has a lumbar transverse process on the left and a small rib on the right (Sts 14aa) (Figures 10.3 and 10.24) (Robinson, 1972). Both the transverse process and the rib are inferiorly inclined, although the rib more so. The rib articulates with an inferiorly

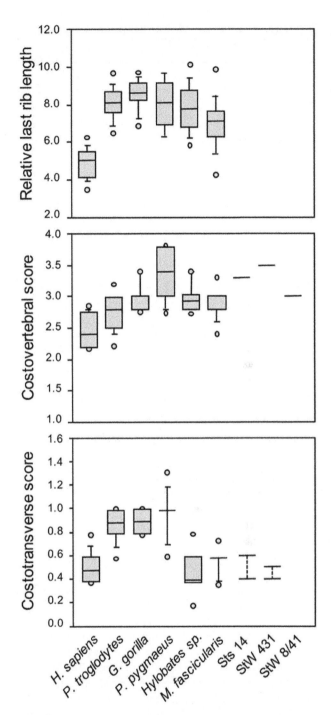

Figure 10.23 Box plots (median, first and third quartiles, and outliers) of relative last rib length, costovertebral score which categorizes configuration of the costovertebral facet position on the vertebral bodies, and costotransverse score which categorizes configuration of the costotransverse facets (see "Methods"). Each of the latter scores averages the inferiormost five thoracic vertebral levels (PLV1-5).

oriented costovertebral facet situated near the supero-inferior center of the vertebral body. There is no costotransverse articulation, as there is no transverse process

Table 10.5   Kendall's τ correlation matrix. Correlations are significant at p < .01 (2-tailed).

	Relative last rib length	Costovertebral score	Costotransverse score
Relative L1 transverse process span	0.21 (n = 122)	not significant	0.24 (n = 119)
Relative last rib length		0.24 (n = 140)	0.25 (n = 141)
Costovertebral score			0.27 (n = 379)

Sts 14 f & aa   Sts 14 g, z & ab

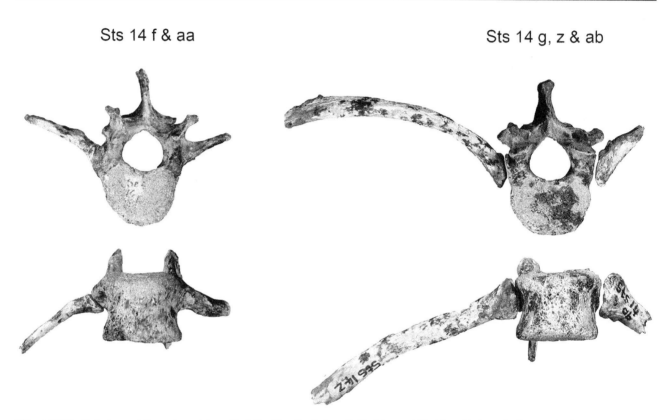

Figure 10.24   Inferiormost thoracic vertebrae Sts 14f with articulating rib Sts 14aa and Sts 14g with its articulating ribs Sts 14 z and Sts 14ab, both shown in superior (top row) and anterior (bottom row) views.

on this vertebra. One level higher, there are remains of both right and left ribs (Sts 14z right and Sts 14ab left) that articulate with PSV7, Sts 14f (Figure 10.24). These are robust ribs, and the better-preserved one (Sts 14z) is incomplete but quite long, showing posterior curvature proximally with a gentle ventral concavity that arcs anteriorly and inferiorly. Although it cannot be quantified, the Sts 14 rib is longer than those usually found in human inferiormost two ribs. This long second-to-last rib is similar to the rib belonging to A.L. 288-1 that is a second- or third-to-last rib (Johanson et al., 1982). Thus it appears that *Australopithecus* did have longer lower ribs than are typical for humans.

The likelihood of long lower ribs in the Sterkfontein hominins is supported by the fact that their last rib length

and first lumbar transverse process span are positively correlated, and it has been widely noted that Sts 14 has long transverse processes (Robinson, 1972; Haeusler et al., 2002). The presence of long lower ribs, in turn, is consistent with the interpretation that *Australopithecus* species also appear to have had relatively wide ilia for their size (see Haeusler and Ruff, Chapter 11, this volume).

Modest inferences about lower rib length also can be made about lower rib length from the morphology of the costovertebral articular surfaces on the vertebrae. Humans have a lower costovertebral score than do nonhuman primates (Mann-Whitney tests, *p* < 0.001 for all comparisons), although the difference is not as pronounced as for lower rib length (Figure 10.23; Table 10.5). This reflects a more rapid transition from having demifacets and ribs that articulate with two vertebral bodies in the middle of the thoracic spine, to a situation in which the rib articulates with only one vertebral body, which typically characterizes the last two ribs. Nonhuman primates tend to have a more gradual transition, reflecting a better-developed articulation between lower ribs and the vertebral column. All three Sterkfontein hominins have high costovertebral facet scores (Figure 10.23), consistent with having relatively long last, or second-to-last, ribs. Again, this is likely related to the adjoining platypelloid pelvis and the elongated lumbar column.

Although the Sterkfontein hominins appear to have had long lower ribs and a more gradual transition from dual single costovertebral articulations than typical for humans, they differ from great apes in having a lower costotransverse scores. Humans, hylobatids, and monkeys have lower scores as well (Figure 10.23, Table 10.5). Sts 14f has only diminutive transverse processes that barely project laterally, and there is no apparent costotransverse contact. There may have been some fibrous contact here, although there is a gap of more than 6–8 between ribs and transverse processes. Stronger contact between tubercles and vertebral transverse processes can plausibly be hypothesized to restrict movement between ribs and the vertebral column, perhaps explaining the differences between great apes and taxa with longer and more flexible lumbar spines. For the Sterkfontein hominins, then, this may suggest that although their lower ribs may have been more robust than those of humans, they may have been capable of a similar pattern of mobility between rib cage and lower thoracic vertebrae.

Although the thoracolumbar series are too incomplete to be quantified using the coding approach taken here, the morphology of costotransverse articulations can be qualitatively compared between the Sterkfontein fossils and other early hominins. In *Australopithecus afarensis* (A.L. 288-1), the preserved long lower rib thought to represent the last or second-to-last rib (Johanson et al., 1982) also lacks evidence of synovial or strong fibrous contact between the rib and vertebral transverse process. *Australopithecus sediba* (Williams et al., 2013) also appears to have large, single costovertebral articulations, but with no strong transverse processes with costotransverse articulations. In *H. erectus* (KNM-WT 15000Y), there are single large costovertebral facets (Walker and Leakey, 1993; Jellema et al., 1993; Haeusler, 2002), but the transverse processes are broken such that no evidence of a possible costotransverse articulation is evident. So, all early hominins appear to have lacked strong costotransverse articulations, perhaps suggesting the possibility of slightly greater movement at the costovertebral joints than in extant great apes.

Given that the thoracolumbar facet transition in early hominins was one level higher relative to the rib cage than is typical for extant hominoids (Haeusler et al.,

2002; Williams et al., 2013, 2016, 2019; Ward et al., 2017, the zygapophyseal facet joints would have restricted rotation between PLV1 and PLV2, whereas the thoracic-like paracoronal orientation of this joint in extant hominoids does not offer this restriction. So, lower rib length would not likely have substantively affected possible rotation or movement at this joint in *Australopithecus*, and the effect of reduced costotransverse contact of the second-to-last rib is uncertain. This combination of features suggests that the inferiormost ribs of humans may have become reduced when the thoracolumbar facet transition shifted inferiorly later in human evolution, perhaps after 1.6 Ma.

When the lowest two ribs of Sts 14 are articulated with the vertebral column, they exhibit an inferior declination (Figure 10.24). However, as none of the ribs is complete, it is not possible to quantify the extent of any anterior declination. Certainly, humans have more anteriorly declined ribs than do nonhuman primates (Jellema et al., 1993; Latimer et al., 2016), and a lateral declination seen in the lowest two ribs of Sts 14 could be related to an overall rib declination and human-like thoracic shape. All of the Sts 14 ribs show a slight torsion along their lengths, typical of what is found in humans and seen in *H. erectus* (KNM-WT 15000; Jellema et al., 1993), *Australopithecus sediba* (MH 2; Schmid et al., 2013; Williams et al., 2019), and *Australopithecus afarensis* (KSD-VP-1/1; Latimer et al., 2016). This torsion is associated with anterior declination of the ribs, and may be related to development of the human-like thoracic kyphosis, given that torsion is not present in humans at birth but develops during adolescence (García-Martínez et al., 2016). Because the great apes have abbreviated their iliocostal space to essentially a single intercostal interval, they do not, and cannot, decline their ribs (Latimer et al., 2016). With this in mind, the obvious declination and associated torsion of the Sterkfontein ribs (also *A. afarensis* and *A. sediba*; Schmid et al., 2013; Latimer et al., 2016) are likely associated with an elongated, flexible lumbar column compared to the great apes.

Although fragmentary, the ribs of Sts 14 suggest an overall human-like thoracic shape, but perhaps with somewhat longer lower ribs. This latter issue likely is owing to the mediolaterally broad upper pelvis in *Australopithecus africanus* (and *Australopithecus afarensis*).

Discussion and conclusions

Taken together, the vertebrae and ribs of the Sterkfontein hominins tell a consistent story. With the exception of StW 572, which exhibits morphology more similar to *Homo*, all other specimens are consistent with being attributed to the same species, likely *Australopithecus africanus*. Overall, all the Sterkfontein specimens resemble other early hominins, with a fundamentally human-like torso, with a series of sinusoidal spinal curvatures similar to those of hominins including *Homo erectus, Australopithecus sediba,* and *A. afarensis.* These characteristics are distinctly human-like, are dissimilar from any other mammal, and are consistent with a fully upright posture. This is evident in the pattern of vertebral wedging, the zygapophyseal facet spacing, the superoinferior zygapophyseal height, and the facet orientations.

Sts 14 also displays anterior extensions of the vertebral bodies like those seen in Scheuermann's kyphosis along with depressions of the vertebral endplates that resemble Schmorl's nodes, both common spinal pathologies seen in humans but not in other hominoids. These two conditions are associated with the sinusoidal spinal curvatures that are necessary for habitual bipedality.

The rib cage of the Sterkfontein hominins although fragmentary appears to indicate declination and torsion of the ribs, again features that are unique to hominins. The Sterkfontein fossils also appear to have longer, thicker lower ribs than is typical for humans, and in this they seem to resemble *Australopithecus afarensis* and *A. sediba* but like these hominins lack costotransverse articulations at the second-to-last ribs. The thoracolumbar transition in the Sterkfontein hominins is like that of all other early hominins, with a facet transition occurring at the second-to-last rib-bearing level, rather than the last rib-bearing level as seen in extant humans and great apes.

In sum, several issues become clear. First, Sts 14 had a long, mobile lumbar spine, a feature directly related to the mechanics of upright walking and running and one markedly dissimilar to that seen in the great apes. Second, the ribs of Sts 14 appear to indicate anteroinferior declination, a rib orientation also not apparent in apes, although more complete lower ribs will provide a more definitive test of this. And finally, Sts 14 displays vertebral morphologies that among primates are only regularly seen in humans and are associated with habitually orthograde posture.

Acknowledgments

The authors would like to thank the curators and staff of the Ditsong Museum of Natural History, University of the Witswatersrand, National Museums of Kenya, National Museum of Ethiopia and Cleveland Museum of Natural History for access to specimens. Funding to conduct this research was provided by the PAST of South Africa, Wenner Gren Foundation, University of the Witswatersrand, University of Missouri Research Board, University of Missouri Research Council, University of Missouri Summer Research Fellowship, Kent State University, and National Science Foundation (BCS 0716244 to CVW).

References

Abitbol, M.M., 1995. Reconstruction of the STS 14 (*Australopithecus africanus*) pelvis. American Journal of Physical Anthropology 96, 143–158.

Alemseged, Z., Spoor, F., Kimbel, W.H., Bobe, R., Geraads, D., Reed, D. and Wynn, J.G., 2006. A juvenile early hominin skeleton from Dikika, Ethiopia. Nature 443, 296.

Allbrook, D.B., 1955. The East African vertebral column: a study in racial variability. American Journal of Physical Anthropology 13, 489–513.

Been, E., Gómez-Olivencia, A., Kramer, P.A., 2012. Lumbar lordosis of extinct hominins. American Journal of Physical Anthropology 147, 64–77.

Benade, M.M., 1980. Thoracic and lumbar vertebrae of African hominids ancient and recent: morphology and functional aspects with specific reference to upright posture, M.A. Thesis, University of the Witwatersrand.

Berge, C., Goularas, D., 2010. A new reconstruction of Sts 14 pelvis (*Australopithecus africanus*) from computed tomography and three-dimensional modeling techniques. Journal of Human Evolution 58, 262–272.

Cook, D.C., Buikstra, J.E., De Rousseau, C.J., Johanson, D.C., 1983. Vertebral pathology in the Afar australopithecines. American Journal of Physical Anthropology 60, 83–102.

D'Anastasio, R., Zipfel, B., Moggi-Cecchi, J., Stanyon, R., Capasso, L., 2009. Possible brucellosis in an early hominin skeleton from Sterkfontein, South Africa. PLoS ONE 4, e6439.

Davis, P.R., 1955. The thoraco-lumbar mortice joint. Journal of Anatomy 89, 370–377.

Digiovanni, B.F., Scoles, P.V., Latimer, B M., 1989. Anterior extension of the thoracic vertebral bodies in Scheuermann's kyphosis. Spine 14, 712–716.

Fischel, A., 1906. Untersuchungen über die Wirbelsäule und den Brustkorb des Menschen. Anatomische Hefte 31, 462–588.

García-Martínez, D., Recheis, W., Bastir, M., 2016. Ontogeny of 3D rib curvature and its importance for the understanding of human thorax development. American Journal of Physical Anthropology 159, 423–431.

Gómez-Olivencia, A., Barash, A., García-Martínez, D., Arlegi, M., Kramer, P., Bastir, M., Been, E., 2018. 3D virtual reconstruction of the Kebara 2 Neandertal thorax. Nature Communications 9, 4387.

Grine, F.E., 2019. The alpha taxonomy of *Australopithecus* at Sterkfontein: The postcranial evidence. Comptes Rendues Paleovol 18, 335–352.

Haeusler, M., Martelli, S.A., Boeni, T., 2002. Vertebrae numbers of the early hominid lumbar spine. Journal of Human Evolution 43, 621–643.

Haeusler, M., Scheiess, R., Boeni, T., 2011. New vertebral and rib material point to modern bauplan of the Nariokotome *Homo erectus* skeleton. Journal of Human Evolution 61, 575–582.

Jellema, L., Latimer, B., Walker, A., 1993. The rib cage. In: Walker, A., Leakey, R.E. (Eds.), The Nariokotome *Homo erectus* skeleton. Harvard University Press: Cambridge, pp. 294–325.

Johanson, D.C., Lovejoy, C.O., Kimbel, W.H., White, T.D., Ward, S.C., Bush, M.E, Latimer, B.M., Coppens, Y., 1982. Morphology of the Pliocene partial hominid skeleton (A.L. 288-1) from the Hadar Formation, Ethiopia. American Journal of Physical Anthropology 57, 403–451.

Lanier, R.R., 1939. The presacral vertebrae of American White and Negro males. American Journal of Physical Anthropology 25, 341–420.

Latimer, B.M., Lovejoy, C.O., Spurlock, L., Haile-Selassie, Y., 2016. The thoracic cage of KSD-VP-1/1. In Haile-Selassie, Y., Su, D. (Eds.), The postcranial anatomy of *Australopithecus afarensis*. Springer: Dordrecht, pp. 143–153.

Latimer, B., Ohman, J.C., Lovejoy, C.O., 1987. Talocrural joint in African hominoids: implications for *Australopithecus afarensis*. American Journal of Physical Anthropology 74, 155–175.

Latimer, B., Ward, C.V., 1993. The thoracic and lumbar vertebrae. In: Walker, A., Leakey, R.E. (Eds.), The Nariokotome *Homo erectus skeleton*. Harvard University Press: Cambridge, pp. 299–293.

Lovejoy, C.O., 2005. The natural history of human gait and posture part 1: spine and pelvis. Gait and Posture 21, 95–112.

Lovejoy, C.O., Johanson, D.C., Coppens, Y., 1982. Elements of the axial skeleton recovered from the Hadar formation: 1974-1977 collections. American Journal of Physical Anthropology 57, 631–636.

Lovejoy, C.O., Suwa, G., Simpson, S.W., Matternes, J.H.,White, T.D., 2009. The great divides: *Ardipithecus ramidus* reveals the postcrania of our last common ancestors with African apes. Science 326, 73, 100–106.

MacLarnon, A., 1993. The vertebral canal. In: Walker, A., Leakey, R.E. (Eds.), The Nariokotome *Homo erectus* skeleton. Harvard University Press: Cambridge, pp. 359–390.

MacLarnon, A.M., Hewitt, G.P., 1999. The evolution of human speech: the role of enhanced breathing control. American Journal of Physical Anthropology 109, 341–363.

Masharawi Y, Rothschild, B., Dar, G., Peleg, S., Robinson, D., Been, E., Hershkovitz, I., 2004. Facet orientation in the thoracolumbar spine. Spine 29, 1755–1763.

McHenry, H.M., 1992. Body size and proportions in early hominids. American Journal of Physical Anthropology 87, 407–431.

Mente, P.L., Spence, H., Aronsson, D.D., 1997. Progression of vertebral wedging in an asymmetrically loaded rat tail model. Spine 22, 1292–1296.

Nakatsukasa, M., Hayama, S., Preuschoft, H., 1995. Postcranial skeleton of a macaque trained for bipedal standing and walking and implications for functional adaptation. Folia Primatologica 64, 1–29.

Pal, G.P., Routal, R.V., 1999. Mechanism of change in the orientation of the articular process of the zygapophyseal joint at the thoracolumbar junction. Journal of Anatomy 195, 199–209.

Palazzo, C., Sailhan, F., Revel, M., 2014. Scheuermannn's disease: an update. Joint Bone Spine 81, 209–214.

Panjabi, M.M., Oxland, T., Takata, K., Goel, V., Duranceau, J., Krag, M., 1993. Articular facets of the human spine. Spine 18, 1298–1310.

Robinson, J.T., 1972. Early hominid posture and locomotion. University of Chicago Press: Chicago.

Rose, M.D., 1975. Functional proportions of primate lumbar vertebral bodies. Journal of Human Evolution 4, 21–38.

Rosenman, B.A., 2008. Triangulating the evolution of the vertebral column in the last common ancestor: thoracolumbar transverse process homology in the Hominoidea. Ph.D. Dissertation, Kent State University.

Russo, G.A., 2010. Prezygapophyseal articular facet shape in the catarrhine thoracolumbar vertebral column. American Journal of Physical Anthropology 142, 600–612.

Russo, G.A., Marsh, D.A., ster, A.D., 2018. Response of the axial skeleton to bipedal loading behaviors in an experimental animal model. The Anatomical Record. doi.org/10.1002/ar.2400.3

Sanders, W. J., 1998. Comparative morphometric study of the australopithecine vertebral series StW H8/H41. Journal of Human Evolution 34, 249–302.

Scheuermann, H.W., 1920. The classic: kyphosis dorsalis juvenilis. Ugeskr Laeger 82, 385–393.

Schmid, P., Churchill, S.E., Nalla, S., Weissen, E., Carlson, K.J., de Ruiter, D.J., Berger, L.R., 2013. Mosaic morphology in the thorax of *Australopithecus sediba*. Science 340, 1234598.

Schultz, A.H., Straus, W.L., 1945. The number of vertebrae in primates. Proceedings of the American Philosophical Society 89, 601–626.

Scoles, P., Latimer, B., DiGiovanni, B., Vargo, E., Bauza, S., Jellema, L., 1991. Vertebral alterations in Scheuermann's kyphosis. Spine 16, 509–515.

Scoles, P.V., Linton, A.E., Latimer, B., Levy, M.E., Digiovanni, B., 1988. Vertebral body and posterior element morphology: the normal spine in middle life. Spine 13, 1082–1086.

Shapiro, L., 1993. Evaluation of "unique" aspects of human vertebral bodies and pedicles with a consideration of *Australopithecus africanus*. Journal of Human Evolution 25, 433–470.

Shinohara, H., 1997. Changes in the surface of the superior articular joint from the lower thoracic to the upper lumbar vertebrae. Journal of Anatomy 290, 461–465.

Singer, K.P., Breidahl, P.D., Day, R.E., 1988. Variations in zygapophyseal joint orientation and level of transition at the thoracolumbar junction. Surgical and Radiological Anatomy 10, 291–295.

Singer, K.P., Willen, J., Breidahl, P.D., Day, R.E., 1989. Radiologic study of the influence of zygapophyseal joint orientation on spinal injuries at the thoracolumbar junction. Surgery, Radiology, Anatomy 11, 233–239.

Stokes, I.A.F., 2002. Mechanical effects on skeletal growth. Journal of Musculoskeletal Neuronal Interactions 2, 277–280.

Tobias, P.V., 1973. A new chapter in the history of the Sterkfontein early hominid site. Journal of the South African Biological Society, 14, 30–44.

Toussaint, M., Macho, G., Tobias, P., Partridge, T., Hughes, A., 2003. The third partial skeleton of a late Pliocene hominin (StW 431) from Sterkfontein, South Africa. South African Journal of Science 99, 215–223.

Veldhuizen, A.G., Baas, P., Webb, P.J., 1986. Observations on the growth of the adolescent spine. Journal of Bone and Joint Surgery 68B, 724–728.

Walker, A., Leakey, R.E., 1993. The Nariokotome Homo erectus skeleton. Harvard University Press: Cambridge.

Ward, C.V., Kimbel, W.H., Johanson, D.C., 2012. New postcranial fossils attributed to Australopithecus afarensis from Hadar, Ethiopia. Journal of Human Evolution 63, 1–51.

Ward, C.V., Latimer, B., 2005. Human evolution and the development of spondylolysis. Spine 30, 1808–1814.

Ward, C.V., Latimer, B., Parker., J., Alander, D.H., Ronan, J., Sanders, C., Holden A., 2007. Radiographic assessment of lumbar facet distance spacing and spondylolysis. Spine 32, E85–E88.

Ward, C.V., Mays, S., Child, S., Latimer, B., 2009. Spondylolysis in an archeological and modern population. American Journal of Physical Anthropology 131, 352–362.

Ward, C.V., Nalley, T.K., Spoor, F., Tafforeau, P., Alemseged, Z., 2017. Thoracic vertebral count and thoracolumbar transition in Australopithecus afarensis. Proceedings of the National Academy of Sciences 114, 6000–6004.

Whitcome, K.K., Shapiro, L.J., Lieberman, D.E., 2007. Fetal load and the evolution of lumbar lordosis in bipedal hominins. Nature 450, 1075.

White, A.A., Panjabi, M.M., 1990. Clinical biomechanics of the spine, 2nd ed. Lippincott: New York.

Williams, S.A., García-Martinez, D., Bastir, M., Meyer, M.R., Nalla, S., Hawks, J., Schmid, P., Churchill, S.E., Berger, L.R. 2017. The vertebrae and ribs of Homo naledi. Journal of Human Evolution 104, 136–154.

Williams, S.A., Meyer, M.R., Nalla, S., García-Martínez, D.G., Nalley, T.K., Eyre, J., Prang, T.C., Bastir, M., Schmid, P., Churchill, S.E., Berger, L.R., 2019. The vertebrae, ribs and sternum of Australopithecus sediba. Paleoanthropology 2018, 156–233.

Williams, S.A., Middleton, E.R., Villamil, C.I., Shattuck, M.R., 2016. Vertebral numbers and human evolution. American Journal of Physical Anthropology 159, S19–S36.

Williams, S.A., Ostrofsky, K.R., Frater, N., Churchill, S.E., Schmid, P., Berger, L.R., 2013. The vertebral column of Australopithecus sediba. Science 340, 1232291–1232295.

Wood, B.A., Quinney, P.S., 1996. Assessing the pelvis of AL 288-1. Journal of Human Evolution 31, 563–568.

Wren, T.L., Ponrartana, S., Aggabau, P.C., Poorghasamians, Ed., Skaggs, D.L., Gilsanz, V., 2018. Increased lumbar lordosis and smaller vertebral cross-sectional area are associated with spondylolysis. Spine 43, 833–838.

Zehnder, S., Ward, C.V., Crow, A., Alander, D., Latimer, B., 2009. Radiographic assessment of lumbar facet distance spacing and pediatric spondylolysis. Spine 34, 285–290.

11

Pelvis

Martin Haeusler and Christopher B. Ruff

The hominin pelvis is a remarkable structure that plays a central role in many critical biological processes, most notably bipedal locomotion and parturition. Here we describe each pelvic element from Sterkfontein and reconstruct the pelvis of the partial skeletons of Sts 14 and StW 431. Appendix I presents member, locus with grid reference, and date of discovery of each specimen (where known).

Descriptions

Sts 14—Partial pelvis

The pelvis of Sts 14 consists of almost the entire right hipbone (Sts 14s), most of the left hipbone (Sts 14r), as well as the bodies and left ala of the two first sacral vertebrae (Sts 14q). The right hipbone was the first element of the Sts 14 skeleton that was discovered. It was chiseled out of a block of breccia (Broom and Robinson, 1947; Broom et al., 1950), leading to eventual preparation of the rest of the known skeletal elements (Robinson, 1972) (see Ward et al., Chapter 4, this volume).

Sts 14s—Right hipbone

Preservation. The right hipbone is missing the anterior superior iliac spine, which has been reconstructed by Robinson (1972) in plaster based on its left counterpart (Figure 11.1). The iliac crest is crushed in its anterior two-thirds. The posterior superior and inferior iliac spines are complete, and the auricular surface is well preserved. On the external aspect, about 3 cm² of the outermost bone layer is missing close to the posterior inferior iliac spine, but otherwise the superficial bone seems to be well preserved. The gluteal lines cannot be discerned, probably due to its subadult age, but the nutrient foramen that usually is associated with the anterior gluteal line is well visible.

On the internal aspect, the iliac fossa is impressed in its anterior third by an indentation with radiating fracture lines. This is responsible for the reduced thickness of the central iliac blade compared to the left ilium. At a distance of 35 mm, just medially from the inferior end of the anterior inferior iliac spine, a second indentation can be found that has produced a step fracture on the internal side of the distal ilium. Further damage occurred opposite to it at the cranial margin of the acetabulum, which is chipped off. The differential diagnosis for these indentations includes carnivore damage (Figure 11.2). In fact, the distance between the upper canines of an extant leopard is in the range of 35 mm (Häusler and Schmid, 1995).

Morphology. The maximum height of the right iliac blade is 75.1 mm measured from the greater sciatic notch to the iliac crest (without the apophysis, which is missing); the minimum width is 45.3 mm from the greater sciatic notch to the anterior iliac margin between the anterior superior and anterior inferior iliac spines (Table 11.1).

The acetabulum measures about 35.5 mm superoinferiorly and 33.5 mm transversely if the above-mentioned damage to its superior margin and the anterior horn is taken into account, where superficial bone has splintered off (Figure 11.3). These acetabular dimensions are significantly smaller than most previous studies proposed although the transverse diameter is similar to the measurement by McHenry (1975b) and the best-fit sphere of Hammond et al. (2013) to the posteroinferior sector of the lunate surface (Table 11.2). Using the approach of Hammond et al. (2013), we therefore also approximated the size of the acetabulum by a best-fit sphere to the lunate surface. For this, the lunate surface was extracted

Martin Haeusler and Christopher B. Ruff, *Pelvis.* In: *Hominin Postcranial Remains from Sterkfontein, South Africa, 1936–1995.* Edited by: Bernhard Zipfel, Brian G. Richmond, and Carol V. Ward, Oxford University Press (2020). © Oxford University Press.
DOI: 10.1093/oso/9780197507667.003.0011

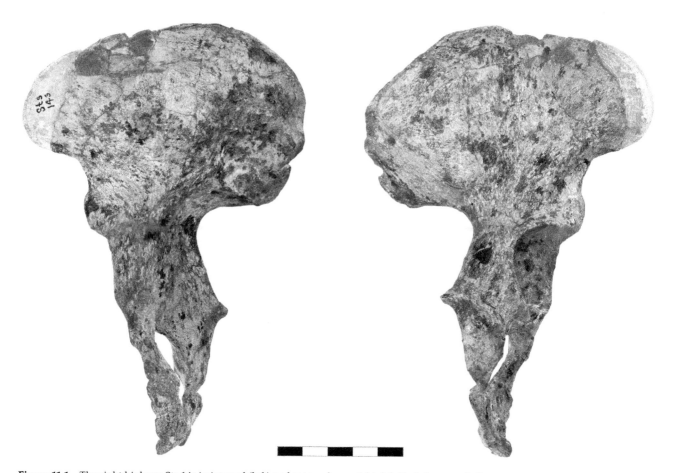

Figure 11.1 The right hipbone Sts 14s in internal (left) and external aspect (right). Scale bar equals 5 cm.

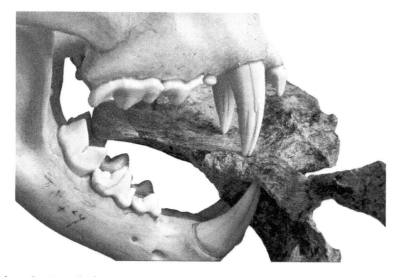

Figure 11.2 The right hipbone Sts 14s in the fangs of a modern leopard demonstrating the close match to the pattern of damage. The crushing of the superior acetabulum, the ventral side of the iliac neck and iliac blade, and perhaps the anterior superior iliac spine and iliac crest might be explained by carnivore damage.

with Meshmixer 3.5 mm from a high-resolution surface scanner-generated 3D model of Sts 14 (acquired by a PT-M4c surface scanner, Polymetrics, Dortmund, Germany). Surface defects, such as the above-mentioned damages, and the convex periphery of the lunate surface were not included. Using Rhinoceros 5.5, a sphere was fitted to the vertex points of the polygonal mesh. Six independent extractions of the lunate surface resulted in a

Table 11.1 Measurements of the Sterkfontein pelvic remains.

Dimension and number of measurement according to Martin and Saller (1957) and Knussmann (1988) where available	Sts 14	Sts 65	StW 431	StW 441	StW 611
Height of iliac blade, from the superior rim of the acetabulum to the iliac crest	75.1*	> 74	105	–	–
Minimum ilium width, from the greater sciatic notch to the anterior iliac margin between the anterior superior and anterior inferior iliac spines	45.3	> 49	60.5	–	–
Acetabulum diameter, superoinferior (22a)	(35.0 –35.5)	–	–	–	–
Acetabulum diameter, transverse (22b)	33.5	–	–	–	–
Acetabulum diameter, best–fit sphere†	35.4	38.5	41.1‡	–	39.6
Auricular surface, superoinferior height	17.8	18.0	23.5	–	–
Auricular surface, distance from anterior margin to posterior inferior iliac spine	24.5	29.3	36.1	–	–
Pubic length, acetabular margin to symphysis	62.7	–	–	–	–
Pubic symphysis, superoinferior height (18)	28.1	–	–	–	–
Pubic symphysis, maximum anteroposterior thickness	12.2	–	–	–	–
Tuberoacetabular sulcus, minimum breadth	17.2	–	–	–	16.0
Mediolateral diameter of lumbosacral articular surface (19)	(26.0)*	–	36.7	–	–
Dorsoventral diameter of lumbosacral articular surface (18)	16.7*	–	21.0	–	–
Maximum anterior breadth of the sacrum (5)	(76.6)*	–	89.5	–	–
Promontory angle (22)	70°	–	68°	–	–
Mediolateral diameter of superior articular facet joint of S1	–	–	13.3	–	–
Craniocaudal diameter of superior articular facet joint of S1	10.3	–	12.5	–	–
Mediolateral diameter of neural canal of S1 (17)	15.8	–	18.8	–	–
Dorsoventral diameter of neural canal of S1 (16)	6.9	–	11.1	–	–
Superoinferior length of the ventral aspect of S1 at midline	18.4*	–	23.6	–	–
First interforaminal breadth (20)	16.2	–	25.9	–	–
Superoinferior length of the ventral aspect of S2 at midline	14.2*	–	20.5	–	–
Superoinferior length of the ventral aspect of S1 and S2 together	33.9*	–	45.7	–	–
Second interforaminal breadth (20)	–	–	19.7	–	–
Mediolateral diameter of neural canal of S3	–	–	17.3	–	–
Dorsoventral diameter of neural canal of S3	3.7	–	4.8	–	–
Length of auricular surface of the sacrum (14)	–	–	43.3	–	–
Breadth of auricular surface of the sacrum (15)	–	–	23.0	–	–

All measurements in mm. Estimated values are in brackets.

* Apophysis or epiphysis is missing

† The best-fit sphere to the entire acetabulum seems to be intermediate between the superoinferior and transverse acetabular diameters. Based on a combined sample of 461 modern humans, chimpanzees, and gorillas, the superoinferior acetabulum diameter can be calculated from the transverse diameter according to the least squares regression formula $Ac_{SI} = 1.011 \times Ac_{TV} + 1.472$ (SEE 1.5 mm, 95% confidence limits ± 2.9 mm; $r^2 = 0.962$; reference sample described in Haeusler and McHenry 2004, 2007).

‡ Mean of StW 431e and StW 431k.

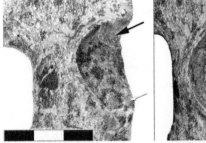
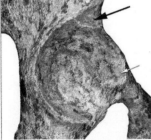

Figure 11.3 A close-up of the right hipbone Sts 14s showing the carnivore damage.

Table 11.2 Comparison of different published measurements of the Sts 14 acetabulum.

Study	Superoinferior height	Transverse width	Best-fit sphere
Present study	35.5	33.5	35.4
Hammond et al. (2013)	–	–	33.1 ± 0.7*
Ruff (2010)	39.0	–	–
McHenry (1975a)	38.7	32.8**	–
Robinson (1972)	39.2	37.2	–

All measurements in mm.

* Hammond et al. (2013) estimated a best-fit sphere to the dorsoinferior segment of the acetabulum

** McHenry (1975b) used only the less compromised transverse diameter to estimate femoral head size

best-fit sphere with a mean diameter of 35.5 mm ± 0.04 (standard deviation). A faint fracture line runs, however, through the posterior acetabulum, leading to a step of 0.2 mm, which might have affected the acetabulum diameter. After virtual reduction of this step by posterior translation of the posteroinferior acetabulum fragment, the diameter of the best-fit sphere slightly reduced to 35.4 mm. This is still more than 2 larger than the estimate by Hammond et al. (2013), which can be explained by a different-quality 3D surface scan, that their estimate was limited to the posteroinferior sectors of the lunate surface only, and the undefined inclusion of the convex periphery of the lunate surface and damaged areas.

The auricular surface is proportionally much smaller relative to the width of the iliac fossa than in modern humans since only the first two sacral vertebrae articulated with the ilium compared with the first three sacral vertebrae in modern humans. The distance from the posterior inferior iliac spine to the anterior margin of the auricular surface measures 24.5 mm, and the superoinferior height of the auricular surface is 17.8 mm. In contrast to the auricular surface, the iliac tuberosity is relatively larger than in modern humans due to the more vaulted posterior region of the iliac crest.

The curvature of the iliac crest is faintly sigmoid when viewed from above (Figure 11.4), which implies

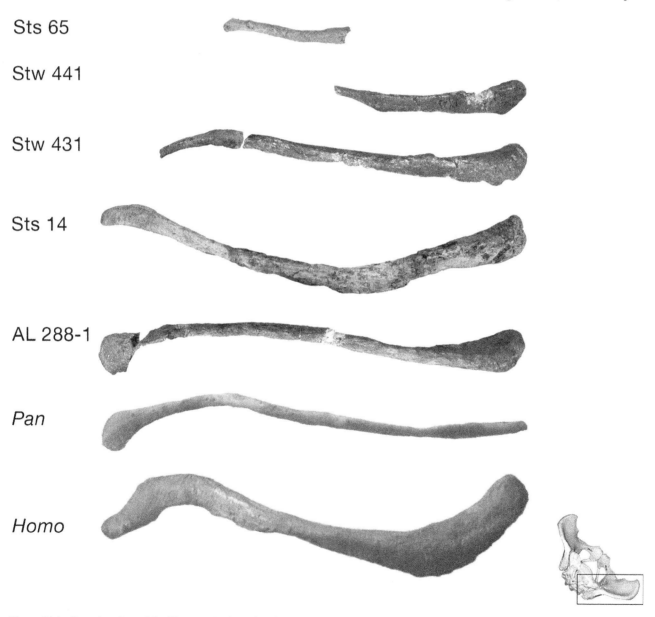

Sts 65

Stw 441

Stw 431

Sts 14

AL 288-1

Pan

Homo

Figure 11.4 Superior view of the iliac crest in *Australopithecus*, *Pan*, and *Homo*, scaled to the same width of the iliac blade. Sts 14 and StW 441 are mirrored from the left side; the iliac crest of StW 431e has been completed with the iliac crest of the mirrored left ilium fragment StW 431j. The distortion of the posterior portion of the A.L. 288-1 ilium has been corrected.

that there is no iliac pillar (Arsuaga, 1981). There is only a faint thickening of the outer table of the ilium parallel to the anterior iliac margin. Although the anterior superior iliac spine is missing, it is obvious from its base as well as from its counterpart on the left side that it was far projecting anterolaterally. The anterior inferior iliac spine is relatively gracile. On the external aspect above the edge of the acetabulum, traces of a ridge that runs to the inferior margin of the anterior inferior iliac spine are visible. The ridge, which is associated with the iliofemoral ligament, is slightly less robust than in StW 431. The area of origin of the reflected head of the rectus femoris muscle cannot be discerned. There is a distinct iliopectineal eminence, and the groove for the iliopsoas muscle is marked. These features are less marked than in StW 431, but they contrast to A.L. 288-1 where the groove for the iliopsoas muscle and the iliopectineal eminence are absent (Stern and Susman, 1983). The anterior horn of the lunate surface of the acetabulum is well developed in Sts 14 as in StW 431, whereas it is diminutive in A.L. 288-1.

The root of the superior pubic ramus is well preserved, but the remainder of the pubis is heavily fractured and the individual pieces are seriously displaced. Most of the medial portion of the pubis including the pubic symphysis and pubic tubercle is missing.

The ischium exhibits a pointed ischial spine at the distal extremity of the wide greater sciatic notch. The lesser sciatic notch also seems to be relatively wider than in modern humans. The entire ischial tuberosity is, however, missing and ischium length cannot be determined—only a "small rounded knob" (Broom et al., 1950) might mark the beginning of the facet for origin of the hamstring muscles. It is therefore not possible to say that ischium length is short compared to that of modern humans as suggested by Robinson (1972). Rather, ischium length was probably markedly longer than in modern humans as indicated by the wide lesser sciatic notch.

Sts 14r—Left hipbone

Preservation: The left hipbone of Sts 14 consists of four main fragments that are joined by plaster (Figure 11.5). One piece includes the anterior ilium with the anterior superior iliac spine, a second the posterior and inferior ilium with the superior portion of the acetabulum, a third the ischium, and a fourth fragment the superior ramus and medial portion of the pubis.

The rough surface of the iliac crest particularly in the region of the anterior superior iliac spine indicates that the apophysis of the iliac crest was at a stage shortly before fusion. At the posterior superior iliac spine and in the region of the iliac tuberosity cancellous bone is exposed. This might suggest that the superficial bone was here still soft and thus particularly susceptible to erosion. At the same time, it could be an indirect sign that the apophysis is missing at these two places.

Many fine crack lines that run perpendicular to the iliac crest cross the external side of the left iliac blade. In their vicinity, small parts of the superficial bone layer have splintered off. This is most extensively the case in the second fragment of the ilium, where on the external side an area of about 5 × 2 cm have flaked off in the region of the posterior superior iliac spine. The internal side of the iliac blade exhibits a finer cracking pattern, and most of the outermost bone layer has splintered off. Weathering thus affected the left hipbone to a much greater extent than the right one but less extensively than the left femur and tibia of Sts 14. It corresponds to weathering stage 3 after Behrensmeyer (1978). According to her observations, the left hipbone must have been lying between 4 and over 15 years on the surface before being covered by sediments. *Morphology.* The iliac fossa is characterized by a gentle bulge, which begins near the anterior superior iliac spine and runs parallel to the iliac crest. A similar swelling is present in StW 431, and its beginning is also indicated in the ilium StW 441, but there is no such bulging in either modern human ilia or those of extant great apes.

The auricular surface is slightly better preserved, but cancellous bone is exposed at its medial and inferior border. It is therefore not possible to provide accurate measurements of its dimensions. The ischium preserves slightly more of its original length than on the right side, although cancellous bone is also here exposed at the whole region of the iliac tuberosity, as has already been mentioned.

The superior pubic ramus has suffered multiple fractures that have been crudely glued with araldite (Figure 11.6). The resulting distortion caused an excessive inward curvature and downward tilt of the superior pubic ramus, which lead to a shortening of the sagittal diameter of the pelvis and a heavily oblique pubic symphysis. The medial portion of the pubis is mostly unfractured, but the region of the pubic tubercle is eroded and cancellous bone is exposed here. The pubic symphysis is slender—its superoinferior height measures 28.1 mm, its maximum anteroposterior thickness 12.2 mm. The joint surface is rough, which is in accordance to the subadult age of the individual. A ventral arc is clearly present and the inferior border of the medial portion of the pubis seems to be concave—although part of it is reconstructed in plaster. These are two features that in modern humans are diagnostic of a female sex (Phenice, 1969). The inferior rami of the pubis and ischium are missing and reconstructed in plaster.

The lunate surface is preserved at the roof of the acetabulum and at the posterior and the inferior sectors (see Figure 11.6). The curvature of these sectors corresponds to best-fit spheres with diameters of 28.8 mm ± 0.7, 34.4 mm ± 0.3, and 35.0 mm ± 1.5. Since in primates the curvature

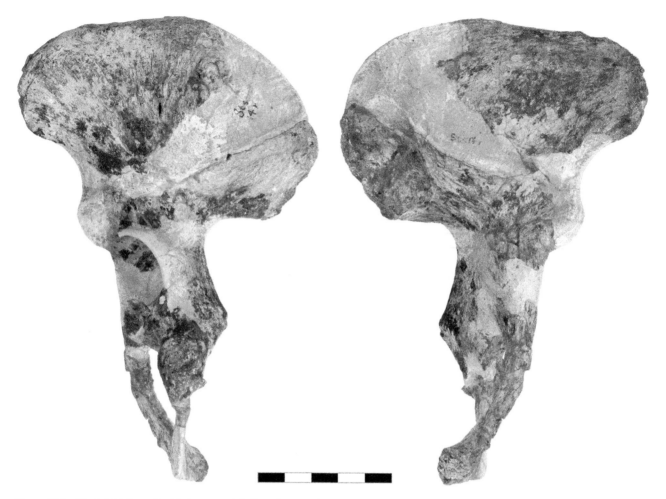

Figure 11.5 The left hipbone Sts 14r in external (left) and internal aspect (right). Scale bar equals 5 cm.

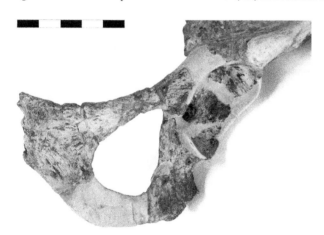

Figure 11.6 The left pubis Sts 14r in external aspect. The ventral arc is well visible. Scale bar equals 5 cm.

of the lunate surface is not homogenous throughout the acetabulum, regression equations of Hammond et al. (2013) based on a human and great ape sample were used for their acetabular sectors 3, 4, and 5–6, respectively, to estimate the best-fit spheres to the entire acetabulum. The

resulting mean diameters were 33.0 mm with a standard error of the estimate (SEE) of 3.3 mm, 36.2 mm (SEE 4.7), and 34.9 mm (SEE 1.9), respectively. This fits well the best-fit sphere to the right acetabulum.

Sts 14q—Sacrum

Preservation. The Sts 14 sacrum preserves the bodies and left ala of the first two sacral vertebrae (Figure 11.7). The two bodies are ventrally not fused due to the subadult age of the specimen (Berge and Gommery, 1999). Obviously, also the second and third vertebral body had not yet united ventrally, which can be inferred from the smooth inferior surface of the second sacral vertebra. The ring apophysis of the lumbosacral articular surface has been lost, but on the sacral ala the lateral epiphyseal plate has already partially fused, and a 14 mm-long portion of it is preserved at the level of the first sacral vertebra. At the level of the second sacral vertebra, the lateral part of the left ala including the auricular surface is broken off and reconstructed in plaster. On the right side, the entire ala together with the right third of the lumbosacral articular

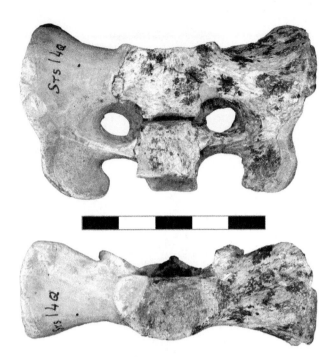

Figure 11.7 The sacrum Sts 14q in pelvic (top) and cranial aspect (bottom). Scale bar equals 5 cm.

surface, the cranial margin of the left first anterior sacral foramen, and the inferior right edge of the second sacral vertebra are reconstructed in plaster.

Due to this damage the transverse diameter of the lumbosacral articular surface cannot be measured accurately, and its sagittal diameter of 16.7 mm is compromised by the loss of the ring apophysis. The auricular surface suffered from both the partially missing lateral epiphyseal plate and the damage at the level of the second sacral vertebra. The articulation with the left hipbone is therefore imperfect.

Morphology. The sacrum appears to be relatively flat compared to modern humans, which is due to the small vertebral bodies and the wide ala sacralis. Similar to modern humans but in contrast to A.L. 288-1, the cranial tip of the ala is well developed.

Sts 65: Right partial hipbone

Preservation. This specimen represents part of a right hipbone that includes most of the ilium and a fragmentary superior pubic ramus (Figure 11.8). It was associated with a fragment of the dorsal part of the body of a lower lumbar vertebra (Robinson, 1972). Unfortunately, this partial vertebra has been lost and is currently not available for study. Most of the iliac crest and the anterior superior iliac spine are broken off in a ragged manner that is suggestive of carnivore damage (Haynes, 1980). The anterior inferior iliac spine as well as the anterior margin of the ilium between these two spines is heavily crushed and broken off. Despite this damage it is still recognizable

that the anterior inferior iliac spine was well developed. The region of the posterior superior iliac spine is well preserved but has been slightly displaced dorsally. It bears a small part of the iliac crest that measures 9.2 in thickness. The posterior inferior iliac spine is intact.

As noted by Robinson (1972), the surface of the hipbone is crossed by many fine cracks and, on the internal side of the ilium, one broad crack running diagonally across the iliac fossa, but the outermost layer of bone is generally well preserved. The weathering thus corresponds to stage 2 after Behrensmeyer (1978) and suggests that the specimen had been lying on the surface between two and six years before it was incorporated into the Sterkfontein cave deposits.

Morphology. The preserved width of the iliac blade measures 99 mm (Table 11.1). The minimum width from the anterior margin of the ilium to the greater sciatic notch is at least 48 mm. The overall pelvic size is thus about 8% larger than that of Sts 14.

Without contact with the rest of the specimen is an approximately 5-cm-long fragment of the middle section of the iliac crest. The thickness of the iliac crest is 6.8 mm at the lateral end of the fragment and 6.7 mm at medial end, but 4.3 mm at the medial third.

On the external aspect, the lateral section of the anterior gluteal line can be palpated over a length of 4 cm until it vanishes above the nutrient foramen approximately in the middle of the iliac blade (Figure 11.9A). Close to the posterior inferior iliac spine the posterior gluteal line is well preserved. Parallel to the anterior margin of the ilium a well-developed buttress extends to the superior margin of the acetabulum. It is, however, not homologous to the iliac pillar of modern humans (Arsuaga, 1981). In contrast to Robinson's (1972) opinion, the region of the origin of the reflected head of the rectus femoris muscle cannot be discerned.

The pubis is heavily crushed and preserves about the proximal half of the superior pubic ramus (Figure 11.9B). The iliopectineal eminence is also severely damaged. Nevertheless, it seems to have been well developed.

Only a small triangular fragment of the iliac section of the joint surface of the acetabulum is preserved. It is surrounded by exposed cancellous bone and bears a small puncture mark in its middle that is probably attributable to a carnivore. The superior acetabular rim is damaged but seems to have been sharp-edged, suggesting osteophytic lipping. Acetabular rim degeneration is also indicated by a 4-wide furrow in the lunate surface that runs parallel to the rim in the region of the anterior inferior iliac spine. It marks the border between the original lunate surface and the osteophytic extension of the acetabular rim.

The size of the acetabulum can be approximated by a best-fit sphere to the lunate surface (see Sts 14). Surface defects, such as the above-mentioned furrow and the

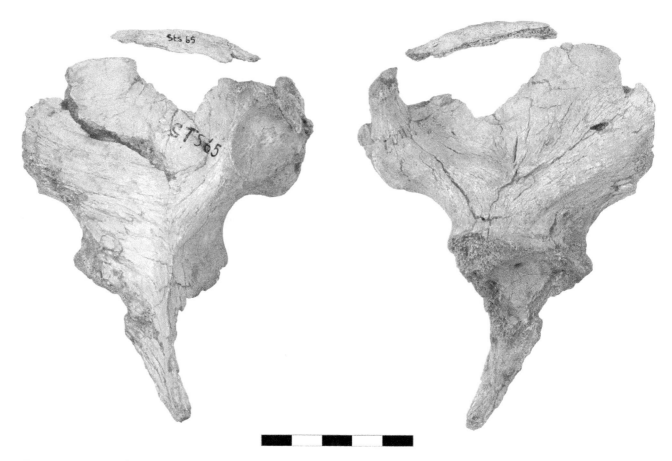

Figure 11.8   The right hipbone Sts 65 in internal (left) and external aspect (right). Scale bar equals 5 cm.

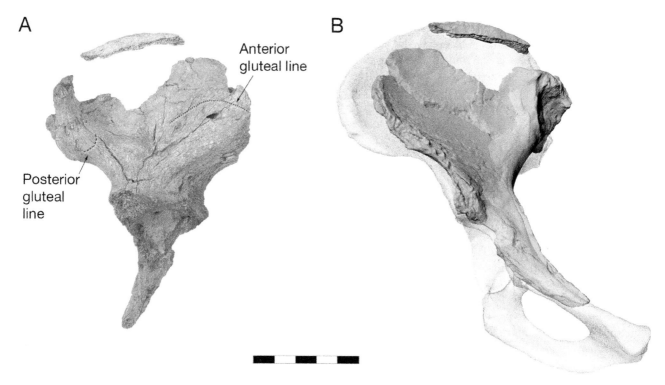

Figure 11.9   A: Muscle markings on the external aspect of Sts 65. B: 3D model of Sts 65 (yellow) superimposed on the reconstructed right hipbone of Sts 14 (transparent; Häusler and Schmid, 1995), ventral view. Scale bar equals 5 cm.

convex periphery of the lunate surface, were not included. Six independent extractions of the lunate surface resulted in a best-fit sphere with a mean diameter of 38.0 mm ± 0.2 (standard deviation). Because a fracture through the roof of the acetabulum might have affected the acetabulum diameter, a sphere was also fitted to the posterior section of the lunate surface only. Its diameter of 38.1 mm ± 0.7 is similar to that of the entire preserved portion of the lunate surface. Based on this, the diameter of a best-fit sphere to the whole acetabulum can be estimated according to the regression equation of Hammond et al. (2013) for acetabular sector 3. This yields a mean diameter of 38.5 mm with a SEE of 3.9 mm.

On the internal aspect, the auricular surface is complete save for 1 cm² at the posterior inferior iliac spine where the superficial bone is missing. The auricular surface is proportionally smaller relative to the width of the iliac fossa than in modern humans because probably only the first two sacral vertebrae articulated with the ilium compared with the first three sacral vertebrae in modern humans. The distance from the posterior inferior iliac spine to the anterior margin of the auricular surface measures 29.3 mm, and the superoinferior height of the auricular surface is 18.0 mm.

The ventral section of the arcuate line, including its continuation as the pubic pecten, is developed to a razor-sharp and very prominent edge, which is unusual in extant hominoids. It is related to the attachment of the iliac fascia and the insertion of the psoas minor muscle. A true tubercle of the psoas minor muscle as in A.L. 288-1 cannot be discerned.

StW 431

The elements of the pelvic girdle

The StW 431 partial skeleton was discovered by the late Alun Hughes and his team under the direction of Philip Tobias in February 1987 (Tobias, 1992). Four additional fragments were later discovered by Kibii and Clarke (2003) among faunal remains from Sterkfontein (see Ward et al., Chapter 4, this volume). In the course of their discovery, each hominid pelvic fossil fragment was given a chronological number. Only later, after realizing that there were no overlapping parts, was it assumed that they all belong to a single individual, which then became known as StW 431 after the number of the first element discovered (Tobias, 1992). The individual bones were identified with letters, and if a specimen consisted of several fragments they were given a second letter (see Table 11.3). Various early publications referred, however, to the original numbers that are actually still written directly on the specimens. They are therefore listed in Table 11.3 together with their new identification numbers and their provenience in the grid system used at Sterkfontein.

During this renaming, the left ilium fragments StW 441/StW 465 were also attributed to the StW 431 skeleton as StW 431i and StW 431w. They are, however, much more gracile than the right ilium and thus obviously belong to a different, smaller individual than the partial skeleton StW 431 (Häusler and Berger, 2001). These fragments are therefore described here under their original identification numbers.

Table 11.3   The elements of the pelvic girdle of StW 431.

ID number	Previous number	Fragment	Grid number and depth
431ea	436a	Part of greater sciatic notch, right	R45 25'6"–26'6"
431eb	436b	Anterior superior iliac spine, right	Q45 21'11"–22'11"
431ec	436c	Iliac crest, right	Q45 22'11"–23'11"
431ed	436d	Anterior inferior iliac spine and part of acetabulum, right	Q45 22'11"–23'11"
431ee	436e	Portion of iliac fossa, right	Q45 23'11"–24'11"
431ef	436f	Pubic body, right	Q46 28'6"–29'6"
431fa/431fb	438 and 444	Acetabulum fragments, right	Q45 22'11"–23'11" and Q45 23'11"–24'11"
431ha	439a	Sacrum	Q45 22'11"–23'11"
431hb	439b	Fragment of right ala sacralis	Q45 24'11"–25'11"
431i	440	Superior pubic ramus, left	Q45 21'11"–22'11"
431j	442	Fragment of left iliac crest	Q45 24'11"–25'11"
431ka	443	Acetabulum fragment, left	Q45 24'11"–25'11"
431kb	–	Fragment of left external ilium	–
431w	S94-1099	Left auricular fragment	Q45 24'11"–25'11" and Q45 21'11"–22'11"
431x	S94-13926	Right posterior superior iliac spine	Q45 24'11"–25'11"
431y	S94-13995	Right auricular fragment	–

The identification numbers StW 431ef, fa/fb, i, kb, w, x, and y (bold) differ from those of the computer version of the Sterkfontein fossil catalogue (see text).

Unfortunately, however, there have also been some cases of misnumbering. Thus, the two fragments of the ischiadic part of the right acetabulum have originally been labeled StW 438 and StW 444. The official Sterkfontein catalogue simply lists now both of them under StW 431ef. Yet, this number also belongs to the right acetabulum and pubic body fragment, which is part of the right ilium StW 431e (the latter originally carried the number StW 436, and its individual parts StW 436a-f became StW 431ea-ef). It is therefore here proposed to label the two fragments of the lower right acetabulum with separate letters, StW 431fa and 431fb, respectively.

To make confusion complete, the fragment of the left superior pubic ramus is marked with ink on the bone as StW 436f instead of the right acetabulum and pubic body fragment. In older excavation logbooks, the left superior pubic ramus fragment was known for some time as StW 440, a number originally given to a specimen identified as an ulna shaft (now: fibula shaft), which was found in a grid square near the StW 431 skeleton. Subsequently, it was marked missing in the fossil logbook and the left pubis came to be known as StW 440. Because the original StW 440 fragment was found in grid square P44 at a depth of 13'5"–14'5" and thus farther away and in a much higher position than the parts of the StW 431 skeleton, the attribution of the left pubis to this skeleton was later questioned

as well, although the note on the left superior pubic ramus itself confirms that it was found in the same grid square and at the same depth as most of the right ilium. Given its morphology, which fits the left acetabulum fragment, there are no reasons to doubt its association with the StW 431 skeleton. We therefore propose the new number StW 431i for the left pubis. This letter becomes available as the left ilium fragment StW 441 clearly belongs to a different, smaller individual than the partial skeleton StW 431. We are aware that this renaming might cause further confusion until the new identification numbers are fully established, but there is no other solution available as most letters of the alphabet are already reserved for other specimens of the StW 431 partial skeleton.

The four additional fragments of the StW 431 skeleton that have been discovered by Kibii and Clarke (2003) among faunal remains are here assigned the numbers StW 431kb, w, x, and y (see Table 11.3).

StW 431e—Right ilium

Preservation. This specimen includes a large section of the right iliac crest, the anterior superior and inferior iliac spines, the middle section of the greater sciatic notch, part of the acetabulum, and the body of the pubis (Figure 11.10). It consists of eight fragments that have been glued

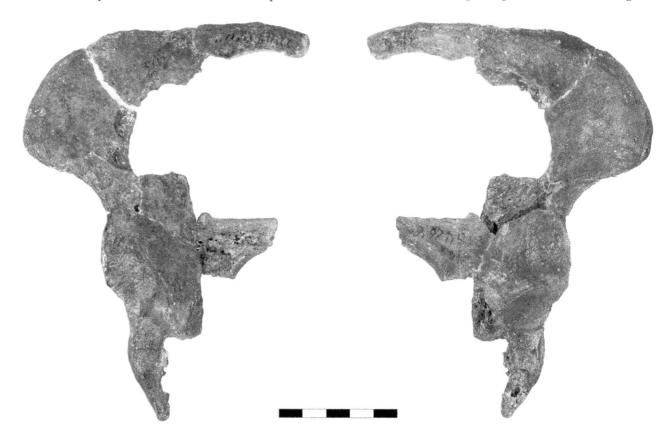

Figure 11.10 The right ilium StW 431e in internal (left) and external aspect (right). Scale bar equals 5 cm.

together. Six of them were originally numbered separately as StW 436a–f (cf. Table 11.3). All fragments fit together closely. Only the alignment of the medialmost cristal fragment allows a little play. Nevertheless, its orientation can be well adjusted through alignment of the curvature of the inner and outer surfaces with the adjacent fragment. There is thus no doubt that as in the other australopithecine ilia the sigmoid curvature of the iliac crest was only faintly developed in comparison to modern humans. Minor defects of the surface bone are present on the anterior iliac crest and around the break lines between the three cristal fragments. Otherwise, the external surface of the ilium is perfectly conserved and there are no signs of weathering or postmortem deformation. The iliac crest is preserved over a stretch of 123.7 mm. The minimum width of the ilium between the notch above the anterior inferior iliac spine and the greater sciatic notch measures 60.5 mm (Table 11.1).

Morphology. The anterior superior iliac spine is very prominent and strongly built. Here we find the thickest part of the iliac crest (15 mm). Going medially, the iliac crest steadily thins, at first slowly and then more markedly after the missing flake of bone between the anterior and the intermediate cristal fragments. This pronounced thinning extends over about 2 cm, approximately, to the center of the intermediate cristal fragment. Here, the inner margin of the iliac crest starts to bend outward.

At the junction to the medialmost cristal fragment, the iliac crest is thinnest at 5 mm and thickens again slightly immediately medial to it. The linea epiphysis marginalis becomes well visible on the ventral aspect. In the middle of the medialmost cristal fragment, a small rugosity can be discerned on the ventral side of the iliac crest that might be related to the attachment of the quadratus lumborum muscle. Further medially, the iliac crest starts to faintly curve dorsally.

On the anterior portion of the iliac crest, a distinct labium internum, linea intermedia, and labium externum can be recognized until the crest reaches its thinnest segment. The labium internum is well developed and bends the iliac crest inward. This feature is identical in StW 441 and possibly indicates a strong pull of the fascia transversalis, the *m. transversus abdominis*, and inguinal ligament. It cannot be established whether Sts 14 also showed this morphology, because its iliac crest is damaged and the apophysis is missing. There are no indications that the iliac crest of A.L. 288-1 is similarly pulled inward, although this is difficult to verify because the anterior superior iliac spine is deformed by a longitudinal crack (Johanson et al., 1982).

The iliac fossa is characterized by a wide bulge that begins at the anterolateral margin of the ilium about 20 below the anterior superior iliac spine and runs parallel to the iliac crest in its anterior sector. A similar swelling, but closer to the iliac crest and less conspicuous, is present on

the undistorted left ilium of Sts 14 (the right ilium is damaged), and its beginning is also indicated in the ilium fragment StW 441. Neither A.L. 288-1 nor modern humans or great apes show a similar bulge, which suggests that it is a unique feature of *Australopithecus* from Sterkfontein. Its functional significance remains unclear, although it might be related to a powerful *m. iliacus*.

On the external aspect of StW 431e, the anterior part of the iliac crest shows some marked enthesophytes. The anterior gluteal line is well developed and joins the iliac crest right at the anterior superior iliac spine (Figure 11.11). This corresponds to a relatively larger and broader anterior portion of the gluteus medius muscle than in modern humans. A similar condition is present in StW 441 and A.L. 288-1 (*contra* Berge, 1994). The remaining course of the anterior gluteal line is not preserved save for the junction of its inferior end with that of the posterior gluteal line, which is represented by a rugosity above the greater sciatic notch just before the bone is broken off. This fits the morphology on the left auricular fragment StW 431w that preserves the posterior gluteal line over a larger stretch. A comparable rugosity can also be found on the A.L. 288-1 hipbone. The course of the inferior gluteal line is not discernible with certainty. Its lateral portion might coincide with the break line below the anterior superior iliac spine.

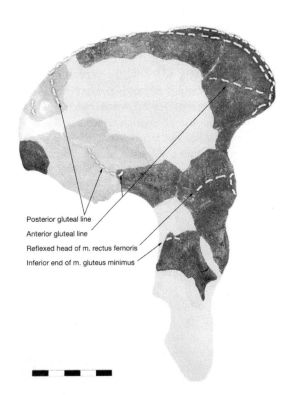

Posterior gluteal line
Anterior gluteal line
Reflexed head of m. rectus femoris
Inferior end of m. gluteus minimus

Figure 11.11 Muscle markings projected on a composite image of the StW 431 right hipbone, external aspect. StW 431j and StW 431w are mirrored from the left side and transparent. Scale bar equals 5 cm.

The anterior inferior iliac spine is remarkably sturdy compared to that of both Sts 14 and A.L. 288-1. It is large superoinferiorly but relatively flat in its dorsoventral dimension compared to that of modern humans. Its marked ventral projection might indicate both a strong pull of the caput rectum of the rectus femoris muscle (superiorly) as well as of the iliofemoral ligament (inferiorly). A depression on its inferior third on the ventral aspect running obliquely from superomedial to inferolateral might be related to the iliopsoas muscle as the marked groove between the anterior inferior iliac spine and the iliopectineal eminence at the cranioventral margin of the acetabulum.

At the dorsal side of the anterior inferior iliac spine a ridge extends from the lower end of the spine toward the craniodorsal margin of the acetabulum. The massive development of this ridge suggests a very strong iliofemoral ligament compared to A.L. 288-1, where such a ridge is only faintly developed. In Sts 14, the cranial margin of the acetabulum is chipped off, and thus the region of attachment of the iliofemoral ligament is damaged as well. Cranial to this ridge and above a shallow groove, another ridge points toward the upper end of the anterior inferior iliac spine of StW 431e. It serves as origin for the *caput reflexum musculi recti femoris*.

The iliopectineal eminence is very sturdy compared to Sts 14. In contrast to the two Sterkfontein specimens, the iliopectineal eminence is completely absent in A.L. 288-1 and a true iliopsoas groove does not occur in this specimen (Stern and Susman, 1983). Medial to the iliopectineal eminence of StW 431, only the ventral sector of the root of the superior pubic ramus is preserved.

The arcuate line is preserved over a stretch of 17 mm just opposite the acetabulum. Although it is marked with longitudinal furrows, it has not formed such a sharp edge as in Sts 65. Its rugose nature suggests, however, that it provided insertion to a strong psoas minor muscle. Nevertheless, a tuberculum musculi psoas minor as in A.L. 288-1 cannot be verified on this short portion of the arcuate line.

The cranial part of the acetabulum with the lunate surface and almost the entire anterior horn is preserved in StW 431e. The anterior horn is well developed and has a broad surface, comparable to that of Sts 14. This contrasts to A.L. 288-1, where the articular surface has no contribution from the pubis (Stern and Susman, 1983; MacLatchy, 1996). The anterior sector of the acetabulum displays a furrow parallel to the superior margin that starts just below the anterior inferior iliac spine. This suggests that, similar to Sts 65, StW 431 was also affected by acetabular rim degeneration in the context of hip osteoarthritis.

The curvature of the preserved lunate surface of StW 431e can be approximated by a best-fit sphere with a diameter of 42.3 mm ± 0.08 (see Sts 65). Because the fracture through the iliopubic eminence might have expanded the lunate surface, another best-fit sphere ignoring the segment of the pubic body fragment was calculated. This sphere has a diameter of 41.4 mm ± 0.3. According to the regression equation of Hammond et al. (2013) for acetabular sector 2, this suggests a best-fit sphere to the entire acetabulum of 41.2 mm (SEE 4.8 mm).

StW 431f—Right acetabulum fragment

Preservation. This specimen consists of two fragments now joined together (Figure 11.12). Their previous numbers are StW 438 and StW 444. They belong to the right body of the ischium and form the inferior part of the acetabulum. The match between the two fragments is good, though inaccurate agglutination resulted in a slight displacement and some expansion of the lunate surface. Superior to the fracture, the superficial bone has flaked away over a short stretch of the lunate surface.

Morphology. The curvature of the preserved lunate surface of StW 431f corresponds to a best-fit sphere with a diameter of 41.1 mm ± 1.6. However, the crack between the two parts might have affected the curvature of the lunate surface. Thus, the best-fit sphere to the lunate surface of the superior fragment has a diameter of 38.3 mm ± 1.5,

Figure 11.12 The right acetabulum fragment StW 431f in lateral (left) and medial aspect (right). Scale bar equals 5 cm.

while the diameter of the corresponding best-fit sphere to the lunate surface of the inferior fragment is 40.2 mm ± 0.9. According to the regression equations of Hammond et al. (2013) for acetabular sectors 5–6 and 7, this suggests best-fit spheres to the entire acetabulum of 38 mm (SEE 1.8) and 37.9 mm (SEE 14.0 mm), respectively.

On the internal surface, this crack widens to more than half a centimeter, such that only a small portion of the surface bone remains on each of the two pieces. Barely 3 mm of the margin of the greater sciatic notch is preserved.

There is no direct contact to the ilium fragment StW 431e. Together with the narrower radius of curvature of the lunate surface compared to StW 431e, this raises the possibility that the right acetabulum fragment StW 431f belongs to a different individual than StW431e. The large standard errors of the estimate associated with the best-fit spheres, however, encompass the value estimated for StW 431e. On the other hand, the width of the lunate surface at the superior extremity of StW 431f is with 17.8 mm almost identical to the 18.0 mm at the inferior extremity of the left acetabulum fragment StW 431k, and also the breadth of the posterior aspect of the ischial body from the acetabular margin to the greater sciatic notch is with 29.8 mm a perfect match to the corresponding dimension in StW 431d if the damage to the greater sciatic notch in that specimen is taken into account. Hence, the overall morphology of the right acetabular fragment StW 431f supports its association with the StW 431 pelvis.

StW 431h—Sacrum

Preservation. The StW 431 sacrum was found in two parts. Specimen StW 431ha comprises the first two-and-a-half sacral vertebral bodies with the left ala and auricular surface, which is almost complete (Figure 11.13). Only the caudal extremity of the auricular surface is slightly damaged. StW 431hb, a fragment of the right sacral ala, is separated from StW 431ha by a gap of about 9 mm on the dorsal side, whereas the union is perfect ventrally and cranially. Only a few small flakes of bone are chipped off along the ventral and cranial fracture line. Moreover, the cranial tips of the transverse processes of both alae, the right superior articular process, and the tips of the median sacral crest are all broken off. Despite this damage it is clear from the remaining morphology that the tip of the left dorsal alar tubercle must have been well developed as in modern humans. This thus contrasts to the condition in A.L. 288-1. *Morphology.* The ventral margin of the lumbosacral articular surface shows mild osteophytic lipping related to a limbus vertebra (Haeusler, 2019) or to brucellosis of the L4–L5 intervertebral disc (D'Anastasio et al., 2009). Osteoarthritic alterations are also found on the left superior articular process of the sacrum. On the ventral aspect, the epiphyses between S1 and S2 and between S2 and

Figure 11.13 The sacrum StW 431h in ventral (top) and cranial aspect (bottom). Scale bar equals 5 cm.

S3 are still open medially, which may suggest an early adult age for StW 431, although there is a great range of variability in modern humans (Schwabe, 1933; Belcastro et al., 2008).

The left auricular surface is broken off at its inferior extremity, but comparison with the auricular surface of the left ilium, StW 431w, demonstrates that less than 1 is missing. It extends over the whole length of the first two-and-a-half vertebral segments, which is more than in Sts 14 and probably Sts 65. As in Sts 14 and A.L. 288-1, the shape of the auricular surface is similar to that of modern humans and contrasts sharply with the narrow joint surface of great apes.

StW 431i—Left superior pubic ramus

Preservation. StW 431i represents the middle part of the left superior pubic ramus. It misleadingly bears the number 436f directly written onto the bone. Medially, the pubic body is broken off along with any pubic tubercle that may have been present, whereas proximally an oblique, probably fresh fracture has cut off the entire iliopectineal eminence. A corresponding oblique fracture is also present on the left acetabulum fragment StW 431k (Figure 11.14). Although the specimen has no direct contact with the acetabulum fragment, the morphology of the arcuate line

Figure 11.14 The left pubis StW 431i (left) and acetabulum StW 431ka (right), ventral aspect. Scale bar equals 5 cm.

Figure 11.15 The left ilium fragment StW 431j, internal (left) and external aspect (right). Scale bar equals 5 cm.

on the internal aspect of the fragment suggests that little bone is missing between the two specimens.

Morphology. On the proximal third of the specimen, the arcuate line is developed to a sharp and prominent crest for the attachment of the pectineal ligament. Further medially, the pecten ossis pubis forms again a prominent crest for the attachment of the lacunar ligament, the inguinal falx (the conjoint aponeurosis of the obliquus internus and transversus abdominis muscles), and the reflected inguinal ligament. The pubic tubercle with the attachment of the inguinal ligament is broken off.

On the inferior margin of the pubic ramus, the obturator sulcus is limited medially by a prominent anterior obturator tubercle. Although the ventral aspect exhibits some oblique wrinkles, the anterior obturator crest is flat and barely prominent, which contrasts strikingly with the condition found in modern humans. It may indicate a relatively weak pull by the pubofemoral ligament. This is surprising, as the well-developed anterior inferior iliac spine and the prominent ridge on the external surface of the ilium above the acetabulum seem to indicate a strong iliofemoral ligament (see StW 431e).

StW 431j—Left ilium fragment

Preservation. StW 431j is a small fragment of about 38 mm of the left posterior iliac crest (Figure 11.15). On the internal side, the epiphyseal line is well visible, which corresponds to the morphology of the right ilium. The crest is thinnest at the medial extremity of the fragment from where it thickens first rapidly and then more slowly until on the lateral third of the crest fragment a pronounced tubercle is reached, which protrudes ventrally. It is possibly related to the attachment of the iliolumbar ligament and the medial end of the quadratus lumborum muscle. It most likely corresponds to a similar, though less marked tubercle on the right iliac crest that is found 12 mm from the medial edge of the StW 431e fragment. In fact, thickness and robusticity of the crest closely correspond to the medialmost end of the right ilium fragment StW 431e if the enthesophyte is not counted, which projects above the crest on the dorsal aspect just opposite to this tubercle. The enthesophyte might be related to the medial margin of the latissimus dorsi muscle. Lateral to the enthesophyte, two distinct facets can be distinguished on the iliac crest that likely are related to the origins of the latissimus dorsi and quadratus lumborum muscles.

Morphology. On the internal side, a discrete ridge can be palpated that runs from the above-mentioned tubercle obliquely inferoposteriorly. It is probably related to the insertion of the ventral sacroiliac ligaments. Concomitant to this faint ridge, the thickness of the iliac blade increases continually toward the inferoposterior margin of the fragment. Rugosities that are found here belong to the beginning of the iliac tuberosity.

Another faint ridge can be discerned on the dorsal side of the iliac blade, which runs from the medial margin of the enthesophyte in inferoposterior direction. This ridge can be identified as the posterior gluteal line and thus suggests that in *A. africanus* the gluteus maximus muscle had an origin on the ilium, as is characteristic for *Homo*. The posterior margin of attachment of the gluteus maximus muscle is probably indicated by a second ridge, which starts just at the medial extremity of the fragment and runs parallel to the first.

StW 431k—Left acetabulum fragment

Preservation. StW 431ka is a large fragment of the inferior left ilium including the superior portion of the acetabulum (Figures 11.14 and 11.16). To the upper margin of the fragment, an approximately 3-cm²-large piece of the external surface of the ilium, StW 431kb, was joined in May 2002. This piece is therefore missing on older casts of the StW 431 skeleton. The upper end of the anterior inferior iliac spine and the entire iliopectineal eminence are broken off. On the internal aspect, the arcuate line is preserved over a straight length of 52 mm. At the transition from the medial to the anterior third of the arcuate line, a 6 mm × 4 mm flake of superficial bone has splintered away. Its anterior third displays longitudinal furrows, similar to the arcuate line on the right ilium.

Morphology. Similar to the left acetabulum StW 431e, StW 431k also displays a furrow parallel to the superior acetabular rim just below the anterior inferior iliac spine implying hip osteoarthritis. The superior sector of the lunate surface shows three punctures. A microscopic examination suggested that the two superomedial impressions were caused before fossilization, perhaps by carnivores. The third puncture is accompanied by a furrow and probably represents a preparation damage.

A best-fit sphere to the lunate surface has a diameter of 41.1 mm ± 0.07. According to the regression equations of Hammond et al. (2013) for acetabular sectors 3–4, this suggests a best-fit sphere to the acetabulum of 41.1 mm (SEE 2.9 mm). This is very similar to that suggested by the right acetabulum fragment StW 431e.

On the dorsal aspect, a faint horizontal ridge can be palpated about 1 cm above the inferior break. It is probably associated with the inferior boundary of the gluteus medius muscle. The remaining morphology of the fragment is identical to that of the right ilium StW 431e.

StW 431w—Left auricular fragment

Preservation. This is a large fragment of the left posterior ilium consisting of three pieces that have been glued together. They fit excellently, but small flakes of bone are chipped off along the fracture lines (Figure 11.17). The fragment measures approximately 89 by 4 mm and comprises the entire auricular joint surface, the medial section of the greater sciatic notch, and both the posterior inferior and superior iliac spines together with the adjacent 18 mm of the iliac crest. The auricular surface articulates perfectly with the sacrum.

Morphology. The distance from the posterior inferior iliac spine to the anterior margin of the auricular surface measures 36.1 mm, and the superoinferior height of the auricular surface is 23.5 mm. The iliac tuberosity is inflated and projects well above the auricular surface. On the external

Figure 11.16 The left acetabulum fragment StW 431ka and kb, external aspect (left) and internal aspect (right). Scale bar equals 5 cm.

Figure 11.17 The left auricular fragment StW 431w, internal aspect (left), and external aspect (right). Scale bar equals 5 cm.

aspect, the inferior end of the posterior gluteal line is well discernible above the greater sciatic notch (see Figure 11.11).

The fragment contacts the left acetabulum fragment StW 431k for a length of 13.0 mm on the external side, while two further small points of contact are on the interior side and the greater sciatic notch. The resulting shape of the greater sciatic notch is relatively narrow and asymmetric, which is reminiscent of males according to modern human standards.

StW 431x—Right posterior superior iliac spine

Preservation. This is a small fragment of the right posterior superior iliac spine (Figure 11.18). It preserves 25.0 mm of posterior iliac crest and 15.0 mm of the adjacent posterior iliac margin.

Morphology. Its morphology is identical to the left posterior superior iliac spine in StW 431w. The maximum thickness of the iliac crest is 14.9 mm.

StW 431y—Possibly right auricular fragment

Preservation. This is an 18.7 × 27.4 × 12.0 mm fragment of the posterior ilium of the right side (Figure 11.19). The concave external aspect preserves 20 of the posterior gluteal surface. On the internal aspect, it seems to include the cranial-most section of the auricular surface and the adjacent iliac tuberosity as well as a small part of the iliac fossa that is set at an angle of 90° to the auricular surface. The auricular surface articulates with the sacrum StW 431h. The small size of the fragment, however, precludes definite identification. Alternatively, it might represent part of the left iliac tuberosity and iliac fossa just cranial of the auricular fragment StW 431w.

Morphology. Little diagnostic morphology is preserved on this specimen.

StW 441: Anterior left ilium

Preservation. This is an 83 × 46 large portion of an anterior left ilium. It consists of three triangular pieces (Figure 11.20).

Figure 11.18 The right posterior superior iliac spine StW 431x, internal aspect (left), and external aspect (right). Scale bar equals 5 cm.

Figure 11.19 The right auricular fragment StW 431y, ventral aspect (left), and dorsal aspect (right). Scale bar equals 3 cm.

The first of them, a 38-long piece of the iliac crest, StW 441, was discovered in September 1986 in grid square Q46 at a depth of 20'10"–21'10". A few months later, in February and March 1987, the left anterior superior iliac spine (without specimen number) was excavated in the adjacent grid R46 at a depth of 20'9"–21'11". A third iliac fragment, StW 465,

Figure 11.20 The left ilium fragment StW 441 in ventral (left) and dorsal aspect (right). Scale bar equals 5 cm.

found four grid squares to the east (R43, 20'22"–21'11") fits neatly between the other two pieces. As the three fragments clearly belong together, the accession number of the first found specimen StW 441 stands for them all.

Between the piece of the anterior superior iliac spine and the adjacent triangular fragment the internal two-thirds of the crest is flaked off over a length of about one centimeter. This leads to only a small contact area at the iliac crest between the two fragments. Their relative orientation in the plane of the iliac blade is therefore subject to some uncertainty. The contact with the medial fragment (originally StW 441) is better and about 10 in width. *Morphology.* The peculiar morphology of the ilium with the projecting anterior superior iliac spine is typical for *Australopithecus.* Though being more delicately built, its shape closely corresponds to the right ilium of StW 431. Because the second and third fragment of StW 441 have been discovered at the same time and in grid squares in the vicinity of the StW 431 partial skeleton, though closer to the top of Member 4, it has been suggested that they all might belong to the same individual, particularly as there are no overlapping parts. However, the dimensions of StW 441 are much smaller; for example, the maximum thickness of the anterior superior iliac spine of StW 441 is 11 mm, much thinner than the 15 mm of StW 431's right ilium. This strongly suggests that they belong to different individuals, particularly as the preserved pelvic elements of StW 431 do not show any indication of asymmetry in size or morphology (Häusler and Berger, 2001). The not fully fused cristal apophysis of StW 441, which is indicated by a linea epiphysis marginalis visible on both the external and internal sides of all three fragments, might suggest a slightly younger age than that of StW 431.

The iliac crest of StW 441 bulges ventrally in the region of the anterior superior iliac spine in the same way as in StW 431 and Sts 14r, which is possibly related to the anterior fibers of the iliacus muscle. On the dorsal aspect,

a prominent pointed tubercle close to the anterior superior iliac spine marks the beginning of the anterior gluteal line, which can be palpated on the medial side of this tubercle for a short stretch up to the fracture margin of the specimen (Figure 11.21). A comparable, though less developed tubercle of the anterior gluteal line is also present in A.L. 288-1 (Johanson et al., 1982) but is absent from both StW 431 and Sts 14. The course of the anterior gluteal line, however, closely corresponds to that in StW 431, Sts 65, and AL 281-1. The gluteus medius muscle therefore seems to have had a larger anterior portion in australopithecines than generally found in modern humans, while the gluteus minimus muscle had a broad area of origin.

A shallow furrow separates the anterior portion of the of the StW 441 iliac crest at the anterior superior iliac spine from the rest of the crest, which thus impresses as a distinct 19 mm-long knob. Such a demarcation is unique to StW 441 and is absent in other australopithecine ilia, including StW 431. The knob bulges ventrally over the iliac fossa, which suggests tension by the fascia iliaca and the inguinal ligament. Another, smaller, knob that is related to the origin of the sartorius muscle lies just below the anterior superior iliac spine.

The iliac crest of StW 441 permits distinction between the muscle attachments of all three abdominal wall muscles. The labium internum starts just medially to the large knob and provides the origin of the transversus abdominis muscle. A small knob in the middle of the StW 441 fragment marks the medial limit of the labium internum. Further medially, the internal margin of the preserved crest remains straight and does not bend outward as in StW 431. The linea intermedia provides origin to the obliquus abdominis internus muscle, while the obliquus abdominis externus muscle inserts on the labium externum. In the middle of the medial fragment StW 441, just opposite to the small knob on the labium internum, a distinct protuberance is visible, which is possibly related to the iliotibial

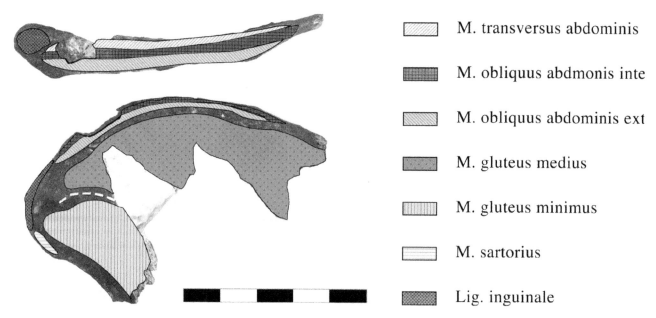

Figure 11.21   Proposed muscle attachments on StW 441, adapted from Häusler and Berger (2001). Scale bar equals 5 cm.

tract. On the medial side of this protuberance, the external portion of the crest thins out abruptly, probably indicating the end of the insertion of the obliquus abdominis externus muscle. A comparable protuberance and abrupt thinning of the labium externum are both lacking from StW 431.

At the inferomedial fracture margin of the specimen, which would be approximately midway between the anterior superior and posterior superior iliac spines, the external table of the iliac blade starts to thicken again. Unfortunately, the corresponding part of the iliac blade is not preserved in StW 431 and Sts 65, but this thickening differs from the morphology of Sts 14 or A.L. 288-1 as well as from that of modern humans or extant great apes. A possible explanation might be a structure homologous to the iliac pillar of humans. The above-mentioned protuberance of the iliac crest in the middle of the medialmost fragment could then be homologized with the cristal tubercle of modern humans, although there is no indication of a cristal tubercle or an iliac pillar in StW 431 or any of the other australopithecines (see also Arsuaga, 1981).

StW 611: Fragment of left ischium

Preservation. This is a fragment of a left ischium that was discovered in grid square O44 at a depth of 13'10"–14'10" (Figure 11.22). It measures 47 mm obliquely superoinferiorly and 35 mm anteroposteriorly. The ischial tuberosity and the ischial spine are broken off, but it preserves part of the ischial portion of the acetabulum with a 31 mm-long segment of the lunate surface. The maximum width of the lunate surface is 15.3 mm. *Morphology.* The diameter of a best-fit sphere to the lunate surface is 39.9 mm ± 0.17. This is smaller than the best-fit

spheres to the lunate surface of Stw 431f and k. According to the regression equations of Hammond et al. (2013) for acetabular sectors 5–6, the best-fit sphere to the entire acetabulum of StW 611 can be estimated to 39.6 mm (SEE 1.9). The width of the dorsal aspect of the fragment at the superior extremity, at 21.8 mm, is also smaller than the corresponding measurement at the inferior extremity of StW 431k, which measures at least 28.0 mm.

The minimum width of the tuberoacetabular sulcus measures 16.0 mm. Relative to acetabulum size, the width of the tuberoacetabular sulcus of StW 611 is thus intermediate between MH2 (*A. sediba*) and the *Australopithecus* sample of Kibii et al. (2011).

Comparative anatomy

Age and sex

Sts 14

Robinson (1972) suggested that Sts 14 belonged to an adult individual. However, the apophyses of the iliac crests and the ischial tuberosities are completely missing, which is most easily seen on the left hipbone (the right iliac crest and ischial tuberosity is damaged). The rough surface at the left anterior superior iliac spine suggests that ossification of the apophysis was probably complete and fusion immanent, which would be equivalent to Risser stage IV (Risser, 1958), if it is possible to equate morphological with radiological findings. In addition, part of the lateral epiphyseal plate of the sacral ala is missing, and the sacral vertebral bodies are unfused (Häusler, 1992, 2001;

Figure 11.22 The ischium StW 611, external aspect (left) and internal aspect (right). Scale bar equals 5 cm.

Berge and Gommery, 1999). The triradiate cartilage of the acetabulum, on the other hand, is completely ossified and no physeal lines are visible. The subadult condition of the sacrum corresponds to a skeletal age of about 16–21 years according to modern human standards (Berge and Gommery, 1999; Bonmatí et al., 2008), whereas Risser stage IV of the iliac apophysis and a female sex indicates about 15 ± 1.2 years (Kotwicki, 2008). The pelvis would thus have reached almost adult size and the remaining growth would mainly be in the vertebral column.

The very wide greater sciatic notch is suggestive of a female sex. The depth of the greater sciatic notch is 46.6% of its width, measured according to Genovés (1959), which is close to the mean of modern human females. This is supported by a female state of the arc composé (composite arch; Genovés, 1959) according to modern human standards, a round shape of the pelvic inlet, a wide subpubic angle, the presence of a ventral arc on the left pubis, and a subpubic concavity (Häusler and Schmid, 1995). A preauricular groove is not present.

Sts 65

The iliac crest apophysis is completely fused and an epiphyseal line cannot be discerned neither in the vicinity of the posterior superior iliac spine nor on the isolated fragment of the middle section of the iliac crest (*contra* Claxton et al., 2016). On the other hand, acetabular rim degeneration implies the presence of hip osteoarthritis (Leunig et al., 2003). This pathology is surprisingly rare in the hominid fossil record and is almost unknown in nonhuman primates (Jurmain, 2000; Haeusler et al., 2015). It means that Sts 65 belonged to an adult individual and perhaps even to one of an advanced age, although

secondary hip osteoarthritis might begin already in young adults.

The greater sciatic notch appears to be slightly narrower than that of Sts 14. Because the lower part of the greater sciatic notch including the ischial spine is not preserved, however, it is not possible to exactly determine its shape. Nevertheless, the anterior margin of the greater sciatic notch is continuous with the anterior margin of the auricular surface, implying a masculine state of the composite arch (arc composé; Genovés, 1959) according to modern human standards.

A faint depression on the posterior preauricular surface was interpreted by Wolpoff (1973) and Claxton et al. (2016) as a preauricular sulcus. The small size and the shallowness of the groove, reflected by an open arc that makes up less than half the circumference of a circle in combination with the absence of a clear border of the depression, is, however, characteristic of a paraglenoid groove that is related to skeletal robusticity (Bruzek, 2002). In modern humans, this paraglenoid groove occurs more often in males, where it is found in up to 40% (Hoshi, 1961; Karsten, 2018).

The larger size of the few comparable dimensions of the Sts 65 hipbone and associated vertebral fragment relative to those of Sts 14, the narrower greater sciatic notch, the shape of the composite arch, and the paraglenoid groove therefore tend to suggest a male sex, although it is unknown whether modern human sex characteristics can be applied to australopithecines (Häusler and Schmid, 1995).

StW 431

The iliac crest apophysis is completely fused although the epiphyseal line is visible on the internal side of StW 431e and StW 431j. Since a faint epiphyseal line may persist throughout life, this suggests that StW 431 was at least a young adult. On the other hand, acetabular rim degeneration is present in the superior sectors of the acetabulum on both sides. This implies hip osteoarthritis, although to a less advanced degree than in Sts 65. StW 431 might therefore well have been an elderly individual.

Although modern human sex characteristics of the hipbone cannot directly be applied to australopithecines (see Häusler and Schmid, 1995), the reconstructed StW 431 hipbone suggests a slightly narrower greater sciatic notch than that of Sts 14 and thus tentatively a male sex. However, because the ischial spine is not preserved, it is not possible to determine the exact shape of the greater sciatic notch.

Pelvic reconstructions

Sts 14 pelvis

Four different reconstructions of the Sts 14 pelvis have been published. Robinson (1972) restored the missing parts with plaster applied directly to the original fossils using visual mirror imaging, glue, and wire. However, he did not supplement the medial portion of the right pubis and he allowed a massive downward tilt of the left pubis. Day (1973) heavily criticized his method and result.

Abitbol's (1995) reconstruction suffered from the same shortcomings. He reconstructed the entire sacrum and the medial portion of the right pubis using casts of Robinson's (1972) reconstruction and added extra plaster to the symphyseal surface of the distorted left pubis to make the symphysis parallel and in the midline. Because the resulting pelvic cavity was platypelloid in all pelvic planes, he postulated a nonrotational birth mechanism with a transverse orientation of the fetal head similar to that described by Tague and Lovejoy (1986) for A.L. 288-1, but unknown in any living mammal.

Haeusler (1992; Häusler and Schmid, 1995) also relied on casts of Robinson's (1972) reconstruction, but he removed all parts that had been added in plaster and separated the various fragments of the pubis along the fracture lines before making a fresh, undistorted reconstruction by mirror imaging. He obtained a round shape of the pelvic canal and the corresponding birth mechanism was surprisingly similar to that of modern human females.

However, the reconstruction of the sacroiliac joint and the resulting posteriorly tilted sacrum and steep orientation of the iliac blades in particular have more recently been criticized by Berge and Goularas (2010). They presented a new virtual three-dimensional model of the Sts 14 pelvis based on computed tomography (CT)

of the original fossils. They reconstructed the sacrum by mirror imaging using five superimposition landmarks. Then they completed the right hipbone with the mirrored anterior superior iliac spine, iliac crest, and pubis from the left side and finally mirrored this composite hipbone to obtain the complete pelvis. Unfortunately, they did not attempt to correct the crushed regions. Because the left (and right) pubis have been pushed into the pelvic cavity during fossilization, the resulting sagittal diameters are too short in Berge and Goularas's (2010) reconstruction and the shape of the pelvic canal too platypelloid. Nevertheless, they concluded that the birth mechanism was essentially human-like except that at the pelvic inlet the fetal head entered in a transverse rather than oblique orientation.

Sts 65 pelvis

Claxton et al. (2016) proposed a virtual reconstruction of a composite pelvis based on Sts 65 and the Sts 14 sacrum. It is, however, questionable, whether this small female sacrum is a suitable substitute for Sts 65. Moreover, a superimposition with the Sts 14 hipbone suggests that the superior public ramus of Sts 65 is only preserved to about half of its length rather than being "nearly complete" (Figure 11.9B). The position and orientation of both the pubic symphysis and the hipbone, therefore, remain unresolved, and they call into question any inferences derived from the reconstruction of Claxton et al. (2016).

StW 431 pelvis

Three different reconstructions of the StW 431 pelvis have been presented so far. Haeusler (2002) based his reconstruction on mirror-image molding using wax and casts of those parts of the pelvis that had been identified by Alun Hughes in 1987. However, only the discovery of the additional fragments StW 431kb, w, s, and y by Kibii and Clarke (2003) provided the crucial information about the sacroiliac joint that allowed correct orientation of the iliac blades.

Based on these new fragments, Kibii and Clarke (2003) restored the missing portion of the right ilium and joined the different fragments using plaster of Paris applied to the original fossils. A virtual 3D reconstruction based on CT scans and mirror-imaging by Haeusler (2006) basically confirmed the findings of Kibii and Clarke (2003). Thus, the StW 431 pelvis seems to be similarly flat to that of extant great apes and flatter than Sts 14 in Berge and Goularas's (2010) reconstruction, with very coronally oriented and laterally flaring iliac blades (see Figure 11.23). This has significant consequences for the lever arms of the

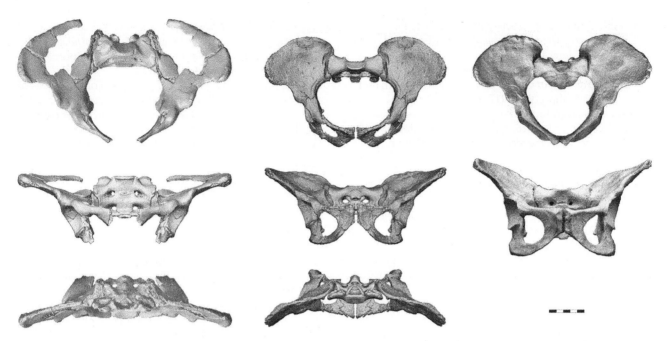

Figure 11.23 Comparison of the reconstruction of the StW 431 pelvis (Haeusler, 2006, left) with that of Sts 14 by Berge and Goularas (2010, middle) and by Häusler (1992, right). Scale bar equals 5 cm. Note the different orientation of the iliac blades and of the pubis.

gluteal muscles and lateral stabilization of the hip during bipedal walking. It remains open whether the divergent morphology of StW 431 and Sts 14 is attributable to the reconstruction of the pelvis, to age differences, sexual dimorphism, or taxonomic differentiation.

Acknowledgments

We would like to thank Carol Ward, Brian Richmond, and Bernhard Zipfel for organizing the Sterkfontein workshop. Access to the fossil collections was granted by Stephany Potze and Francis Thackeray (Ditsong National Museum of Natural History, Pretoria), Bernhard Zipfel, Lee Berger, Beverly Kramer and the late Phillip Tobias (University of the Witwatersrand, South Africa), and Metasebia Endalamaw, Yared Assefa, and Mamitu Yilma (National Museum of Ethiopia, Addis Ababa). We thank the two anonymous reviewers for comments that helped to improve the manuscript. Additional financial support for MH was provided by the Swiss National Science Foundation Grants Nos. 31003A_156299 and 31003A_176319.

References

Abitbol, M.M., 1995. Reconstruction of the Sts 14 (*Australopithecus africanus*) pelvis. American Journal of Physical Anthropology 96, 143–158.

Arsuaga, J.L., 1981. Iliac angular measurements in *Australopithecus*. Journal of Human Evolution 10, 293–302.

Behrensmeyer, A.K., 1978. Taphonomic and ecologic information from bone weathering. Paleobiology 4, 150–162.

Belcastro, M.G., Rastelli, E., Mariotti, V., 2008. Variation of the degree of sacral vertebral body fusion in adulthood in two European modern skeletal collections. American Journal of Physical Anthropology 135, 149–160.

Berge, C., 1994. How did the australopithecines walk? A biomechanical study of the hip and thigh of *Australopithecus afarensis*. Journal of Human Evolution 26, 259–273.

Berge, C., Gommery, D., 1999. Le sacrum de Sterkfontein Sts 14 Q (*Australopithecus africanus*): nouvelles données sur la croissance et sur l'âge osseux du spécimen (hommage à R. Broom et J.T. Robinson). Comptes Rendus de l'Académie des Sciènces de Paris, Sciences de la Terre et des Planètes 329, 227–232.

Berge, C., Goularas, D., 2010. A new reconstruction of Sts 14 pelvis (*Australopithecus africanus*) from computed tomography and three-dimensional modeling techniques. Journal of Human Evolution 58, 262–272.

Bonmatí, A., Arsuaga, J.L., Lorenzo, C., 2008. Revisiting the developmental stage and age-at-death of the "Mrs. Ples" (Sts 5) and Sts 14 specimens from Sterkfontein (South Africa): do they belong to the same individual? Anatomical Record 291, 1707–1722.

Broom, R., Robinson, J.T., 1947. Further remains of the Sterkfontein ape-man, *Plesianthropus*. Nature 160, 430–431.

Broom, R., Robinson, J.T., Schepers, G.W.H., 1950. Sterkfontein ape-man, *Plesianthropus*. Transvaal Museum Memoir 4, 1–117.

Bruzek, J., 2002. A method for visual determination of sex, using the human hip bone. American Journal of Physical Anthropology 117, 157–168.

Claxton, A.G., Hammond, A.S., Romano, J., Oleinik, E., DeSilva, J.M., 2016. Virtual reconstruction of the *Australopithecus africanus* pelvis Sts 65 with implications for obstetrics and locomotion. Journal of Human Evolution 99, 10–24.

D'Anastasio, R., Zipfel, B., Moggi-Cecchi, J., Stanyon, R., Capasso, L., 2009. Possible brucellosis in an early hominin skeleton from Sterkfontein, South Africa. PLoS ONE 4, e6439.

Day, M.H., 1973. Locomotor features of the lower limb in hominids. In: Zuckerman, S. (Ed.), The concepts of human evolution. Academic Press: London, pp. 29–51.

Genovés, S., 1959. L'estimation des différences sexuelles dans l'os coxal; différences métriques et différences morphologiques. Bulletins et Memoires de la Société d'Anthropologie de Paris 10, 3–95.

Haeusler, M., 2002. New insights into the locomotion of *Australopithecus africanus* based on the pelvis. Evolutionary Anthropology 11, 53–57.

Haeusler, M., 2006. 3D-computer assisted reconstruction of the Stw 431 (*Australopithecus africanus*) pelvis. American Journal of Physical Anthropology Suppl. 42, 99.

Haeusler, M., 2019. Spinalpathologies in fossil hominins. In: Been, E., Gomez-Olivencia, A., Kramer, P. (Eds.), Spinal evolution: morphology, function, and pathology of the spine in hominoid evolution. Springer: Berlin, pp. 213–245.

Haeusler, M., Landis, S., Zipfel, B., 2015. Hip joint osteoarthritis in the MLD 46 (*Australopithecus africanus*) proximal femur. American Journal of Physical Anthropology Suppl. 60, 157.

Haeusler, M., McHenry, H.M., 2004. Body proportions of *Homo habilis* reviewed. Journal of Human Evolution 46, 433–465.

Haeusler, M., McHenry, H.M., 2007. Evolutionary reversals of limb proportions in early hominids? Evidence from KNM-ER 3735 (*Homo habilis*). Journal of Human Evolution 53, 383–405.

Hammond, A.S., Plavcan, J.M., Ward, C.V., 2013. Precision and accuracy of acetabular size measures in fragmentary hominin pelves obtained using sphere-fitting techniques. American Journal of Physical Anthropology 150, 565–578.

Häusler, M., 1992. Rekonstruktion des Beckens von Sts 14 (*Australopithecus africanus*). Anthropologisches Institut und Museum der Universität Zürich-Irchel.

Häusler, M., 2001. New insights into the locomotion of *Australopithecus africanus*: Implications of the partial skeleton Stw 431 (Sterkfontein, South Africa). Ph.D. Dissertation, Universität Zürich.

Häusler, M., Berger, L.R., 2001. Stw 441/465: a new fragmentary ilium of a small-bodied *Australopithecus africanus* from Sterkfontein, South Africa. Journal of Human Evolution 40, 411–417.

Häusler, M., Schmid, P., 1995. Comparison of the pelves of Sts 14 and AL 288-1: implications for birth and sexual dimorphism in australopithecines. Journal of Human Evolution 29, 363–383.

Haynes, G., 1980. Evidence of carnivore gnawing on Pleistocene and recent mammalian Bbones. Paleobiology 6, 341–351.

Hoshi, H., 1961. On the preauricular groove in the Japanese pelvis with special reference to the sex difference. Okajimas Folia Anatomica Japonica 37, 259–269.

Johanson, D.C., Lovejoy, O.C., Kimbel, W.H., White, T.D., Ward, S.C., Bush, M.E., Latimer, B.M., Coppens, Y., 1982. Morphology of the Pliocene partial skeleton (A.L. 288-1) from the Hadar Formation, Ethiopia. American Journal of Physical Anthropology 57, 403–451.

Jurmain, R.D., 2000. Degenerative joint disease in African great apes: an evolutionary perspective. Journal of Human Evolution 39, 185–203.

Karsten, J.K., 2018. A test of the preauricular sulcus as an indicator of sex. American Journal of Physical Anthropology 165, 604–608.

Kibii, J.M., Churchill, S.E., Schmid, P., Carlson, K.J., Reed, N.D., de Ruiter, D.J., Berger, L.R., 2011. A partial pelvis of *Australopithecus sediba*. Science 333, 1407–1411.

Kibii, J.M., Clarke, R.J., 2003. A reconstruction of the Stw 431 *Australopithecus* pelvis based on newly discovered fragments. South African Journal of Science 99, 225–226.

Knussmann, R., 1988. Anthropologie: Handbuch der vergleichenden Biologie des Menschen. Zugleich 4. Aufl. des Lehrbuchs der Anthropologie, begründet von Rudolf Martin. Fischer: Stuttgart.

Kotwicki, T., 2008. Improved accuracy in Risser sign grading with lateral spinal radiography. European Spine Journal 17, 1676–1685.

Leunig, M., Beck, M., Woo, A., Dora, C., Kerboull, M., Ganz, R., 2003. Acetabular rim degeneration: a constant finding in the aged hip. Clinical Orthopedics and Related Research 417, 201–207.

MacLatchy, L.M., 1996. Another look at the australopithecine hip. Journal of Human Evolution 31, 455–476.

Martin, R., Saller, K., 1957. Lehrbuch der Anthropologie 1, 433–476.

McHenry, H.M., 1975a. A new pelvic fragment from Swartkrans and the relationship between robust and gracile australopithecines. American Journal of Physical Anthropology 43, 245–261.

McHenry, H.M., 1975b. Biomechanical interpretation of the early hominid hip. Journal of Human Evolution 4, 343–355.

Phenice, T.W., 1969. A newly developed visual method of sexing the os pubis. American Journal of Physical Anthropology 30, 297–302.

Risser, J.C., 1958. The iliac apophysis; an invaluable sign in the management of scoliosis. Clinical Orthopedics 11, 111–119.

Robinson, J.T., 1972. Early hominid posture and locomotion. University of Chicago Press: Chicago.

Ruff, C.B., 2010. Body size and body shape in early hominins—implications of the Gona pelvis. Journal of Human Evolution 58, 166–178.

Schwabe, R., 1933. Rückbildung der Bandscheibe im menschlichen Kreuzbein. Virchows Archiv 287, 651–713.

Stern, J.T., Susman, R.L., 1983. The locomotor anatomy of Australopithecus afarensis. American Journal of Physical Anthropology 60, 279–317.

Tague, R.G., Lovejoy, C.O., 1986. The obstetric pelvis of A.L. 288-1 (Lucy). Journal of Human Evolution 15, 237–256.

Tobias, P.V., 1992. New researches at Sterkfontein and Taung with a note on Piltdown and its relevance to the history of palaeo-anthropology. Transactions of the Royal Society of South Africa 48, 1–14.

Wolpoff, M.H., 1973. Posterior tooth size, body size, and diet in South African gracile australopithecines. American Journal of Physical Anthropology 39, 375–393.

12

Femur

Jeremy M. DeSilva and Mark W. Grabowski

Fossil femora from Sterkfontein, South Africa have featured prominently in the history of human evolutionary thought and the recognition of *Australopithecus* as a bipedal hominin (Broom, 1938; Le Gros Clark, 1947; Broom and Robinson, 1949; Tobias, 1998). Femora discovered at the Sterkfontein Type Site by Robert Broom were described and functionally interpreted by Robinson (1972). Additional work by C. Owen Lovejoy (Lovejoy and Heiple, 1970; Heiple and Lovejoy, 1971; Lovejoy et al., 1973) helped frame the still-ongoing debate about hip function in *Australopithecus*. Since that time, many additional fossil femora have been recovered at Sterkfontein (Appendix I). While this is the first thorough description of these femora, others have commented on the state of preservation and morphology of these bones. Most notably, Harmon (2009a) detailed the preservation and morphology of the proximal femora and provided a thorough comparative analysis. Likewise, DeGusta (2004) partially described the femora, though only in relation to the Bouri femur from the Middle Awash, Ethiopia. Here, we provide a detailed description of the Sterkfontein hominin femora, and additional comparative and functional interpretations of these fossils.

Descriptions

Felid femora

In a study of the MLD 46 femur, Reed et al. (1993) noted how similar fragmentary hominin femora can be to fossil felids and developed a methodology for distinguishing them. Most notably, the most inferior aspect of the femoral head is rounded off in hominin femora but presents a hook-like anatomy in felids. Additionally, the fovea capitis is located more posteriorly and is not as centrally located in felid femoral heads compared with hominins. There are several proximal femora originally identified as hominin that are more likely felids (Reed et al., 1993; DeSilva, 2011). These include StW 30a, StW 30b, and StW 31 and are therefore not described in anatomical detail (see Appendix I for locus). The following specimens are all identified as hominin.

Sts 14—Left proximal femur

Preservation. Sts 14t is a poorly preserved left femoral neck and proximal shaft, preserving about 200 mm of the total length of the femur (Figure 12.1; Table 12.1). The bone is crushed and consists of glued-together fragments presumably held in the position of their discovery. The proximal most ~80 mm of the bone is badly cracked and held in place with glue but appears to have been recovered and preserved with the fragments in proper anatomical position. However, the more distal pieces are terribly distorted and result in the shaft bowing anteriorly in the sagittal plane and laterally in the coronal plane. Cortical bone along the shaft is crushed and flaking. The femoral head and posteromedial aspect of the neck are not preserved and have been reconstructed with plaster. Robinson (1972) reports that this reconstruction was made by R. Broom and based on part on a natural cast of the head preserved in breccia surrounding the fossil. Most of the femoral neck is present, but there is significant cracking and abrasion to the surface. The greater and lesser trochanters are sheared away. It is difficult to ascertain any shaft anatomy due to the extensive damage to the fossil.

Morphology. The femoral head is reconstructed with plaster to be ~28.8 mm superoinferiorly and ~29 mm

Jeremy M. DeSilva and Mark W. Grabowski, *Femur.* In: *Hominin Postcranial Remains from Sterkfontein, South Africa, 1936–1995.* Edited by: Bernhard Zipfel, Brian G. Richmond, and Carol V. Ward, Oxford University Press (2020). © Oxford University Press.
DOI: 10.1093/oso/9780197507667.003.0012

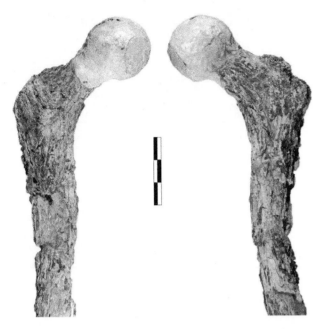

Figure 12.1 Sts 14t in posterior (left) and anterior (right) view. Notice that the femoral head and part of the neck is reconstructed with plaster and there is severe damage to the femoral shaft. Scale bar in cm.

anteroposteriorly. Robinson (1972) reports that the plaster head is 31 mm—a measurement that can be replicated by measuring the head diagonally in the anteroinferior to posterosuperior direction. Based on the acetabulum of Sts 14, the head is likely to have been ~30.8 mm (Ruff, 2010). The neck is superoinferiorly tall: 22.6 mm superoinferiorly and approximately 14.2 mm anteroposteriorly, though the neck appears crushed and this is likely a minimum value. The femoral neck appears long, though it is not directly measurable given the absence of the head/neck junction. However, given that the head and neck reconstruction were based on a natural cast in the breccia, a morphological neck length of 27.9 mm can be estimated. Robinson (1972) estimates the neck-shaft angle to be 118°. The trochanteric fossa is detectable just medial to the damaged surface for the greater trochanter. A portion of the intertrochanteric crest is palpable between the damaged trochanters. An obturator externus groove is not detectable. Robinson (1972) reports a faint linea aspera, spiral line, and gluteal ridge, but the specimen is tremendously damaged, fragmented, and distorted, so these features are unclear.

Table 12.1 Proximal femur and midshaft measurements. All measurements in mm.

	Head diameter	Neck height (SI)	Neck breadth (AP)	Morphological neck length	Subtrochanteric ML	Subtrochanteric AP	Midshaft ML	Midshaft AP
Sts 14	30.8	22.6	14.2 (min)	27.9	–	–	–	–
StW 25	30.2	20.2	–	–	–	–	–	–
StW 99/100	(39.1)	27.1	19.1	44.0	34.6	26.1	28.2	24.1
StW 121	–	–	–	–	–	–	25.2	22.8
StW 311	35.7	26.9	19.3	–	–	–	–	–
StW 361	29.1	–	–	–	–	–	–	–
StW 392	31.5	19.2	–	–	–	–	–	–
StW 403	(32.7)	24.4	16.7	–	–	–	–	–
StW 479	(32.7)	23.5	17.4	32.6	–	–	–	–
StW 501	33.0	23.4	17.3	–	–	–	–	–
StW 522	30.4	20.6	15.2	31.0	–	–	–	–
StW 527	32.2	–	–	–	–	–	–	–

Note: (femoral head diameters) in parentheses are taken from Ward et al. (2015) who used sphere-fitting techniques to approximate this value on eroded femoral heads.

StW 25—Right proximal femur

Preservation. StW 25 is a fragmentary right femoral head and partial neck, primarily preserving the posterior part of the bone (Figure 12.2; Table 12.1). Preserved overall is 39 mm mediolaterally and 30.7 mm superoinferiorly. The anterior aspect of the neck and part of the head are sheared away. The superior part of the femoral head is

StW 25   StW 311   StW 361   StW 392   StW 527

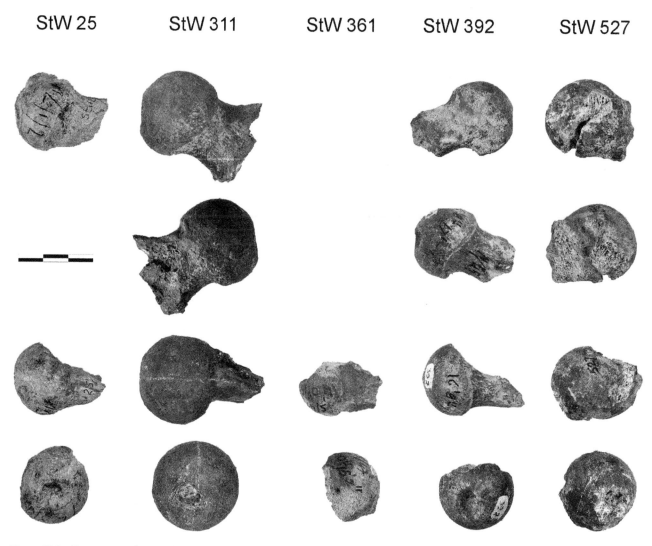

Figure 12.2 Fragmentary femoral heads of (from left to right): StW 25, StW 311, StW 361, StW 392, and StW 527. Fossils are positioned (from top to bottom) in anterior, posterior, superior, and medial views. As discussed in the text, StW 311 is from Member 5 and may be from *A. robustus*, whereas the other fossils all derive from Sterkfontein Member 4. Scale bar in cm.

cracked and a 14.1 mm mediolateral fragment of the head is displaced superiorly by matrix, artificially expanding the head size. The crack extends medially and inferiorly under the fovea capitis. There is some erosion to the posterior surface of the head, including a 12.2 mm-long, ~3 mm-wide gash. The break in the femoral neck is obliquely oriented.

Morphology. The head is roughly 30.2 mm superoinferiorly. The anteroposterior head diameter is 30.4 mm. Failure to account for the displaced fragment of bone on the superior head would result in a superoinferior head diameter closer to 32.0 mm. The fovea is large, deep, and slightly offset posteriorly and inferiorly. It is 6.9 mm anteroposteriorly and 5.9 mm superoinferiorly. The neck height is 20.2 mm superoinferiorly. Superiorly, the ~13 mm mediolateral of preserved neck slopes gently from the head. Posteriorly, a lip of articular bone from the head extends

onto the neck and grades smoothly into it. The cross-section of the neck reveals thick inferior and thin superior cortex.

StW 99/100—Right proximal femur

Preservation. StW 99/100 is a large right femur preserving the proximal 253 mm of the head, neck, and shaft (Figure 12.3; Table 12.1). It consists of three pieces that have been reattached: (1) a femoral head; (2) 67 mm superoinferiorly of the neck and trochanteric region; and (3) the remaining ~200 mm of the shaft. The first two pieces conjoin cleanly at the head-neck junction. The second and third pieces conjoin cleanly anteriorly, but there is a large gap separating them posteriorly. There is erosion around the perimeter of the head; the articular surface of the head is only preserved anteromedially. The neck is quite well preserved. The greater trochanter

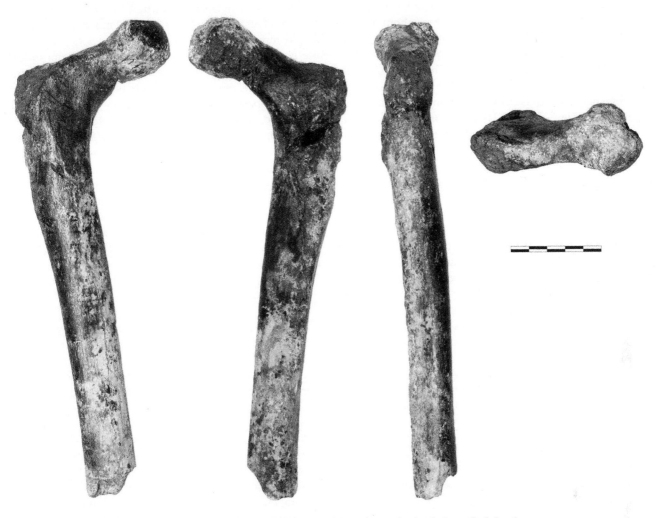

Figure 12.3 StW 99/100 in anterior (bottom left), lateral (bottom right), and superior (top) views. Scale bar in cm.

is sheared away. The intertrochanteric crest is not preserved; there is a large gap between the lesser and greater trochanters where the larger two fragments of the fossil have been conjoined. Anteriorly, there is a small (~8 mm) triangular-shaped piece of bone missing where the larger two fragments of the fossil conjoin. The shaft is reasonably well preserved, though there is some cortical flaking along the posterior aspect of the distal shaft. Anteriorly, there is a longitudinal crack running the length of the distal shaft piece.

Morphology. The minimum anteroposterior femoral head diameter is 35.5 mm, though this is an underestimate given damage to the posterior surface of the head. Sphere-fitting techniques yield a femoral head diameter of roughly 39.1 mm (Ward et al., 2015), slightly larger than the estimated FHD of 38 mm (Ruff and Higgins, 2013) and slightly smaller than the 39.7 mm estimate (DeSilva et al., 2013). The neck is anteroposteriorly compressed and is 27.1 mm superoinferiorly and 19.1 mm anteroposteriorly. Anteriorly, the neck is smooth with no

evidence for an intertrochanteric line. The posterior neck is smooth and laterally preserves some of the trochanteric fossa. Just medial to the fossa is a palpable obturator externus groove that continues inferomedially across the neck of the femur. The morphological neck length is an estimated 44 mm. While there is no intertrochanteric line, a weak spiral line forms along the junction of the inferior femoral neck and medial shaft and merges with the lesser trochanter medially. A prominent pectineal line arches inferolaterally from the lesser trochanter and continues ~44 mm until it meets the gluteal line. Laterally, there is a prominent third trochanter/hypotrochanteric fossa complex which continues distally as a line that meets and then runs parallel with the pectineal line. This entire V-shaped region bounded by the pectineal line medially and the gluteal line laterally bulges from the posterior surface of the bone. Just inferior to the lesser trochanter, the subtrochanteric dimensions of the bone are 34.6 mm mediolaterally and 26.1 mm anteroposteriorly. Distally, the pectineal line and gluteal line merge into a prominent linea aspera. The shaft is mediolaterally expanded throughout and at the point of break in the distal shaft, the bone is 28 mm mediolaterally and 24.3 mm anteroposteriorly. Based on estimates of total length, the midshaft mediolateral width is between 28 mm and 28.4 mm; the anteroposterior midshaft breadth is between 23.8 mm and 24.4 mm. In the lateral view, the shaft is slightly anteriorly bowed. The cortical thickness at the point of break in the shaft is 5.5 mm anteriorly, 6.6 mm posteriorly, 6.9 mm medially, and 7.1 mm laterally. Relative to the linea aspera, the head and neck of StW 99 are anteriorly angled.

StW 121—Right femoral shaft

Preservation. StW 121 is a distal right femur, preserving 204.1 mm from slightly above the midshaft to just superior to where the condyles would have been (Figure 12.4; Table 12.1). The break near midshaft is obliquely angled so that more of the posterior surface than anterior surface is preserved, and shows little signs of wear on the broken surface. The posterolateral aspect of the surface near the break is slightly worn. The cortical bone of the shaft is generally well preserved, with some localized erosion. The distal end of the shaft is very worn, particularly on the anterior portion, where none of the patellar surface is preserved. The epicondyles and condyles have not been preserved. Because the anterior surface is worn this gives the distal end an angled appearance when viewed from the side.

Morphology. The shaft is cylindrical, with notable extension toward the posterior side for the linea aspera and posterolateral side. There is a strong linea aspera that begins below approximately midshaft and divides into distinct medial and lateral supracondylar lines. In anatomical view there is a strong lateral bow of the shaft. The minimum mediolateral width of the cortex mediolaterally is 8.9 mm at the proximal break. At approximately midshaft the mediolateral width is 25.2 mm, anteroposterior width is 22.8 mm. The mediolateral width of the distal end, which is notably worn on the lateral side is 42.5 mm. The width of the shaft at the distal end at the last portion of the preserved surface is 20.1 mm. The cortex measures 9.3 mm at the medial side and 9.6 mm at the lateral side.

StW 311—Right proximal femur

Preservation. StW 311 is a right femoral head and neck (Figure 12.2; Table 12.1). The head is perfectly preserved. Only ~26 mm mediolateral of the neck is present, terminating in a jagged break. The posterior surface of the neck is well preserved; the anterior is mostly sheared away exposing some trabecular bone. The inferomedial portion of the neck is still present. The neck is broken diagonally,

roughly at the neck-shaft junction through the intertrochanteric crest. Preserved length is 52 mm mediolaterally and 49 mm superoinferiorly.

Morphology. The femoral head diameter is 35.7 mm superoinferiorly and 35.6 mm anteroposteriorly. The fovea capitis is large, measuring 10.3 mm both superoinferiorly and anteroposteriorly. The head is large and globular—it is pinched off from the neck, except anteriorly, where it grades smoothly from the articular surface of the head onto the neck. The neck is 26.9 mm superoinferiorly and an estimated 19.3 mm anteroposteriorly. The preserved bone along the superior aspect of the neck slopes inferiorly from the head and angles superiorly, qualitatively suggesting a relatively short femoral neck. The posterior neck is smooth and terminates inferiorly with a small elevation that is likely the most proximal part of the lesser trochanter.

StW 361—Proximal femur

Preservation. StW 361 is a poorly preserved head and fragmentary neck of a likely right subadult femur (Figure 12.2; Table 12.1). The anterior aspect of the femoral head is preserved, though there are patches of weathering anteriorly and superiorly; the posterior femoral head has been sheared away, exposing internal trabecular bone. A very small portion of the inferior neck is preserved. Overall, StW 361 measures 35.4 mm mediolaterally.

Morphology. The superoinferior height of the femoral head is 29.1 mm. Anteriorly, a prominent epiphyseal line is present indicating recent fusion of the capital femoral epiphysis and subadult status of this individual. Posteriorly, there is considerable damage, but the fovea can be identified and it is positioned inferior and posterior to the midpoint of the head.

StW 392—Right proximal femur

Preservation. StW 392 is a right partial femoral head and neck, preserving most of the head and posterior portion of the neck extending laterally 23.3 mm (Figure 12.2; Table 12.1). A small portion of cortex (6.5 mm) has flaked from the lateral extent of the neck posteriorly, but otherwise the cortex is very well-preserved. The anterior aspect of the head and neck are sheared away, exposing trabecular networks. Total preserved dimensions are ~43 mm mediolaterally and ~37 mm superoinferiorly.

Morphology. The head is 31.5 mm superoinferiorly. It is offset from the neck inferiorly, but grades into the neck superoposteriorly. The fovea is offset posteriorly and somewhat inferiorly and is large, measuring 7.8 mm superoinferiorly and 10.1 mm anteroposteriorly. There is a faint epiphyseal line detectable posteriorly and inferiorly, suggesting that this femur belonged to a young adult.

Figure 12.4 StW 121 in (from left to right) anterior, posterior, and lateral views. Scale bar in cm.

The neck height superoinferiorly is 19.2 mm. Inferiorly, the neck is strongly convex, and pillar-like. There is a small piece of cortex preserved anteroinferiorly, allowing for a 13.7 mm minimum anteroposterior neck width, though this likely underestimates the true neck width. The oblique break through the neck reveals thick cortex inferiorly and thinner cortex superiorly.

StW 403—Right proximal femur

Preservation. StW 403 is a partial right femoral head and neck preserving 44.2 mm mediolaterally and 55.2 mm superoinferiorly (Figure 12.5). The entire inferior and much of the anterior aspect of the head is damaged, revealing trabecular bone and making any caliper-based femoral head diameter measurements suspect. Superiorly and posteriorly, the head is well preserved. The neck is

well preserved but is sheared medial to the intertrochanteric crest. Anteriorly, there is a noticeable oval impression at the head-neck junction that may be carnivore damage (Harmon, 2009a).

Morphology. The preserved anteroposterior head diameter is 31.1 mm, which is a minimum. Fitting a sphere to the head results in a more likely 32.7 mm (Ward et al., 2015). The fovea is slightly offset posteriorly and is 8.8 mm anteroposteriorly and 7.6 mm superoinferiorly. The superior rim of the fovea is elevated slightly above the articular surface of the femoral head. Posteriorly and somewhat superiorly, the head grades directly into the neck, without an obvious distinction between the two. The neck height is 24.4 mm and width 16.7 mm, making the neck superoinferiorly tall. Posteriorly, there is a very small indentation along the neck that may be the medial extent of the obturator externus groove.

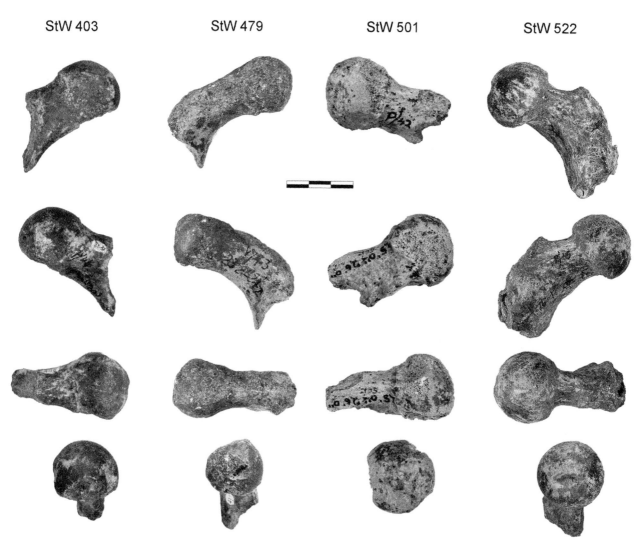

StW 403   StW 479   StW 501   StW 522

Figure 12.5 Femoral heads and necks from Sterkfontein Member 4 (from left to right): StW 403, StW 479, StW 501, and StW 522. Fossils are positioned (from top to bottom) in anterior, posterior, superior, and medial views. Scale bar in cm.

StW 448—Right femoral shaft

Preservation. StW 448 is roughly the distal third of a right femoral shaft, preserving 111.1 mm (Figure 12.6; Table 12.1). Proximally, where the shaft is broken, two spikes of shaft are located anteriorly and posteriorly. The surface of the bone has a series of cracks running superoinferiorly on the anterior and medial sides. There is a strong distinction between the cortical and trabecular bone on the distal break.

Morphology. There is a strong linea aspera that gives the shaft a distinctly elliptical shape and distally separates into slight lateral and medial supracondylar lines. The mediolateral width at the proximal end is 22.8 mm; anteroposterior width is 22.3 mm. The cortex at the proximal end is 7.7 mm laterally and 7.1 mm medially.

StW 479—Right proximal femur

Preservation. StW 479 is a partial right femoral head, neck, and part of the proximomedial shaft preserving 56 mm mediolaterally and 63 mm superoinferiorly (Figure 12.5; Table 12.1). The anterior, superior, and inferior aspect of the femoral head is eroded and the superior neck is eroded, exposing underlying trabecular bone. The fossil is sheared along the intertrochanteric crest, preserving only the most proximal tip of the lesser trochanter. The greater trochanter is not preserved.

Figure 12.6 StW 448 in (from left to right) anterior, medial, and superior views. Notice in superior view the well-developed linea aspera. Scale bar in cm.

Cortical and trabecular bone are exposed at this break in the neck-shaft.

Morphology. The maximum anteroposterior width of the femoral head is 28.3 mm, which is a minimum head diameter. Sphere-fitting techniques estimate a diameter of 32.7 mm (Ward et al., 2015). The fovea is offset posteriorly and inferiorly and is 8.8 mm anteroposteriorly and 5.3 mm superoinferiorly. There is a faint epiphyseal line along the head-neck junction posteriorly, suggesting this femur belonged to a young adult. Posteriorly and superiorly, the head grades onto the neck without a clear distinction between the two. There is damage to the superior neck, though a small preserved portion of cortex allows for an estimated superoinferior height measurement of 23.5 mm. The anteroposterior width is 17.4 mm. Anteriorly, the neck is smooth with no evidence of an intertrochanteric line at the neck-shaft junction. Posteriorly, there is a slight but palpable depression marking the groove made by the tendon of the obturator externus muscle. A raised tubercle at the midline of the femoral neck is likely the quadrate tubercle. Using this landmark to delineate the likely path of the intertrochanteric crest, the morphological neck length of the femur is 32.6 mm. A raised pillar of bone at the inferior neck-shaft junction demarcates the proximal portion of the lesser trochanter.

StW 501—Left proximal femur

Preservation. StW 501 is a fragmentary and eroded left femoral head and neck preserving 53 mm mediolaterally and 48 mm superoinferiorly (Figure 12.5; Table 12.1). The cortical shell of the head is gone posteriorly and eroded anteriorly and superiorly at the head-neck junction, exposing underlying trabecular bone. There is pitting damage around the fovea capitis. Posteriorly, there is considerable damage along the neck, with only some cortex preserved superiorly. Anteriorly, a small (17 mm x 12 mm) piece of cortical bone has flaked away exposing underlying trabecular bone, and a crack runs mediolaterally from the head to this damaged cortical shell. The total neck length preserved is 32.6 mm.

Morphology. The minimum preserved head diameter is 32.5 mm; with an estimated actual size of ~33 mm. The fovea appears to be more centrally located on this specimen compared with others and is approximately 7.5 mm anteroposteriorly and 7.9 mm superoinferiorly. Enough of the neck is preserved to estimate the superoinferior height of 23.4 mm. The width is approximately 17.3 mm, estimated because of damage to the anterior neck. The superior cortext of the neck is exposed and is notably thin.

StW 522—Left proximal femur

Preservation. StW 522 is a left proximal femur preserving the complete head, most of the neck, and a small portion of the proximal shaft (Figure 12.5; Table 12.1). The greater trochanter and lateral portion of the proximal shaft are not preserved at all; underlying trabecular bone is exposed. The lesser trochanter has been sheared away, exposing some underlying trabecular bone. Only the most proximomedial portion of the femoral shaft is preserved. There is abrasion caused either by carnivores or perhaps from mechanical preparation along the anterior portion of the head and neck. The total preserved amount of bone is 59.5 mm mediolaterally and 62.1 mm superoinferiorly.

Morphology. The femoral head is quite globular and strongly pinched off from the femoral neck except posteriorly and superiorly where they grade together. The articular surface of the femoral head, as viewed superiorly, is in a roughly neutral orientation. The femoral head is 30.9 mm anteroposteriorly and 30.4 mm superoinferiorly. The fovea is offset inferior and posterior and is 8.7 mm anteroposteriorly and 5.1 mm superoinferiorly. Superiorly, the notch of the femoral neck is preserved, including the far medial rim of the trochanteric fossa and the insertion region of the obturator externus muscle. The minimum neck height is 20.6 mm superoinferiorly; width is 15.2 mm anteroposteriorly. Posteriorly, there is a small portion of the intertrochanteric crest preserved, proximal and medial to which is a distinct obturator externus groove, which is 2.8 mm wide superoinferiorly. The morphological neck length from the lateral rim of the femoral head to the intertrochanteric crest is ~31 mm.

StW 527—Proximal femur

Preservation. StW 527 is a fragmentary and badly damaged femoral head, possibly from the left side (Figure 12.2; Table 12.1). There is a matrix filled crack that separates the medial two-thirds of the bone from the rest of the head. The two pieces remain physically connected, but they have shifted relative to one another. There is erosion at the junction between these two pieces as well, exposing some trabecular struts. Some cortex has flaked away from the rim of the fovea. Only a very small portion of the head-neck junction is present. Part of the neck is sheared exposing a network of trabecular bone. Overall, 38.6 mm mediolateral is preserved.

Morphology. The femoral head diameter, taken superoinferiorly on the larger preserved fragment, is 32.2 mm. The fovea is positioned inferior and posterior and is 9.2 mm anteroposteriorly and 7.5 mm superoinferiorly.

TM 1513—Left distal femur

Preservation. TM 1513 is 73.2 mm of the distal end of the left femur, preserved to about 50 mm above the patellar surface where the shaft is broken off at an angle to produce a sharp point of bone on the lateral side (Figure 12.7;

Table 12.1). The lateral aspect of the bone has been sheared away, removing the lateral epicondyle and a small portion of the lateral patellar surface. A small amount of abrasion is present on the top of the medial condyle. The surface cortex shows a fair amount of abrasion and pitting with sections of smooth cortical bone remaining. The popliteal surface is generally well preserved.

Morphology. The adductor tubercle is small but sharp and pronounced. Medially, just inferior to the medial epicondyle, there is a distinct groove running along the rim of the articular surface of the medial condyle. The condyles are asymmetrical with the medial side significantly larger than the lateral. The articular surfaces of both are well defined. The anteroposterior depth of the medial condyle is 45 mm; the lateral is damaged anteriorly but is likely about the same, estimated here to be 44 mm. The maximum mediolateral width of the medial condyle is 22 mm. The maximum width of the lateral condyle is 21.3 mm. The intercondylar breadth is 13 mm and the height of the intercondylar notch taken from a line joining the most the inferior point of the medial and lateral condyles to the top of the notch is 12.2 mm. The medial wall of the intercondylar notch indents slightly onto the medial condyle and marks the insertion for the posterior cruciate ligament. Along the posterior aspect of the lateral wall of the intercondylar notch is a distinct pit for the anterior cruciate ligament. The patellar surface depth is 2.6 mm in depth and 22.9 mm in height. There is some distinction between the cortical and trabecular bone at the break, with the cortex measuring 6.2 mm at its widest point.

Sts 34—Right distal femur

Preservation. Sts 34 is 69.3 mm of the distal end of a right femur, preserved to about 35.5 mm above the patellar surface where the shaft is cleanly broken off at an oblique angle with a greater amount of anterior than posterior surface preserved (Figure 12.8; Table 12.1). Preservation is good, with the exception of some surface pitting and a chip taken out of the surface superior to the patella surface. A large portion of the posterior surface of the medial condyle is missing and there is erosion to the posterior surface of the lateral condyle. The popliteal surface shows signs of abrasion and pitting.

Morphology. Posterolaterally, there is a small tubercle for plantaris superior to a larger insertion for the lateral head of gastrocnemius. Along the inferior rim of the lateral surface is a groove for the popliteus. Medially, there is a slight adductor tubercle. Both medial and lateral epicondyles are pronounced. The bicondylar breadth is 62.7 mm. The condyles are asymmetrical with the medial side significantly larger than the lateral. There is a subtle groove on the medial side of the distal side separating the patellar from condylar surface. The articular surfaces are not well

Femur

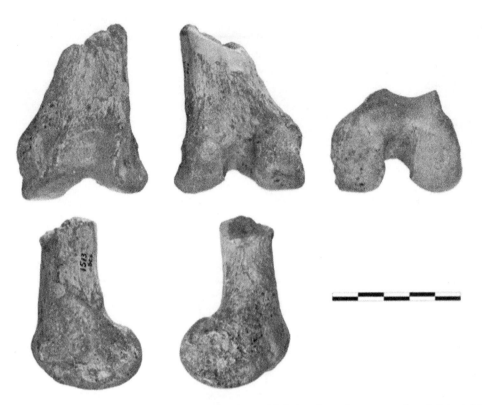

Figure 12.7 TM 1513 in (top row from left to right) anterior, posterior, and inferior views; (bottom row from left to right) lateral and medial views. Scale bar in cm.

Table 12.2 Distal femur measurements. All measurements in mm.

	Biepicondylar breadth	Lat. condyle height[a]	Lat. condyle width	Med. condyle height	Med. condyle width	Intercondylar height	Intercondylar width
TM 1513	55.7	44.0	21.3	45.0	22.0	12.2	13.0
Sts 34	62.7	45.1 (min)	19.6 (est)	51.7 (est.)	21.9	16.0	16.5
StW 129	–	47.3	20.8 (min)	–	–	–	–
StW 318	–	41.0	18.5	–	–	–	–

[a] Includes lateral patellar lip

defined. The anteroposterior depth of the condyles cannot be determined due to breakage, though the lateral condyle from the eroded posterior surface to the rim of the lateral patellar lip is 45.1 mm. Robinson (1972) estimates the medial condylar height to be 51.7 mm. The maximum mediolateral width of the medial condyle is 21.9 mm. The width of the lateral condyle is 19.3 mm minimum. The intercondylar breadth is 16.5 mm. The height of the intercondylar notch is an estimated 16 mm. A palpable indentation along the medial edge of the intercondylar notch delineates the insertion for the posterior cruciate ligament. The patellar surface depth is 5.5 mm and 24.2 mm in height measured from its most distal to most proximal point.

StW 129—Left distal femur

Preservation. StW 129 is a left lateral condyle and distal portion of the femoral shaft. It consists of two parts that have been neatly glued together: a fragment of the shaft and the remaining distal shaft and partial lateral condyle (Figure 12.9; Table 12.1). The lateroposterior portion of the distal condyle is missing. The medial condyle is sheared away. There are three deep grooves running lengthwise on the popliteal surface, as well as at the break between the two portions. The anterior surface is well preserved with a portion of the patellar surface.

Morphology. There is a distinct attachment for the lateral head of the gastrocnemius and a groove for the tendon of popliteus. The lateral epicondyle is distinct and well

219

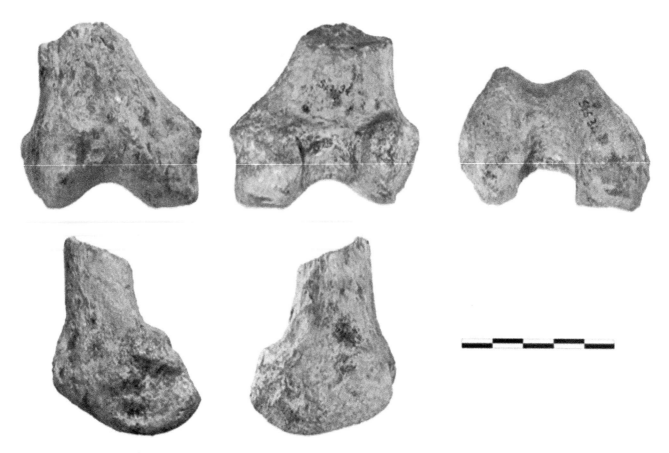

Figure 12.8 Sts 34 in (top row from left to right) anterior, posterior, and inferior views; (bottom row from left to right) medial and lateral views. Scale bar in cm.

preserved. The anteroposterior depth of the lateral condyle is 47.3 mm. The lateral condyle width is 20.8 mm mediolaterally, though there is some erosion making this a minimum. The superoinferior height of the articular surface of the condyle is 37.5 mm measured from the separation of the condylar and patellar articular surface demarcated by a shallow incurving of the lateral margin. In the distal view, the lateral patellar lip rises strongly from the patellar surface. The cortex measures 3.5 mm at the lateral side. Overall dimension of the specimen is 59.1 mm.

StW 318—Right distal femur

Preservation. StW 318 is a fragment of the lateral distal end of the right femur that includes the lateral condyle and epicondyle and more than half of the lateral portion of the patellar surface (Figure 12.9; Table 12.1). All portions are fairly well preserved, with some wear on the lateral edge of the surface with exposed trabecular bone. The most proximal portion is represented by a shaft of bone. The lateral side including the epicondyle is well-preserved.
Morphology. Laterally, there are deep pits for the popliteus and lateral head of the gastrocnemius muscle. The

anteroposterior height of the preserved lateral edge from the lateral patellar surface to the lateral condyle is 41 mm. The maximum mediolateral width of the lateral condyle is 18.5 mm, though the lateral margin is somewhat worn. The superoinferior height of the articular surface of the condyle is 24.7 mm measured from the separation of the condylar and patellar articular surface demarcated by a shallow incurving of the lateral margin—the total height of the lateral condyle including the lateral patellar lip is 41 mm. In the inferior view, the lateral patellar lip rises above the patellar surface. The cortex measures 1.5 mm at the lateral side. The overall length dimension of this specimen is 52.4 mm.

Comparative morphology and interpretation

Assignation to Member 4 or 5

Most of the femora discussed in this chapter were recovered from deposits clearly identified as Member 4. The proper geological context of two femora—StW 99/100 and StW 311—are discussed in more detail here. StW 99/100 is listed in the fossil catalogue in the hominin

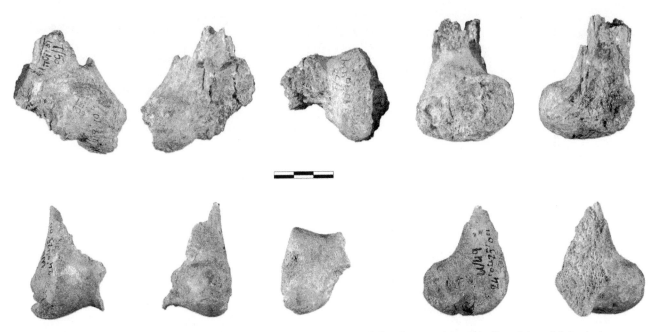

Figure 12.9 Distal femora StW 129 (top row) and StW 318 (bottom row). Both fossils are positioned in (from left to right) anterior, posterior, inferior, lateral, and medial views. Scale bar in cm.

vault at the Evolutionary Studies Institute as deriving from Member 5, though we treat it here as coming from Member 4. StW 99/100 was recovered from grid W/45, depth 6'10"–7'10" [Appendix I]. This grid falls squarely within the boundary of Member 4 as determined by Kuman and Clarke (2000). Furthermore, Moggi-Cecchi et al. (2006) describe craniodental material from the same grid and similar depth—StW 116—and another fossil from similar depth and nearby grid, StW 104. StW 104 and 116 are from juveniles and are therefore not associated with StW 99/100 (an adult). However, these fossils are described as coming from Member 4, and are included in the *A. africanus* hypodigm (Moggi-Cecchi et al., 2006), suggesting that StW 99/100 was also recovered from Member 4.

Following Kuman and Clarke (2000), there is one fossil (StW 311) that derives from Member 5 East and could belong to early *Homo* or to *A. robustus*. While it remains possible that StW 311 is *Homo*, morphologically, it is quite similar to femora from Swartkrans that have been assigned to the robust australopiths. These include SK 82, SK 97, SK 14024, SKX 19, SKX 3121, and SKT1/LB-2. In a comparison of the femoral head to femoral neck height of hominin femora, StW 311 is most like three *A. robustus* femora, making this assignation reasonable (Figure 12.10). Like other *A. robustus* femora (and unlike early eastern African *Homo* femora), StW 311 has an anteroposteriorly compressed (or superoinferiorly tall) femoral neck (Figure 12.11). Additionally, the ratio of the head diameter to neck breadth of StW 311 is outside the range of Sterkfontein Member 4 femora, and within the *A. robustus* sample

(Figure 12.12). For these reasons, we find it reasonable to tentatively assign StW 311 to *A. robustus*.

Mixed assemblage at Sterkfontein Member 4?

Clarke (1988, 2008, 2013) has posited the existence of a second, "pre-*Paranthropus*" species at Sterkfontein, and he has resurrected *Australopithecus prometheus* with the discovery of StW 573 (Granger et al., 2015). Variation in proximal femoral anatomy has informed this discussion of anatomical variation in the Sterkfontein assemblage. Partridge et al. (2003) noted that the Jacovec Cavern femur StW 598 possesses a small head and a long femoral neck, similar to the anatomy found in StW 99/100, and similar to the anatomy found in the Swartkrans *A. robustus* femora SK 82 and SK 97. However, StW 522 appears different in possessing a more globular femoral head and a short femoral neck.

Harmon (2009a) qualitatively assessed size and shape variation in the Sterkfontein femora and found that while size variation can easily be encompassed in a single variable species, shape variation is not as easily encompassed within a single taxon. With a larger sample that included all of the Sterkfontein femora, she found that StW 99 was an outlier and that it was most different from StW 522. It remains unclear how to interpret these findings, but several possibilities exist.

The femora at Sterkfontein may be sampling a single, highly variable, quite dimorphic species. The work of Harmon (2009a) would appear to challenge this possibility. Alternatively, given the at least 600,000 years of

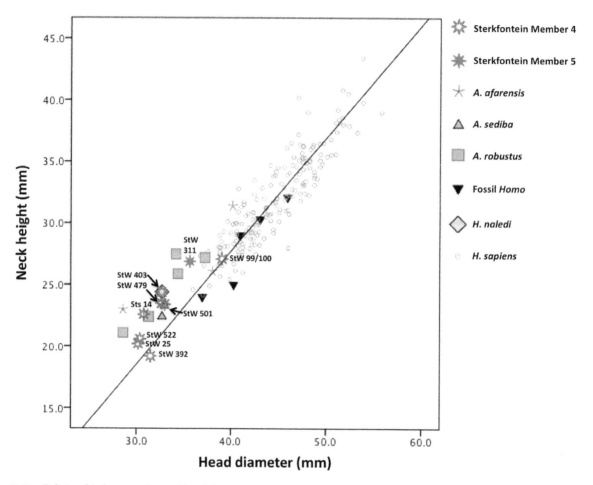

Figure 12.10 Relationship between femoral head diameter (x-axis) and femoral neck height (y-axis) in humans and fossil hominins. The line drawn through the data is a RMA regression characterizing the human sample. Ruff and Higgins (2013; Chapter 16, this volume) remarked on the relatively superoinferiorly tall femoral necks in the australopiths. Here, we agree that for a given head diameter, the Sterkfontein femora have relatively tall (superoinferiorly) necks, though some (StW 392 and StW 99/100) fall precisely on the modern human regression line. Note the position of fossil *Homo*, directly on the human trend line. Additionally, note the position of StW 311 (Sterkfontein Member 4) which is within a cluster of three *A. robustus* femora from Swartkrans, South Africa. Data from Table 1 and DeSilva et al., 2013; Ward et al., 2015.

sediment accumulation at Sterkfontein (Pickering and Kramers, 2010), the fossil assemblage may simply be picking up a single evolving lineage; high variation being the result of time-averaging. This effect may be exacerbated by the inclusion of a Member 5 East fossil (StW 311) into previous analyses. It could also be that, as Clarke (1998, 2013) has proposed, the Sterkfontein Member 4 assemblage consists of two species. The apparent morphological differences between StW 99/100 and StW 522 would be consistent with this finding, though identifying which morph is best assigned to *A. africanus* and which morph to *A. prometheus* will require the comparison of each of these isolated elements to the extraordinarily complete template skeleton StW 573.

However, one final possibility that we currently support is that the eroded head of StW 99/100 has artificially inflated the apparent differences between StW

99/100 and StW 522. An eroded head and damaged intertrochanteric region make it difficult to measure the morphological neck length—we place the value ~44 mm, slightly lower than the 46 mm in Partridge et al. (2003). Additionally, their estimate of 36 mm for the femoral head diameter is clearly too small (as is Harmon's (2009a) value of 34.6 mm). Using a sphere-fitting technique, the femoral head diameter of StW 99/100 is between 38 mm (Ruff et al., Chapter 16, this volume) and 39.1 mm (Ward et al., 2015). Using these values, the difference between StW 99/100 and StW 522 is not nearly as extreme. The long femoral neck may simply be because StW 99/100 is a larger individual than StW 522. When examined in the context of modern human femoral variation (Figure 12.13), the neck length: head diameter ratio of StW 99/100 and StW 522 is easily encompassed within a single variable taxon. Thus, while there may very well be two

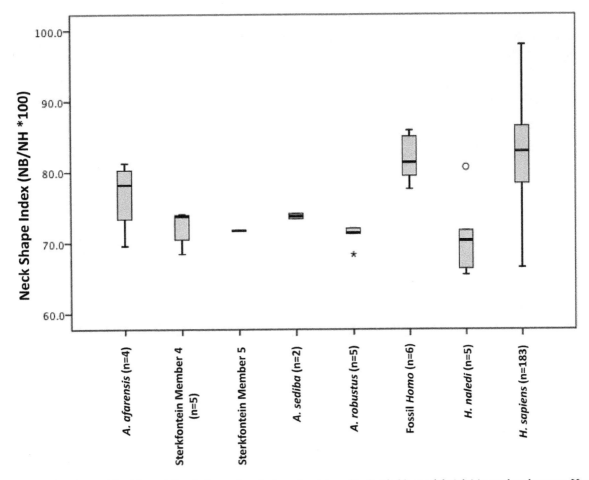

Figure 12.11 There is considerable variation in femoral neck shape (neck breadth divided by neck height) in modern humans. However, in general, australopiths have been characterized as having a more anteroposteriorly compressed (or superoinferiorly tall) femoral neck relative to modern humans and fossil *Homo* (though see *H. naledi*). While there is some overlap, the Sterkfontein Member 4 femora and slightly more anteroposteriorly compressed than those from *A. afarensis*. Box plots illustrate the median value (black bars), 25th to 75th percentiles (gray boxes), range of values (whiskers), and outliers (circles). Data from Table 12.1 and DeSilva et al., 2013; Ward et al., 2015.

taxa in Sterkfontein Member 4, the proximal femora from Sterkfontein Member 4 are morphologically more homogeneous than originally thought.

Functional anatomy

While Lovejoy's hypothesis that australopiths had a bipedal gait indistinguishable from that found in modern humans is often presented in the context of *A. afarensis* (e.g., Lovejoy, 1988), it originally was based on his observations of the Sterkfontein hip joint from Sts 14 and the distal femora Sts 34 and TM 1513 (Lovejoy and Heiple, 1970; Heiple and Lovejoy, 1971; Lovejoy et al., 1973). Compared with modern humans, Sts 14 possessed a long femoral neck and a small (reconstructed) femoral head. This combination of a long femoral neck and small femoral head, relative to humans today, is present in all of the Sterkfontein Member 4 femora (Figure 12.13; Ruff et al., Chapter 16, this volume). Instead of

regarding these anatomical differences as evidence for functionally different gait, Lovejoy et al. (1973) presented a biomechanical analysis in which the long femoral neck increased the mechanical advantage of the lesser gluteals (*Mm. gluteus medius* & *minimus*) during a single-legged stance phase of bipedal walking. Because these muscles contracted with less force due to their insertion farther from the hip joint, the joint reaction force would be less in australopiths, explaining the relatively small femoral head. Furthermore, in the distal femur (Heiple and Lovejoy, 1971), *Australopithecus* possessed key anatomies for human-like bipedal gait, including a high bicondylar angle (14–15°), an elevated lateral lip adaptive for patellar retention, and an elliptical lateral condyle for dissipating the high loads incurred on the knee during upright walking.

While most regard the Sterkfontein femoral material as good evidence for bipedalism in early hominins, and have for some time (Le Gros Clark, 1947; Broom and

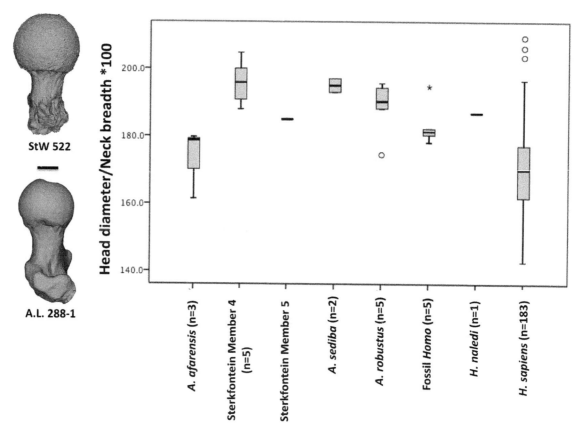

StW 522

A.L. 288-1

Figure 12.12 Left: StW 522 (top) is compared with A.L. 288-1 (bottom) in superior view. The scale bar is 10 mm. Note that while the breadth of the femoral neck is quite similar in the two fossils, StW 522 has a much larger femoral head. Note as well that the expansion of the femoral head is directed anteriorly (to the left in the image). To the right, the relative size of the femoral head to the neck breadth is quantified. Notice that *A. afarensis* is modern human-like in these proportions while the Sterkfontein Member 4 femora are quite different and possess a more anteroposteriorly compressed neck relative to the head size. We hypothesize that this morphology would result in greater transverse plane rotation at the hip joint in *A. africanus* compared with *A. afarensis*. Box plots illustrate the median value (black bars), 25th to 75th percentiles (gray boxes), range of values (whiskers), and outliers (circles). Data from Table 1 and DeSilva et al., 2013; Ward et al., 2015.

Robinson, 1949; Washburn and Patterson, 1951), there is evidence in the femur that the Sterkfontein hominins may have walked in a biomechanically different manner than humans today. Ruff and Higgins (2013) hypothesize that the subtle differences in cortical bone distribution in the femoral neck and the superoinferiorly reinforced femoral neck (Figure 12.10) found in the Sterkfontein fossils is consistent with a small, but functionally significant, degree of lateral displacement of the body mass over the stance leg during gait. Subtle mediolateral sway of the body during gait would likely have been energetically less efficient, as would the, on average, shorter legs estimated for the Sterkfontein australopiths (Holliday, 2012; Pontzer, 2012).

Furthermore, attempts have been made to anatomically and functionally compare the Sterkfontein fossils to other hominin assemblages, particularly from Koobi Fora, Kenya, Swartkrans, and earlier australopiths from Hadar. Using StW 99, 311, 501, 522, and MLD 46 as representatives of *A. africanus* Harmon (2009b) found that these

femora were generally similar to those from Swartkrans, Koobi Fora, and Hadar, but that *A. africanus* had quite low neck-shaft angles (especially StW 99 and MLD 46). McHenry and Berger (1998) noticed that while many of the isolated forelimb fossils from Sterkfontein were quite large, the femora were generally small (only StW 99 and 431 were large enough to make the "medium" category). Green et al. (2007) quantified these observations using a resampling approach and agreed that the hindlimb joints (including the hip) of *A. africanus* were relatively smaller than that found in *A. afarensis*, signifying a potential difference in substrate use between the two australopith species. Tardieu (1981) suggested that the distal femora of *A. africanus* (TM 1513 and Sts 34) were quite human-like and similar to those from the Hadar 333 locality but different from the A.L. 129 and A.L. 288 femora. McHenry (1986) disagreed and found the distal femora of *A. africanus* and *A. afarensis* to be quite similar. Lovejoy (2007) agrees and has regarded all australopith distal femora

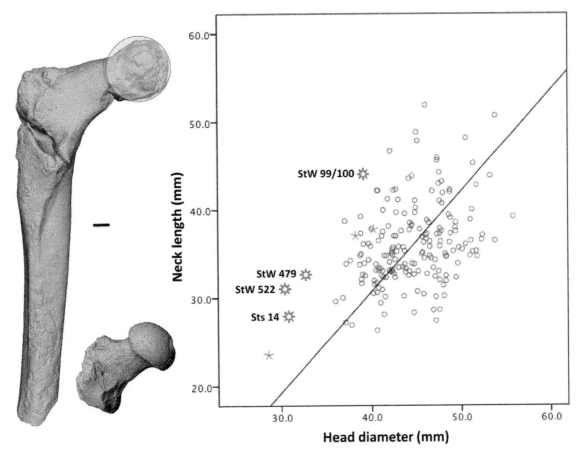

Figure 12.13 Left: StW 99/100 is mirrored to reflect the right side and a sphere has been positioned to estimate the actual (39.1 mm) diameter of the femoral head based on a sphere-fitting technique applied to the preserved anterior surface (Ward et al., 2015). Below StW 99/100 is the much smaller femur StW 522. The scale bar is 10 mm. Visual inspection (as presented in Partridge et al., 2003) would suggest StW 99/100 has a relatively longer morphological neck length than StW 522. To the right is the range of neck length to femoral head diameter in modern humans ($R^2 = 0.12$; $P < 0.001$). Notice that all of the Sterkfontein Member 4 femora (red stars) have similarly long necks given the size of the femoral head and present the same level of variation one might find in a modern human assemblage. The shortest neck belongs to Sts 14. Notably, both the head diameter and the neck length are approximations measured from plaster reconstructions based on impressions of these elements left in the fossil breccia. It is likely that these measurements—especially the neck length—are not entirely accurate. Note as well that the Sterkfontein femora generally have longer necks than those from Hadar (n = 3; blue asterisks).

as possessing the key anatomies for human-like bipedal locomotion.

Sterkfontein and Hadar comparison

While we agree with others (e.g., McHenry, 1986) that the distal femora of *A. afarensis* and *A. africanus* are quite similar, we note here unrecognized and potentially functionally relevant differences in proximal femoral anatomy between fossils assigned to *A. afarensis* and those typically assigned to *A. africanus*. These are itemized here:

1. Compared with *A. afarensis*, the Sterkfontein Member 4 femora have slightly longer femoral necks relative to the size of the femoral head (Figure 12.13; Table 12.3). Sample sizes remain low in

both collections, but we find this an observation of interest, and of potential functional and evolutionary significance.

2. The proximal femora from Sterkfontein that are complete enough to preserve the anterior neck-shaft junction (n = 2: StW 99/100 and StW 479) lack an intertrochanteric line. This rugosity, thought to signify extended hip posture and the tightening of the iliofemoral ligament, is present to varying degrees in all adult Hadar femora (n = 4). A study on human femora found that those with relatively longer necks were more likely to lack an intertrochanteric line, supporting a relationship between neck length and intertrochanteric line development (DeSilva et al., 2006), and consistent with the observation that *A. africanus* femora have

Table 12.3 Comparative anatomy of hominin proximal femur.

	Neck length/head diameter*100	Head diameter/neck breadth*100	Neck breadth/neck height*100	Midshaft AP/ML*100
Modern human (n=183)	81.0 ± 10.3 (58.1–112.7)	170.7 ± 12.0 (142.7–209.3)	82.4 ± 5.8 (66.5–98.0)	109.7 ± 5.6[a]
Sterkfontein Member 4	101.2 ± 9.0 (n = 4; 90.6–112.5)	195.8 ± 6.8 (n = 5; 187.9–204.7)	72.1 ± 2.6 (n = 5; 68.4–73.9)	86.6 ± 3.6 (n = 2; 85.4–90.5)
Sterkfontein Member 5 (StW 311)	–	185.0	71.8	–
A. afarensis	90.9 ± 7.8 (n = 3; 82.2–96.9)	173.3 ± 10.3 (n = 3; 161.5–179.7)	76.8 ± 5.1 (n = 4; 69.6–81.2)	108.2 (n = 1)
A. sediba	–	195.0±2.8 (n=2; 193.0–197.0)	73.8 ± 0.51 (n = 2; 73.4–74.1)	–
A. robustus	95.9±6.6 (n = 2; 91.2–100.7)	188.6 ± 8.4 (n = 5; 174.5–195.6)	71.0 ± 1.5 (n = 5; 68.3–72.1)	–
Fossil Homo	91.8 ± 9.7 (n = 3; 81.4–100.7)	183.4 ± 6.5 (n = 5; 178.1–194.7)	81.8 ± 3.3 (n = 6; 77.6–85.9)	89.3 ± 14.4 (n=12; 68.1–109.4)
H. naledi	103.7 (n = 1)	186.9 (n = 1)	70.9 ± 6.0 (n = 5; 65.5–80.7)	109.5 (n = 1)

[a] Data from Aiello and Dean (1990).

longer necks than those assigned to A. afarensis. However, the functional link between these variables is currently unclear.

3. The neck shape index (anteroposterior breadth/superoinferior height) is slightly lower in A. africanus than in A. afarensis (Figure 12.11)—in other words, the neck is more anteroposteriorly compressed (or superoinferiorly tall) in A. africanus. This difference is even more apparent when the anteroposterior breadth of the femoral neck is compared with the diameter of the femoral head (Figure 12.12; See also Figure 12.13 for comparison of femoral head diameter to neck length). In A. africanus, the head is considerably anteroposteriorly larger than the femoral neck, and there is no overlap whatsoever between the Sterkfontein and Hadar populations for this variable. The head of A. africanus extends more anteriorly than posteriorly, suggesting that A. africanus possessed slightly different hip mechanics (i.e., A. africanus may have been capable of more transverse plane rotation at the hip joint than was A. afarensis), or may reflect use of different substrates.

4. While the shape of the femoral midshaft is known from only one fossil (A.L. 288-1) and possesses a roughly circular shape, the two femoral midshafts from A. africanus (StW 99/100 and StW 121) are mediolaterally broad (Figure 12.14), as found in some early Homo fossil and most H. erectus femora (Ruff, 1995; though see Ruff et al., Chapter 16, this volume for another interpretation of the midshaft of StW 121).

In general, these findings—that the proximal femur of A. afarensis differs in subtle ways from A. africanus—are consistent with findings from the upper limb (McHenry and Berger, 1998; Green et al., 2007) and the foot (Harcourt-Smith and Aiello, 2004) that A. afarensis and A. africanus may have possessed slightly different walking mechanics (see also Harmon, 2009b). However, it remains noteworthy that there have been no differences found in the morphology of the distal femur in these two taxa (McHenry and Berger, 1998; Lovejoy, 2007; Squyres, 2016; but see Tardieu, 1981).

Internal anatomy

The internal structural anatomy of the Sterkfontein proximal femur (head and neck) has been studied using both conventional computed tomography (CT) scanning and micro-CT scanning approaches (see Ruff et al., Chapter 16, this volume for a more in-depth analysis of Sterkfontein femoral internal anatomy). Chirchir et al. (2015) found that the femoral heads from Sterkfontein possess, on average, higher bone volume fraction (i.e., the highest density of trabecular bone per unit volume) than extant humans and apes or other fossil hominins. The sampled fossils were StW 403 (36.6% bone volume/total volume, or BV/TV), StW 311 (47.5% BV/TV), and StW 527 (52.7% BV/TV). More recently, Ryan et al. (2018) discovered human-like orientation of the trabecular bone in femoral heads StW 99, 311, 392, 403, 479, and 501, indicating that A. africanus loaded their hips in a human-like way. Ruff and Higgins (2013) used conventional CT scans to quantify the

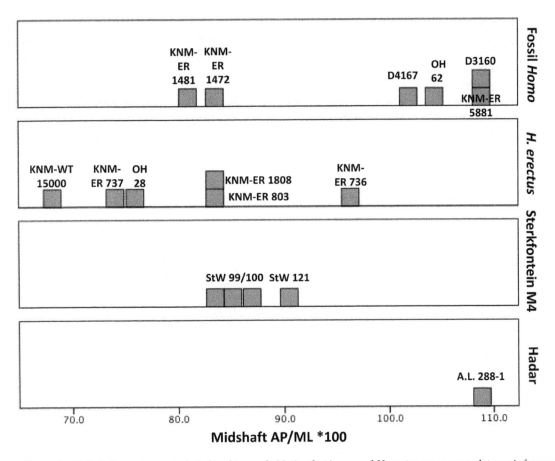

Figure 12.14 Femoral midshaft dimensions are plotted in this graph. Notice that in general *H. erectus* possesses a platymeric femoral midshaft (Ruff, 1995). Additionally, the femora from Sterkfontein Member 4 have a similarly anteroposteriorly compressed femoral midshaft. The three estimates for StW 99/100 are based on the different positions of the midshaft using three different overall length estimates of the bone. The platymeric midshaft of *A. africanus* is in contrast to A.L. 288-1, which possesses an anteroposteriorly expanded midshaft.

distribution of cortical bone in the Sterkfontein hominin femoral necks, including: StW 99, 311, 403, 479, 501, and 522. They found human-like asymmetrical distribution of cortical bone at the neck-shaft junction. Only StW 99 possessed thickened superior bone outside the range typically found in modern humans. At the midneck, most of the specimens were in the human range (StW 99, 311, 403, 522). StW 479 was just outside the human range, while StW 501 was ape-like in possessing quite thick superior cortex. More recently, Claxton (2018) used higher resolution micro-CT techniques to quantify the cortical bone structure of the femoral necks and found Sterkfontein femora (StW 99, 311, 479, 501) to possess human-like ratios of superior to inferior cortex. However, the magnitude of superior cortex alone was thick and ape-like, inconsistent with predictions generated from human-like loading models of the hip joint (Claxton, 2018).

Size variation

Femoral head diameter is often used as a means for estimating body mass in fossil taxa. Fortunately, there are 11 proximal femoral head diameters that can be measured in the Sterkfontein collection. One of these, already discussed, is from Member 5 (StW 311). The other 10 are from Member 4, and may be from *A. africanus*. The smallest of these (StW 361) is from a juvenile and measures 29.1 mm. The largest (StW 99/100) is approximately 39.1 mm. Interestingly, the other eight are quite similar in size, ranging from 30.2 mm to 33.0 mm. It is difficult to know how to interpret this assemblage. It is of note that Pickering et al. (2004) has identified carnivore damage on many of these femora including definitive crenulation and tooth marks on StW 129 and 367; possible tooth marks on StW 25, 403, and 479; and possible crenulation on StW 99, 121, 501, and 522. It is possible that there was a sex bias favoring smaller, presumably female, individuals in the accumulation of hominin remains at Sterkfontein (Lockwood, 1999).

Femoral length

Because a complete femur is not yet known from Sterkfontein Member 4 or 5, regression-based approaches have been used to estimate leg length from fragmentary fossils. However, these estimates have varied widely. Sts 14 is generally regarded as one of the shorter femora, with estimates ranging from between 276 mm (Steudel-Numbers and Tilkens, 2004) and 295 mm (McHenry, 1991) up to 310 mm (Broom et al., 1950). Lovejoy and Heiple (1970) estimated Sts 14 to have had a 280 mm leg length. Based on the femoral head diameters, StW 392 and StW 25 have yielded femoral length estimates of 311 mm and 320 mm, respectively (McHenry, 1991). Vančata (1994) calculated femur lengths of 325 mm and 336 mm for the distal femora Sts 34 and TM 1513, though Steudel-Numbers and Tilkens (2004) estimate a lower length of 276 mm for Sts 34. Estimates vary the most for StW 99/100, ranging from 362 mm (Vančata, 1994) and 380 mm (McHenry, 1991) to as high as 433.5 mm (Holliday, 2012).

Conclusion

While a complete femur from Sterkfontein Member 4 or 5 remains elusive, enough fragmentary remains have been found to characterize the general femoral anatomy of these hominins. While some have postulated that the differences between StW 99/100 and StW 522 are evidence for taxonomic diversity, we regard these as large and small versions of the same general morphology. Multiple taxa may have been present at Sterkfontein; we just do not see evidence for it in the femora. However, while the distal femur of *A. africanus* is similar to that found in *A. afarensis*, the proximal femur is not. Sterkfontein Member 4 femora have relatively longer and more compressed femoral necks, and a large femoral-head-to-femoral-neck ratio. Whether these subtle morphological differences reflect differences in substrate use, reflect differences in locomotor kinematics, or result in functional equivalency remains unclear.

Acknowledgments

We are grateful for B. Zipfel and C. Ward for the opportunity to contribute to this volume. Thank you to the curators and colleagues who provided access to the hominin femora discussed in this chapter. Thanks to F. K. Manthi, E. Mbua, and the National Museum of Kenya. The Evolutionary Studies Institute and University of the Witwatersrand Fossil Access Committee permitted our study of fossil material housed at the ESI in Johannesburg, South Africa. Thank you to S. Potze for her assistance at the Ditsong Museum of Natural History (Pretoria, South Africa). We are grateful to R. Meindl and O. Lovejoy (Kent State University), L. Jellema (Cleveland Museum of Natural History), M. Morgan (Peabody Museum of Archaeology and Ethnology), and M. Wolpoff (University of Michigan) for access to human femora measured in this study. Funding for this research was provided by The George Washington University's Selective Excellence Program, Boston University, and Dartmouth College.

References

Aiello, L., Dean, C., 1990. An introduction to human evolutionary anatomy. Amsterdam: Academic Press.

Broom, R., 1938. Further evidence on the structure of the South African Pleistocene anthropoids. Nature 142, 897–899.

Broom, R., Robinson, J.T., 1949. The lower end of the femur of *Plesianthropus*. Annals of the Transvaal Museum 21, 181–182.

Broom, R., Robinson, J.T., Schepers, G.W.H., 1950. Sterkfontein ape-man *Plesianthropus*. Transvaal Museum Memoires 4: Pretoria.

Chirchir, H., Kivell, T.L., Ruff, C.B., Hublin, J.J., Carlson, K.J., Zipfel, B., Richmond, B.G., 2015. Recent origin of low trabecular bone density in modern humans. Proceedings of the National Academy of Science 112, 366–371.

Clarke, R.J., 1988. A new *Australopithecus* cranium from Sterkfontein and its bearing on the ancestry of *Paranthropus*. In: Grine, F.E. (Ed.), Evolutionary history of the "robust" Australopithecines. Aldine de Gruyter: New York, pp. 285–292.

Clarke, R.J., 2008. Latest information on Sterkfontein's *Australopithecus* skeleton and a new look at *Australopithecus*. South African Journal of Science 104, 443–449.

Clarke, R.J., 2013. *Australopithecus* from Sterkfontein Caves, South Africa. In: Reed, K., Fleagle, J.G., Leakey R.E. (Eds.), The paleobiology of *Australopithecus*. Springer: Dordrecht, pp. 105–123.

Claxton, A., 2018. A re-assessment of femoral neck cortical thickness. Ph.D. Dissertation, Boston University.

DeGusta, D., 2004. Pliocene hominid postcranial fossils from the Middle Awash, Ethiopia. Ph.D. Dissertation, University of California, Berkeley.

DeSilva, J.M., 2011. A shift toward birthing relatively large infants early in human evolution. Proceedings of the National Academy of Science 108, 1022–1027.

DeSilva, J.M., Holt, K.G., Churchill, S.E., Carlson, K.J., Walker, C.S., Zipfel, B., Berger, L.R., 2013. The lower limb and mechanics of walking in *Australopithecus sediba*. Science 340, 1232999.

DeSilva, J.M., Shoreman, E., MacLatchy, L.M., 2006. A fossil hominoid proximal femur from Kikorongo Crater, southwestern Uganda. Journal of Human Evolution 50, 687–695.

Granger, D.E., Gibbon, R.J., Kuman, K., Clarke, R.J., Bruxelles, L., Caffee, M.W., 2015. New cosmogenic burial ages for Sterkfontein Member 2 *Australopithecus* and Member 5 Oldowan. Nature 522, 85–88.

Green, D.J., Gordon, A.D., Richmond, B.G., 2007. Limb-size proportions in *Australopithecus afarensis* and

Australopithecus africanus. Journal of Human Evolution 52, 187–200.

Harcourt-Smith, W.E.H., Aiello, L.C., 2004. Fossils, feet and the evolution of human bipedal locomotion. Journal of Anatomy 204, 403–416.

Harmon, E., 2009a. Size and shape variation in the proximal femur of *Australopithecus africanus*. Journal of Human Evolution 56, 551–559.

Harmon, E., 2009b. The shape of the early hominin proximal femur. American Journal of Physical Anthropology 139, 154–171.

Heiple, K.G., Lovejoy, C.O., 1971. The distal femoral anatomy of *Australopithecus*. American Journal of Physical Anthropology 35, 75–84.

Holliday, T.W., 2012. Body size, body shape, and the circumscription of the genus *Homo*. Current Anthropology 53, S330–S345.

Kuman, K., Clarke, R.J., 2000. Stratigraphy, artefact industries and hominid associations for Sterkfontein, Member 5. Journal of Human Evolution 38, 827–847.

Le Gros Clark, W.E., 1947. Observations on the anatomy of the fossil Australopithecinae. Journal of Anatomy 81, 300–333.

Lockwood, C.A., 1999. Sexual dimorphism in the face of *Australopithecus africanus*. American Journal of Physical Anthropology 108, 97–127.

Lovejoy, C.O., 1988. Evolution of human walking. Scientific American 259, 118–125.

Lovejoy, C.O., 2007. The natural history of human gait and posture: Part 3. The knee. Gait and Posture 25, 325–341.

Lovejoy, C.O., Heiple, K.G., 1970. A reconstruction of the femur of *Australopithecus africanus*. American Journal of Physical Anthropology 32, 33–40.

Lovejoy, C.O., Heiple, K.G., Burstein, A.H., 1973. The gait of *Australopithecus*. American Journal of Physical Anthropology 38, 757–779.

McHenry, H.M., 1986. The first bipeds: a comparison of the *A. afarensis* and *A. africanus* postcranium and implications for the evolution of bipedalism. Journal of Human Evolution 15, 177–191.

McHenry, H.M., 1991. Femoral lengths and stature in Plio-Pleistocene hominids. American Journal of Physical Anthropology 85, 149–158.

McHenry, H.M., Berger, L.R., 1998. Body proportions in *Australopithecus afarensis* and *A. africanus* and the origin of the genus *Homo*. Journal of Human Evolution 35, 1–22.

Moggi-Cecchi, J., Grine, F.E., Tobias, P.V., 2006. Early hominid dental remains from Members 4 and 5 of the Sterkfontein Formation (1966-1996 excavations): catalogue, individual associations, morphological descriptions and initial metrical analysis. Journal of Human Evolution 50, 239–328.

Partridge, T.C., Granger, D.E., Caffee, M.W., Clarke, R.J., 2003. Lower Pliocene hominid remains from Sterkfontein. Science 300, 607–612.

Pickering, T.R., Clarke, R.J., Moggi-Cecchi, J., 2004. Role of carnivores in the accumulation of the Sterkfontein Member 4 hominid assemblage: a taphonomic reassessment of the complete hominid fossil sample (1936-1999). American Journal of Physical Anthropology 125, 1–15.

Pickering, R., Kramer, J.D., 2010. Re-appraisal of the stratigraphy and determination of new U-Pb dates for the Sterkfontein hominin site, South Africa. Journal of Human Evolution 59, 70–86.

Pontzer, H., 2012. Ecological energetics in early *Homo*. Current Anthropology 53, S346–S358.

Reed, K.E., Kitching, J.W., Grine, F.E., Jungers, W.L., Sokoloff, L., 1993. Proximal femur of *Australopithecus africanus* from member 4, Makapansgat, South Africa. American Journal of Physical Anthropology 92, 1–15.

Robinson, J.T., 1972. Early hominid posture and locomotion. University of Chicago Press: Chicago.

Ruff, C.B., 1995. Biomechanics of the hip and birth in early *Homo*. American Journal of Physical Anthropology 98, 527–574.

Ruff, C.B., 2010. Body size and body shape in early hominins—implications of the Gona pelvis. Journal Human Evolution 58, 166–178.

Ruff, C.B., Higgins, R., 2013. Femoral neck structure and function in early hominins. American Journal of Physical Anthropology 150, 512–525.

Ryan, T.M., Carlson, K.J., Gordon, A.D., Jablonski, N., Shaw, C.N., Stock, J.T., 2018. Human-like hip joint loading in *Australopithecus africanus* and *Paranthropus robustus*. Journal of Human Evolution 121, 12–24.

Squyres, N., 2016. Shape variation in the distal femur of modern humans and fossil hominins. Ph.D. Dissertation, Johns Hopkins University.

Steudel-Numbers, K.L, Tilkens, M.J., 2004. The effect of lower limb length on the energetic cost of locomotion: implications for fossil hominins. Journal of Human Evolution 47, 95–109.

Tardieu, C., 1981. Morpho-functional analysis of the articular surfaces of the knee-joint in primates. In: Chiarelli, A.B., Corruccini, R.S. (Eds.), Primate evolutionary biology. Springer Verlag: Berlin, pp. 68–80.

Tobias, P.V., 1998. Ape-like *Australopithecus* after seventy years: was it a hominid? Journal of the Royal Anthropological Institute 4, 283–308.

Vančata, V., 1994. New estimates of femoral length in early hominids. Anthropologie 32, 269–272.

Ward, C.V., Feibel, C.S., Hammond, A.S., Leakey, L.N., Moffett, E.A., Plavcan, J.M., Skinner, M.M., Spoor, F., Leakey, M.G., 2015. Associated ilium and femur from Koobi Fora, Kenya, and postcranial diversity in early *Homo*. Journal of Human Evolution 81, 48–67.

Washburn, S.L., Patterson, B., 1951. Evolutionary importance of the South African "man-apes." Nature 167, 650–651.

13

Tibia and fibula

Kristian J. Carlson, Bernhard Zipfel, and William L. Jungers

Taken as a whole, the Sterkfontein Cave hominin fossil assemblage, particularly from Member 4 (Partridge, 1978; Kuman and Clarke, 2000), provides a unique opportunity to examine morphological variability at select times during the course of human evolution. The suggestion of two contemporaneous species of *Australopithecus* within Member 4 has been made largely on craniodental material (Clarke, 1985, 1989, 1994, 2012; 2013; Lockwood and Tobias, 2002; Fornai, 2009; but see Fornai et al., 2015), while others recognize the presence of only a single species, *Australopithecus africanus*, again based largely on craniodental material (Grine, 2013). Studies of hominin postcranial material from Sterkfontein, on the other hand, have been seldom recruited in these taxonomic debates (e.g., Berge et al., 2007), instead tending to focus more on comparative morphology and function (Lovejoy and Heiple, 1970; Heiple and Lovejoy, 1971; Robinson, 1972; Vrba, 1979; Clarke and Tobias, 1995; McHenry and Berger, 1998; Deloison, 2003; Green et al., 2007; Harmon, 2009; Zipfel et al., 2009; DeSilva and Devlin, 2012; Ruff and Higgins, 2013; Zeininger et al., 2016; Dowdeswell et al., 2017; Su and Carlson, 2017; Pickering et al., 2018; Ryan et al., 2018). Tibiae and fibulae, in particular, have rarely served as centerpieces in these types of functional investigations (but see Berger and Tobias, 1996; DeSilva, 2009; Zipfel and Berger, 2009; Barak et al., 2013; Tommy, 2018; Desilva et al., 2019). As a result, the extent of morphological variability expressed by hominin tibiae and fibulae from Sterkfontein remains unassessed, largely because these elements have not been subjected to a careful systematic study.

This chapter provides the initial formal descriptions of tibiae and fibulae assigned to hominin taxa from Sterkfontein Members 4 and 5 (Appendix I). Carnivores were a factor in the accumulation of the Sterkfontein hominin assemblage, particularly in Member 4 (Pickering et al., 2004), but this was not evaluated in descriptions. We do not include specimens from the StW 573 skeleton from Member 2 (Clarke, 1998), as they have been systematically described elsewhere (see Heaton et al., 2019. We provide qualitative and quantitative comparisons to other hominin fossils, focusing on australopith tibiae, in order to contextualize the extent of the observed variability exhibited by the Sterkfontein fossils. Our comparisons focus on answering specific questions, including whether there are previously unrecognized associations among the Sterkfontein fossils and whether there is evidence for more than one functional constellation of features in these elements (i.e., multiple "morphs").

Descriptions

StW 396—Right proximal tibia

Zipfel and Berger (2009) described StW 396; whenever possible, we attempt to replicate their measurements, as well as to provide supplementary measurements and description. Only when there is a discrepancy do we note their earlier measurements and provide additional description.

We suggest that StW 396 may be associated with StW 567, which is a right distal tibia also retaining evidence of an epiphyseal line (Figure 13.1). Thus, while both specimens exhibit similar developmental states, they also appear to exhibit similar preservation conditions and proportionate articular dimensions (Figure 13.1, Table 13.1). The recovery of StW 396 from Member 4 deposits versus the recovery of StW 567 from Member 5 deposits, however, casts some doubt on such an association.

Kristian J. Carlson, Bernhard Zipfel, and William L. Jungers, *Tibia and fibula*. In: *Hominin Postcranial Remains from Sterkfontein, South Africa, 1936–1995.* Edited by: Bernhard Zipfel, Brian G. Richmond, and Carol V. Ward, Oxford University Press (2020). © Oxford University Press.
DOI: 10.1093/oso/9780197507667.003.0013

Figure 13.1   StW 396 is a partial right proximal tibia illustrated in posterior (top left), medial (top right), and superior views (bottom).

Preservation. StW 396 (Figure 13.1) is a partial right proximal tibia of a young adult retaining most of the medial tibial condyle and parts of the medial sides of the tibial plateau and proximal shaft. An epiphyseal line is still visible medially and posteriorly. The specimen has a maximum length of 46.3 mm from its proximal- to distal-most points, which correspond to what remains of the medial intercondylar tubercle and the medial proximal shaft, respectively. The specimen does not appear to exhibit any plastic deformation.

The specimen is broken and missing below a supero-lateral to inferomedial oriented plane, which runs through the medial region of the surface of the lateral condyle and passes beneath the medial condyle and tibial plateau. As a result of the break being internally rotated relative to the sagittal plane, comparatively more of the posterior surface of the specimen is retained than the anterior surface. Trabecular bone with matrix infill is exposed over the entire distal break. A second more proximal break located on the medial surface isolates a triangular portion of the metaphysis from the rest of the specimen, but it exhibits no anatomical displacement and refits cleanly. This more proximal break connects the distal break to the epiphyseal line, but it does not appear to extend superior to the epiphyseal line.

The margin of the medial condyle is eroded antero-medially and posteriorly, with small isolated areas of abrasion in between. Centrally, in the posterior half of the medial condyle, there is a circular area of pitting, which has been suggested as reflective of a degenerative condition (Zipfel and Berger, 2009). The intercondylar eminence, including parts of both medial and lateral intercondylar tubercles, is preserved, although apices of both tubercles are broken off. The lateral condyle is represented by only a small portion of its medial-most region. Other areas of the tibial plateau show additional minor weathering or abrasion.

Morphology. Due to the proximal location of the distal-most break, a distinct medullary cavity is absent and only trabecular bone and intervening matrix infill is present. Moreover, cortical thickness of the preserved shaft is not well-defined as a result. Thus, functionally informative cross-sections of the shaft, or local thickness measurements of the proximal diaphysis, are unavailable. An external circumference measurement of the shaft also is not possible because of the oblique orientation and proximal location of the break.

Dimensions of the tibial plateau or those of either tibial condyle cannot be measured due to their incompleteness, except for mediolateral breadth of the medial condyle (estimated as 23.8: Table 13.2). The medial condyle is relatively intact, exhibiting a markedly depressed central region that results in a concave surface anteroposteriorly

Table 13.1 Tibial metrics used in RMA regression. All measurements in mm.

Group	n	Medial condyle ML width (left)	Medial condyle ML width (right)	Talar facet ML width of anterior margin (left)	Talar facet ML width of anterior margin (right)
Homo sapiens	16	30.7 (2.5) [16]	30.9 (2.3) [16]	30.7 (2.4) [16]	30.4 (2.6) [16]
Pan troglodytes	15	24.8 (1.8) [15]	25.2 (1.8) [15]	24.3 (2.1) [15]	24.5 (2.1) [15]
Pan paniscus	14	24.5 (1.7) [13]	24.7 (1.0) [12]	23.6 (1.6) [13]	23.4 (1.3) [12]
Gorilla gorilla	10	34.9 (4.9) [9]	34.5 (5.3) [9]	36.7 (4.9) [9]	35.6 (4.7) [9]
Gorilla beringei	20	33.5 (4.8) [13]	35.5 (4.8) [12]	33.3 (4.1) [17]	33.5 (3.6) [16]
Pongo abelii	10	22.4 (2.7) [9]	22.8 (2.6) [10]	21.0 (2.2) [8]	20.9 (2.7) [10]
Pongo pygmaeus	10	22.9 (3.1) [10]	24.0 (3.2) [9]	22.1 (2.7) [10]	22.7 (3.2) [9]
Papio anubis	15	16.9 (1.6) [12]	17.7 (1.9) [15]	18.1 (2.0) [13]	18.6 (2.2) [15]
StW 396/567		–	23.8	–	21.5
StW 514/515		–	20.2	–	20.1
KNM-KP 29285		–	27.3	–	27.9
A.L. 129-1b		–	20.0	–	–
A.L. 288-1aq/1ar		–	19.5	–	21.2
A.L. 330-6		–	26.3	–	–
A.L. 333x-26		–	26.6	–	–

Cells contain mean values, one standard deviation in parentheses, and the sample size preserving measurements in brackets. All measurements taken on original fossils.

and mediolaterally. The central depression gradually shallows until reaching the edge of the articular surface. The medial condyle is generally elliptical, with a major axis in the anteroposterior direction.

Insufficient area of the lateral condyle surface is retained in StW 396 to characterize its form. It also cannot be ascertained whether the lateral condyle occupies the same horizontal plane as the medial condyle. Medial and lateral intercondylar tubercles are broken at their apices, with the break on the lateral tubercle also extending approximately one-third down its posterior side. Zipfel and Berger (2009: 72) describe the intercondylar tubercles as "high" with a line connecting the estimated locations of their apices angled 30° clockwise from the mediolateral plane. The distance between the tubercle bases is estimated as 7.7 mm (see Berger and Tobias, 1996), and the intercondylar eminence breadth at the same location is 11.1 mm wide.

The anterior intercondylar fossa is missing its anterior half, but its preserved medial and lateral borders are well delineated. The medial border preserves a smoothened area immediately posterior to the break but anterior to where the border begins to ascend proximally onto the intercondylar tubercle. This suggests the medial meniscus had a well-defined attachment for its anterior horn. The posterior intercondylar fossa is 14 mm deep anteroposteriorly and is relatively horizontal in orientation. There is roughening on StW 396 at the attachment point of the posterior cruciate ligament, while the surrounding surface in this area of the posterior intercondylar fossa is relatively smooth.

Medially, the tibial plateau sits on an inflated metaphysis such that the proximal shaft approaches the tibial plateau as a gradual, sloped junction. Few muscular attachments on the tibial plateau and proximal shaft are preserved on StW 396. The insertion of *m. semimembranosus* is a subtly defined, oval-shaped depression on the posteromedial side of the tibial plateau. Its largest dimension is proximodistal (minimum estimate = 10.3 mm), while anteroposteriorly it is shorter (maximum estimate = 9 mm)

Table 13.2 Distal tibial metrics. All measurements in mm.

Left Tibiae

Group	Total n	Talar facet anterior width (left)	Talar facet central width (left)	Talar facet posterior width (left)	Talar facet medial AP length (left)	Talar facet central AP length (left)	Talar facet lateral AP length (left)
Homo sapiens	16	30.7	28.1	25.9	22.8	27.6	23.7
		(2.4)	(2.0)	(2.4)	(2.2)	(2.9)	(2.3)
		[16]	[16]	[16]	[16]	[16]	[16]
Pan troglodytes	15	24.3	19.6	18.2	19.6	21.0	17.6
		(2.1)	(2.2)	(1.5)	(1.7)	(1.6)	(2.1)
		[15]	[15]	[15]	[15]	[15]	[15]
Pan paniscus	14	23.6	19.1	16.7	17.7	18.6	17.7
		(1.6)	(1.5)	(1.1)	(1.1)	(1.0)	(0.9)
		[13]	[13]	[13]	[13]	[13]	[13]
Gorilla gorilla	10	36.7	27.7	22.5	26.3	25.9	23.0
		(4.9)	(4.1)	(3.1)	(4.1)	(3.1)	(2.6)
		[9]	[9]	[9]	[9]	[9]	[9]
Gorilla beringei	20	33.3	26.5	22.9	25.4	24.7	19.9
		(4.1)	(3.6)	(3.3)	(3.0)	(3.0)	(3.4)
		[17]	[17]	[17]	[17]	[17]	[17]
Pongo abelii	10	21.0	18.5	16.8	14.7	17.9	13.7
		(2.2)	(2.2)	(2.5)	(1.5)	(1.9)	(1.8)
		[8]	[8]	[8]	[8]	[8]	[8]
Pongo pygmaeus	10	22.1	20.0	18.9	14.8	19.1	13.3
		(2.7)	(3.0)	(3.8)	(1.6)	(2.4)	(1.9)
		[10]	[10]	[10]	[10]	[10]	[10]
Papio anubis	15	18.1	14.4	13.3	13.8	15.4	12.8
		(2.0)	(1.4)	(1.5)	(1.9)	(1.2)	(1.3)
		[13]	[12]	[13]	[13]	[13]	[13]
StW 181		–	–	19.7	–	–	–
StW 358		18.4	–	17.1	16.7	16.9	16.8
StW 389		23.0	20.4	19.6	16.2	20.2	22.6
KNM-ER 1481		29.8	26.4	23.4	22.3	23.6	18.0
KNM-ER 1500		24.6	20.3	18.6	17.6	20.0	18.6
KNM-ER 2596		23.2	17.5	16.2	15.6	18.9	16.2
A.L. 333-6		21.8	19.6	18.8	17.8	21.9	17.7
A.L. 333-7		–	22.3	20.0	20.8	24.7	–
A.L. 545-3		17.0	16.9	16.6	16.2	20.1	19.5
KSD-VP-1/1e		26.7	23.0	22.1	20.6	20.4	27.4

Right tibiae

Group	Total n	Talar facet anterior width (right)	Talar facet central width (right)	Talar facet posterior width (right)	Talar facet medial AP length (right)	Talar facet central AP length (right)	Talar facet lateral AP length (right)
Homo sapiens	16	30.4	27.9	26.0	22.9	27.3	24.4
		(2.6)	(2.1)	(2.1)	(1.8)	(2.7)	(2.3)
		[16]	[16]	[16]	[16]	[16]	[16]
Pan troglodytes	15	24.5	19.8	18.7	19.5	20.4	17.3
		(2.1)	(1.8)	(1.5)	(1.8)	(1.7)	(1.5)
		[15]	[15]	[15]	[15]	[15]	[15]
Pan paniscus	14	23.4	19.2	16.9	17.5	18.9	17.1
		(1.3)	(1.4)	(0.6)	(1.4)	(1.6)	(1.0)
		[12]	[12]	[12]	[12]	[12]	[12]

(continued)

Table 13.2   Continued

Right tibiae

Group	Total n	Talar facet anterior width (right)	Talar facet central width (right)	Talar facet posterior width (right)	Talar facet medial AP length (right)	Talar facet central AP length (right)	Talar facet lateral AP length (right)
Gorilla gorilla	10	35.6 (4.7) [9]	27.2 (4.2) [9]	22.0 (3.6) [9]	25.5 (4.0) [9]	24.8 (3.1) [9]	22.4 (3.1) [9]
Gorilla beringei	20	33.5 (3.6) [16]	27.1 (3.5) [16]	23.3 (3.0) [16]	25.7 (3.0) [16]	24.8 (2.6) [16]	20.2 (3.3) [16]
Pongo abelii	10	20.9 (2.7) [10]	19.1 (2.1) [10]	17.5 (1.8) [10]	15.1 (1.7) [10]	18.1 (2.2) [10]	15.0 (1.9) [10]
Pongo pygmaeus	10	22.7 (3.2) [9]	20.8 (3.4) [9]	19.7 (4.4) [9]	14.8 (1.7) [9]	18.9 (2.7) [9]	14.0 (1.7) [9]
Papio anubis	15	18.6 (2.2) [15]	15.1 (1.6) [15]	13.8 (1.7) [15]	13.9 (1.8) [15]	15.3 (1.4) [15]	12.9 (1.6) [15]
StW 515		20.1	–	–	–	–	–
StW 567		21.5	20.4	15.9	14.5	22.0	23.5
KNM-KP 29285		27.9	25.1	23.3	19.9	23.0	23.4
KNM-WT 15000		33.4	27.8	23.1	23.2	28.4	27.7
A.L. 288-1ar		21.2	17.4	17.2	14.4	14.8	13.3
U.W. 88-21		22.9	19.0	19.7	18.5	19.3	16.8
U.W. 88-97		21.0	17.2	18.6	17.6	19.6	17.9

Cells contain mean values, one standard deviation in parentheses, and the sample size preserving measurements in brackets. Dashes in cells indicate missing measurements on fossils due to damage. All measurements were taken on original fossils, except those for U.W. 88-21 and U.W. 88-97 that were measured on casts. All measurements on extant distal tibiae followed the protocol of DeSilva (2009) and were taken with calipers by one of us (KC) in order to eliminate interobserver error. In a small reliability study incorporating 10% of the extant comparative sample, the second set of proximal measurements was within 3.7% of the original set. For distal measurements, the second set of mediolateral breadth measurements was within 2.9% of the original set, and the second set of anteroposterior length measurements was within 6.2% of the original set, except for the lateral-most dimension that was slightly higher. None of the second set of measurements were consistently higher or lower than the original sets.

and eventually tapers posteriorly as an equally subtle groove. Abrasion of the medial margin of the medial condyle obscures the precise superior border, requiring estimation of the proximodistal dimension of this depression. Evidence of a soleal line and the arrangement of a popliteus attachment are not preserved due to the location of the distal break. What remains of the attachment area of popliteus on the posterior surface is smooth. There is no evidence of a nutrient foramen on the preserved shaft.

StW 514—Right proximal tibia

Berger and Tobias (1996) described StW 514. They noted that it is associated with StW 515, a partial right distal tibia (Figure 13.2, Table 13.1). Berger and Tobias (1996) refer to StW 514 and StW 515 as StW 514a and StW 514b, respectively. We use the former set of specimen numbers, as these correspond to the original catalogue entries. Whenever possible, we attempt to replicate measurements of Berger

and Tobias (1996), as well as report supplementary measurements and additional description. Only when there is a discrepancy do we note their earlier measurements and emphasize their description.

Preservation. StW 514 is a partial right proximal tibia of an adult (Figure 13.2) that preserves parts of both tibial condyles, the tibial plateau, and the proximal shaft. It has a maximum length of 66.7 mm from its proximal- to distal-most points, which correspond to the medial border of the lateral condyle and the anterolateral shaft, respectively. This specimen does not appear to exhibit any plastic deformation.

A vertically oriented break isolates much of the medial condyle and its supporting tibial plateau from the remainder of the specimen, but there is no apparent displacement of either portion. In superior view, the break passes diagonally through the anterior intercondylar area and posteromedially through the lateral border of the medial condyle. It extends distally into the shaft

StW 514

StW 515

Figure 13.2 StW 514 is a right proximal tibia (top) and StW 515 is an associated distal tibia (bottom). Specimens are illustrated from left to right in anterior, posterior, lateral, medial and superior views.

until it meets a second, obliquely oriented break through the shaft. Anteriorly and superiorly, the vertical break has well-defined margins with a maximum gap of approximately 0.8. Posteriorly, the vertical break is continuous with a sizable, wedge-shaped missing section of the tibial plateau and medial condyle that exposes underlying trabecular bone. In the superior view, this wedge-shaped section is narrowest centrally and progressively widens mediolaterally until it intersects where the posterior surface apparently would be if it were retained.

The lateral-most regions of the lateral condyle and tibial plateau have been sheared off by a second, less extensive vertical break that is oriented anterolaterally to posteromedially (Figure 13.2). Thus, form and orientation of the largely missing proximal fibular facet cannot be assessed. The more proximal of the two breaks transecting the shaft is oriented obliquely such that its most proximal position occurs on the posteromedial shaft. Near this point, there is a small amount of localized crushing along the superiormost extent of the break. The break proceeds inferiorly onto the anterior and posterior surfaces reaching its most distal position at the inferior border of the tibial tuberosity. Also, on the anterior surface, a small triangular section of the cortex has spalled off the tibial plateau slightly inferior to the medial condyle (Figure 13.2).

Distally, the shaft is jaggedly broken and missing below an oblique plane that roughly parallels the orientation of the more proximal break transecting the shaft. A small vertically oriented notch in the shaft is centered about the distal break, interrupting the distal extension of the tibial tuberosity. Otherwise, this break does not result in anatomical displacement of either portion. The distalmost break exposes the medullary cavity, thus facilitating reasonable cortical thickness measurements of the shaft. Two small fragments (16.4 mm and 10.9 mm in maximum dimensions) are among those comprising StW 514, but while they preserve external cortical surface, they do not retain sufficient diagnostic features to include them further.

Morphology. The distal break exposes general thickness of the elliptical diaphyseal shaft, plus dimensions of the medullary cavity. While a true cross-section of the diaphysis could be derived via radiographic or CT imaging to more precisely characterize its shape and thickness, here only qualitative assessment is provided. The shaft cross-section has a more mediolaterally narrowed anterior than posterior border (i.e., sharper border anteriorly). Below the tibial plateau, the shaft is triangular with surfaces facing anteromedially, laterally, and posteriorly. Maximum thickness of the shaft at the distal-most break occurs anteriorly (6.6 mm) and less so posteriorly (5.8 mm), while minimum thickness occurs posterolaterally (2.4 mm) and less so anteromedially (2.7 mm). The anteroposterior breadth of the external shaft at the proximal-most point of the break is estimated as 26.5 mm, while mediolateral breadth of the external shaft is 19.0 mm. The external circumference above the distal break and in a transverse plane through the inferior part of the tibial tuberosity is 80 mm. Insufficient shaft remains to characterize either shaft curvature or the extent of tibial retroflexion.

The dimensions of the tibial plateau are 31.7 mm anteroposteriorly (measured from below the medial border of the lateral condyle posteriorly to below the anterior intercondylar area anteriorly) and 52.2 mm mediolaterally. The medial condyle is estimated as 20.2 mm in mediolateral breadth (Table 13.2), while its anteroposterior dimension cannot be measured due to damage on its posterior margin. The lateral condyle is estimated as 19.2 mm in mediolateral breadth and 26.8 mm in anteroposterior dimension. Berger and Tobias (1996) report a similar mediolateral breadth as estimated here (about 20 mm), but they report a substantially smaller anteroposterior dimension of 16.8 mm. The latter measure cannot be accurate but rather would appear to be an error in the original description by Berger and Tobias (1996).

The medial condyle is palpably depressed in its central region, but otherwise it is generally flattened in mediolateral and anteroposterior dimensions. The lateral condyle of StW 514, on the other hand, is anteroposteriorly convex and mediolaterally flattened. It is also relatively circular in shape. The articular surface of the lateral condyle is slightly tilted from the horizontal in StW 514 such that its lateral surface is slightly inferior to its medial surface.

The medial and lateral condyles appear to occupy approximately the same horizontal plane in anterior and posterior views (Figure 13.2). The intercondylar tubercles are sheared off in a horizontal plane at or just above their bases. The remaining edges are worn smooth, making it impossible to characterize height or form of the intercondylar tubercles. The intervening region of the intercondylar eminence is also broken off and worn. While we estimate the distance between the preserved portions of the tubercle bases as 5 mm, which is less than the 6 mm estimated by Berger and Tobias (1996), it is not possible to characterize the precise direction of the ridge connecting the intercondylar tubercles. Bases of the tubercles in StW 514 appear to be converging where the intercondylar eminence is broken away posteriorly. Intercondylar eminence breadth is evaluated relative to the overall breadth of the tibial plateau (52.2 mm), or to the breadth of the medial or lateral condyles (20.2 mm and 19.2 mm, respectively).

While the anterior intercondylar fossa is relatively complete, due to the incompleteness of the posterior intercondylar fossa and damage in this region, it is not possible to assess whether a notch indicating a ligamentous connection of the posterior horn of the lateral meniscus to the intercondylar eminence is present or absent. It also is not possible to assess an attachment of the posterior cruciate ligament in this area. The attachment for the anterior cruciate ligament is represented as a poorly delineated roughened surface rather than a pit. Where the posteromedial margin of the lateral condyle intersects the lateral edge of the posterior intercondylar fossa, what remains of its border is short and appears to be relatively vertically oriented.

While the tibial plateau is missing posterolaterally, the remaining metaphysis is inflated such that the proximal shaft approaches the tibial plateau as a more gradually sloped junction, as in modern humans, and unlike the sharply angled shelf-like junction that characterizes apes.

A vertically oriented tibial tuberosity is positioned lateral to the diaphyseal midline, with a sharper lateral margin and a rounder medial margin. Distal to the tuberosity, the subcutaneous border remains flattened with no groove.

Proximally, above the tuberosity and in the same sagittal plane, but distal to the intracapsular area, a depression and groove in the tibial surface are visible. At the proximal border of the depression, the superomedial to inferolateral orientation of the groove is more ape-like than modern human-like (Berger and Tobias, 1996).

Lateral to the tibial tuberosity, the shaft is slightly hollowed and laterally facing, while medially the shaft is flattened and faces anteromedially. There is a visible, vertically oriented roughening for the pes anserinus attachment inferior to the tuberosity and bordering the medial side of the subcutaneous border. It is truncated by the distal break. We would not characterize this attachment as "clearly present as a marked depression," as others have done (Berger and Tobias, 1996: 344).

The insertion of *m. semimembranosus* is ovoid in shape with prominent superior, anterior, and inferior margins, The well-defined depression on StW 514 is proximodistally larger on its anterior border (7.8 mm) and tapers posteriorly (maximum estimate = 7.5 mm) until its form is interrupted by a missing region of the plateau.

Posteriorly on the metaphysis, the proximal attachment area of *m. popliteus* is hollowed out inferior to the posterior surface of the lateral condyle and tibial plateau. Lateral to this, the soleal line is well marked as a raised ridge extending superolaterally to inferoposteriorly from a break through the lateral portion of the tibial plateau to the distal break truncating the shaft. The shaft is broken proximal to where the soleal line would reach the superior margin of the origin of *m. flexor digitorum longus*.

The insertion of the iliotibial band is marked on the anterolateral face of the lateral condyle by a small 7.6 mm (proximodistal) by 6.0 mm (mediolateral) ovoid pit rather than a smooth or protruding (Gerdy's) tubercle. Inferior to this pit on the metaphysis, a horizontal ridge extending approximately 7.2 mm in anteroposterior length separates the proximal-most attachments of *m. extensor digitorum longus* (above) and *m. tibialis posterior* (below). Extending inferiorly from the anterior border of this horizontal ridge until reaching the distal break through the shaft is a distinct vertical ridge indicating the interosseous membrane attachment. A superiorly facing nutrient foramen is located 0.7 mm anterior to the vertical ridge at the level of the tibial tuberosity. This vertically oriented ridge separates the origins of *m. tibialis anterior* (anterior to ridge) and *m. tibialis posterior* (posterior to ridge). The interosseous membrane attachment is noticeably closer to the medial border of the tibial tuberosity than it is to the soleal line.

StW 181—Left distal tibia

Preservation. StW 181 (Figure 13.3, Table 13.2) is a partial left distal tibia of an adult that is 29.3 mm in length from its proximal- to distal-most points, which correspond to the posterolateral shaft and posterolateral rim of the talar facet, respectively. The specimen is broken proximally and missing through the metaphysis, or just superior to it. The malleolus is broken off and missing inferior to the plane of the talar facet, exposing trabecular bone in medial view. The talar facet is broken in a coronal plane and missing slightly anterior to its deepest point in the medial or lateral view. It is not possible to estimate anteroposterior or mediolateral curvature of the shaft based on the remaining portion.

In general, little of the cortical shell of the diaphysis remains anteriorly, medially, and posteriorly. Anteriorly, missing cortical bone from what remains of the shaft and metaphysis exposes underlying trabecular bone at a coronally oriented break. A vertically oriented groove through the exposed trabecular bone on the anterior surface extends distal to the talar facet, resulting in a rounded notch missing from the talar facet. Cortical bone is missing or weathered from the remainder of the metaphyseal surface, often exposing underlying trabecular bone. Most of the medial and posterior surfaces of the shaft are not preserved, particularly superior to the metaphysis.

Laterally, the shaft retains cortical bone only on its posterior and inferior regions, both of which exhibit mild weathering of the surface. The area on the lateral surface bordering the superior edge of the remaining distal fibular facet exhibits minor abrasion, as well as pitting along its posterior border. The posterior margin of this overall area is modestly ridged, also with a heavily abraded surface.

Morphology. A medullary cavity is absent due to the distal location of the break and cortical thickness of the preserved shaft cannot be accurately measured. Thus, functionally informative cross-sections of the shaft, or local thickness measurements thereof, are unavailable. The external circumference of the distal shaft also is not measurable.

Based on what remains of the metaphysis (only its posterior surface), it appears to be posteriorly bulbous, resembling a modern human-like condition rather than a non-human ape-like condition. The metaphysis extends a maximum of 4.7 mm beyond the posterior border of the talar facet. An anteroposterior offset of the talar facet on the distal metaphysis, however, is not possible to estimate because of the missing anterior region.

There is an unusual, inferiorly opening triangular pit in the metaphysis, possibly a vascular foramen, bordering the posteromedial corner of the talar facet. Mediolaterally the pit is 3.5 mm, while anteroposteriorly it is 1.4 mm. Its maximum anteroposterior dimension is on its medial edge. The posterior border of the pit is roughly parallel with the course of the more lateral part of the posterior rim of the talar facet, resulting in a slight anterior indentation of the talar facet in this region.

Overall shape of the talar facet cannot be determined because of damage. Insufficient shaft remains to estimate shaft angle relative to the talar facet in the anterior view, or to estimate anterior tilt in the lateral view. At the posteromedial border of the talar facet, a subtle lip exists where a section of the articular surface descends inferiorly onto the base of the otherwise missing malleolus. Thus, only mediolateral breadth along the posterior border can be reliably measured (19.7 mm: Table 13.2), while other mediolateral and all anteroposterior dimensions of the talar facet are not measurable because of the missing areas.

There is a subtle anteroposterior-oriented trochlear keel in the mediolateral midpoint of the talar facet. In its anterocentral region, the remaining talar facet provides evidence suggestive of what may have been a pronounced "beak" at the intersection of the keel and the anterior rim of the talar facet, although any direct evidence of a central "beak" in the anterior margin of the talar facet has been removed. Where the keel intersects the posterior rim of the talar facet, an inferiorly directed projection, or beak, is absent. Rather, there is only a slight swelling at this intersection. In lateral view, the talar facet is anteroposteriorly concave. The precise anteroposterior location of its deepest point cannot be definitively assessed, however, due to the extent of the anteriorly missing articular surface.

The posterior surface of the diaphysis appears mediolaterally convex superior to the plafond, transitioning into what appears to be a more flattened surface superior to the metaphysis.

The lateral surface of the shaft is anteroposteriorly flattened in the region of the distal fibular facet, not

StW 181

StW 358

StW 389

StW 567

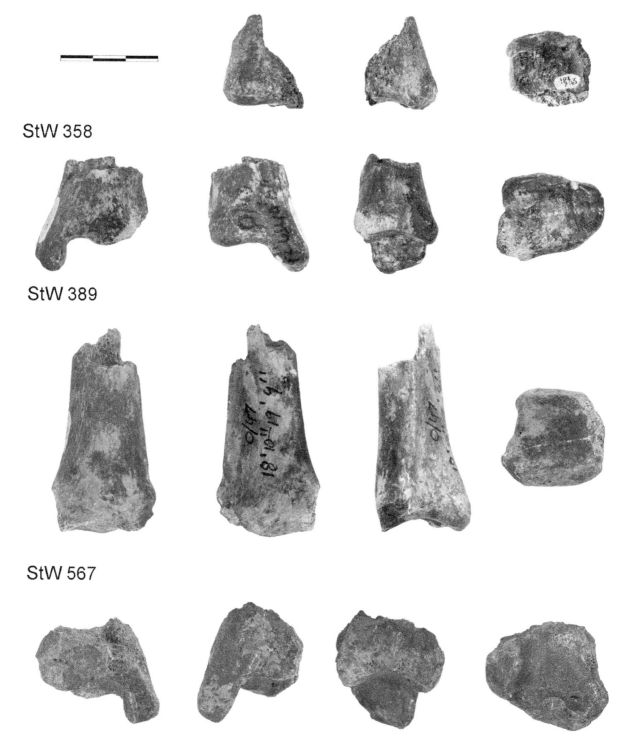

Figure 13.3 Distal tibiae in each row are illustrated from left to right in anterior (not shown for StW 181), posterior, lateral, and inferior views (anterior down).

being at all convex or concave, and is entirely absent superior and anterior to a plane running superoposterior to anteroinferior through the talar facet. Remnants of a vertical ridge along the lateral surface terminate as a distinct tubercle for the posterior tibiofibular ligament. This tubercle has a maximal anteroposterior thickness at its inferior termination (5.0 mm), which extends inferiorly to occupy the same transverse plane as the most superior part of the intact distal fibular facet.

The distal fibular facet has a superoinferior dimension of 2.8 mm, which is a minimum estimate since the superior and inferior margins of the facet diverge toward the missing anterior region of the facet. The remaining facet has an anteroposterior dimension (10.0 mm), which must be considered a minimum due to the missing anterior portion. The distal fibular facet of StW 181 is distinctly separated from its talar facet by a rounded ridge.

StW 358—Left distal tibia

StW 358 has been associated with StW 363, a left talus, on the grounds that the specimens articulate (Fisk and Macho, 1992). We suggest that StW 358 also may be associated with StW 356, a partial left fibular shaft. We propose this association based on proximity of the recovery locations of the specimens (e.g., adjacent excavation squares), as well as on their sampling individuals of apparently similar body sizes and preservation conditions.

Preservation. StW 358 is a partial left distal tibia of an adult that is 33.0 mm in proximodistal length (Figure 13.3, Table 13.2). Proximally, the specimen is broken through the diaphysis above the metaphyseal expansion. The break is slightly irregular and oriented approximately in a transverse plane. At the break, a small section of the anterior cortical shell (10 mm in mediolateral breadth and 3 mm in superoinferior height) has been pushed inward relative to the anterior surface of the shaft. A second, larger section of the cortical shell (13 mm in mediolateral breadth and 11 mm in superoinferior height) has been separated and pulled out from the posteromedial surface of the shaft, such that its anterior margin protrudes slightly beyond the plane of the medial surface of the shaft. Anteriorly and posterolaterally, there are a couple of additional bone flakes missing from the shaft surface at the break. A crack through the talar facet extends from its anteromedial border to the midline of the lateral border. Glue is visible in the depth of the lateral half of the crack. It is not possible to estimate anteroposterior or mediolateral curvature of the shaft based on the remaining portion.

Anterolaterally, the surface of the metaphysis is damaged and missing, exposing trabecular bone where the distal fibular facet and the anterolateral margin of the talar facet would be. Relatively little abrasion, weathering, or surface damage is evident elsewhere on the specimen, except on the malleolar surface. A strip of the surface lateral and anterior to the distal tip of the malleolus is abraded, with small abraded areas also evident along the medial surface near its inferior face, and at the inferiormost point on the posterolateral surface.

Morphology. A medullary cavity is absent at the proximal break, instead exposing only trabecular bone. Cortical thickness of the preserved shaft tapers inferiorly because of the presence of trabecular bone, meaning a representative cross-section of the shaft, or local thickness measurements thereof, are unavailable because of the distal location of the break. The external circumference of the shaft immediately below the proximal break is 68 mm.

The anterior side of the shaft is subtly, mediolaterally convex, transitioning to a slightly undulating surface at the metaphysis. The posterior side of the shaft is mediolaterally convex, facing posteriorly at its proximal-most extent, but gradually flattening at the metaphysis.

The metaphysis is only slightly anteroposteriorly expanded relative to the shaft and the talar facet. The talar facet appears relatively centered on the metaphysis such that anterior and posterior expansions of the metaphysis are essentially equivalent. Insufficient shaft is preserved to reliably measure the angle of the shaft relative to the talar facet in anterior or lateral views. The maximum anteroposterior dimension of the metaphysis (estimated as 23.8 mm) is lateral to the midline.

There is a medial "squatting facet" resembling the form illustrated by Singh (1959: Figure 11). Its lateral-most extent is missing due to damage. Presence or absence of a lateral "squatting facet" and the dimensions of the distal fibular facet cannot be assessed to damage in the respective areas.

Even with the anterolateral portion of the talar facet partially missing, it is still clear that the talar facet is relatively trapezoidal in overall shape. It has an anterior border that is mediolaterally broader than the posterior border, and a fibular border that is anteroposteriorly broader than the malleolar border (Table 13.2). Mediolateral breadths of the talar facet are 18.4 mm (estimated) and 17.1 mm, anteriorly and posteriorly, respectively (Table 13.2). While anterior breadth is estimated due to damage, central mediolateral breadth is not considered reliably measurable because of damage in the same region. Anteroposterior lengths of the facet are 16.8 mm (estimated), 16.9 mm, and 16.7 mm, laterally, centrally, and medially, respectively (Table 13.2). Laterally, anteroposterior breadth is estimated due to damage. Medially, anteroposterior breadth is a maximum measurement because of the lack of a distinct boundary between the border of the talar facet and the "squatting facet" that continues onto the anterior surface of the metaphysis. The talar facet continues inferiorly and perpendicularly onto the lateral surface of the malleolus at an abrupt transition. Posterior to the anteroposterior midpoint of this vertical part of the articular surface, the articular surface faces laterally and extends inferiorly a maximum of 7.2 mm below the horizontal portion of the talar facet. Anterior to the anteroposterior midpoint of the malleolus, the inferior continuation of the articular surface descends more inferiorly onto the lateral surface of the malleolus and faces anterolaterally rather than strictly laterally. Anterior to the anteroposterior midpoint, the articular surface extends inferiorly a maximum of 9.6 mm below the horizontal portion of the talar facet.

In inferior view, the talar facet is mildly mediolaterally convex because of a central trochlear keel. Along the anterior border of the talar facet, and in anterior view, there is a distinct inferiorly projecting swelling at its junction with the trochlear keel. Along the posterior border of the talar facet, and in posterior view, there is an inferiorly projecting swelling also at its junction with the keel. The keel is more distinct anteriorly than posteriorly, and especially compared to its central region. It diverges subtly from the midline, being instead oriented posterolateral to anteromedial. In lateral view, the talar facet is anteroposteriorly concave, with its deepest point (2.7 mm) being approximately centrally located.

The medial side of the shaft, while damaged, appears relatively flattened and gradually expands medially as it transitions onto the surface of the malleolus. In the medial view, the malleolus visually dominates the rest of the distal end, largely obscuring the talar facet beneath an only weakly anteroposteriorly expanded metaphysis. The anteroposterior dimension of the malleolus is greatest at its base (16.8 mm), near its junction with the talar facet. The malleolus narrows toward its inferior surface. The mediolateral dimension of the malleolus is greatest in the posterior half of its base (9.6 mm). The greatest mediolateral breadth of the malleolus occurs in its distal half, after which it rapidly narrows to the inferior surface of the malleolus.

In inferior view, the malleolus gradually narrows in mediolateral breadth from its anteroposterior midpoint until its sharp anterior border, where it meets the anterolaterally facing articular facet that also angles anterior toward this border.

The maximum superoinferior dimension of the malleolus projecting inferior to the plane of the talar facet (12.6 mm) is relatively more anterior rather than posterior. This asymmetry creates an inferiorly, blunt-pointed malleolus rather than a squared-off malleolus in medial view. The inferior surface of the malleolus is slightly uneven, but without distinct ridges marking attachments for tibial contributions to the deltoid ligament.

Posteriorly, a gentle swelling protrudes from the medial surface of the metaphysis immediately above the talar facet, and is paired with what appears to be an even

more protuberant region evident more laterally. The two subtle swellings are 4.5 mm and intervene between the impressions for tendons of *m. tibialis posterior* and *m. flexor hallucis longus*. These paired swellings effectively create a shallow intervening furrow running proximodistally and slightly medial to the inferiorly projecting junction of the trochlear keel and the posterior border of the talar facet.

The groove for *m. tibialis posterior* is 16.3 mm in length, proceeding from the posteromedial side of the metaphysis and descending inferomedially onto the posterior surface of the malleolus. The groove is 6.5 mm in mediolateral breadth, and <1.0 mm in depth.

At the proximal break, the lateral side of the shaft is not readily distinguishable from the posterior side. The interosseous crest dominates the lateral side of the preserved shaft, already occupying an anterolateral position at the proximal break. It continues inferiorly along the anterolateral margin of the shaft, but its termination cannot be assessed because of damage to this area. Specifically, it is not possible to characterize its vertical position relative to the missing distal fibular facet, nor can a maximum anteroposterior thickness of the interosseous crest be reliably estimated.

StW 389—Left distal tibia

Preservation. StW 389 is a partial left distal tibia of an adult (Figure 13.3, Table 13.2) that is 63.8 mm in length on the posterolateral shaft and the posterior rim of the talar facet, respectively. Proximally, the specimen is broken through the diaphysis. The surface of the break is relatively even but tilted anteriorly from a transverse plane such that its distal-most point is anteromedial on the shaft and its proximal-most point is posterolateral on the shaft. Posteriorly and posterolaterally at the break, a small section of the full thickness of the cortical shell protrudes 10.4 mm proximally from the remainder of the break. Relatively little abrasion, weathering, or surface damage is evident on the specimen, with perhaps the exception of the lateral surface immediately proximal to the distal fibular facet. It is not possible to estimate anteroposterior or mediolateral curvature of the shaft based on the remaining portion.

Distally, the malleolus is missing, having been cleanly sheared off in a transverse plane immediately inferior to the point where the talar facet continues onto the lateral surface of the malleolus, and also parallel to the vertical plane of the medial surface of the shaft. This results in an inferiorly projecting lip on the medial border of the talar facet. The central part of the lip has been abraded, creating a concave border. Trabecular bone is exposed on the medial surface beginning 11 mm–13.5 mm superior to the talar facet and continuing down to the level of the facet, or slightly more distally at the posteromedial corner of the facet. In the anteromedial region, the break through the malleolus has left a portion of the articular surface behind, creating a small ridge (<1 mm) projecting inferior to the talar facet. The break has left slightly more of an inferiorly projecting ridge in the posteromedial region. The medial-most section of the posterior rim of the talar facet (estimated length is 6 mm) is missing due to damage. *Morphology.* The proximal break exposes thickness of the elliptical shaft and dimensions of the medullary cavity. A cross-section could be estimable, including its shape and orientation, but it is not reported here. External circumference of the shaft just below the break is 66 mm. At the break, the mediolateral diameter of the external shaft is 19.4, while anteroposteriorly it is 20.8 mm. A medullary cavity is present and is elliptical in shape with its major axis running anterolaterally to posteromedially (10.5 mm in maximum internal diameter) and its minor axis (8.5 mm in maximum internal diameter) running perpendicularly. Maximum (21.1 mm) and minimum (17.8 mm) external diameters of the shaft are similarly oriented. Shaft thicknesses at the proximal break are largest anteriorly (4.8 mm) and posteromedially (5.9 mm) and smallest medially (3.5 mm) and posterolaterally (4.2 mm).

The anterior side of the shaft is mediolaterally convex, transitioning to becoming more flattened at the metaphysis. The posterior surface of the shaft is mediolaterally convex, facing posterolaterally at its proximal surface but gradually flattening into a posteriorly facing surface nearer the metaphysis. The metaphysis is expanded relative to the talar facet.

The metaphysis is moderately bulbous and expanded relative to the shaft, with an anteroposterior dimension 28.7 mm in the midline, while the talar facet is 21.3 mm in the same dimension. The angle of the shaft to the talar facet in anterior or lateral view is not reliably estimated due to the majority of the shaft being missing.

The talar facet is offset on the metaphysis such that, in inferior view, more of the metaphysis extends anteriorly beyond the facet than posteriorly. There is a "squatting facet" (Singh, 1959), extending anteriorly over the rim of the talar facet onto the anteroinferior surface of the metaphysis. It has a mediolateral dimension estimated as 12.5 mm due to damage along the medial part of the anterior rim of the talar facet, and a maximum superoinferior dimension of 3.4 mm. This "squatting facet" is more mediolaterally centered about the trochlear keel of the talar facet than is the example illustrated by Singh (1959: Figure 10.b), and a distinct medial "squatting facet" is not apparent.

The talar facet is trapezoidal in shape, with the anterior border being mediolaterally broader than the posterior border, and the fibular border being anteroposteriorly broader than the malleolar border. Mediolateral breadths of the talar facet are 23.0 mm, 20.4 mm, and 19.6 mm,

anteriorly, centrally, and posteriorly, respectively (Table 13.2). Central breadth was measured in a coronal plane just anterior to the posterior border of the distal fibular facet. Anteroposterior lengths of the talar facet are 22.6 mm, 20.2 mm, and 16.2 mm (estimated), laterally, centrally, and medially, respectively (Table 13.2). Central length was measured along the trochlear keel, while the medial dimension represents a minimum estimate due to damage on the medial part of the anterior rim.

From the inferior view, the talar facet is mediolaterally flattened except for a gently sloping trochlear keel extending along its anteroposterior midline. From the posterior view, the junction between the posterior rim of the facet and the trochlear keel is marked by a relatively laterally positioned gentle inferiorly projecting swelling. From the lateral view, the talar facet is concave anteroposteriorly, with its deepest point (4.3 mm) being centrally located.

The medial surface of the shaft is relatively flat, facing anteromedially at its proximal surface. The medial surface gradually becomes more medially facing nearer the metaphysis. The malleolus form cannot be assessed as it is missing below the level of the talar facet. Likewise, missing or damaged malleolar and surrounding regions do not permit assessment of the presence or absence of a groove for the tendon of *m. tibialis posterior*. There is no evidence of a distinct groove for the *m. flexor hallucis longus* tendon on the posterior surface of the remaining metaphysis.

A distinct lateral surface can be differentiated from the posterior surface only immediately above the metaphysis. In this region, the fibular notch is mediolaterally concave, and bounded anteriorly and posteriorly by distinct ridges. Anteriorly, the ridge is a prominent interosseous crest that occupies a lateral position on the shaft toward the anterolateral surface of the metaphysis, terminating 2.2 mm above the superior margin of the distal fibular facet. The crest gradually thickens in its anteroposterior dimension until reaching a 4.1 mm maximum at its inferior termination, where the interosseous crest takes the form of a small vertically oriented tubercle for attachment of the anterior tibiofibular ligament, perhaps indicating the presence of an inferiorly thickened interosseous membrane just above the talocrural joint (Stern, 2003). The attachment for the posterior tibiofibular ligament is evident as a second (smaller) tubercle, but unlike the proximodistal orientation of the anterior talofibular ligament attachment, is inclined in an anterosuperior to posteroinferior direction. Maximum distance (inferior-most) between the tubercles in a transverse plane is 20 mm.

The presence of a vertical gap between the termination of the interosseous crest at the anterior tubercle and the distal fibular facet that is located distal to it appears to resemble the chimpanzee condition rather than the modern human condition since in the latter the interosseous crest appears to end anterior to or extend inferior to the superior margin of the distal fibular facet. In other words, the anteroposterior position of the distal fibular facet on the StW 389 metaphysis is shifted anterior to an anteroposterior midpoint of the lateral surface rather than the more evenly distributed facet about the anteroposterior midpoint of the lateral surface exhibited by modern humans.

The distal fibular facet of StW 389 measures 13.5 mm and 5.0 mm in anteroposterior and superoinferior dimensions, respectively. The distal fibular facet is distinguished from the talar facet by a subtle, rounded ridge. The distal fibular facet is more obliquely than perpendicularly oriented to the longitudinal axis of the shaft.

StW 515—Right distal tibia

StW 515 (Figure 13.2, Tables 13.1 and 13.2) is likely associated with StW 514, proximal tibia.

Preservation. StW 515 is a partial right distal tibia of an adult (Figure 13.2) that is 36.9 mm in length from its proximal- to distal-most points, which correspond to the posterior shaft and the broken inferior surface of the malleolus, respectively. Multiple breaks, particularly medially and posteriorly, have removed diagnostic and functionally informative features of its surface. Internal differentiation between cortical, subchondral, and trabecular bone boundaries is difficult at breaks exposing internal areas due to the preservation conditions of the specimen, nor are trabecular bone networks distinct. The specimen is hypermineralized such that it is heavier than expected for its size. It is not possible to estimate anteroposterior or mediolateral curvature of the shaft based on the remaining portion.

Proximally, the specimen is unevenly broken through its diaphysis just above the beginning of the metaphyseal expansion. Anteriorly, most of the surface of the shaft has flaked off from the proximal break and is missing nearly to the talar facet. Medially, the remaining shaft is moderately abraded, while anterolaterally a vertical strip of cortex is spalled off parallel and anterior to the interosseous crest extending distally until the horizontal plane of the distal fibular facet. Posteriorly, an obliquely oriented break passes posterosuperiorly to anteroinferiorly through the metaphysis removing much of its surface along with the posterior half of the talar facet. The most proximal point of this break is located posterolaterally on the distal shaft, from which point it gradually extends distally onto the lateral and medial sides of the specimen. On the lateral surface, this break has removed all evidence of the posterior margin of the talar facet and the posterior tubercle. Only the most anterior area of the distal fibular facet remains.

Medially, a break extends inferiorly and anteriorly through the malleolus, leaving only the anterior portion of the malleolus in place. The inferior tip of the remaining portion of the malleolus is broken off and missing. The break through the metaphysis removes the talar facet posterior to its deepest point (in the lateral view), making it impossible to assess the relative anteroposterior location of this deepest point in the talar facet. The lateral half of the anterior edge of the talar facet has been worn away.

Morphology. A medullary cavity is absent, instead being replaced by trabecular bone due to the distal location of the break. Because of the extensive mineralization that obscures cortical and trabecular bone boundaries, the nature of the internal structure is difficult to determine. Thus, a functionally meaningful cross-section of the shaft, or local thickness measurements thereof, are unavailable. Measurement of an external circumference on the distal shaft also is not possible.

Damage to the anterior surface does not permit an accurate assessment of its surface morphology; however, it is clear that the metaphysis is bulbous relative to the shaft and the talar facet. In inferior view, the metaphysis projects beyond the anterior border of the talar facet despite the missing cortical bone on the anterior surface of the shaft.

Posteriorly, the shaft appears mediolaterally convex, although the posterior profile of the metaphysis is not possible to precisely characterize because of damage in the region. It is also not possible to evaluate the extent of any offset of the talar facet on the metaphysis. Finally, an insufficient amount of the shaft is preserved to estimate angles of the shaft relative to the talar facet in anterior or lateral view.

There is no evidence of a medial "squatting facet" or smooth groove for the talar neck (cf. Singh, 1959), while the presence or absence of a lateral "squatting facet" cannot be determined because this area of the anterior margin of the talar facet is damaged.

Overall shape of the talar facet cannot be characterized because of damage to the specimen. Thus, anteroposterior and mediolateral dimensions of the talar facet cannot be measured because of damaged and missing areas, except for mediolateral breadth, which is 20.1 anteriorly (Table 13.2).

In inferior view, the talar facet appears relatively flattened mediolaterally but is interrupted by a palpable trochlear keel that appears to be oriented in an anteroposterior direction. Anterior and posterior intersections of the keel and talar facet margins are missing, although on the retained portion of the talar facet the base of the anterior intersection is swollen, hinting at the probable existence of some form of a rounded or sharp "beak."

In lateral view, the talar facet is anteroposteriorly concave. Its deepest point cannot be assessed since the facet is missing posteriorly and is damaged along its anterior margin. Only a small anteroposteriorly concave section of the external surface is preserved superior to the largely missing distal fibular facet. In this remaining portion, a distinct border between the distal fibular facet and the talar facet is absent. Rather, only a change in orientation of the articular surfaces distinguishes between the two facets.

Medially, the shaft is subtly, anteroposteriorly convex, transitioning seamlessly onto the medial surface of the malleolus. The malleolus extends 9.8 mm (minimum measurement) inferior to the plane of the talar facet. Laterally, an indication of the form of the interosseous crest is ill-defined because of abrasion on the lateral surface near the distal fibular facet. Its terminal point, and therefore presence of a distinct tubercle for the anterior tibiofibular ligament, cannot be definitively characterized. Nonetheless, there is a small portion of the distal fibular facet retained immediately inferior to the apparent termination of the interosseous crest.

The preserved portion of the distal fibular facet is 4.4 mm wide anteroposteriorly, while superoinferiorly it is 5.2 mm. Based solely on the preserved portion, the facet is oriented superolaterally to inferomedially on the metaphysis and is oblique to the lateral surface of the distal shaft.

StW 567—Right distal tibia

StW 567 may be associated with proximal tibia StW 396.

Preservation. StW 567 (Figure 13.3, Tables 13.1 and 13.2) is a partial right distal tibia of a young adult that is 40.1 mm in length from its proximal- to distal-most points, which correspond to the posteromedial shaft and the inferior surface of the malleolus, respectively. An epiphyseal line is visible anteriorly and anterolaterally. Proximally, the specimen is broken through the distal shaft. The surface of the break is irregular, with few distinct trabeculae and trabecular spaces visible due to the matrix infill. The break is tilted from a transverse plane such that its inferior-most point lies on the anterior surface of the metaphysis and its superior-most point lies on the posteromedial side of the shaft. It is not possible to estimate anteroposterior or mediolateral curvature of the shaft based on the remaining portion.

Orange-reddish matrix still adheres to the external surface of the fossil, particularly laterally and posteriorly. This color of matrix is typical in Member 5 (R. Clarke, June 2018, pers. comm.). A small concentration of matrix is centrally located on the posterior surface of the distal shaft and fills an eroded area on the posterior aspect of the malleolus. There is not enough of the posterior surface retained to evaluate distinct morphological features. The entire lateral surface of the metaphysis is abraded

such that the distal fibular facet cannot be discerned. The matrix fills a majority of the exposed trabecular spaces. At the intersection of the trochlear keel and the anterior border of the talar facet, abrasion of the inferiorly projecting surface exposes underlying trabecular bone. A small elliptical pit is visible on the anterolateral surface of the talar facet, measuring 2.8 mm anteroposteriorly and 3.8 mediolaterally. It has a depth of approximately 1 mm–2 mm. Matrix is visible at the bottom of this pit.

Morphology. A medullary cavity is absent, instead being filled by trabecular bone. Cortical bone of the preserved shaft has begun an inferior taper due to the presence of trabecular bone. Thus, a functionally meaningful cross-section of the shaft, or local thickness measurements thereof, are unavailable due to the distal location of the break. Measurement of an external circumference on the distal shaft also is not possible.

A profile of the anterior surface of the distal shaft is not possible due to damage. The anterior surface of the metaphysis appears relatively bulbous and expanded relative to the shaft and talar facet. Posteriorly, the metaphysis appears as if it would have been relatively mediolaterally flattened, but this is ultimately uncertain given the amount of missing external surface in the area. There is no distinct groove for *m. tibialis posterior* or *m. flexor hallucis longus* either on the metaphysis or on the malleolus, although parts of these surfaces remain obscured by adhering matrix.

The talar facet is offset on the metaphysis such that the latter projects beyond the former more anteriorly than posteriorly. Insufficient shaft is preserved to estimate angles of the shaft to the talar facet in the anterior or lateral view.

The anteroposterior dimension of the metaphysis is 33.1 mm in the midline, while the talar facet is estimated as 22.4 mm in the same dimension due to damage on the anterior border of the talar facet. The maximum anteroposterior dimension of the metaphysis is lateral to the midline. There does not appear to be a medial or lateral "squatting facet" (Singh, 1959), although damage at the intersection of the trochlear keel and the anterior border of the talar facet would have removed evidence if it were relatively small.

The talar facet is trapezoidal in shape, with the anterior border being mediolaterally broader than the posterior border, and the fibular border being anteroposteriorly broader than the malleolar border. The medial portion of the anterior border is pinched inward, shortening the anteroposterior dimension of the talar facet in this region. Mediolateral breadths of the talar facet are 21.5 mm, 20.4 mm, and 15.9 mm, anteriorly, centrally, and posteriorly, respectively (Table 13.2). Anteroposterior lengths of the facet are 23.5 mm, 22 mm (estimated), and 14.5 mm, laterally, centrally, and medially, respectively (Table 13.2).

In inferior view, the talar facet is flattened medio-laterally but interrupted in the midline by a gently sloping, anteroposteriorly oriented trochlear keel. At the anterior and posterior intersections of the keel and the talar facet margins, inferiorly projecting swellings are present. While the anterior intersection is abraded, it is relatively "beak"-like, especially when compared to the posterior intersection. In posterior view, the intersection between the posterior rim of the facet and the trochlear keel is marked by a relatively laterally positioned gentle inferiorly projecting swelling. In the lateral view, the talar facet is anteroposteriorly concave, with its deepest point (4.8 mm) being centrally located

The talar facet extends inferiorly onto the lateral surface of the malleolus. In the posterior half of this continuation, the articular surface extends inferiorly a maximum of 8.9 mm and is laterally facing. Anterior to the anteroposterior midpoint of this continuation, the articular surface gradually extends more inferiorly onto the lateral surface of the malleolus and transitions to facing more anterolaterally. In the anterior half of this distal continuation, the articular surface extends inferiorly a maximum of 15.9 mm below the talar facet.

Medially, a profile of the distal shaft cannot be assessed due to damage. More inferiorly, the medial side of the malleolus is flattened, with its surface facing relatively medial. The medial surface of the malleolus exhibits several superiorly facing foramina, possibly due to the non-adult developmental state. The anteroposterior dimension of the malleolus (22.8 mm) is greatest near its junction with the talar facet, and narrows toward its distal end. The greatest mediolateral dimensions of the malleolus are posteriorly rather than anteriorly situated, and inferiorly rather than superiorly situated, reaching a maximum mediolateral breadth (12.0 mm) approximately halfway between the plane of the talar facet and the inferiormost point of the malleolus. The narrower anterior border of the malleolus is relatively rounded, while in medial view the inferior projection of the malleolus is blunt-pointed rather than squared-off.

StW 356—Left fibular shaft

StW 356 may be associated with StW 358, a partial left distal tibia. We tentatively propose this association based on proximity of the recovery locations of the specimens (adjacent excavation squares), as well as their sampling apparently similar body sizes and preservation conditions.

Preservation. StW 356 is a partial left fibula (Figure 13.4) that is 227 mm in length, consisting of four conjoining shaft fragments. Neither articular end is preserved, as the specimen has an oblique superolateral to inferomedial break through the proximal shaft and a transverse break through the distal shaft. The proximal fragment is about

Figure 13.4 StW 356 is a left fibula shaft that is possibly associated with StW 358, a partial left distal tibia. The reassembled shaft is illustrated from left to right in anterior, medial and posterior views. Proximal is up.

one-third longer than the cumulative length of the three distal conjoined fragments. The shaft of the composite is slightly misaligned and exhibits minor surface damage at the distal of its two refit breaks. The proximal fragment refits to the composite fragment relatively securely on the medial side, but the refit is more tenuous anteriorly and laterally due to existing damage. A medullary cavity is not exposed at either broken end, suggesting their proximity to articular ends.

A nutrient foramen is visible on the posteromedial surface of the proximal-most fragment 13.5 mm superior from where the composite group (three) of more distal fragments refits. Proximal to the nutrient foramen there is 117 mm of the shaft preserved, while 110 mm of the shaft is preserved distal to the nutrient foramen. A groove for peroneal tendons is not palpable on the distal surface of the shaft, nor is the beginning of an expansion for the lateral malleolus distinct. A malleolar fossa is not preserved. Despite the absence of the fibular head or the malleolus, a majority of the shaft appears to be preserved (morphology exists hinting at proximity to articular structures).

Morphology. An irregular break separates the proximal fragment from the composite, approximately 12 mm–13 mm inferior to the nutrient foramen. The fracture exposes a relatively rounded cross-section. The external contour of the cross-section is unlike the anteroposteriorly elongated external contour of a cross-section typical of the mid-diaphyseal region of the modern human fibula. The medullary space exposed by the break is also approximately circular, with a diameter of 2.8 mm. At the level of the nutrient foramen, mediolaterally the external diameter of the shaft is 11.6 mm, while anteroposteriorly the external diameter of the shaft is 11.5 mm. Periosteal circumference of the shaft at the level of the nutrient foramen is 38 mm.

The shaft is relatively straight with subtle laterally directed flaring at both its proximal-most and distal-most breaks. Anteriorly, the shaft is sharpened proximally by a well-defined interosseous crest. Superior to the nutrient foramen the interosseous crest is present as a single anteriorly positioned ridge, creating a triangular shaft with rounded posteromedial and posterolateral corners. This single ridge divides into medial and lateral lips approximately 39 mm proximal to the nutrient foramen. Medial and lateral lips continue inferiorly diverging on the anterior surface until the medial lip gradually dissipates on the distal shaft and the lateral lip bifurcates 72 mm distal to the nutrient foramen. One branch of the lateral lip continues inferiorly onto the anterior surface for a brief extent before disappearing, while the other more laterally positioned branch wraps around the shaft onto its lateral surface. Ultimately, the continuation of the lateral lip onto the lateral surface of the shaft extends to the distal break as a well-defined ridge at the posterolateral border of the shaft. Thus, the proximal portion of the subcutaneous surface appears to face anterolaterally.

At the proximal break, a medial crest is distinct at the posteromedial corner of the shaft, separating medial and posterior surfaces. The defined medial crest descends inferiorly on the shaft where it begins to migrate onto the anterior surface 11 mm proximal to the nutrient foramen. The medial crest gradually dissipates as it approaches the similarly fading medial lip on the anterior surface of the composite.

Posteriorly, the shaft is flattened proximally, rounded in the mid-diaphyseal region, and flattens again distally. Both of the flattened sections appear to face more posterolaterally than posteriorly, but this is difficult to assess with certainty because of missing articular ends. A distinct ridge at the posterolateral border is visible at the proximal-most break. It gradually descends inferiorly onto the posterior surface of the shaft, and dissipates superior to the nutrient foramen, approximately at the same point where the medial crest begins migrating toward the medial lip on the anterior surface of the shaft.

Comparative anatomy

The Sterkfontein tibiae were compared to a sample of extant hominoid and fossils hominin tibiae (Table 13.3).

Proximal Tibia

The plateaus of both proximal tibiae from Sterkfontein (StW 396 and StW 514) (Figures 13.1 and 13.2) sit on an inflated metaphysis such that the proximal shaft approaches the tibial plateau as a gradual, sloped junction, as exhibited by modern humans rather than the sharply angled, shelf-like junction that characterizes chimpanzee tibiae (Aiello and Dean, 1990; Ward et al., 2001). In this, they resemble tibiae attributed to *Australopithecus anamensis* (KNM-KP-29285), the larger *A. afarensis* specimens (A.L. 330-6, A.L. 333-42, A.L. 333x-26), *A. sediba* (U.W. 88-64/78), and *Paranthropus boisei* (KNM-ER 1500). StW 514 is slightly less inflated, like the smaller *A. afarensis* specimens A.L. 129-1b and A.L. 288-1aq. All of these specimens exhibit even less metaphyseal flare than the tibia possibly attributable to *Homo erectus* (KNM-ER 1481), suggesting that body size may be a factor in the extent of this morphology, although the general similarity between StW 396 and the comparatively smaller StW 514 suggests that body size should not be considered as the only relevant factor.

In the only specimen for which it can be evaluated, StW 514, the tibial condyles occupy the same horizontal plane in anterior and posterior views (Figure 13.2). This morphology contrasts with the condition seen in tibiae of modern humans and from some attributed to *A. afarensis* (A.L. 330-6, A.L. 333-42, A.L. 333x-26) and *H. erectus* (KNM-ER 1481). Instead, StW 514 more closely resembles the tibiae of chimpanzees and other early hominin specimens, *A. anamensis* (KNM-KP-29285), *A. afarensis* (A.L. 129-1b), *A. sediba* (U.W. 88-64/78), and *P. boisei* (KNM-ER 1500).

The intercondylar eminence in both Sterkfontein tibiae is relatively wider that of *A. afarensis* (A.L. 129-1b, A.L. 288-1aq, A.L. 330-6, A.L. 333-42, A.L. 333x-26), is a reasonable match for that of *A. sediba* (U.W. 88-64/78), and that of StW 514 is especially narrower than in *A. anamensis* (KNM-KP-29285) or *H. erectus* (KNM-ER 1481). Given that the latter specimens are relatively large, relative intercondylar eminence breadth might appear related to body size, but the variation within the Sterkfontein and *A. afarensis* samples calls such a conclusion into question.

The deep, horizontal posterior intercondylar fossa of StW 396 (Figure 13.1) resembles the modern human condition rather than the more vertically oriented and anteroposteriorly shortened non-human ape condition. It appears slightly less posteriorly sloped than the posterior intercondylar fossa of StW 514, although this specimen is

Table 13.3 Comparative sample[1] used in these analyses.

		Source	Description
KNM-KP 29285	Right partial tibia	NMK	Leakey et al. (1995); Ward et al. (2001)
KNM-ER 1481	Left partial tibia	NMK	Day et al. (1975)
KNM-ER 1500	Left partial tibia	NMK	Day et al. (1976); Dominguez-Rodrigo et al. (2013)
KNM-ER 2596	Left partial tibia	NMK	DeSilva and Papakyrikos (2011)
KNM-WT 15000	Right tibia	NMK	Walker and Leakey (1993)
A.L. 129-1b	Right proximal tibia	NME	Johanson and Coppens (1976)
A.L. 288-1aq	Right proximal tibia	NME	Johanson et al. (1982)
A.L. 330-6	Right proximal tibia	NME	Ward et al. (2012)
A.L. 333x-26	Right proximal tibia	NME	Lovejoy et al. (1982)
A.L. 333-42	Left proximal tibia	NME	Lovejoy et al. (1982)
A.L. 288-1ar	Right distal tibia	NME	Johanson et al. (1982)
A.L. 333-6	Left distal tibia	NME	Lovejoy et al. (1982)
A.L. 333-7	Left distal tibia	NME	Lovejoy et al. (1982)
A.L. 333-96	Left distal tibia	NME	Lovejoy et al. (1982)
A.L. 545-3	Left distal tibia	NME	Ward et al. (2012)
KSD-VP-1/1e	Left tibia	NME	Haile-Selassie et al. (2010)
U.W. 88-64/78	Right proximal tibia	ESI	DeSilva et al. (2019)
U.W. 88-24	Left proximal tibia	ESI	DeSilva et al. (2019)
U.W. 88-21	Right distal tibia	ESI	DeSilva et al. (2019)
U.W. 88-97	Right distal tibia	ESI	DeSilva et al. (2019)
Homo sapiens	n = 16 (both tibiae)	WITS	–
Gorilla beringei	n = 20 (both tibiae)	NMNH, RMCA	–
Gorilla gorilla	n = 10 (both tibiae)	NMNH, RMCA	–
Pan paniscus	n = 14 (both tibiae)	RMCA	–
Pan troglodytes	n = 15 (both tibiae)	NMNH	–
Papio anubis	n = 15 (both tibiae)	NMNH, RMCA	–
Pongo abelii	n = 10 (both tibiae)	NMNH	–
Pongo pygmaeus	n = 10 (both tibiae)	NMNH	–

Abbreviations are as follows: ESI = Evolutionary Studies Institute, University of the Witwatersrand, NME = National Museum of Ethiopia, NMK = National Museums of Kenya, NMNH = National Museum of Natural History (Smithsonian), RMCA = Royal Museum of Central Africa, WITS = Dart Collection, School of Anatomical Sciences, University of the Witwatersrand. [1]Comparative extant samples are comprised of males and females (*H. sapiens* = 8 males, 8 females; *G. beringei* = 11 males, 9 females; *G. gorilla* = 7 males, 3 females; *P. paniscus* = 7 males, 7 females; *P. troglodytes* = 7 males, 8 females; *P. anubis* = 8 males, 3 females, and 4 unknown; *P. abelii* = 3 males, 7 females; *P. pygmaeus* = 2 males, 8 females). Comparative fossil samples include specimens attributed to *Australopithecus anamensis* (KNM-KP 29285), *A. afarensis* (A.L. and KSD specimens), *A. sediba* (U.W. specimens), *Paranthropus boisei* (KNM-ER 1500), *Homo erectus* (KNM-ER 1481 and KNM-WT 15000), or an unattributed hominin (KNM-ER 2596).

damaged in this area. StW 396 appears less sloped than in *A. anamensis* (KNM-KP 29285), and slightly more sloped than in one *A. afarensis* tibia (A.L. 288-1aq), but it bears a closer resemblance to that of others assigned to *A. afarensis* (A.L. 129-1b, A.L. 330-6, and A.L. 333x-26), *P. boisei* (KNM-ER 1500), and *H. erectus* (KNM-ER 1481). The posterior intercondylar fossa of StW 396 is also anteroposteriorly more elongated compared to smaller tibiae assigned to *A. afarensis* (A.L. 129-1b, A.L. 288-1aq), but resembles the condition of larger tibiae assigned to *A. anamensis* (KNM-KP-29285) and the larger *A. afarensis* specimens (A.L. 330-6, A.L. 333x-26). This variation is unlikely allometric, however, as the *H. erectus* tibia (KNM-ER 1481) has a similarly elongate posterior intercondylar fossa but is even larger in size.

The margins of the anterior intercondylar fossae of StW 396 and StW 514 are well-delineated and rounded, like those of other *Australopithecus* fossils, *A. afarensis*

(KNM-KP-29285) and *A. afarensis* (A.L. 129-1b, A.L. 288-1aq, A.L. 330-6, A.L. 333-42, A.L. 333x-26), and they differ from the raised margin exhibited by *H. erectus* (KNM-ER 1481).

Medial condylar shape varies in shape and contours within the Sterkfontein sample, as it does across early hominins. The medial condyle of StW 396 is generally elliptical, with a major axis in the anteroposterior direction, similar to that of *A. anamensis* (KNM-KP 29285), *A. afarensis* (A.L. 129-1b, A.L. 288-1aq, A.L. 330-6, A.L. 333-42, A.L. 333x-26), *A. sediba* (U.W. 88-64/78), and *H. erectus* (the non-adult KNM-WT-15000). StW 514, though, exhibits a more circular medial condyle that is more non-human ape-like than modern human-like. Also, in StW 396, the medial condyle has a marked central depression that becomes shallower until it reaches the edge of the articular surface, as exhibited by the A.L. 333x-26 tibia assigned to *A. afarensis*. In StW 514, however, the depression is contained centrally and is flatter, as in *A. anamensis* (KNM-KP 29285), most *A. afarensis* (A.L. 129-1b, A.L. 288-1aq, A.L. 330-6, A.L. 333-42), *A. sediba* (U.W. 88-64/78), and *H. erectus* (KNM-ER 1481). The Sterkfontein medial condyles are similar, though in having a low notch for the anterior horn of medial meniscus as do *A. anamensis* (KNM-KP-29285) and *A. afarensis* (A.L. 129-1b, A.L. 288-1aq, A.L. 330-6, A.L. 333-42, A.L. 333x-26), differing from the more raised margin exhibited by *H. erectus* (KNM-ER 1481).

The lateral condyle of StW 514 is relatively circular in shape with a mediolaterally flattened articular surface. It shares an anteroposterior convexity with a tibia attributed to *A. afarensis* (A.L. 129-1b), but exhibits more convexity than others attributed to *A. anamenis* (KNM-KP-29285), *A. afarensis* (A.L. 288-1aq, A.L. 330-6, A.L. 333-42, A.L. 333x-26), *A. sediba* (U.W. 88-24, U.W. 88-64/78), and *H. erectus* (KNM-ER-1481). The articular margin of the StW 396 lateral condyle has a posteromedial corner that slopes toward the posterior intercondylar fossa, similar to the condition exhibited in some tibiae attributed to *A. anamensis* (KNM-KP-29285), *A. afarensis* (A.L. 288-1aq, A.L. 330-6), and *H. erectus* (KNM-ER 1481), and unlike the more well-defined and sharper margin exhibited by other tibiae attributed to *A. afarensis* (A.L. 129-1b, A.L. 333x-26). In StW 514, though, this margin is almost vertically oriented, as in chimpanzees and *A. afarensis* (A.L. 129-1b, A.L. 330-6, A.L. 333x-26) and unlike in modern humans, *A. anamensis* (KNM-KP-29285), *A. africanus* (StW 396), or *H. erectus* (KNM-ER 1481).

A ligamentous connection for the posterior horn of the lateral meniscus cannot be definitively determined in StW 396, but a distinct facet in the posterior intercondylar fossa marks the attachment site of the posterior horn of the medial meniscus, despite a sharply defined posterolateral margin of the medial condyle. This configuration recalls the arrangement exhibited by tibiae assigned to *A. anamensis* (KNM-KP-29285) and *A. afarensis* (A.L. 288-1aq, A.L. 330-6, A.L. 333-42, A.L. 333x-26), but to a lesser extent that of *H. erectus* (KNM-ER 1481) which exhibits a smoother posterolateral margin.

Tibial tuberosity morphology can only be assessed for StW 514 (Figure 13.2). The rounded medial and rounded sharp lateral margins of the tibial tuberosity resemble the morphology seen in tibiae attributed to *A. anamensis* (KNM-KP-29285), *A. afarensis* (A.L. 129-1b, A.L. 288-1aq, A.L. 330-6, A.L. 333x-26), and *H. erectus* (KNM-ER 1481), but not that of *P. boisei* (KNM-ER 1500) that has more equally rounded lateral and medial margins. Proximal to the tuberosity and distal to the intracapsular area there is a depression and a relatively superomedial to inferolateral groove that resemble that those on tibiae attributed to *A. afarensis* (A.L. 129-1b, A.L. 288-1aq, A.L. 333x-26), while this groove appears more transversely oriented in other

hominin tibiae of *A. afarensis* (KSD-VP-1/1e) and *A. anamensis* (KNM-KP-29285). Distal to the tuberosity and medial to its sharp anterior margin StW 514 lacks the distinct groove seen in tibiae of chimpanzees and *A. afarensis* (A.L. 129-1b, A.L. 288-1aq, though to a lesser extent A.L. 333x-26), instead more closely resembling tibiae of *A. anamensis* (KNM-KP-29285) and *H. erectus* (KNM-ER 1481).

The insertion site for *m. semimembranosus* of both StW 396 and StW 514 ultimately recalls the form exhibited by tibiae attributed to *A. afarensis* (A.L. 288-1aq, A.L. 330-6). The attachment site for *m. semimembranosus* of StW 396 is as not as well-defined as on StW 514 or other tibiae attributed to *A. afarensis* (A.L. 288-1aq, A.L. 330-6, A.L. 333x-26), though, perhaps because this specimen may represent an earlier developmental stage.

The strong soleal line of StW 514 resembles that of tibiae attributed to *A. anamensis* (KNM-KP-29285) and *A. afarensis* (A.L. 129-1b, A.L. 288-1aq, A.L. 333x-26). The horizontal ridge separating attachment sites of *m. extensor digitorum longus* and *m. tibialis posterior* apparent on StW 514 is not apparent on tibiae attributed to *A. afarensis* that preserve the region (A.L. 129-1b, A.L. 288-1aq, A.L. 333x-26). The attachment site for the interosseus membrane in StW 514 is offset toward the margin of the tibial tuberosity rather than the soleal line, as in *A. anamensis* (KNM-KP-29285) and a tibia attributed to *A. afarensis* (A.L. 129-1b), and unlike on other tibiae attributed to *A. afarensis* (A.L. 288-1aq, A.L. 333x-26) where it is more equidistant.

Distal tibia

External surfaces of the distal tibial diaphysis vary within the Sterkfontein sample, as they do among other early hominins. The posterior aspect of the diaphysis is flat in StW 358 but is more transversely convex in StW 181, StW 389, and StW 515. A flat posterior diaphysis is also apparent on two tibiae attributed to *A. afarensis* (A.L. 333-6 and A.L 545-3) as well as one from *A. sediba* (U.W. 88-97), but less so in a tibia attributed to *A. anamensis* (KNM-KP-29285, three other *A. afarensis* (A.L. 288-1ar, A.L. 333-7, KSD-VP-1/1e), another *A. sediba* (U.W. 88-21), *P. boisei* (KNM-ER 1500), and *H. erectus* (KNM-ER-1481).

The anterior surface of the diaphysis shows a different pattern of variation. In StW 389, the diaphysis is transversely convex and becomes flatter at the metaphysis, but not as much as seen in StW 358 or *P. boisei* (KNM-ER 1500) but is similar to that of *A. afarensis* (A.L. 288-1ar, KSD-VP-1/1e, A.L. 333-6, A.L. 545-3), *A. sediba* (U.W. 88-21, U.W. 88-97), and *H. erectus* (KNM-ER 1481).

StW 358 is the only Sterkfontein tibia preserving tendinous grooves. On this specimen *m. tibialis posterior* and *m. flexor hallucis longus* tendon grooves are are separated by a broad eminence, resembling the human condition rather than exhibiting the narrower, sharper-ridged separation that characterizes one *A. afarensis* (A.L. 288-1ar) and also chimpanzees. The groove for *m. tibialis posterior* resembles the form of that exhibited on tibiae attributed to *A. anamensis* (KNM-KP 29285), *A. afarensis* (A.L. 288-1ar, A.L. 333-6, A.L. 545-3), and *A. sediba* (U.W. 88-97), although its margins are less sharply defined than those of the *A. afarensis* specimens.

Metaphyseal flaring is considerable in all early hominins, including StW 389, StW 515, and StW 567, as well as *A. anamensis* (KNM-KP 29285), *A. afarensis* (A.L. 288-1ar, KSD-VP-1/1e, A.L. 333-6, A.L. 545-3), *A. sediba* (U.W. 88-21, U.W. 88-97), and *H. erectus* (KNM-ER-1481, KNM-WT-15000), but it appears uniquely minimal in StW 358. The maximum anteroposterior dimension of the StW 358, 389, 515, and 567 metaphysis is lateral to the midline, also seen in some *A. afarensis* (A.L. 288-1ar, KSD-VP-1/1e, A.L. 545-3), one *A. sediba* (U.W. 88-97), and *P. boisei* (KNM-ER-1500), but it is unlike the position in other tibiae attributed to *A. afarensis* (A.L. 333-6) or *A. sediba* (U.W. 88-21) that exhibit a maximum dimension approximately on the midline, or those attributed to *A. anamensis* (KNM-KP-29285) and *H. erectus* (KNM-ER-1481) which are relatively uniform in anteroposterior dimensions.

The maximum inferior projection of the malleolus is situated farther anterior rather than posterior in StW 567. This creates a blunt-pointed malleolus rather than a squared-off malleolus in medial view unlike StW 515 and other hominins (e.g., A.L. 545-3). In this manner, StW 567 resembles other tibiae attributed to *A. africanus* (e.g., StW 358) or to *A. anamensis* (KNM-KP 29285), *A. afarensis* (e.g., A.L. 288-1ar, A.L. 333-6), *A. sediba* (e.g., U.W. 88-21, U.W. 88-97), *P. boisei* (e.g., KNM-ER-1500), and *H. erectus* (e.g., KNM-ER-1481). The inferior surface of the malleolus is slightly uneven, but without distinct ridges marking attachments for tibial contributions to the deltoid ligament.

The anterior border of the malleolus is sharper in StW 358 than StW 567, which is more more blunted as it is in *A. anamensis* (KNM-KP 29285), *A. afarensis* (A.L. 288-1ar, A.L. 333-6, A.L. 545-3), *A. sediba* (U.W. 88-21, U.W. 88-97), and *H. erectus* (KNM-ER 1481).

The malleolus becomes mediolaterally broader along its distal half in StW 358, StW 515, and StW 567, *A. anamensis* (KNM-KP 29285), *A. afarensis* (e.g., A.L. 288-1ar, A.L. 333-6, A.L. 545-3), and *A. sediba* (e.g., U.W. 88-21). A more distal maximum breadth of the malleolus appears to characterize chimpanzees, unlike the more even breadth apparent in the modern human malleolus and exhibited by *A. sediba* (U.W. 88-97) and *H. erectus* (KNM-ER 1481).

In the medial view, the malleolus tapers distally in StW 358 and StW 567 like malleoli of *A. anamensis* (KNM-KP 29285), *A. afarensis* (A.L. 288-1ar, A.L. 333-6), *A. sediba* (U.W. 88-21, U.W. 88-97), *P. boisei* (KNM-ER-1500), and *H. erectus* (KNM-ER-1481). This tapering is also seen in the articular extension of the talar facet onto the lateral surface of the malleolus in most of these specimens, although the inferior margin of the facet is more horizontal, or squared-off, on one *A. afarensis* tibia (A.L. 545-3).

In inferior view, the StW 567 malleolus is mediolaterally narrower in its anterior region (6.4 mm maximum), with a relatively blunted anterior border similar to the condition exhibited by tibiae attributed to *A. afarensis* (e.g., A.L 288-1ar, A.L. 333-6, A.L. 545-3), *A. sediba* (e.g., U.W. 88-21, U.W. 88-97), and *H. erectus* (e.g., KNM-ER-1481), and unlike the more sharply converging articular and medial surfaces exhibited by tibiae attributed to *A. africanus* (e.g., StW 358) and *P. boisei* (e.g., KNM-ER-1500).

The distal talofibular facet extends inferior to the interosseus crest in all Sterkfontein tibiae that preserve it (except perhaps StW 181), as in tibiae of *A. afarensis* (A.L. 288-1ar, A.L. 333-6, A.L. 545-3) and *A. sediba* (U.W. 88-21). The facet on StW 181 appears inclined only slightly superolateral to inferomedial, resembling the more vertical condition exhibited on modern human tibiae and those attributed to one *A. afarensis* (A.L. 288-1ar), *A. sediba* (U.W. 88-21), and *H. erectus* (KNM-ER 1481). The talofibular facet is oriented more obliquely on StW 389, StW 515, and StW 567, as non-human apes and those on *A. anamensis* (KNM-KP-29285) and most *A. afarensis* tibiae (KSD-VP-1/1e, A.L. 333-6, A.L. 333-7, A.L. 545-3). In the anterior or posterior view, the proximal end of the facet in StW 181 is raised above the metaphyseal surface as in *A. anamensis* (KNM-KP-29285), but unlike the more flush morphology exhibited by *A. afarensis* (A.L. 288-1ar, A.L. 333-6, A.L. 545-3) or *A. sediba* (e.g., U.W. 88-21).

The tubercles for attachment of the posterior and anterior talofibular ligaments of StW 181 and StW 389 are more prominent than on tibiae attributed to *A. anamensis* (KNM-KP-29285), some *A. afarensis* (A.L. 288-ar, A.L. 333-6), *A. sediba* (U.W. 88-21, U.W. 88-97), and *H. erectus* (KNM-ER 1481), but in overall prominence and form are similar to another *A. afarensis* (A.L. 545-3) and *P. boisei* (KNM-ER 1500). The thickened attachment for the anterior talofibular ligament may indicate the presence of a human-like inferiorly thickened interosseus membrane (Stern, 2003).

In lateral view, the talar facet is situated further to the posterior than anterior side of the metaphysis in StW 389 and StW 567, as it is in *A. afarensis* (e.g., A.L. 288-1ar and A.L. 545-3) and in the offset *P. boisei* (KNM-ER-1500). In contrast, StW 358 has a more centrally positioned talar facet, like tibiae of *A. anamensis* (KNM-KP-29285), *A. afarensis* (e.g., KSD-VP-1/1e and A.L. 333-6), *A. sediba* (e.g., U.W. 88-21 and U.W. 88-97), and *H. erectus* (e.g., KNM-ER-1481).

Also in the lateral view, the talar facet of StW 358, StW 389, and StW 567 is anteroposteriorly concave, with its deepest point being centrally located, as in two *A. afarensis* (A.L. 333-6, A.L. 545-3), *P. boisei* (KNM-ER1500), and *H. erectus* (KNM-ER 1481). This morphology contrasts with the more posteriorly situated deepest point along the talar facet in other tibiae attributed to *A. afarensis* (A.L. 288-1ar, KSD-VP-1/1e), or the relatively more anteriorly situated deepest point in tibiae attributed to *A. anamensis* (KNM-KP 29285) and *A. sediba* (U.W. 88-21, U.W. 88-97).

In distal view, the tibial facet of StW 389 is less trapezoidal than that of StW 567 or one *A. afarensis* (A.L. 545-3), but more than StW 358 and most *A. afarensis* tibiae (A.L. 288-1ar, KSD-VP-1/1e, A.L. 333-6), *A. sediba* (U.W. 88-21, U.W. 88-97), and *P. boisei* (KNM-ER-1500), but also less rectangular than seen in A. *anamensis* (KNM-KP 29285) and *H. erectus* (KNM-ER 1481).

Trochlear keeling is mildest in StW 181 and slightly less mild in StW 389 than in StW 358, StW 515, and StW 567, more closely resembling milder keeling present in some tibiae attributed to *A. afarensis* (A.L. 333-7, A.L. 545-3) and *A. sediba* (U.W. 88-97). The more strongly keeled morphology with stronger anterior and posterior swellings of the other Sterkfontein specimens is reminiscent of that seen in *A. anamensis* (KNM-KP-29285), other *A. afarensis* (A.L. 288-1ar, A.L. 333-6, KSD-VP-1/1e), *A. sediba* (U.W. 88-21), and to a lesser extent *P. boisei* (KNM-ER-1500) and *H. erectus* (KNM-ER-1481).

At the anterior and posterior intersection of the keel and the talar facet margins, inferiorly projecting swellings are present in StW 358, StW 515, and StW 567. The anterior swelling forms a "beak"-like morphology, especially when compared to the milder posterior intersection, resembling the morphology of tibiae attributed to *A. anamensis* (KNM-KP 29285), *A. afarensis* (A.L. 288-1ar, A.L. 333-6, KSD-VP-1/1e), one *A. sediba* (U.W. 88-21), *P. boisei* (KNM-ER-1500), and *H. erectus* (KNM-ER-1481). This condition differs from the flatter condition exhibited on StW 181 and StW 389, *A. afarensis* (A.L. 333-7, A.L. 545-3), and one *A. sediba* tibiae (e.g., U.W. 88-97). The relative lateral position of the projecting posterior swelling of StW 358, StW 389, and StW 567 is also apparent on tibiae attributed to *A. anamensis* (KNM-KP 29285), *A. afarensis* (A.L. 288-1ar, A.L. 333-6, KSD-VP-1/1e), and *A. sediba* (e.g., U.W. 88-21). The "squatting facet" on StW 389 is broadly similar to a facet on *A. anamensis* (KNM-KP-29285), *A. afarensis* (A.L. 288-1ar), and *A. sediba* (U.W. 88-21, U.W. 88-97).

Fibula

The proximal portion of the subcutaneous surface of StW 356 appears to face anterolaterally, somewhat intermediate to the lateral orientation characteristic of modern humans and the anterior orientation described in most distal fibulae of *A. afarensis* (Stern and Susman, 1983).

Discussion

There is reasonable evidence supporting an association of the partial proximal tibia, StW 396, and the partial distal tibia, StW 567, as attributable to the same individual. While we can never know with certainty, there are multiple lines of evidence supporting such an association. The two fossils exhibit at least some proximal and distal linear dimensions that would fit comfortably within the extant anthropoid sample comprised of proximal and distal dimensions measured on the same tibiae. The predicted mediolateral breadth of the medial condyle for StW 396 underestimates the observed value by 8.0%, while the predicted values for StW 514, KNM-KP 29285, and A.L. 288-1aq overestimate observed values by 1.4, 5.1, and 10.8%, respectively (Figure 13.5; Table 13.4). Thus, in terms of the amount of difference between the predicted and observed values, the StW 396 and StW 567 pairing also fits within the range of discrepancies exhibited by these known hominin fossil pairings. While it may be noteworthy that the predicted value of the mediolateral breadth of the medial condyle on StW 396 uniquely underestimates the observed value, and each of the predicted values for the other three hominins overestimates their respective observed values, it is worth keeping in mind that the former is not skeletally mature while the latter three individuals are. The presence of an epiphyseal line in StW 396 (Figure 13.1) and StW 567 (see also Barak et al., 2013: Figures 2f and 4) suggests that this individual could still have been undergoing some indeterminant (but likely minor) amount of longitudinal postcranial growth. Growth could factor into the contrasting overestimates in adults versus an underestimate in the non-adult hominin. Finally, the external appearance and preservation conditions of StW 396 and StW 567 also appear to support an association between these two specimens.

In addition to a potential association between StW 396 and StW 567, other associations within the Sterkfontein tibiae and fibulae have been proposed. StW 514 and StW 515 comprise a second pair of associated proximal and distal tibiae fragments, respectively (Berger and Tobias, 1996). There is no reason to preclude this association in our view. A partial distal tibia, StW 358, has been associated with StW 363, a left talus, by Fisk and Macho (1992). We suggest that StW 358 also may be associated with the partial left fibular shaft, StW 356, based on proximity of the recovery locations of these two specimens (adjacent excavation squares), as well as based on their sampling individuals of apparently similar body sizes and preservation conditions.

The association of StW 396 and StW 567 provides an informative comparator for the associated proximal tibia, StW 514, and distal tibia, StW 515. In other words, the knee and ankle joints of two individuals attributed to *A. africanus* can be compared. Previous studies of StW 567 (Zipfel and Berger, 2009) and StW 514 (Berger and Tobias,

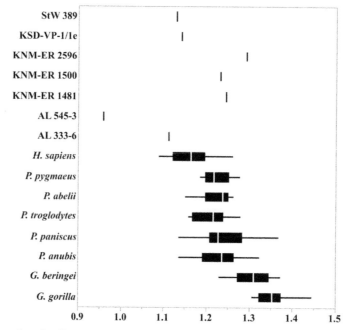

Standardized ML breadth of anterior margin of talar facet (left)

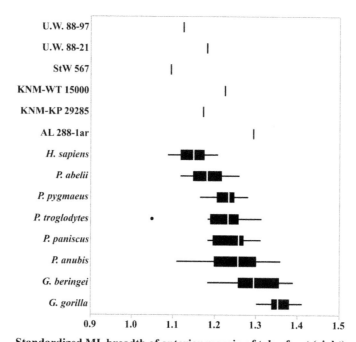

Standardized ML breadth of anterior margin of talar facet (right)

Figure 13.5 Mediolateral breadth of the anterior margin of the talar facet size—standardized by the geometric mean of mediolateral breadths and anteroposterior lengths. Side-specific comparisons are made (left on top; right on bottom). One left distal tibia (StW 358) retaining five of the six talar dimensions used for standardizing the dimension of interest (see Table 13.2) was omitted. When using these five dimensions, however, to standardized ML breadth of the anterior margin of the talar facet for StW 358, it generates a standardized value of 1.0710 for the specimen. This value is just below the observed modern human range for left tibiae, and below all other hominin left distal tibiae except that of A.L. 545-3.

Table 13.4 Reduced major axis regression results [log10 (mediolateral breadth of the medial condyle) regressed on log10 (mediolateral breadth of the anterior margin of the talar facet)]. Where y = medial condyle and x = talar facet. Statistical analysis performed using PAST version 3.20. Data log10 transformed to normalize, resulting in non-significant (p > 0.05) Shapiro-Wilk (W) tests.

	Model[a]	R^2	p
Extant right tibiae (n = 98)	y = (1.0265)x + (-0.026441)	0.906	<0.001
95% bootstrapped CI (N = 1999)	Slope = (0.95983, 1.0916)		
Extant left tibiae (n = 96)	y = (1.0199)x + (–0.025587)	0.898	<0.001
95% bootstrapped CI (N = 1999)	Slope = (0.95686, 1.0819)		

Using right side equation	Observed ML breadth of anterior margin of talar facet	Observed ML breadth of medial condyle	Predicted ML breadth of medial condyle
StW 396+567 (r)	21.5	23.8	21.9 (–8.0%) (estimates using 95% slope CI = 17.9 –26.8)
StW 514+515 (r)	20.1	20.2	20.5 (1.4%)
KNM-KP 29285 (r)	27.9	27.3	28.7 (5.1%)
A.L. 288-1aq+1ar (r)	21.2	19.5	21.6 (10.8%)

[a] In the model, the predictor variable (x) is the log10 of mediolateral breadth of the anterior margin of the talar facet (TF). The predicted value for StW 396 underestimates the observed value by 8.0%, while the predicted values for StW 514, KNM-KP 29285, and A.L. 288-1aq overestimate observed values by 1.4, 5.1, and 10.8%, respectively.

1998) have highlighted the modern human-like morphology of the former and the non-human ape-like morphology of the latter. In the present study, shapes and forms of the medial and lateral condyles provide the clearest support for this interpretation (Table 13.5). In fact, medial condyle form of StW 396 is similar to that of other hominin tibiae attributed to *A. anamensis*, *A. afarensis*, *A. sediba*, and *H. erectus* (Table 13.5). By comparison, the rounded medial condyle shape of StW 514 is unlike that of any other hominin tibia studied here, while the anteroposteriorly convex lateral condyle appears to resemble only that of A.L. 129-1b (Table 13.5). In a quantitative investigation of lateral condyle contours, Organ and Ward (2006) also noted this similarity in StW 514 and A.L. 129-1b lateral condyles, and further suggested relatively similar contours of lateral condyles on A.L. 333x-26, A.L. 333-42, and the damaged A.L. 288-1q. In other features, such as relative superoinferior height of the proximal metaphysis and morphology of the anterior intercondylar fossa, the two proximal tibiae from Sterkfontein bear strong resemblances to one another, to other hominins (tibiae attributed to *A. anamensis* and *A. afarensis*), and ultimately to modern humans as well (Table 13.5).

Despite StW 515 being heavily damaged, comparison of the distal tibiae in the two Sterkfontein pairings highlights additional commonalities. Both StW 515 and

StW 567 appear to exhibit an anteroposteriorly inflated distal metaphysis, as does another distal tibia from Sterkfontein preserving this region (StW 389) (Table 13.5). This feature, which is found in the earliest representatives of the genus *Australopithecus* (*A. anamensis*: see Ward et al., 2001), appears to be characteristic of habitual bipeds, as it would theoretically dissipate weight-bearing loads through the distal tibia more effectively than a non-inflated metaphysis by expanding the volume of bone (cortical and trabecular) through which these loads are transferred. Interestingly, StW 358 is alone among the hominin tibiae examined here in lacking an inflated distal metaphysis (Table 13.5). The functional significance for bipedalism of a non-inflated tibial metaphysis is unclear, although it its perhaps noteworthy that its absence characterizes non-human ape distal tibiae. Both StW 515 and StW 567 appear to exhibit a relatively oblique orientation of the distal fibular facet, as does StW 389 (Table 13.5). Stern and Susman (1983) observed that a relatively oblique orientation of the distal fibular facet characterizes non-human apes more than it characterizes modern humans who instead tend to have a more vertically oriented joint. This configuration of the distal tibiofibular joint may permit greater range of motion in the former. It is worth noting that StW 181 is alone among the distal tibiae from Sterkfontein in exhibiting the modern

Table 13.5 Summary of descriptive tibial comparisons.

	Trait	Similar to	Dissimilar to
StW 396	Medial condyle depression	*A. afarensis* (A.L. 333x-26)	*A. anamensis* *A. afarensis* *A. africanus* (StW 514) *A. sediba* *H. erectus*
	Medial condyle shape	*A. anamensis* *A. afarensis* *A. sediba* *H. erectus*	*A. africanus* (StW 514)
	Relative intercondylar eminence distance	*A. anamensis* *A. africanus* (StW 514) *A. sediba* *H. erectus*	*A. afarensis*
	Anterior intercondylar fossa soft tissue attachments	*A. anamensis* *A. afarensis* *A. africanus* (StW 514)	*H. erectus*
	Posterior intercondylar fossa soft tissue attachments	*A. anamensis* *A. afarensis* *H. erectus*	*A. afarensis*
	Posterior intercondylar fossa orientation	*P. boisei* *H. erectus*	*A. anamensis* *A. afarensis* *A. africanus* (StW 514)
	Metaphysis height	*A. anamensis* *A. afarensis* *A. africanus* (StW 514) *A. sediba* *P. boisei* *H. erectus*	None
	m. semimembranosus insertion	*A. afarensis* *A. africanus* (StW 514)	
StW 514	Medial condyle depression	*A. anamensis* *A. afarensis* *A. sediba* *H. erectus*	*A. afarensis* (A.L. 333x-26) *A. africanus* (StW 396)
	Medial condyle shape	None	*A. anamensis* *A. afarensis* *A. africanus* (StW 396) *A. sediba* *H. erectus*
	Lateral condyle form	*A. afarensis* (A.L. 129-1b)	*A. anamensis* *A. afarensis* *A. sediba* *H. erectus*
	Similar horizontal plane of condyles	*A. anamensis* *A. afarensis* (A.L. 129-1b) *A. sediba* *P. boisei*	*A. afarensis* *H. erectus*

Table 13.5 Continued

	Trait	Similar to	Dissimilar to
	Relative intercondylar eminence distance	*A. anamensis* *A. africanus* (StW 396) *A. sediba* *H. erectus*	*A. afarensis*
	Anterior intercondylar fossa soft tissue attachments	*A. anamensis* *A. afarensis* *A. africanus* (StW 396)	*H. erectus*
	Posterior intercondylar fossa orientation	*A. afarensis*	*A. anamensis* *A. africanus* (StW 396) *H. erectus*
	Metaphysis height	*A. anamensis* *A. afarensis* *A. africanus* (StW 396) *A. sediba* *P. boisei* *H. erectus*	None
	Tibial tuberosity form	*A. anamensis* *A. afarensis* *H. erectus*	*P. boisei*
	m. semimembranosus insertion	*A. afarensis* *A. africanus* (StW 396)	
	Soleal line form	*A. anamensis* *A. afarensis*	
	Interosseous membrane attachment position	*A. anamensis* *A. afarensis* (A.L. 129-1b)	*A. afarensis*
StW 181	Posterior swelling of central keel on talar facet	*A. afarensis* *A. sediba* (U.W. 88-97) *P. boisei* *H. erectus*	*A. anamensis* *A. afarensis* *A. africanus* *A. sediba* (U.W. 88-21)
	Posterior shaft form	*A. anamensis* *A. afarensis* *A. africanus* *A. sediba* (U.W. 88-21) *P. boisei* *H. erectus*	*A. afarensis* *A. africanus* (StW 358) *A. sediba* (U.W. 88-97)
	Tibiofibular tubercle form (posterior only)	*A. afarensis* (A.L. 545-3) *A. africanus* (StW 389) *P. boisei*	*A. anamensis* *A. afarensis* *A. sediba* *H. erectus*
	Distal fibular facet orientation	*A. afarensis* (A.L. 288-1ar) *A. sediba* (U.W. 88-21) *H. erectus*	*A. anamensis* *A. afarensis* *A. africanus*
StW 358	Talar facet shape	*A. afarensis* *A. sediba* *P. boisei*	*A. anamensis* *A. afarensis* (A.L. 545-3) *A. africanus* *H. erectus*

(continued)

Table 13.5   Continued

Trait	Similar to	Dissimilar to
Anterior 'beak' of central keel on talar facet	*A. anamensis* *A. afarensis* *A. africanus* (StW 567) *A. sediba* (U.W. 88-21) *P. boisei* *H. erectus*	*A. afarensis* *A. africanus* (StW 389) *A. sediba* (U.W. 88-97)
Posterior swelling of central keel on talar facet	*A. anamensis* *A. afarensis* *A. africanus* *A. sediba* (U.W. 88-21)	*A. afarensis* *A. africanus* (StW 181) *A. sediba* (U.W. 88-97)
Position of greatest depth of talar facet in lateral view	*A. afarensis* *A. africanus* *P. boisei* *H. erectus*	*A. anamensis* *A. afarensis* *A. sediba*
Position of talar facet on metaphysis	*A. anamensis* *A. afarensis* *A. sediba* *H. erectus*	*A. afarensis* *A. africanus* *P. boisei*
Inflated metaphysis	*P. boisei*	*A. anamensis* *A. afarensis* *A. africanus* *A. sediba* *H. erectus*
Position of metaphysis maximum anteroposterior dimension	*A. afarensis* *A. africanus* *A. sediba* (U.W. 88-97) *P. boisei*	*A. anamensis* *A. afarensis* (A.L. 333-6) *A. sediba* (U.W. 88-21) *H. erectus*
Posterior metaphysis surface	*A. afarensis* (A.L. 333-6)	*A. afarensis* *A. africanus* (StW 181) *A. sediba* *P. boisei* *H. erectus*
Medial malleolus medial profile	*A. anamensis* *A. afarensis* *A. africanus* (StW 567) *A. sediba* *P. boisei* *H. erectus*	*A. afarensis* (A.L. 545-3)
Medial malleolus anterior border	*P. boisei*	*A. afarensis* *A. africanus* (StW 567) *A. sediba* *H. erectus*
Medial malleolus mediolateral breadth	*A. anamensis* *A. afarensis* *A. africanus* *A. sediba* (U.W. 88-21)	*A. sediba* (U.W. 88-97) *H. erectus*
Groove for *m. tibialis posterior*	*A. anamensis* *A. afarensis* *A. sediba* (U.W. 88-97)	None

Table 13.5 Continued

	Trait	Similar to	Dissimilar to
	Posterior shaft form	*A. afarensis* *A. sediba* (U.W. 88-97)	*A. afarensis* *A. africanus* *A. sediba* (U.W. 88-21) *P. boisei* *H. erectus*
StW 389	Talar facet shape	*A. afarensis* (A.L. 545-3) *A. africanus* (StW 567)	*A. anamensis* *A. afarensis* *A. sediba* *P. boisei* *H. erectus*
	Anterior 'beak' of central keel on talar facet	*A. afarensis* *A. sediba* (U.W. 88-97)	*A. anamensis* *A. afarensis* *A. africanus* *A. sediba* (U.W. 88-21) *P. boisei* *H. erectus*
	Posterior swelling of central keel on talar facet	*A. anamensis* *A. afarensis* *A. africanus* *A. sediba* (U.W. 88-21)	*A. afarensis* *A. africanus* (StW 181) *A. sediba* (U.W. 88-97)
	Position of greatest depth of talar facet in lateral view	*A. afarensis* *A. africanus* *P. boisei* *H. erectus*	*A. anamensis* *A. afarensis* *A. sediba*
	'Squatting facet' form	*A. anamensis* *A. afarensis* (A.L. 288-1ar) *A. sediba* (U.W. 88-97)	None
	Position of talar facet on metaphysis	*A. afarensis* *A. africanus* (StW 567) *P. boisei*	*A. anamensis* *A. afarensis* *A. africanus* (StW 358) *A. sediba* *H. erectus*
	Inflated metaphysis	*A. anamensis* *A. afarensis* *A. africanus* *A. sediba* *H. erectus*	*A. africanus* (StW 358) *P. boisei*
	Position of metaphysis maximum anteroposterior dimension	*A. afarensis* *A. africanus* *A. sediba* (U.W. 88-97) *P. boisei*	*A. anamensis* *A. afarensis* (A.L. 333-6) *A. sediba* (U.W. 88-21) *H. erectus*
	Tibiofibular tubercle form	*A. afarensis* (A.L. 545-3) *A. africanus* (StW 181) *P. boisei*	*A. anamensis* *A. afarensis* *A. sediba* *H. erectus*
	Distal fibular facet anteroposterior position	*A. afarensis* *A. africanus* (StW 515) *A. sediba* (U.W. 88-21)	None

(continued)

Table 13.5   Continued

	Trait	Similar to	Dissimilar to
	Distal fibular facet orientation	A. anamensis A. afarensis A. africanus	A. afarensis (A.L. 288-1ar) A. africanus (StW 181) A. sediba (U.W. 88-21) H. erectus
StW 515	Inflated metaphysis	A. anamensis A. afarensis A. africanus A. sediba H. erectus	A. africanus (StW 358) P. boisei
	Medial malleolus mediolateral breadth	A. anamensis A. afarensis A. africanus A. sediba (U.W. 88-21)	A. sediba (U.W. 88-97) H. erectus
	Distal fibular facet anteroposterior position	A. afarensis A. africanus (StW 389) A. sediba (U.W. 88-21)	None
	Distal fibular facet orientation	A. anamensis A. afarensis A. africanus	A. afarensis (A.L. 288-1ar) A. africanus (StW 181) A. sediba (U.W. 88-21) H. erectus
StW 567	Talar facet shape	A. afarensis (A.L. 545-3) A. africanus (StW 389)	A. anamensis A. afarensis A. sediba P. boisei H. erectus
	Anterior 'beak' of central keel on talar facet	A. anamensis A. afarensis A. africanus (StW 358) A. sediba (U.W. 88-21) P. boisei H. erectus	A. afarensis A. africanus (StW 389) A. sediba (U.W. 88-97)
	Posterior swelling of central keel on talar facet	A. anamensis A. afarensis A. africanus A. sediba (U.W. 88-21)	A. afarensis A. africanus (StW 181) A. sediba (U.W. 88-97)
	Position of greatest depth of talar facet in lateral view	A. afarensis A. africanus P. boisei H. erectus	A. anamensis A. afarensis A. sediba
	Position of talar facet on metaphysis	A. afarensis A. africanus (StW 389) P. boisei	A. anamensis A. afarensis A. africanus (StW 358) A. sediba H. erectus
	Inflated metaphysis	A. anamensis A. afarensis A. africanus A. sediba H. erectus	A. africanus (StW 358) P. boisei

Table 13.5   Continued

Trait	Similar to	Dissimilar to
Position of metaphysis maximum anteroposterior dimension	*A. afarensis* *A. africanus* *A. sediba* (U.W. 88-97) *P. boisei*	*A. anamensis* *A. afarensis* (A.L. 333-6) *A. sediba* (U.W. 88-21) *H. erectus*
Medial malleolus medial profile	*A. anamensis* *A. afarensis* *A. africanus* (StW 358) *A. sediba* *P. boisei* *H. erectus*	*A. afarensis* (A.L. 545-3)
Medial malleolus anterior border	*A. afarensis* *A. sediba* *H. erectus*	*A. africanus* (StW 358) *P. boisei*
Medial malleolus mediolateral breadth	*A. anamensis* *A. afarensis* *A. africanus* *A. sediba* (U.W. 88-21)	*A. sediba* (U.W. 88-97) *H. erectus*
Distal fibular facet orientation	*A. anamensis* *A. afarensis* *A. africanus*	*A. afarensis* (A.L. 288-1ar) *A. africanus* (StW 181) *A. sediba* (U.W. 88-21) *H. erectus*

Note that when specimen numbers follow a specific taxon, this indicates a single example from among the multiple specimens that were inspected for the respective taxon. When reference is made to multiple examples for a respective taxon, specimen numbers are not provided in the table but can be found in the description text.

human-like orientation (Table 13.5). StW 181 is also alone among the Sterkfontein distal tibiae in exhibiting a relatively flattened posterior margin of the talar facet, which is also common among modern human tibiae. StW 515, similar to StW 389, exhibits a distal fibular facet that extends anteriorly and passes inferior to the terminal end of the interosseous crest; the same configuration is evident in *A. afarensis* and *A. sediba* (Table 13.5)(see also DeSilva et al., 2013). This feature of the distal tibiofibular joint is also more commonly observed in non-human apes than modern humans, and may indicate less entrapment of the lateral malleolus by a more superior attachment of the interosseous membrane, thus also enabling more range of motion in the joint.

Overall shape of the talar facet is similar in StW 389 and StW 567, being relatively more trapezoidal and less modern human-like, than in StW 358, which is more squared, and thus relatively more modern human-like (Table 13.5). DeSilva (2009) has argued that an expanded mediolateral breadth of the anterior margin of the talar facet would enable dissipation of peak compressive loads across this part of the talar facet during ankle dorsiflexion. He reasoned that dorsiflexed ankle posture (and high compressive loads) is common during chimpanzee

vertical climbing, and thus, a relatively wide anterior margin of the talar facet is a useful osteological correlate of vertical climbing. Like DeSilva (2009) who noted relatively narrower size-standardized anterior margins of the talar facet in hominins compared to African apes (hominins usually fall within the range of observed human variation), we also observed these patterns in our hominin sample (Figure 13.5). However, we do not agree with DeSilva (2009) that this is an indication of australopiths being generally non-adept climbers, although we do agree that hominins were likely vertically climbing in a kinematically distinct manner from chimpanzees. Other analyses of talar facet shape in human foragers who engage in climbing for selectively advantageous reasons (procurement of highly valued food resources) have shown that a mediolaterally expanded anterior margin of the talar facet is not necessarily an accurate indicator of proficient human climbing capability (Venkataraman et al., 2013 a,b). In our extant comparative sample, which included mountain gorillas, orangutans, and a generally more behaviorally and taxonomically diverse sample than other similar studies (DeSilva, 2009; Venkataraman et al., 2013 a,b), it is clear that relative mediolateral breadth of the anterior margin of the talar facet is not a useful

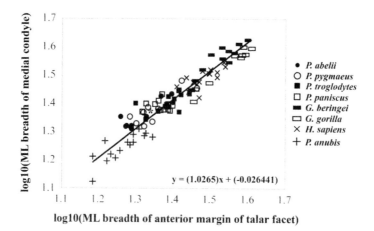

Figure 13.6 Bivariate plot of mediolateral breadth of the anterior margin of the talar facet regressed on mediolateral breadth of the medial condyle. The reported RMA regression equation based on 98 anthropoids has an $r^2 = 0.906$, and a $p < 0.001$ (see Table 13.5). An asterisk indicates the projected position of the paired StW 396 (1.3766) and StW 567 (1.3324) fossil dimensions.

osteological correlate of vertical climbing frequency or capability. Rather, we suggest it may be an indicator of habitually dorsiflexed ankle postures over a greater extent of the entire positional behavior repertoire. In extant non-human apes, especially African apes, this would include behavioral modes such as quadrupedalism on terrestrial and arboreal substrates (Hunt et al., 1996; D'Août et al., 2002; Thorpe and Crompton, 2006; Raichlen et al., 2009; Finestone et al., 2018) in addition to vertical climbing, each of which would benefit from dissipating large compressive loads across the anterior margin of the talar facet in a more dorsiflexed ankle. In early hominins, which usually fall within the range of observed modern human variability (Figure 13.5), this feature is less able to exclusively differentiate their behavioral repertoire from that of modern humans (potentially greater vertical climbing frequency in early hominins) because it presumably reflects additional factors that drive variability in bipeds (possibly substrate or terrain variability). In fact, the "squatting facets" and possibly the "beaking" that are observed in some Sterkfontein distal tibiae may be a more accurate osteological correlate of extreme dorsiflexion of the ankle during climbing than is the relative shape of their talar facet, as they are unlikely reflective of squatting per se.

Conclusions

Among the tibiae and fibulae from Sterkfontein, there is a clear range of variation expressed in several functionally relevant features. The distal tibia, StW 358, uniquely exhibits an absence of metaphyseal expansion beyond the talar facet. If the fibular shaft, StW 356, is associated with StW 358, it could offer additional functional insights into the locomotor repertoire of this individual that

might help contextualize this apparently unique trait. Likewise, the paired StW 514/515 and StW 389/567 also can offer glimpses of greater functional differences in knee joint configurations and ankle joint configurations in two individuals derived from the Sterkfontein assemblage, and especially highlight the unique combination of medial and lateral condyle form exhibited by StW 514. Additional morphological and functional evaluation of the ranges of variation expressed in many of the features exhibited by these Sterkfontein fossils, along with systematic comparisons to new discoveries and additional material not considered here (see Jungers et al., 2009), will undoubtedly further elucidate the potential existence of distinct "morphs" in the Sterkfontein hominin fossil assemblage.

Acknowledgments

We would like to thank the other organizers of the Sterkfontein Workshop—Drs. Carol Ward and Brian Richmond—for the invitation to participate and contribute this chapter, and Dr. Charles Lockwood for his efforts in initiating the workshop. We also wish to thank the Evolutionary Studies Institute at the University of the Witwatersrand (South Africa) and the Ditsong National Museum of Natural History (South Africa) for granting access to the fossils being described. We are grateful to the National Research Foundation (NRF) and the Department of Science and Technology (DST) of South Africa, as well as the University of Southern California, for providing funds to support this research.

For granting permission to study the comparative Ethiopian fossils, we would like to thank the government of Ethiopia, the Cultural Heritage Collection and Laboratory

Directorate of Ethiopia, the Authority for Research and Conservation of Cultural Heritage (ARCCH), and the National Museum of Ethiopia, as well as Mr. Desalegn Abebaw Andualem, Mr. Yared Assefa, Dr. Yohannes Haile-Selassie, and Dr. William Kimbel. We also would like to thank Metasebia Endalamaw for her assistance while studying the Ethiopian fossils. For granting permission to study the comparative Kenyan fossils, we would like to thank the government of Kenya, the National Museum of Kenya in Nairobi, as well Dr. Fredrick Manthi and Dr. Job Kibii. We also thank Francis Ndiritu for his assistance while studying Kenyan fossils.

For granting permission to study the comparative extant material, we would like to thank the School of Anatomical Sciences at the University of the Witwatersrand, the National Museum of Natural History (Smithsonian), and the Royal Museum of Central Africa. We are grateful to the following individuals who facilitated this research: Mr. Brendon Billings, Dr. Emmanuel Gilissen, Mr. Wim Wendelen, and Mr. Darrin Lunde.

References

Aiello, L.C., Dean, C., 1990. An introduction to human evolutionary anatomy. Academic Press: New York.

Barak, M.M., Lieberman, D.E., Raichlen, D., Pontzer, H., Warrener, A.G., Hublin J-J., 2013. Trabecular evidence for a human-like gait in *Australopithecus africanus*. PLoS One 8, e0077687.

Berge, C., Clarke, R.J., Goularas, D., Kibii, J.M., 2007. The *Australopithecus* pelvic morphology (StW 431 and Sts 14): the taxonomic evidence. Journal of Morphology 268, 1048–1049.

Berger, L.R., Tobias, P.V., 1996. A chimpanzee-like tibia from Sterkfontein, South Africa and its implications for the interpretation of bipedalism in *Australopithecus africanus*. Journal of Human Evolution 30, 343–348.

Clarke, R.J., 1985. *Australopithecus* and early *Homo* in southern Africa. In: Delson, E. (Ed.), Ancestors: the hard evidence. Alan R. Liss: New York, pp. 171–177.

Clarke, R.J., 1989. A new *Australopithecus* cranium from Sterkfontein and its bearing on the ancestry of *Paranthropus*. In: Grine, F.E. (Ed.), Evolutionary history of the "robust" australopithecines. Aldine de Gruyter: New York, pp. 285–292.

Clarke, R.J., 1994. Advances in understanding the craniofacial anatomy of South African early hominids. In: Corruccini, R.S., Ciochon, R.L. (Eds.), Integrative paths to the past: paleoanthropological advances in honor of F. Clark Howell. Prentice-Hall: Englewood Cliffs, pp. 205–222.

Clarke, R.J., 1998. First ever discovery of a well-preserved skull and associated skeleton of *Australopithecus*. South African Journal of Science 94, 460–463.

Clarke, R.J., 2012. A brief review of history and results of 40 years of Sterkfontein excavations. In: Reynolds, S.C., Gallagher, A. (Eds.), African genesis: perspectives on hominin evolution. Cambridge University Press: Cambridge, pp. 120–141.

Clarke, R.J., 2013. *Australopithecus* from Sterkfontein Caves, South Africa. In: Reed, K., Fleagle, J., Leakey, R. (Eds.), The paleobiology of *Australopithecus*. Vertebrate paleobiology and paleoanthropology. Springer: Dordrecht, pp. 105–123.

Clarke, R.J., Tobias, P.V., 1995. Sterkfontein Member 2 foot-bones of the oldest South African hominid. Science 269, 521–524.

D'Août, K., Aerts, P., De Clercq, D., De Meester, K., Van Elsacker, L., 2002. Segment and joint angles of hind limb during bipedal and quadrupedal walking of the bonobo (*Pan paniscus*). American Journal of Physical Anthropology 119, 37–51.

Day, M.H., Leakey, R.E.F., Walker, A.C., Wood, B.A., 1975. New hominids from East Rudolf, Kenya, I. American Journal of Physical Anthropology 42, 461–476.

Day, M.H., Leakey, R.E.F., Walker, A.C., Wood, B.A., 1976. New hominids from East Turkana, Kenya. American Journal of Physical Anthropology 45, 369–436.

Deloison, Y., 2003. Fossil foot bones anatomy from South Africa between 2.4 and 3.5 My old. Interpretation in relation to the kind of locomotion. Revue Biometrie Humaine et Anthropologie 21, 189–230.

DeSilva, J.M., 2009. Functional morphology of the ankle and the likelihood of climbing in early hominins. Proceedings of the National Academy of Sciences 106, 6567–6572.

DeSilva, J.M., Carlson, K.J., Claxton, A., Harcourt-Smith, W., McNutt, E., Sylvester, A.D., Walker, C., Zipfel, B., Churchill, S.E., Berger, L.R., 2019. The anatomy of the lower limb skeleton of *Australopithecus sediba*. PaleoAnthropology 2019, 357–405.

DeSilva, J.M., Devlin, M.J., 2012. A comparative study of the trabecular bony architecture of the talus in humans, non-human primates, and *Australopithecus*. Journal of Human Evolution 63, 536–551.

DeSilva, J.M., Holt, K.G., Churchill, S.E., Carlson, K.J., Walker, C.S., Zipfel, B., Berger, L.R., 2013. The lower limb and mechanics of walking in *Australopithecus sediba*. Science 340, 1232999.

DeSilva, J.M., Papakyrikos, A., 2011. A case of valgus ankle in an early Pleistocene hominin. International Journal of Osteoarchaeology 21, 732–742.

Domínguez-Rodrigo, M., Pickering, T.R., Baquedano, E., Mabulla, A., Mark, D.F., Musiba, C., Bunn, H.T., Uribelarrea, D., Smith, V., Diez-Martin, F., Pérez-González, A., Sánchez, P., Santonja, M., Barboni, D., Gidna, A., Ashley, G., Yravedra, J., Heaton, J.L., Arriaza, M.C., 2013. First partial skeleton of a 1.34-million-year-old *Paranthropus boisei* from Bed II, Olduvai Gorge, Tanzania. PLoS One 8, e0080347.

Dowdeswell, M.R., Jashashvili, T., Patel, B.A., Lebrun, R., Susman, R.L., Lordkipanidze, D., Carlson, K.J., 2017. Adaptation to bipedal gait and fifth metatarsal structural properties in *Australopithecus, Paranthropus,* and *Homo.* Comptes Rendues Palevol 16, 585–599.

Finestone, E.M., Brown, M.H., Ross, S.R., Pontzer, H., 2018. Great ape walking kinematics: implications for hominoid evolution. American Journal of Physical Anthropology 166, 43–55.

Fisk, G.R., Macho, G.A., 1992. Evidence of a healed compression fracture in a Plio-Pleistocene hominid talus from Sterkfontein, South Africa. International Journal of Osteoarchaeology 2, 325–332.

Fornai, C., 2009. Testing the second australopithecine species hypothesis for the South African site of Sterkfontein: Geometric morphometric analysis of maxillary molar teeth. M.S. Thesis, University of the Witwatersrand.

Fornai, C., Bookstein, F.L., Weber, G.W., 2015. Variability of *Australopithecus* second maxillary molars from Sterkfontein Member 4. Journal of Human Evolution 85, 181–192.

Green, D.J., Gordon, A.D., Richmond, B.G., 2007. Limb-size proportions in *Australopithecus afarensis* and *Australopithecus africanus.* Journal of Human Evolution 52, 187–200.

Grine, F.E., 2013. The alpha taxonomy of *Australopithecus africanus.* In: Reed, K., Fleagle, J., Leakey, R. (Eds.), The paleobiology of *Australopithecus.* Vertebrate paleobiology and paleoanthropology. Springer: Dordrecht, pp. 73–104.

Haile-Selassie, Y., Latimer, B.M., Alene, M., Deino, A.L., Gilbert, L., Melillo, S.M., Saylor, B.Z., Scott, G.R., Lovejoy, C.O., 2010. An early *Australopithecus afarensis* postcranium from Woranso-Mille, Ethiopia. Proceedings of the National Academy of Sciences 107, 12121–12126.

Harmon, E., 2009. Size and shape variation in the proximal femur of *Australopithecus africanus.* Journal of Human Evolution 56, 551–559.

Heaton, J.L., Pickering, T.R., Crompton, R.H., Stratford, D., Carlson, K.J., Jashashvili, T., Beaudet, A., Bruxelles, L., Kuman, K., Clarke, R.J., 2019. The long limb bones of the StW 573 *Australopithecus* skeleton from Sterkfontein Member 2: descriptions and proportions. Journal of Human Evolution 133, 167–197.

Heiple, K.G., Lovejoy, C.O., 1971. The distal femoral anatomy of *Australopithecus.* American Journal of Physical Anthropology 35, 75–84.

Hunt, K.D., Cant, J.G.H., Gebo, D.L., Rose, M.R., Walker, S.E., Youlatos, D., 1996. Standardized descriptions of primate locomotor and postural modes. Primates 37, 363–387.

Kuman, K., Clarke, R.J., 2000. Stratigraphy, artefact industries and hominid associations for Sterkfontein, Member 5. Journal of Human Evolution 38, 827–847.

Johanson, D.C., Coppens, Y., 1976. A preliminary anatomical diagnosis of the first Plio/Pleistocene hominid discoveries in the Central Afar, Ethiopia. American Journal of Physical Anthropology 45, 217–234.

Johanson, D.C., Lovejoy, C.O., Kimbel, W.H., White, T.D., Ward, S.C., Bush, M.E., Latimer, B.M., Coppens, Y., 1982. Morphology of the Pliocene partial hominid skeleton (A.L. 288-1) from the Hadar Formation, Ethiopia. American Journal of Physical Anthropology 57, 403–451.

Jungers, W.L., Larson, S.G., Harcourt-Smith, W., Morwood, M.J., Sutikna, T., Rokhus Due Awe, Djubiantono, T. 2009. Descriptions of the lower limb skeleton of *Homo floresiensis.* Journal of Human Evolution 57, 538–554.

Leakey, M.G., Feibel, C.S., McDougall, I., Walker A., 1995. New four-million-year-old hominid species from Kanapoi and Allia Bay, Kenya. Nature 376, 565–571.

Lockwood, C.A., Tobias, P.V., 2002. Morphology and affinities of new hominin cranial remains from Member 4 of the Sterkfontein Formation, Gauteng Province, South Africa. Journal of Human Evolution 42, 389–450.

Lovejoy, C.O., Heiple, K.G., 1970. A reconstruction of the femur of *Australopithecus africanus.* American Journal of Physical Anthropology 32, 33–40.

Lovejoy, C.O., Johanson, D.C., Coppens, Y., 1982. Hominid lower limb bones recovered from the Hadar formation: 1974-1977 collections. American Journal of Physical Anthropology 57, 679–700.

McHenry, H.M., Berger, L.R., 1998. Body proportions in *Australopithecus afarensis* and *A. africanus* and the origin of the genus *Homo.* Journal of Human Evolution 35, 1–22.

Organ, J.M., Ward, C.V., 2006. Contours of the hominoid lateral tibial condyle with implications for *Australopithecus.* Journal of Human Evolution 51, 113–127.

Partridge, T.C., 1978. Re-appraisal of lithostratigraphy of Sterkfontein hominid site. Nature 275, 282–287.

Pickering, T.R., Clarke, R.J., Moggi-Cecchi, J., 2004. Role of carnivores in the accumulation of the Sterkfontein Member 4 hominid assemblage: a taphonomic reassessment of the complete hominid fossil sample (1936-1999). American Journal of Physical Anthropology 125, 1–15.

Pickering, T.R., Heaton, J.L., Clarke, R.J., Stratford, D., 2018. Hominin hand bone fossils from Sterkfontein Caves, South Africa (1998-2003 excavations). Journal of Human Evolution 118, 89–102.

Raichlen, D.A., Pontzer, H., Shapiro, L.J., Socko, M.D., 2009. Understanding hind limb weight support in chimpanzees with implications for the evolution of primate locomotion. American Journal of Physical Anthropology 138, 395–402.

Robinson, J.T., 1972. Early hominid posture and locomotion. Chicago University Press: Chicago.

Ruff, C.B., Higgins, R., 2013. Femoral neck structure and function in early hominins. American Journal of Physical Anthropology 150, 512–525.

Ryan, T.M., Carlson, K.J., Gordon, A.D., Jablonski, N., Shaw, C.N., Stock, J.T., 2018. Human-like bipedal gait in *Australopithecus africanus* and *Paranthropus robustus.* Journal of Human Evolution 121, 12–24.

Singh, I., 1959. Squatting facets on the talus and tibia in Indians. Journal of Anatomy 93, 540–550.

Smith, R.J., 2009. Use and misuse of the reduced major axis for line-fitting. American Journal of Physical Anthropology 140, 476–486.

Stern Jr., J.T., 2003. Essentials of gross anatomy. Stony Brook University: Stony Brook.

Stern Jr., J.T., Susman, R.L., 1983. The locomotor anatomy of *Australopithecus afarensis*. American Journal of Physical Anthropology 60, 279–317.

Su, A., Carlson, K.J., 2017. Comparative analysis of trabecular bone structure and orientation in South African hominin tali. Journal of Human Evolution 106, 1–18.

Thorpe, S.K.S., Crompton, R.H., 2006. Orangutan positional behaviour and the nature of arboreal locomotion in Hominoidea. American Journal of Physical Anthropology 131, 384–401.

Tommy, K.A., 2018. A comparative study of trabecular architecture in the distal tibia of humans, hominoids, and australopiths from South Africa. M.Sc. Thesis, University of the Witwatersrand.

Venkataraman, V.V., Kraft, T.S., DeSilva, J.M., Dominy, N.J., 2013a. Phenotypic plasticity of climbing-related traits in the ankle joint of great apes and rain forest hunter-gatherers. Human Biology 85, 309–328.

Venkataraman, V.V., Kraft, T.S., Dominy, N.J., 2013b. Tree climbing and human evolution. Proceedings of the National Academy of Sciences 110, 1237–1242.

Vrba, E.S., 1979. A new study of the scapula of *Australopithecus africanus* from Sterkfontein. American Journal of Physical Anthropology 51, 117–130.

Walker, A., Leakey, R.E.F., (Eds.), 1993. The Nariokotome *Homo erectus* skeleton. Harvard University Press: Cambridge.

Ward, C.V., Kimbel, W.H., Harmon, E.H., Johanson, D.C., 2012. New postcranial fossils of *Australopithecus afarensis* from Hadar, Ethiopia (1990-2007). Journal of Human Evolution 63, 1–51.

Ward, C.V., Leakey, M.G., Walker A., 2001. Morphology of *Australopithecus anamensis* from Kanapoi and Allia Bay, Kenya. Journal of Human Evolution 41, 255–368.

Zeininger, A., Patel, B.A., Zipfel, B., Carlson, K.J., 2016. Trabecular architecture in the StW 352 fossil hominin calcaneus. Journal of Human Evolution 97, 145–158.

Zipfel, B., Berger, L.R., 2009. Partial hominin tibia (StW 396) from Sterkfontein, South Africa. Palaeontologia Africana 44, 71–75.

Zipfel, B., DeSilva, J.M., Kidd, R.S., 2009. Earliest complete hominin fifth metatarsal –implications for the evolution of the lateral column of the foot. American Journal of Physical Anthropology 140, 532–545.

14

Tarsals

Tea Jashashvili, Matthew W. Tocheri, Kristian J. Carlson, and Ronald J. Clarke

Twelve hominin tarsal bones were recovered during the excavations of Alun Hughes and Phillip Tobias from 1980 to 1988 (non-inclusive). These include one calcaneus and five tali from Member 4, as well as a partial tarsus from Member 2 comprised of an associated calcaneus, talus, navicular, and three cuneiforms (Appendix 1). Recent evaluations of Sterkfontein stratigraphy and depositional history underscore the complexity of the caves (Pickering and Herries, Chapter 3, this volume; see also Clarke, 2002, 2006, 2007, 2008; Bruxelles et al., 2014; Granger et al., 2015). Such complexity is particularly crucial to consider when differentiating Member 4 and 5 deposits (e.g., see Kuman and Clarke, 2000) since hominin fossils found in Member 4, or adjacent to Member 5, are usually attributed to *Australopithecus*.

Previous comparative studies of the Sterkfontein Member 4 tarsals have tended to differ in their morphological and locomotor interpretations, often due, in part, to contrasting approaches that were adopted. For example, several authors have analyzed different aspects of the partial calcaneus from Member 4, StW 352, attributed to *Australopithecus africanus* (Deloison, 1993, 2003, 2004; Zipfel et al., 2011; Clarke, 2013; Prang, 2016b; McNutt et al., 2017; McNutt et al., 2018; DeSilva et al., 2019). Based on evaluations of external morphology, Deloison (2003) noted modern human-like features, while Prang (2015a) noted *Pan*-like features. Zeininger et al. (2016) investigated trabecular structure beneath articular surfaces of the StW 352 calcaneus, noting both African ape-like and modern human-like similarities, which was suggested to indicate more varied calcaneal loading than typical of modern humans. Deloison (1993, 2003) provided an initial description of external morphology for Member 4 tali, recognizing features indicative of four different locomotor groups. Others (DeSilva and Devlin, 2012; Su and

Carlson, 2017) evaluated internal trabecular structure in some of these tali, with one of the studies (Su and Carlson, 2017) emphasizing a range of loading patterns expressed among them. Most recently, DeSilva et al. (2019) offered some basic external comparisons of the Sterkfontein tarsals to those of other hominins.

Previous evaluations of the StW573 partial foot from Member 2 have produced conflicting interpretations. Clarke and Tobias (1995) proposed a modern human-like rearfoot, emphasizing external morphology of the talus (StW 573a), paired with a more ape-like forefoot, emphasizing external morphology of the navicular and cuneiforms. Deloison (2003) also described the StW 573a talus as modern human-like, while others have highlighted the ape-like nature of its angular measurements and trochlear shape (Harcourt-Smith, 2002; Kidd and Oxnard, 2005). Less disagreement exists regarding the navicular (StW 573b), however, as there seems to be a consensus on its primitive and more African ape-like external morphology (Clarke and Tobias, 1995; Deloison, 2003; Kidd and Oxnard, 2005). The medial cuneiform (StW 573c) has most frequently been described as modern human-like rather than ape-like in its external morphology (Harcourt-Smith, 2002; McHenry and Jones, 2006; Gill et al., 2015), but this is not a consensus view (see Clarke and Tobias, 1995). The intermediate cuneiform (StW 573e) appears to exhibit a mixture of ape-like (long anteroposterior and wide dorsoplantar dimensions) and modern human-like characters (presence of an L-shaped facet for the medial cuneiform) (DeSilva et al., 2019). Likewise, the lateral cuneiform (StW 573g) has been described as exhibiting a mixture of ape-like (presence of a plantarly projected hamulus) and modern human-like characters (flattened navicular articulation) (DeSilva et al., 2019). Finally, DeSilva et al. (2019) have even suggested recently that the StW 573h calcaneus

Tea Jashashvili, Matthew W. Tocheri, Kristian J. Carlson, and Ronald J. Clarke, *Tarsals*. In: *Hominin Postcranial Remains from Sterkfontein, South Africa, 1936–1995.* Edited by: Bernhard Zipfel, Brian G. Richmond, and Carol V. Ward, Oxford University Press (2020). © Oxford University Press.
DOI: 10.1093/oso/9780197507667.003.0014

should not be associated with the remainder of the tarsus, instead suggesting it was from a "monkey" based on its "markedly mediolaterally concave cuboid facet typical of cercopithecoids" (p. 45). However, below we present data against this contention.

The primary aims of this chapter are, first, to provide systematic descriptions of the entire selection of Sterkfontein hominin tarsals, and, second, to undertake comparative morphological evaluations compared to those of hominins and extant hominoids to provide additional data to that of Deloison (2003). For specimens that have been initially described elsewhere, here we emphasize additional morphological details that complement those provided in the previously published descriptions. The goal of this systematic treatment is to further current debates generated by conflicting results of earlier studies. In addition, previous investigations of hominin fossils from Sterkfontein Member 4 emphasized morphological diversity present in the assemblage, suggesting that it may encompass multiple species (Clarke 1994; Lockwood

and Tobias, 2002; Clarke, 2013). Thus, an additional goal of the chapter is to assess the extent of morphological variability exhibited by Member 4 tarsals, especially the tali, and contextualize it in this debate.

Descriptions

Calcanei

StW 352—Right calcaneus

Preservation. This specimen consists of most of the anterior three-quarters of a right calcaneus measuring 51.2 in maximum length (anteroposteriorly), 44.9 mm mediolaterally, and 24.6 mm dorsoplantarly (Figure 14.1, Table 14.1). Posteriorly, the tuberosity is obliquely broken off from the posterior aspect of the peroneal trochlea to where the medial and plantar surfaces begin to flare medially. The remainder of the bone is in reasonably good condition, other than exhibiting some damage to the plantomedial

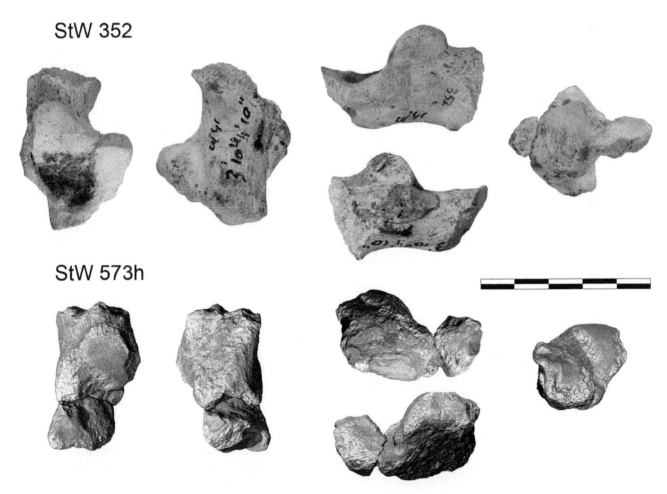

StW 352

StW 573h

Figure 14.1 Sterkfontein calcanei. The top row illustrates StW 352 in dorsal, plantar, medial (top) and lateral (bottom), and anterior views from left to right; while the bottom row illustrates 3D surface renderings of StW 573f in dorsal, plantar, medial (top) and lateral (bottom), and anterior views from left to right. Scale bar in cm.

Table 14.1 Calcaneus measurements. All measurements in mm.

	StW 352	StW 573f
1. Anteroposterior length	40.6	37.5*
2. Maximum mediolateral width	44.9	…
3. Dorsoplantar length	24.7	28.8
4. Anterior talar facet length	24.7*	…
5. Anterior talar facet width	11.0	…
6. Anterior talar facet depth	5.8	…
7. Posterior talar facet length	22.6	22.9*
8. Posterior talar facet width	16.2*	14.1
9. Posterior talar facet height	6.2	5.9
10. Mediolateral cuboid facet width	14.8	…
11. Dorsoplantar cuboid facet width	18.4	…
12. Cuboid facet depth	…	3.9*

Measurement definitions are: 1 = length between most posterior point of the posterior talar facet and most anterior point of the cuboid facet; 2 = width between most medial point and most lateral point; 3 = maximum dorsoplantar length posteriorly just before calcaneal tubercle; 4 = length of anterior talar facet measured along its longitudinal axis; 5 = width measured posteriorly between most medial point and most lateral point of the anterior talar facet; 6 = depth from the deepest point of the anterior talar facet to a horizontal plane through the highest points of the anterior talar facet; 7 = length of posterior talar facet measured along its longitudinal axis; 8 = width between the most elevated medial and lateral points of the posterior talar facet (in lateral view); 9 = height between highest point on the lateral rim of the posterior talar facet and a horizontal line defined mediolateral lowest points on the posterior talar facet; 10 = length measured dorsally between medial and lateral sides of the cuboid facet; 11 = length measured medially between plantar and dorsal sides of the cuboid facet; 12 = distance between the deepest point of the cuboid facet and a horizontal line defined amongst the mediolateral highest points of the cuboid facet.
*estimated value.

portion of the cuboid facet, the lateral margin of the posterior talar facet, and the dorsomedial edge of the anterior talar facet. The posterior margin of the sustentaculum tali is missing cortical bone from its surface. The posterior talar facet is well preserved, but it is eroded slightly along its lateral edge.

Morphology. The posterior talar facet is oriented slightly oblique relative to the long axis of the bone. Overall, this facet is saddle shaped, with a strong convexity anteroposteriorly and a shallow concavity mediolaterally. The posterior half of the anterior talar facet is well preserved, but only the most lateral aspects of the anterior half are preserved. Nonetheless, it is clear that this facet extends along the entire length of the sustentaculum tali and the anterior portion of the bone, and that it is moderately concave. The preserved portion of the cuboid facet appears relatively flat and probably would not have extended considerably further plantomedially based on the anterior limit of the pronounced sustentacular sulcus, which extends all the way from the posterior part of the sustentaculum tali to the plantar edge of the cuboid facet medially. Other

non-articular features of note include a prominent anterior plantar tubercle and a massive peroneal trochlea, but see McNutt et al. (2017) who demonstrate that the massiveness of the peroneal trochlea is partly inflated due to matrix that has infiltrated a crack in it (~5 mm).

StW 573f—Left calcaneus

Preservation. This specimen consists of two fragments that join together at the neck (Figure 14.1, Table 14.1). Maximum dimensions of the articulated aggregate are: 41.2 mm anteroposteriorly, 22.9 mm mediolaterally, and 28.3 mm dorsoplantarly. Most of the preserved surfaces are badly eroded on both fragments, except for much of the cuboid facet and the anteriormost part of the anterior talar facet. The tuberosity is transversely broken off and missing immediately behind the posterior talar facet. The remaining anterior portion of the bone is missing its plantar aspect.

Morphology. The preserved surface of the posterior talar facet is moderately convex anteroposteriorly and flat mediolaterally, and its orientation is slightly oblique to the long axis of the bone. What remains of the anterior talar facet is relatively flat. The cuboid facet is missing its dorsal and lateral borders, but what remains gives the appearance of being markedly concave with an evidently central position for the cuboid beak. This concavity was observed by Clarke (1998) as being 'bowl-shaped like that of a chimpanzee, indicating mobility, whereas that of a human has a notch that receives a keel on the articular surface of the cuboid to provide greater stability.' Deloison (2003) also commented on this anatomy, referring to the cuboid articulation as being in the form of a cone and very similar to that of the chimpanzee, signifying a much greater freedom of movement than that of Homo sapiens. She also, significantly, observed that the angle between the plane of the cuboid articulation and the base of the bone measures 130 degrees, with the cuboid articulation facing anteroinferiorly, whereas that of the ape faces directly anteriorly. She wondered whether this indicated the inception of a plantar vault to the StW 573 foot. The anatomy and size of the calcaneum and the fact that it fits perfectly onto the StW 573 talus demonstrates that there is no justification for the declaration by DeSilva et al. (2019) that they 'do not agree that the partial calcaneus belongs with StW 573 and identify it instead as a monkey based on the markedly mediolaterally concave cuboid facet typical of cercopithecoids.' They did not refer to the fact that, as observed by Clarke (1998) and Deloison (2003), it is also typical of chimpanzee. Furthermore, even a very large modern Papio calcaneus is considerably smaller than that of StW 573 and does not share its overall morphology. In addition, the cuboid facet of cercopithecoids faces anteriorly and not inferoanteriorly.

Tali

StW 88—Right talus

Preservation. This talus is complete except for the supero-medial quadrant of its head (Figure 14.2, Table 14.2). The specimen has a maximum anteroposterior dimension of 44.4 mm. The plantar surface of the head is damaged, with otherwise only slight abrasion marks on the plantar margin of the lateral malleolar facet. The sulcus tali exhibits some preparation scratches.

Morphology. The trochlea is only slightly concave medio-laterally with medial and lateral rims that are roughly parallel in superior view. The trochlea extends anteriorly onto the medial side of the neck. The medial malleolar facet also extends anteriorly and laterally onto the neck, and it is noticeably concave in its anteroposterior dimension. The dorsal half of the lateral malleolar facet is flat superiorly, but its more plantar portion flares laterally, creating an overall dorsoplantarly oriented concavity on the facet. The groove for the tendon of *m. flexor hallucis longus* is slightly concave and oblique in orientation rather than being channel-like and more vertical in orientation.

The anterior and middle calcaneal facets are confluent and together are slightly convex laterally and moderately concave medially. The anterior facet is continuous with the facet for contact with the tibiocalcaneal ligament that is 7.1 mm wide. The posterior calcaneal facet is concave mediolaterally and ovoid shaped overall. The talar neck is stout and expanded transversely, measuring 16.6 mm in length anteroposteriorly. The mediolateral axis of the head shows only a minimal degree of torsion (17.0°) relative to a plane through the trochlear rims. The sulcus tali is well marked and deeply excavated, being 5.6 mm in depth dorsoplantarly.

StW 102—Right talus

Preservation. This talus preserves most of its body with considerable damage and erosion on its lateral side (Figure 14.2, Table 14.2). Its head and most of its neck are broken off with only a small lateral portion of its neck remaining. Its maximum dimensions are 44.2 mm anteroposteriorly, 33.2 mm mediolaterally, and 27.4 mm dorsoplantarly. The trochlear facet is badly eroded distolaterally, but key aspects of its shape are still discernable. The medial tubercle is broken, and the lateral tubercle is eroded. The lateral malleolar facet is badly eroded and damaged. The anterior part of medial malleolar facet is broken, and the entire medial side of the body is moderately abraded.

Morphology. The trochlea is moderately curved medio-laterally with medial and lateral rims that are not parallel. Instead, the trochlea is considerably wider anteriorly than it is posteriorly (Table 14.2). A distinct horizontal groove marks the posterior limit of the trochlea. The groove for the tendon of *m. flexor hallucis longus* is slightly oblique in its orientation. The sulcus tali is marked and deeply excavated. The medial malleolar facet appears to have extended anteriorly and laterally onto the neck, at least based on what is preserved of its plantar portion. Although the lateral malleolar facet is badly damaged, its retained morphology suggests that its plantar portion flared laterally creating an overall concavity dorsoplantarly. The posterior calcaneal facet is markedly concave mediolaterally, flattened anteroposteriorly, and ovoid overall.

StW 347—Left talus

Preservation. This specimen preserves its head, neck, and the anterior one-third of its body (Figure 14.2, Table 14.2). Maximum dimensions of the specimen are 32.1 mm anteroposteriorly, 28.7 mm mediolaterally, and 18.0 mm dorsoplantarly. While surfaces of the head and neck are well preserved, the body preserves only the anterior portions of the trochlea and the facet for the medial malleolus.

Morphology. The mediolateral axis of the head shows a considerable degree of torsion (40.0°) relative to a plane through the trochlear rims. The navicular facet is convex and oval shaped and narrows toward its plantar aspect. The preserved part of the trochlea suggests that it was saddle shaped with only a slight concavity mediolaterally. The medial malleolar facet extends anteriorly onto the neck, but it is only slightly concave both anteroposteriorly and dorsoplantarly. Viewed plantarly, the anterior calcaneal facet extends from the lateral aspect of the talar head until the point at which the body is broken immediately anterolateral to the sulcus tali. The facet is slightly convex anterolaterally and moderately concave postero-medially. The preserved sulcus tali is deeply excavated and well marked.

StW 363—Left talus

Preservation. This specimen is preserved in two pieces that have been glued together (Figure 14.3, Table 14.2). The crack separating them traverses the length of the trochlea and runs along the anterior margin of the posterior calcaneal facet. Maximum dimensions of the articulated aggregate are 42.2 mm anteroposteriorly, 30.3 mm mediolaterally, and 21.2 mm dorsoplantarly. Damaged and badly eroded areas of the surface include the medial and plantar portions of the head and neck, medial and lateral edges of the posterior calcaneal facet, the medial tubercle, the groove for *m. flexor hallucis longus*, the anterolateral edge of the anterior calcaneal facet, as well as the plantar aspect of the lateral malleolar facet.

StW 88

StW 102

StW 347

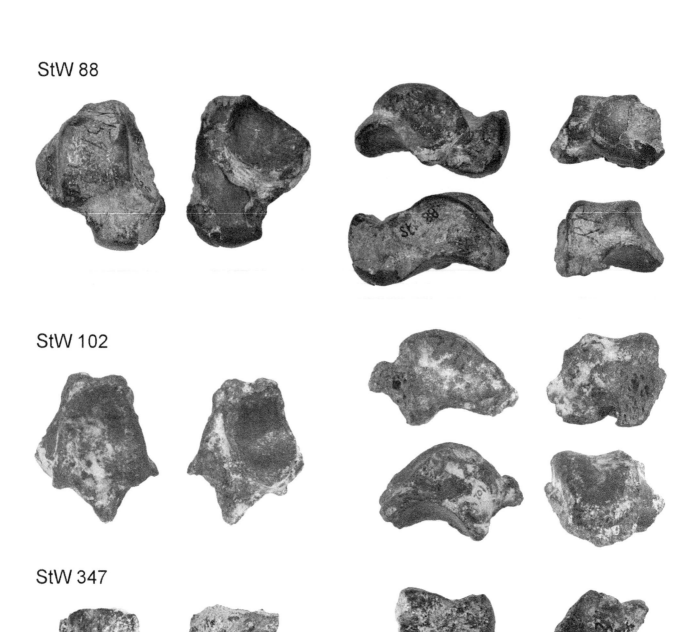

Figure 14.2 Sterkfontein tali. The top row illustrates StW 88 in dorsal, plantar, lateral (top) and medial (bottom), and anterior (top) and posterior (bottom) views from left to right; the middle row illustrates StW 102 in dorsal, plantar, medial (top) and lateral (bottom), and anterior (top) and posterior (bottom) views from left to right; and the bottom row illustrates StW 347 in dorsal, plantar, medial (top) and lateral (bottom), and anterior views from left to right. Scale bar in cm.

Table 14.2 Talus measurements. Linear measurements in mm; angular measurements in degrees.

	StW 88	StW 102	StW 347	StW 363	StW 486	StW 573a
1. Talar length	44.4	...	...	42.2*		41.3
2. Talar neck length	16.6	...	14.8	16.3	16.6	16.0
3. Mid-trochlear width	19.2	19.6	19.3*	19.0	20.8	20.1*
4. Anterior trochlear width	22.2	...	21.2	21.9	22.3	24.3
5. Posterior trochlear width	15.8	14.5	...	15.1	...	14.7
6. Talar width	32.9	30.1	...	30.3*	...	30.8*
7. Talar head height	24.5	...	21.4	21.3*	25.2*	25.0
8. Talar head width	17.8	...	16.5	14.2*	16.7*	17.0
9. Plantar facet length	23.8	24.9*	...	20.3*	...	25.1*
10. Plantar facet width	17.3	18.3	...	12.0	...	14.3
11. Talofibular lateral projection	8.6	8.0	...	8.9*	11.9	6.5*
12. Talar neck angle	32.5	32.0*	24.9*	32.9	27.4	31.8
13. Talar head torsional angle	17.0	...	...	35.3	39.2	26.5

Measurements are adapted from Gebo and Schwartz (2006).

*estimated value.

Morphology. The trochlea is slightly curved mediolaterally. The medial and lateral rims are not quite parallel as the trochlea is slightly wider anteriorly than it is posteriorly. The trochlea extends onto the neck almost as far anteriorly as the anterior border of the medial malleolar facet, which is slightly concave anteroposteriorly and fairly flattened dorsoplantarly. The lateral malleolar facet extends lateral from the trochlear margin, especially along its inferior extent, but the surface is relatively flat overall. The head shows a moderate degree of torsion relative to the trochlea. The preserved portions of the navicular facet indicate that this facet was convex mediolaterally and slightly convex dorsoplantarly. The middle and anterior calcaneal facets are smoothly confluent. The middle calcaneal facet is moderately concave, while the anterior one is transversely convex. The posterior calcaneal facet is moderately concave mediolaterally, relatively flattened anteroposteriorly, and ovoid in shape. The sulcus tali is deep and well marked.

StW 486—Right talus

Preservation. Most of the plantar portion of this talus is missing except for the lateral part of the posterior calcaneal facet (Figure 14.3, Table 14.2). Posterior portions of the medial and central body are also missing. Its maximum dimensions are 35.3 mm anteroposteriorly, 36.2 mm mediolaterally, and 25.9 mm dorsoplantarly. Only the lateral malleolar facet is preserved in its entirety.

Morphology. The trochlea is markedly concave mediolaterally, with medial and lateral rims that appear to have been relatively parallel based on their preserved portions. The anterior border of the trochlea is notched centrally. The medial malleolar facet extends anteriorly onto the neck (just slightly beyond the medial aspect of the trochlea) and is only very slightly concave anteroposteriorly and relatively flat dorsoplantarly. The lateral malleolar facet is flattened dorsally and gently extends laterally for the initial two-thirds of its surface until it more noticeably flares laterally near its inferior margin. This creates a slight dorsoplantar concavity in the facet. Relative to a plane through the trochlear rims, the mediolateral dimension of the head shows a moderate degree of torsion (39.2°). The preserved portions of the navicular facet are convex mediolaterally and slightly less so dorsoplantarly.

StW 573a—Left talus

Preservation. This specimen consists of two adjoining fragments that are missing dorsoposterior and plantar portions of the lateral side of the body (Figure 14.3, Table 14.2). A large sagittally oriented crack runs roughly through the middle of the bone where the two pieces abut. Maximum dimensions of the articulated aggregate are 43.3 mm anteroposteriorly, 29.3 mm mediolaterally, and 20.9 mm dorsoplantarly. Damaged and badly eroded areas include the posterior region of the lateral rim of the trochlea, posterior and plantar aspects of the lateral malleolar facet, and the lateral aspect of the posterior calcaneal facet.

Morphology. The trochlea is saddle shaped and moderately curved mediolaterally. The lateral rim is elevated slightly more superiorly than the medial rim. Despite damage to the lateral rim posteriorly, it appears that the rims are not quite parallel in the anteroposterior direction, making the trochlea slightly wider anteriorly than posteriorly. Medially, the trochlear surface extends anteriorly onto the neck approximately to the same extent

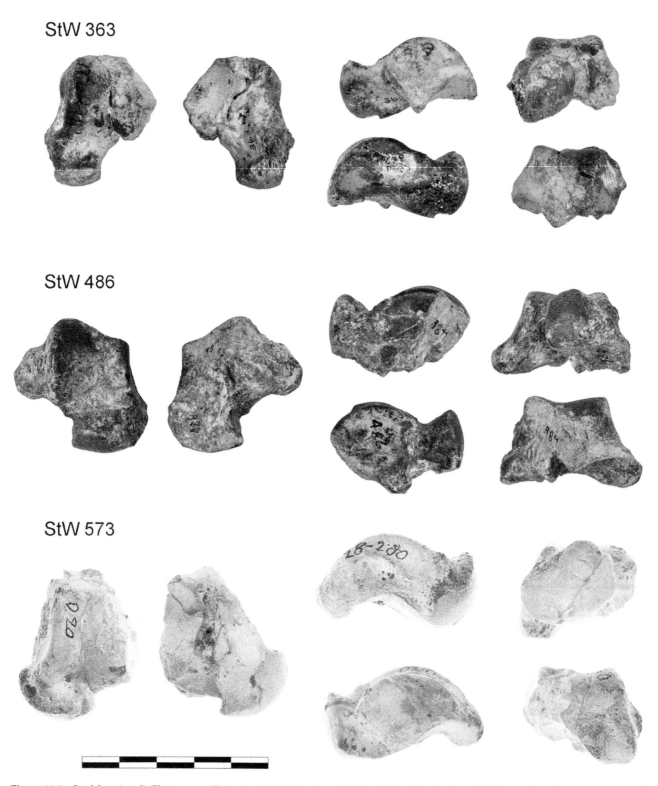

Figure 14.3 Sterkfontein tali. The top row illustrates StW 363 in dorsal, plantar, lateral (top) and medial (bottom), and anterior (top) and posterior (bottom) views from left to right; the middle row illustrates StW 486 in dorsal, plantar, medial (top) and lateral (bottom), and anterior (top) and posterior (bottom) views from left to right; and the bottom row illustrates StW 573a in dorsal, plantar, medial (top) and lateral (bottom), and anterior (top) and posterior (bottom) views from left to right. Scale bar in cm.

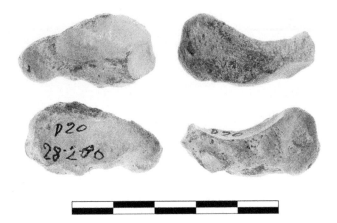

Figure 14.4 Sterkfontein navicular. The top row illustrates StW 573b in anterior and dorsal views from left to right, while the bottom row illustrates posterior and plantar views from left to right. Scale bar in cm.

Table 14.3 Navicular measurements. All measurements in mm.

	StW 573b
1. Talar facet width	22.8
2. Talar facet height	14.4*
3. Talar facet depth	4.3
4. Navicular mediolateral length	33.7
5. Medial cuneiform facet mediolateral width	12.8
6. Medial and intermediate cuneiform ridge height	11.3
7. Intermediate and lateral cuneiform ridge height	8.0
8. Intermediate cuneiform facet dorsal width	12.7
9. Lateral cuneiform facet lateral height	9.0
10. Lateral cuneiform facet plantar width	7.8
11. Lateral cuneiform facet dorsal width	8.4

Measurement definitions are: 1 = mediolateral width of the talar facet; 2 = dorsoplantar height of the talar facet; 3 = length between deepest point on the talar facet and a horizontal line defined by talar facet width; 4 = maximum mediolateral length of navicular; 5 = width between most medial and most dorsolateral points of the medial cuneiform facet; 6 = dorsoplantar height of the ridge defined by the lateral side of medial and medial side of intermediate cuneiform facets; 7 = dorsoplantar height of the ridge defined by the lateral side of intermediate and medial side of lateral cuneiform facets for; 8 = width between most dorsomedial and most dorsolateral points for the intermediate cuneiform facet; measurements 9-11 are taken on the lateral cuneiform facet as described for the other cuneiform facets.
*estimated value.

as the anterior border of the medial malleolar facet. The latter facet is slightly concave anteroposteriorly and dorsoplantarly. Only the dorsoanterior portion of the lateral malleolar facet is present; it is flat and gently extends laterally toward its preserved plantar aspect. The mediolateral dimension of the head displays a moderate degree of torsion relative to a plane through the trochlear rims. The navicular facet is moderately convex mediolaterally and only slightly convex dorsoplantarly. The anterior calcaneal facet is markedly large and wide in its anterior portion, narrowing considerably toward the medial tubercle. It is moderately convex anteriorly and moderately concave posteriorly. The posterior calcaneal facet is incomplete, but its preserved portion indicates that it is moderately concave mediolaterally and relatively flattened anteroposteriorly. A deeply excavated and well-marked sulcus tali is visible.

Navicular

StW 573b—Left navicular

Preservation. This specimen is reasonably intact and well-preserved except for its dorsolateral portion, which is eroded and missing its dorsal border and the surface of the talar facet. Its maximum dimensions are 16.6 mm anteroposteriorly, 33.7 mm mediolaterally, and 18.9 mm dorsoplantarly (Figure 14.4; Table 14.3).
Morphology. The preserved portion of the talar facet indicates that the facet was ovoid in shape and almost twice as wide mediolaterally relative to its length dorsoplantarly. The facet is moderately concave mediolaterally and only slightly concave dorsoplantarly. A distinct small, flat, and ovoid-shaped cuboid facet is present plantolaterally. The cuboid facet is positioned closer to the lateral cuneiform facet than to the talar facet. The anterior side is

dominated by the cuneiform articular surfaces comprised of three continuous facets. Of these, the medial cuneiform facet is the largest, being slightly concave medially and transitioning to slightly convex laterally, and about twice as broad mediolaterally as it is long dorsoplantarly. The intermediate cuneiform facet is roughly triangular in shape with a plantar apex, slightly convex dorsoplantarly, and slightly concave in its most lateral aspect. On the lateral cuneiform facet, the plantar border is slightly wider than its dorsal border, and the medial and lateral borders are gently curved plantarly. Overall, the lateral cuneiform facet shows a slight concavity mediolaterally and dorsoplantarly. The tuberosity is anteroposteriorly long relative to its dorsoplantar width, which is widest anteriorly. The medial (tuberosity) portion of the bone is more than twice as wide dorsoplantarly than is the lateral portion.

Cuneiforms

StW 573c—Left medial cuneiform

Preservation. This specimen is well preserved overall except for a few areas of surface erosion and chipping (Figure 14.5, Table 14.4). A large crack runs anteroposteriorly through the central part of the bone. Maximum dimensions of the specimen are 16.6 mm anteroposteriorly, 13.4 mm mediolaterally, and 25.9 mm dorsoplantarly. All facets are in reasonably good condition.

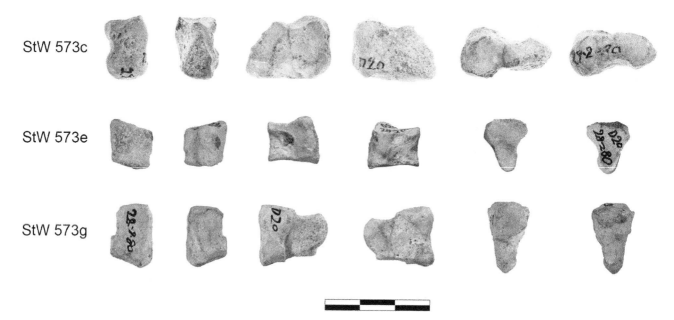

StW 573c

StW 573e

StW 573g

Figure 14.5 Sterkfontein cuneiforms. The top row illustrates the StW 573c medial cuneiform in plantar, dorsal, lateral, medial, posterior, and anterior views from left to right; the middle row illustrates the StW 573e intermediate cuneiform in dorsal, plantar, medial, lateral, anterior, and posterior views from left to right; and the bottom row illustrates the StW 573g lateral cuneiform in dorsal, plantar, medial, lateral, posterior, and anterior views from left to right. Scale bar in cm.

Table 14.4 Medial cuneiform measurements. Linear measurements in mm, angular measurements in degrees.

	StW 573c
1. Navicular facet width	11.1
2. Navicular facet height	14.9*
3. Navicular facet depth	1.5
4. MT1 facet height	21.3
5. MT1 facet middle width	6.4
6. MT1 facet plantar width	9.8
7. Plantar width	16.6
8. Dorsal width	7.6
9. Intermediate cuneiform facet posterior width	2.9
10. Intermediate cuneiform facet posterior height	6.4
11. Intermediate cuneiform facet dorsal height	6.4*
12. Intermediate cuneiform facet dorsal length	8.1
13. MT2 facet width	4.8
14. Anteroposterior angle of torsion	7.4

Measurement definitions are: self-explanatory for 1-13; 14 = angle of torsion between dorsoplantar axes of navicular and MT1 facets.
*estimated value.

Morphology. The facet for the hallucal metatarsal faces anteriorly, but it is slightly inclined medially. Its surface is slightly convex dorsally, while plantarly it begins as moderately concave on the medial side and becomes more convex on the lateral side. The navicular facet faces posteriorly and is triangular in shape with a dorsally positioned apex and a rounded plantarly positioned base. It is moderately concave dorsoplantarly and slightly concave mediolaterally. The intermediate cuneiform facet occupies a small portion of the lateral surface posteriorly and extends distally along the lateral surface dorsally. Posterior and dorsal portions of the facet are roughly orthogonal to one another, but both are relatively flat. A small facet for the second metatarsal occupies the dorsoanterior corner of the lateral surface. It is continuous with the intermediate cuneiform facet posteriorly and extends toward the edge of the hallucal metatarsal facet anteriorly.

StW 573e—Left intermediate cuneiform

Preservation. This specimen is well preserved except for its dorsal surface, which is eroded (Figure 14.5, Table 14.5). Maximum dimensions are 12.1 mm anteroposteriorly, 12.4 mm mediolaterally, and 15.1 mm dorsoplantarly. Small chips are missing along the dorsal borders of the navicular and medial cuneiform facets, the dorsomedial corner of the second metatarsal facet, and the middle of the medial border of the navicular facet.

Morphology. This specimen is shaped like a triangular wedge. The facet for the second metatarsal occupies the entire anterior side and is dorsally broad and plantarly narrow; the facet is slightly convex dorsally and slightly concave plantarly. The posterior side is entirely occupied by the navicular facet, which is roughly triangular in outline, slightly concave dorsoplantarly, and flattened mediolaterally. The lateral side displays a relatively flat, triangular-shaped facet for the lateral cuneiform that

Table 14.5 Intermediate cuneiform measurements (in mm).

	StW 573e
1. Navicular facet dorsoplantar length	15.1
2. Navicular facet mediolateral width	11.2
3. MT2 facet dorsoplantar length	13.1*
4. MT2 facet mediolateral width	12.4
5. Dorsal anteroposterior medial width	12.1
6. Dorsal anteroposterior middle width	11.0
7. Dorsal anteroposterior lateral width	10.0
8. Plantar anteroposterior width	11.6
9. Medial cuneiform facet dorsal anteroposterior width	12.1
10. Lateral cuneiform facet dorsal anteroposterior width	6.4

Measurement definitions are: self-explanatory for 1-9; 9 = anteroposterior width of the dorsal margin of the medial cuneiform facet; and 10 = anteroposterior width of the dorsal margin of the lateral cuneiform facet.
*estimated value.

Table 14.6 Lateral cuneiform measurements (in mm).

	StW 573g
1. Navicular facet dorsoplantar height	9.5
2. Navicular facet mediolateral width	8.2
3. Navicular facet dorsoplantar height	20.8
4. MT3 facet dorsoplantar height	16.3
5. MT3 facet dorsal mediolateral width	11.7
6. MT3 facet middle mediolateral width	6.8
7. MT3 facet plantar mediolateral width	2.8
8. Dorsal anteroposterior lateral length	17.8
9. Dorsal anteroposterior medial length	13.7
10. Plantar anteroposterior length	12.0
11. Cuboid facet width	7.7
12. Cuboid facet height	7.1
13. MT4 facet width	4.5
14. MT4 facet height	6.2
15. Intermediate cuneiform facet width	5.1
16. Intermediate cuneiform facet height	5.2
17. MT2 facet width	4.4
18. MT2 facet height	5.2

Measurement definitions are self-explanatory.

extends along the posterior three-quarters of the dorsal part. The medial cuneiform facet occupies a small portion of the medial surface posteriorly and extends anteriorly and dorsally along the medial surface. Posterior and dorsal portions of this facet are both relatively flat and approximately orthogonal to one another.

StW 573g—Left lateral cuneiform

Preservation. This specimen is well preserved except for a small chip missing from the medioplantar corner of its navicular facet (Figure 14.5, Table 14.6). A large crack runs anteroposteriorly through the center of the bone. Its maximum dimensions are 17.9 mm anteroposteriorly, 12.0 mm mediolaterally, and 20.8 mm dorsoplantarly.

Morphology. This specimen is wide dorsally and narrow plantarly with large grooves occurring centrally on both its medial and lateral sides. It is slightly longer dorsoplantarly than anteroposteriorly. A relatively flattened facet for the third metatarsal occupies the entire anterior side and is dorsally broad and plantarly narrow. A rectangular-shaped facet for the fourth metatarsal is set obliquely to the anterior side dorsally. This facet is continuous with an hourglass-shaped facet for the cuboid extending posteriorly along nearly the entire dorsal aspect of the lateral side. The navicular facet is relatively flat and quadrangular shaped, but with a slightly longer lateral than medial dimension. The medial side displays two distinct, flattened facets for the intermediate cuneiform and the second metatarsal along its dorsal aspect separated by a shallow groove. The more posterior, triangular-shaped facet is for the intermediate cuneiform, while the more anterior, rhomboid-shaped facet is for the second metatarsal. The non-articular plantar portion of the bone forms a hook-like process.

Comparative anatomy

Materials and methods

With the exception of talar comparative data, all measurements were collected on either right or left tarsals of 20 adult extant hominoids curated in the Department of Comparative Anatomy of the National Museum of Natural History, Paris, France (NMNH); the Royal Museum of Central Africa, Tervuren, Belgium (RMCA); and the Raymond A. Dart Collection of Human Skeletons at the University of the Witwatersrand, Johannesburg, South Africa (DART) (Dayal et al., 2009). Common chimpanzees and lowland gorillas are free-ranging individuals from Congo, Gabon, and Cameroon. The modern human sample is derived from several South African populations (Soto, Vendi, Xosa, Pondo, and Zulu). Renderings of all extant hominoid tarsals were produced from image data sets generated from medical computed tomography (CT) scanning in different hospitals. More information on scanning protocols and facilities (hospitals) has been published elsewhere (Jashashvili et al., 2015; Dowdeswell et al., 2017; Patel et al., 2018 a,b). Talar comparative data were collected from extant hominoid tali curated in the aforementioned collections as well as by one of us (MWT) from specimens in the American Museum of Natural History (AMNH), the Cleveland Museum of Natural History (CMNH), and the Academy of Natural Sciences in Philadelphia (ANSP), Royal Belgian Institute of Natural Sciences (RBINS), and the Smithsonian's National Museum of Natural History (USNM). The comparative talus sample includes either right or left tali from

53 common chimpanzees, 71 lowland gorillas, and 62 modern humans. Specifically, in addition to data generated from the 20 tali per group subjected to medical CT scanning like other tarsals, data were generated from an additional 33 common chimpanzees, 51 lowland gorillas, and 42 modern humans subjected to laser scanned using a NextEngine model scanner to create surface renderings (Tocheri, 2009). More information on acquisition protocols of the laser scanning has been published elsewhere (Tocheri, 2009; Dunn et al., 2014). Hominin tarsals in the comparative sample were laser scanned by one of us (MWT) and have been used in several previous investigations (Jungers et al., 2009).

Anatomical terminology adheres to the terminology used for Hadar hominin tarsals (Latimer et al., 1982; Ward et al., 2012). All linear measurements are reported in millimeters (mm), and were taken by one of us (TJ) on original Sterkfontein fossils using dial calipers. All angular measurements are reported in degrees (°), and were taken by one of us (TJ) on digital renderings of Sterkfontein fossils produced from image data acquired by high resolution computed tomography.

Linear measurements for extant fossils were collected on surface rendering of bones in Amira software [Amira 6.4.0, © 1999–2017 FEI, SAS] as the distance between two landmarks defining a measurement (Tables 14.1–14.6). Comparative specimens were side-matched to corresponding sides of Sterkfontein elements (right: calcaneus and talus, or left: navicular and cuneiforms) by mirroring opposite-side specimens. For calcanei and tali, for which both right and left sides are represented in the Sterkfontein sample, all left-side elements (including fossils) were mirrored and illustrated as right tali.

Boxplots (depicting median, upper and lower quartiles, and outliers) were used to compare extant samples and fossil specimens. Discriminant analyses are performed using common covariance on size-adjusted variables (adjusted by the geometric mean). Canonical details are calculated from overall pooled within-group covariance matrices and fossils were projected into respective discriminant spaces. In addition, joint shape was evaluated using the square root of corresponding facet areas to eliminate the effect of size.

Results and discussion

Calcaneus

Overall, the StW 352 and StW 573h calcanei exhibit a number of similarities despite the relatively poor and incomplete condition of the latter (Table 14.7). Absolute magnitudes of linear measurements on both calcanei tend to be more similar to those of gorillas and chimpanzees

than they are to the larger values on modern human calcanei (Tables 14.1 and 14.7). Dimensions of the anterior talar facet fall with the range of gorilla values, while posterior talar facet length and cuboid mediolateral and dorsoplantar facet widths are more comparable to those of chimpanzees (Table 14.7). Depth of the cuboid facet of StW 573h falls within the upper ends of chimpanzee, gorilla, and modern human ranges, with the latter two groups demonstrating non-significant differences (Table 14.7).

StW 352 and StW 573h exhibit similarities to the U.W. 88-99 calcaneus attributed to *Australopithecus sediba*, except in calcaneal dorsoplantar height and posterior talar facet height where StW 352 exhibits both a greater dorsoplantar height and a taller posterior talar facet than U.W. 88-99 (Table 14.7). The peroneal tubercle is preserved in several fossil hominin calcanei (StW 352, StW 573h, A.L. 333-8, A.L. 333-55, U.W. 88-99) (Gebo and Schwartz, 2006; DeSilva et al., 2019), but in at least one of the Sterkfontein calcanei, StW 352, the tubercle has been distorted via diagenetic processes (McNutt et al., 2017).

The shape of the anterior talar facet surface of StW 352 is similar to that of U.W. 88-99, where both resemble the mean shapes of gorillas and modern humans (Figure 14.5. Posterior talar facet shapes of StW 352 and StW 573h most resemble those of U.W. 88-99 among the hominin calcanei and chimpanzees among the extant groups (Figure 14.5). The Sterkfontein calcanei exhibit posterior talar facet shapes that do not particularly resemble that of A.L.333-8, which demonstrates a more modern human-like shape (Figure 14.5). Moreover, the posterior talar facet of StW 352 has a high radius of curvature similar to that of U.W. 88-99. While it also resembles the posterior talar facet of OH 8, it does not appear particularly flattened as is the modern human condition (Zipfel et al. 2011; Prang 2016 a,b; DeSilva et al., 2019) (Figure 14.5). The general direction in which the posterior talar facet faces relative to the long axis of the calcaneal body results in a narrow angle in StW 352 and U.W. 88-99 (Figure 14.6), which is more similar to the configuration exhibited by African apes rather than the greater discrepancy in directions exhibited by modern humans.

The calcaneal discriminant analysis has a 3.64% misclassification rate among the extant specimens. Specifically, two gorillas are misclassified as chimpanzees. When projecting the StW 352 partial calcaneus into discriminant space, it is predicted with 0.91 probability to be a chimpanzee. StW 573h was not complete enough to project into the discriminant space. The first canonical variable accounts for 86.9% of variance, and separates modern humans from gorillas and chimpanzees. StW 352 is positioned within chimpanzee morphospace by the first canonical variable (Figure 14.6, Table 14.7). The second canonical variable accounts for 13.10% of variance,

Table 14.7   Comparative calcaneal data. All measurements in mm.

	Humans (N = 20)	Pan (N = 18)	Gorilla (N = 17)	A.L.333-8	A.L.333-55	U.W. 88-99	OH 8	Can1	Can2
1. Anteroposterior articular length	47.89 (3.42) 42.11–53.97	32.13 (2.45) 27.00–37.52	40.04 (4.24) 34.90–49.79	…	…	42.3	…	–0.10	–0.60
2. Maximum mediolateral width	49.82 (2.96) 43.75–53.52	36.87 (3.85) 30.91–45.79	46.27 (6.24) 38.88–60.68	…	…	34.2	…	–0.72	–0.53
3. Dorsoplantar length	38.25 (3.40) 33.25–45.07	20.06 (1.88) 16.69–22.96	26.57 (4.07) 21.17–39.56	38.1[a]	38.8[a]	30.5	…	0.13	–0.12
4. Anterior talar facet length	33.64 (3.10) 28.02–39.50	20.66 (2.36) 15.68–26.13	25.92 (3.24) 20.87–31.91	…	…	27.6	25.0	–0.24	–0.61
5. Anterior talar facet width	7.37 (1.50) 4.41–10.35	6.58 (0.86) 4.90–8.22	10.55 (2.38) 6.87–14.89	…	…	11.0*	9.7	–0.96	0.44
6. Anterior talar facet depth	5.87 (1.30) 2.38–7.92	4.23 (0.88) 2.09–5.81	5.68 (1.73) 3.10–9.29	…	…	5.8	4.5	–0.96	–0.85
7. Posterior talar facet length	29.34 (2.26) 25.69–32.92	20.18 (1.94) 16.36–23.79	25.67 (2.60) 21.81–31.20	29.8[1]	26.5*	20.1	19.9*	–0.34	0.54
8. Posterior talar facet width	20.82 (2.07) 16.99–25.44	14.73 (1.51) 11.60–18.21	19.26 (2.35) 15.27–23.58	24.4[a]	24.5[a]	17.2	12.5*	–0.53	0.07
9. Posterior talar facet height	4.99 (0.84) 2.48–6.47	5.16 (0.67) 4.31–6.70	5.98 (1.05) 4.18–8.35	6.2	5.6	4.6	4.3	–1.20	–0.92
10. Mediolateral cuboid facet width	21.23 (2.24) 17.17–25.20	17.43 (1.29) 15.07–20.37	20.91 (2.12) 16.50–25.36	…	…	…	16.8	–1.01	–0.73
11. Dorsooplantar cuboid facet width	25.20 (2.00) 20.87–28.31	17.56 (1.37) 14.14–19.34	21.71 (2.30) 17.21–24.92	…	…	…	12.9	0.00	0.00
12. Cuboid facet depth	3.53 (1.09) 1.75–5.13	4.09 (0.62) 3.12–5.49	3.23 (0.73) 2.35–4.50	…	…	…	2.1	…	…

[a]Measurements from Latimer and colleagues (1982).

Measurements in cells of extant taxa correspond to mean and one standard deviation (parenthetical) in the top row and range (minimum-maximum) in the bottom row. Can 1 and Can 2 represent standardized scoring coefficients of the discriminant analysis.

*estimated value.

and separates gorillas from chimpanzees and modern humans. StW 352 is positioned within an area of overlap in all three groups by the second canonical variable.

Talus

Absolute magnitudes of Sterkfontein talar dimensions (talar length, mid-trochlear width, and talar width) are broadly comparable, and most tend to fall between the mean and lower end of modern human ranges (Table 14.8). Talar neck angles of all Sterkfontein tali, however, are noteworthy since they all fall outside observed modern human variation, and instead are collectively more similar to those of gorillas and chimpanzees (Table 14.8). Torsional angles of the talar heads of StW 88 and StW 573a are similar to those of gorillas and chimpanzees, while other tali are more modern human-like (Figure 14.7; Table 14.8). The shape of the talar head articulation on Sterkfontein tali resembles those of gorillas and chimpanzees, while plantar facet shape is intermediate between those of gorillas and modern humans (Figure 14.7; Table 14.8). In both articular shapes, StW 363 is different from other Sterkfontein tali, falling consistently among chimpanzees. Moreover, in absolute magnitudes, StW 363 exhibits smaller talar head and plantar facet dimensions compared to other Sterkfontein tali. StW 363 could belong subadult individuals because it textural porosity (Tumarkin-Deratzian, 2009).

Most absolute magnitudes of Sterkfontein talar dimensions are similar to those of U.W. 88-99. Torsional angle of the talar heads of StW 88 and StW 573a are similar to torsional angles of TM-1517, A.L. 288-1, KNM-ER 1464, KNM-ER 1476, U.W. 88-99, and OH 8 (Figure 14.7; Table 14.8). A trapezoidal trochlea shape (narrow posteriorly and broad anteriorly) is exhibited by all Sterkfontein tali, as well as those of A.L. 288-1, OH 8, and U.W. 88-98. This configuration also characterizes African apes. The talofibular lateral projection is well marked in StW 486, unlike other Sterkfontein tali that are less marked and more comparable to KNM-ER 1464, KNM-ER 1467, and TM 1517 (Figure 14.7; Table 14.8). The groove for the tendon of *m. flexor hallucis longus* in all Sterkfontein tali is nearly as well marked as those of African apes, except for StW88 which exhibits a modern human-like less evident groove. All Sterkfontein tali, again except StW 88, exhibit an elevated medial trochlear rim in comparison to the lateral rim. StW 88 exhibits a

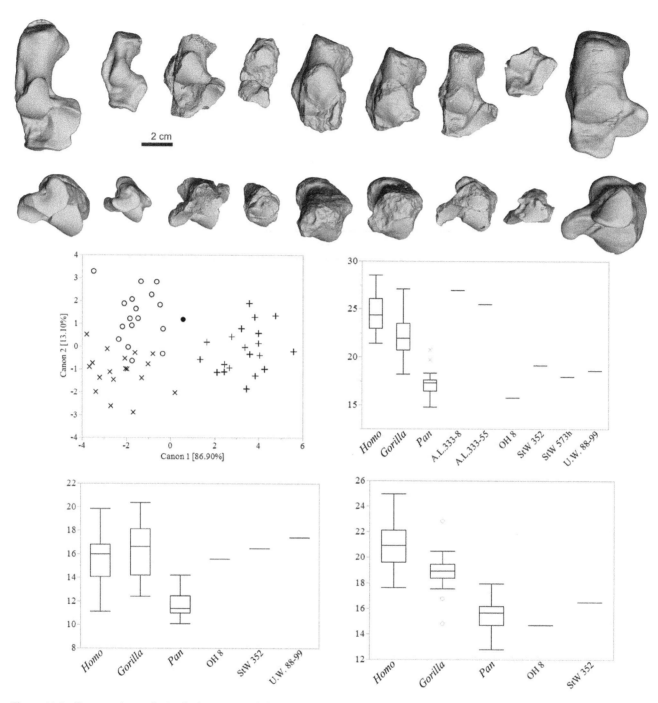

Figure 14.6 Comparative analysis of calcanei. From left to right, calcaneal renderings are illustrated in dorsal (top row) and anterior (second row) views for *Gorilla, Pan*, StW 352, StW 573f, A.L. 333-8, A.L. 333-55, U.W. 88-99, OH 8, and modern humans. Discriminant analysis results are illustrated in the third row (left) for modern humans (crosses), *Gorilla* (open circles), *Pan* (x-marks), and StW 352 (filled circle). Box plots in the third row (right) illustrate posterior talar facet shape (i.e., calculated as the square root of the product of posterior talar facet length and width), while in the bottom row box plots illustrate anterior talar facet shape (i.e., calculated as the square root of the product of anterior talar facet length and width) and cuboid facet shape (i.e., calculated as the square root of the product of cuboid facet mediolateral and dorsoplantar widths) on the left and right, respectively.

Table 14.8 Comparative talar data (in mm or degrees).

	Humans (N = 62)	Pan (N = 53)	Gorilla (N = 71)	SKX42695	TM-1517	A.L.288-1	A.L.333-75	A.L.333-147
1. Talar length	53.71 (5.65) 39.54–62.61	44.03 (3.03) 36.79–55.04	55.74 (7.01) 37.74–69.10			37.3	…	46.8
2. Talar neck length	19.88 (2.25) 12.89–24.18	19.71 (1.58) 16.25–22.97	23.15 (3.02) 14.27–31.69			12.2	…	15.0
3. Mid-trochlear width	25.67 (2.77) 18.96–30.31	16.55 (1.43) 13.70–20.52	21.96 (3.28) 16.59–29.69	18.90		17. 8	…	24.1
4. Anterior trochlear width	27.04 (2.85) 20.04–32.27	20.14 (1.61) 17.08–24.41	28.75 (3.55) 20.17–35.63			20.3	…	22.3
5. Posterior trochlear width	19.62 (2.74) 13.99–24.94	13.29 (1.75) 10.50–17.89	18.13 (3.32) 10.42–24.27			16.4	…	15.3
6. Talar width	39.62 (3.79) 30.26–45.73	34.08 (2.64) 27.19–42.65	47.85 (6.95) 25.28–60.73			27.8	…	36.9
7. Talar head height	29.93 (3.41) 20.20–37.92	20.77 (1.81) 15.74–25.98	29.00 (3.72) 18.43–36.83	23.1	27.7	20.2	25.08	26.6*
8. Talar head width	19.86 (2.74) 12.06–24.35	13.89 (1.82) 9.49–19.35	16.88 (2.77) 10.89–22.81			14.1	16.25	18.2*
9. Plantar facet length	29.89 (2.97) 22.00–34.82	22.14 (1.79) 17.67–27.69	29.23 (4.42) 20.40–38.93			20.8	…	26.7
10. Plantar facet width	19.58 (2.15) 15.12–23.65	11.35 (1.01) 9.01–14.06	18.38 (2.75) 14.15–25.73			9.3	…	16.0
11. Talofibular lateral projection	10.66 (1.78) 7.12–15.42	10.14 (1.23) 7.36–13.19	13.43 (2.13) 9.02–19.61			7.3	…	8.5
12. Talar neck angle	16.70 (2.00) 11.20–19.70	33.03 (4.05) 24.90–40.70	36.52 (4.00) 28.50–42.90	35.0	32.0	32.4	…	…
13. Talar head torsional angle	38.27 (3.22) 30.10–53.20	27.38 (2.82) 21.60–34.30	22.99 (3.58) 15.20–30.20	30.0	26.1	28.6	…	…

	KNM-ER 813	KNM-ER 1464	KNM-ER 1476	KNM-ER 5428	U.W. 88-98	OH 8	Can 1	Can 2
1. Talar length	50.1	49.2	45.2	…	41.6	…	0.12	0.23
2. Talar neck length	17.3	13.8	…	15.0	14.8	13.3	0.33	0.32
3. Mid-trochlear width	25.2	25.0	20.4	33.9	18.3	19.7	**-0.51**	0.40
4. Anterior trochlear width	26.2	26.5	21.5	36.4	21.1	20.3	0.11	**-0.56**
5. Posterior trochlear width	22.5	20.8	18.7	25.8	16.4	13.9*	-0.06	**-0.14**
6. Talar width	38.1*	41.8	34.8	46.0	30.3	28.2	0.43	**-0.46**
7. Talar head height	…	28.4	25.8	36.6	24.0	24.5	**-0.57**	-0.20
8. Talar head width	…	20.2	…	24.0	19.5	17.1	-0.09	0.30
9. Plantar facet length	…	29.9	26.9	…	25.2*	…	**-0.31**	0.15
10. Plantar facet width	19.0	16.9	14.6	23.1	14.4	15.3	**-0.46**	**-0.59**
11. Talofibular lateral projection	8.7	11.0	8.7	9.5	7.5	7.2	0.00	0.00
12. Talar neck angle	14.7*	23.7	30.2	20.0	31.3	33.5	…	…
13. Talar head torsional angle	47.8*	24.0	31.2	39.0	22.0	26.0	…	…

Measurements in cells of extant taxa correspond to mean and one SD (parenthetical) in the top row and range (i.e., min-max) in the bottom row.
Can 1 and Can 2 represent standardized scoring coefficients of the discriminant analysis.
*estimated value.

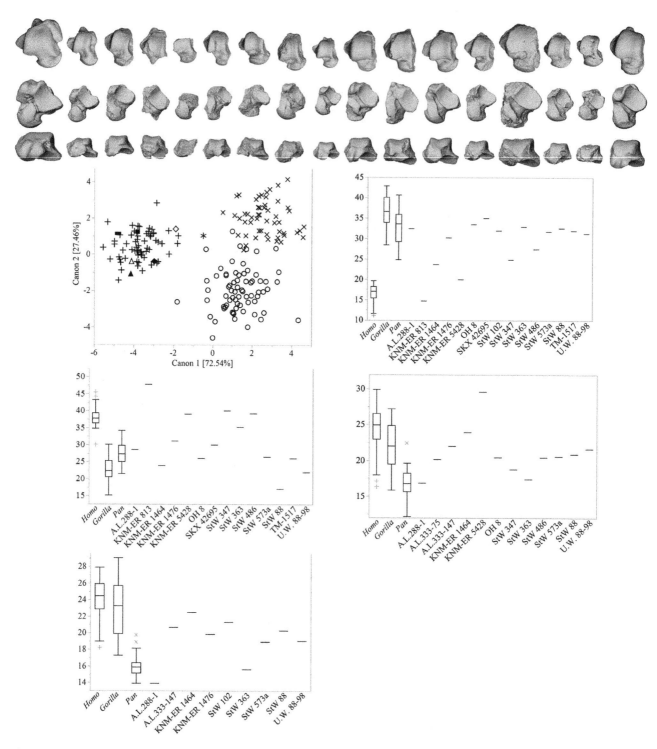

Figure 14.7 Comparative analysis of tali. From left to right, talar renderings are illustrated in dorsal (top row), plantar (second row), and posterior (third row) views for *Gorilla, Pan*, StW 88, StW 102, StW 347, StW 363, StW486, StW 573a, A.L. 288-1, A.L. 333-147, KNM-ER 813, KNM-ER 1464, KNM-ER 1476, KNM-ER 5428, U.W. 88-98, OH 8, and modern humans. Discriminant analysis results are illustrated in the fourth row (left) for modern humans (crosses), *Gorilla* (open circles), *Pan* (x-marks), StW 573a (open triangle), StW363 (star), StW 88 (filled diamond), A.L. 288-1 (open diamond), A.L. 333-147 (filled rectangle), KNM-ER 1464 (filled triangle), and U.W. 88-98 (filled square). Box plots in the fourth row illustrate talar neck angle (right), while those in the fifth row illustrate talar head torsional angle (left) and talar head shape (right) (i.e., calculated as the square root of the product of talar head height and width). Box plots in the sixth row illustrate plantar facet shape (i.e., calculated as the square root of the product of plantar facet height and width).

flattened trochlea, like those of modern humans, that is also comparable to those of A.L. 333-147 and KNM-ER 5428.

The talus discriminant analysis has a 3.23% misclassification rate among the extant specimens. Specifically, one gorilla is misclassified as a modern human and five gorillas are misclassified as chimpanzees (Table 14.8). Fossils (StW 573a, StW 88, AL. 288-1, AL. 33-147, KNM-ER 1464 and U.W. 88-98) were projected into discriminant space and predicted with 1.0 probability to be modern human, while StW 363 was predicted with 0.41 probability to be a human, with 0.39 probability to be a chimpanzee, and with 0.20 probability to be a gorilla. The first canonical variable accounts for 72.54% of variance, and separates modern humans from gorillas and chimpanzees. All fossils are either very close to modern humans in morphospace (StW 88, StW 573a, KNM-ER 1464, A.L.288-1, and A.L.333-147) or between modern humans and African apes (StW 363) (Figure 14.7; Table 14.8). The second canonical variable accounts for 27.46% of variance and separates gorillas from chimpanzees and modern humans.

Navicular

StW 573b is absolutely small for most measurements, and tends to fall consistently closer to mean chimpanzee values than those of other extant groups. Absolute length of StW 573b is similar to that of OH 8, and smaller than A.L. 333-47 and A.L. 333-36. A laterally narrow and medially broad navicular body characterizes all fossils in the study and is similar to the configuration exhibited by African apes as opposed to the more quadrangular shape exhibited by modern humans (Figure 14.8). The area of the lateral cuneiform facet on StW573b is similar to that on both modern human and chimpanzee naviculars, while areas of medial and intermediate cuneiform facets, as well as the talar facet, are most similar to those of chimpanzees. StW 573b, like A.L. 333-47 and A.L. 33-36, has a moderately projecting tuberosity that is somewhat intermediate between those of modern humans and African ape.

Concavity of the talar facet is not informative, as African apes and modern humans do not exhibit significant differences in this feature (Table 14.9). Shape of the medial cuneiform facet on StW 573b is similar to that on A.L. 333-47 and A.L. 333-36, whereas OH 8 appears to have a smaller medial cuneiform facet (Figure 14.8, Table 14.9). Shape of the intermediate cuneiform facet is similar to that of OH 8, whereas it is smaller than those of A.L.333-47 and A.L.333-36, which tend to resemble gorillas and chimpanzees (Figure 14.8, Table 14.9). Shape of the lateral cuneiform facet is again similar to that of OH 8, and similar to the configuration of the intermediate cuneiform facet; it is also smaller than those of A.L. 333-47 and A.L. 333-36 (Figure 14.8, Table 14.9). StW 573b,

A.L. 333-47, and A.L. 333-36 lateral cuneiform facets are flattened like those of modern humans rather than being more convex like those of African apes.

The navicular discriminant analysis has a 8.2% misclassification rate among extant specimens. Specifically, three gorillas are misclassified as chimpanzees and two chimpanzees are misclassified as gorillas (Figure 14.7). Fossils (StW 573b, A.L. 333-36, and A.L. 333-47) were projected into discriminant space and predicted with 0.99 % probability to be chimpanzees, while OH 8 was predicted with 0.93 probability to be a chimpanzee. The first canonical variable accounts for 84.89% of variance, and separates modern humans from gorillas and chimpanzees. All fossils, except OH 8, are either closer to gorillas or chimpanzees than to modern humans in morphospace (Figure 14.7, Table 14.9). The second canonical variable accounts for 15.11% of variance, and separates A.L. 333-36, A.L. 333-47, and OH 8 from extent groups and StW 573b.

Medial cuneiform

All fossils share similar most dimensions to African apes (Table 14.10). StW 573c is particularly similar to chimpanzees in its absolute dimensions. Shapes of the navicular facet and the first metatarsal facet of StW 573c are similar to those of chimpanzees, while the angle between these facets is modern human-like (Figure 14.8; Table 14.10).

StW 573c and OH 8 exhibit a narrower navicular facet than is exhibited by A.L. 333-28, while StW 573c exhibits a taller navicular facet compared to those of A.L. 333-28 and OH 8 (Figure 14.8; Table 14.10). StW 573c, like A.L. 333-28, exhibits a convex facet for the first metatarsal, resembling the African ape condition. This differs from OH 8 and modern humans, which exhibit a flatter facet for the first metatarsal (Figure 14.8). The presence of an attachment area of *m. peroneus longus* on StW 573c, as well as on A.L.333-28, resembles the modern human condition, as an attachment is not present on the medial cuneiforms of African apes (Susman et al., 1984). StW 573c, A.L. 333-28, and OH 8 exhibit L-shaped facets for the intermediate cuneiform, which is similar to the modern human-like condition (Figure 14.8).

The medial cuneiform discriminant analysis has a 1.64% misclassification rate among extant specimens. One chimpanzee is misclassified as a modern human. Two fossils (StW 573c and OH 8) were projected into discriminant space and predicted with 1.0 probability to be modern human. The first canonical variable accounts for 72.79% of variance and separates gorillas from chimpanzees and modern humans. StW 573c and OH 8 plot in the morphospace occupied by modern humans (Figure 14.8; Table 14.10). The second canonical variable accounts for 27.21% of variance and separates chimpanzees from gorillas and modern humans.

Table 14.9 Comparative navicular data. All measurements in mm.

	Modern humans (N = 20)	Pan (N = 21)	Gorilla (N = 20)	StW 573b	A.L.333-36	A.L.333-47	OH 8	Can 1	Can 2
1. Talar facet width	29.10 (1.73) 25.92–33.00	22.81 (2.40) 19.13–27.75	29.34 (2.83) 25.23–34.78	22.8	26.1[a]	24.1[a]	23.8	**−0.02**	0.78
2. Talar facet height	20.58 (1.93) 17.15–23.86	14.15 (1.90) 10.19–16.88	16.42 (2.04) 13.57–20.87	14.4*	16.5[1]	16.5[a]	15.8	0.50	**−0.09**
3. Talar facet depth	4.43 (0.62) 2.93–5.76	4.04 (0.61) 2.89–5.38	4.78 (0.60) 3.51–5.95	4.3	5.8	5.8	3.8	**−0.24**	**0.01**
4. Navicular mediolateral length	39.28 (3.17) 32.48–44.07	33.20 (2.81) 29.54–38.62	41.49 (4.23) 35.42–49.22	33.7	37.0[a]	36.3[a]	31.4	**−0.24**	**−0.12**
5. Medial cuneiform facet mediolateral width	24.74 (3.12) 18.47–29.08	19.58 (1.87) 17.23–23.81	25.68 (3.42) 20.43–33.86	12.8	13.5	12.4	11.6	**−0.32**	0.82
6. Medial and intermediate cuneiform ridge height	16.09 (2.83) 12.53–22.23	10.45 (1.41) 8.29–13.35	12.70 (3.47) 8.30–20.51	11.3	11.0	11.4	7.6	0.16	**−0.12**
7. Intermediate and lateral cuneiform ridge height	14.29 (1.85) 9.10–17.19	8.44 (1.74) 5.72–12.58	9.85 (2.52) 4.95–15.08	8.0	13.3	13.4	8.2	0.31	0.15
8. Intermediate cuneiform facet dorsal width	15.78 (1.67) 12.64–19.38	10.76 (1.25) 8.33–12.86	13.81 (1.78) 11.05–17.05	12.7	13.4	12.1	11.2	**−0.13**	0.97
9. Lateral cuneiform facet lateral height	9.26 (1.59) 6.71–11.79	9.08 (1.12) 7.43–11.97	11.62 (1.70) 7.55–15.79	9.0	10.0	11.9	8.7	**−0.69**	0.57
10. Lateral cuneiform facet plantar width	8.52 (1.37) 6.15–10.93	8.42 (1.74) 5.56–12.52	10.18 (1.88) 5.85–13.10	7.8	7.1	7.7	6.5	**−0.46**	0.34
11. Lateral cuneiform facet dorsal width	11.23 (1.41) 8.73–13.50	10.33 (1.85) 6.67–14.74	12.21 (1.82) 9.89–18.03	8.4	10.7	9.4	8.6	0.00	0.00

[a]Measurements from Latimer and colleagues (1982).

Measurements in cells of extant taxa correspond to mean and one SD (parenthetical) in the top row and range (i.e., minimum-maximum) in the bottom row. Can 1 and Can 2 represent standardized scoring coefficients of the discriminant analysis.

*estimated value.

Intermediate cuneiform

StW 573e exhibits absolutely small dimensions of the navicular facet, which is triangular shaped, as well as absolutely small dimensions of the facet for the second metatarsal. These conditions are most similar to those exhibited by chimpanzees (Figure 14.9; Table 14.11). Flattening of the facet for the second metatarsal exhibited by StW 573e, however, is modern human-like. StW 573e also exhibits African ape-like dorsal and plantar dimensions of the intermediate cuneiform (Figure 14.9; Table 14.11).

StW 573e and OH 8 exhibit similarly small absolute dimensions of the navicular facet, while the facet of the second metatarsal on both fossils is also small and flattened (Figure 14.9). StW 573e, having absolutely larger medial and lateral cuneiform facet lengths compared to those of OH 8, is more similar to African apes in this regard (Table 14.11). StW 573e and OH 8 are similar in their concave and triangular-shaped facet for the navicular, and in their relatively short anteroposterior length.

The intermediate cuneiform discriminant analysis has a 0% misclassification rate among extant specimens. Both fossils were projected into discriminant space and StW 573e was predicted with 0.99 probability to be a gorilla, while OH 8 predicted with 0.86 probability to be a gorilla and with 0.14 probability to be a chimpanzee. The first canonical variable accounts for 85.03% of variance, and separates modern humans from gorillas and chimpanzees. StW 573e and OH 8 plot between modern humans and gorillas in morphospace (Figure 14.9; Table 14.11). The second canonical variable accounts for 14.97% of variance, and separates gorillas from chimpanzees and modern humans.

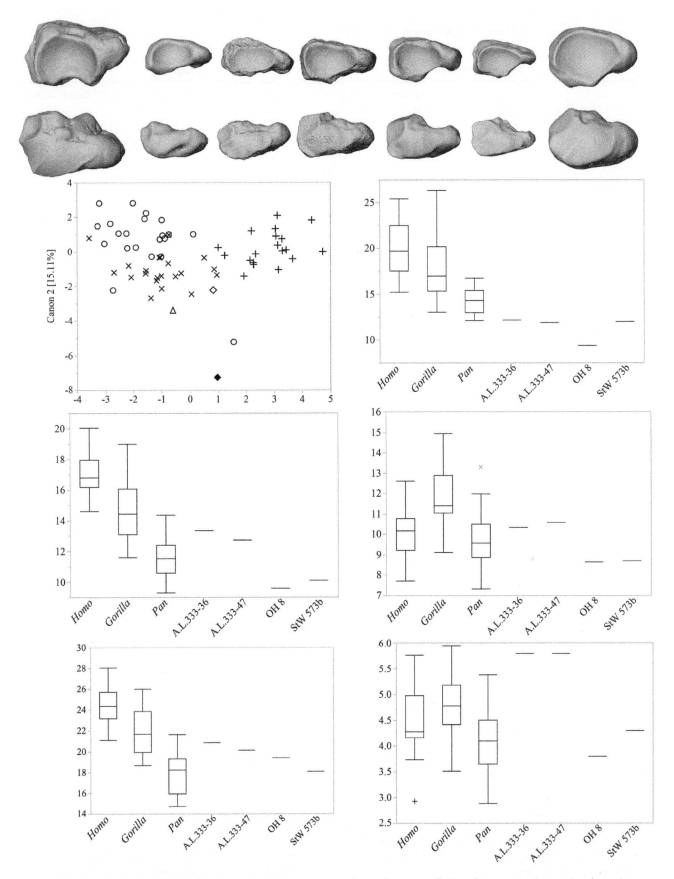

Figure 14.8 Comparative analysis of naviculars. From left to right, navicular renderings are illustrated in posterior (top row) and anterior (second row) views for *Gorilla, Pan*, StW 573b, A.L. 333-36, A.L. 333-47, OH 8, and modern humans. Discriminant analysis results are illustrated in the fourth row (left) for modern humans (crosses), *Gorilla* (open circles), *Pan* (x-marks), StW 573b (open triangle), A.L. 333-36 (filled circle), A.L. 333-47 (filled diamond), and OH 8 (open diamond). Box plots in the third row illustrate medial cuneiform facet shape (right) (i.e., calculated as the square root of the product of medial cuneiform facet mediolateral width and height of ridge formed by medial and lateral cuneiforms). Box plots in the fourth row illustrate intermediate cuneiform facet shape (left) (i.e., calculated as the square root of the product of dorsal intermediate cuneiform facet width and height of ridge formed by lateral and intermediate cuneiforms), and lateral cuneiform facet shape (right) (i.e., calculated as the square root of the product of lateral cuneiform facet height and dorsal width). Box plots in the fifth raw illustrate talar facet shape (left) (i.e., calculated as square root of product of talar facet length and width) (left) and talar facet depth (right).

Table 14.10   Comparative medial cuneiform data. Linear measurements in mm, angular data in degrees.

	Modern humans (N = 20)	Pan (N = 21)	Gorilla (N = 20)	StW 573c	A.L.333-28	OH 8	Can 1	Can 2
1. Navicular facet width	14.40 (1.16) 2.24–16.82	10.86 (1.26) 8.31–13.95	11.91 (1.42) 9.78–13.95	11.1	13.3	10.3	0.62	**−0.01**
2. Navicular facet height	21.20 (2.27) 17.68–25.72	16.65 (1.78) 13.85–19.81	20.97 (3.13) 15.37–28.97	14.9*	13.0	11.9	**−0.54**	0.47
3. Navicular facet depth	1.76 (0.69) 0.70–3.11	1.32 (0.48) 0.32–2.39	1.43 (0.42) 0.54–2.56	1.5	1.4	1.5	0.37	0.85
4. MT1 facet height	27.98 (1.84) 25.26–31.54	20.91 (1.67) 17.97–24.06	23.40 (2.98) 18.99–30.49	21.3	…	19.9	**−0.18**	0.47
5. MT1 facet middle width	12.55 (0.91) 10.96–14.26	8.79 (1.28) 5.47–10.56	10.71 (2.24) 6.52–15.21	6.4	9.8	8.7	**−0.08**	1.33
6. MT1 facet plantar width	13.34 (1.56) 10.06–15.36	12.26 (1.46) 8.90–15.03	13.60 (2.15) 10.42–18.81	9.8	12.0	9.4	0.15	**−0.15**
7. Plantar width	25.26 (1.75) 22.19–29.26	20.09 (1.84) 17.20–24.60	20.14 (2.17) 16.96–24.58	16.6	20.4	13.8	**−0.49**	**−0.47**
8. Dorsal width	14.24 (1.37) 12.02–17.34	10.59 (1.05) 8.10–12.07	16.85 (1.90) 13.37–21.27	7.6	…	9.0	0.99	0.41
9. Intermediate cuneiform facet posterior width	4.61 (1.97) 1.90–8.79	2.73 (0.74) 1.63–4.09	3.94 (1.54) 1.60–8.19	2.9	…	1.8	0.11	1.71
10. Intermediate cuneiform facet posterior height	11.40 (1.44) 8.93–14.56	6.03 (1.98) 1.85–9.55	8.89 (2.81) 2.91–13.56	6.4	…	6.1	**−0.10**	1.54
11. Intermediate cuneiform facet dorsal height	6.60 (1.00) 4.93–8.10	6.09 (0.92) 4.68–8.12	7.30 (1.37) 5.11–10.68	6.4*	..	4.2	0.68	0.23
12. Intermediate cuneiform facet dorsal length	10.76 (2.21) 6.92–15.49	7.96 (1.56) 4.97–11.41	7.52 (1.60) 4.56–10.70	8.1	…	6.8	**−0.03**	0.65
13. MT2 facet width	4.91 (1.43) 2.70–7.88	4.66 (0.64) 3.31–6.19	5.61 (1.23) 4.00–8.01	4.8	…	3.0	0.47	0.39
14. Anteroposterior angle of torsion	7.81 (2.75) 3.58–13.40	16.98 (5.13) 7.59–24.71	21.44 (4.32) 14.75–32.86	7.4*	…	5.2	…	…

Measurements in cells of extant taxa correspond to mean and one SD (parenthetical) in the top row and range (i.e., min-max) in the bottom row. Can 1 and Can 2 represent standardized scoring coefficients of the discriminant analysis.
*estimated value.

Lateral cuneiform

In all absolute measurements, StW 573g generally resembles those of African apes more than modern humans (Table 14.12). Shapes of the facets on StW 573g for the cuboid, third metatarsal, and navicular are also similar to those on African apes (Figure 14.10).

StW 573g and A.L.333-74 exhibit a number of similarities in absolute magnitudes of dimensions. For example, narrow anteroposterior lengths of StW 573g and A.L.333-74 resemble one another, as well as those of African apes (Figure 14.10, Table 14.12). The hamulus associated with *m. tibialis posterior* and *m. peroneus longus* is present on StW 573g and A.L. 333-74, and is like those observed on African ape lateral cuneiforms. StW 573g and A.L. 333-74 exhibit flattened navicular and third metatarsal facets; however, this is a modern human-like feature rather than an African ape-like feature.

The lateral cuneiform discriminant analysis has a 0% misclassification rate among extant specimens. During this discriminant analysis, we excluded measurements

of the fourth metatarsal facet since gorillas and modern humans variably lack them. StW 573g and OH 8 were projected into discriminant space and predicted with 1.0 probability to be a chimpanzee (StW 573e) and with 0.83 probability to be a chimpanzee (OH 8) and 0.16 probability to be a modern human (OH 8). The first canonical variable accounts for 88.62% of variance, and separates modern humans from gorillas and chimpanzees. StW 573g plots within gorilla and chimpanzee morphospace (Figure 14.10, Table 14.12). The second canonical variable accounts for 11.38% of variance, and separates StW 573g and OH 8 from the extant groups.

Discussion and conclusions

Qualitative evaluation and morphometric comparisons of the Sterkfontein Member 2 and Member 4 hominin tarsals underscore a mosaic of African ape-like and modern human-like features. The StW 352 calcaneus plots with chimpanzee in discriminant function space, largely based

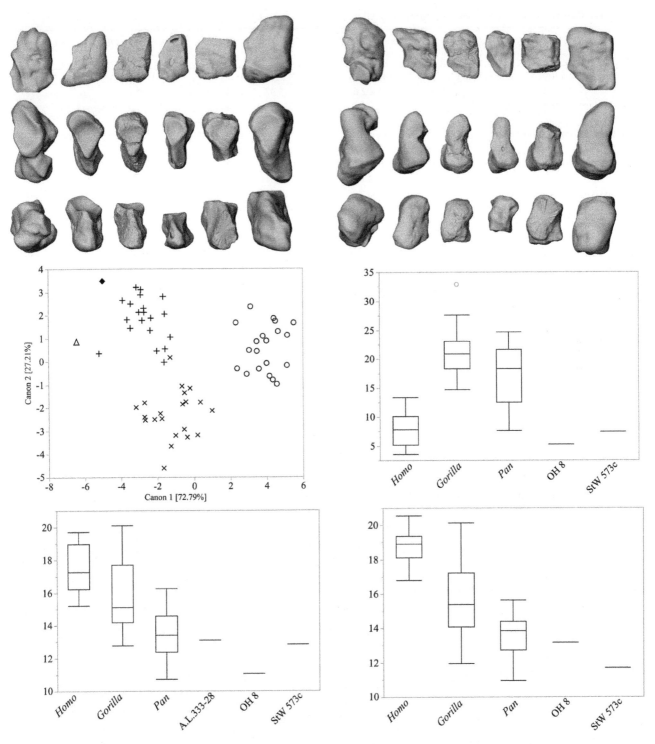

Figure 14.9 Comparative analysis of medial cuneiforms. Within the top row, medial cuneiform renderings are grouped into medial (left column) and lateral (right column) views. Within each column, from left to right, renderings are illustrated for *Gorilla, Pan*, StW 573c, OH 8, A.L.333-28, and modern humans. The second row illustrates posterior (left column) and anterior (right column) views, while the third row illustrates dorsal (left column) and plantar (right column) views. Discriminant analysis results are illustrated in the fourth row (left) for modern humans (crosses), *Gorilla* (open circles), *Pan* (x-marks), StW 573c (open triangle), and OH 8 (filled diamond). Box plots in the fourth row illustrate the angle between navicular and MT1 facets (right), while those in the fifth row illustrate navicular facet shape (left) (i.e., calculated as the square root of the product of navicular facet height and width) and first metatarsal facet shape (right) (i.e., calculated as the square root of the product of first metatarsal facet height and width).

Table 14.11 Comparative intermediate cuneiform data. All measurements in mm.

	Modern humans (N = 20)	Pan (N=18)	Gorilla (N = 18)	StW 573e	OH8	Can 1	Can 2
1. Navicular facet dorsoplantar length	19.44 (2.03) 15.34–22.65	13.35 (1.26) 11.01–15.57	16.73 (2.07) 14.42–21.78	15.1	13.1	0.38	**–0.08**
2. Navicular facet mediolateral width	14.62 (1.25) 12.85–17.38	10.58 (1.14) 8.57–12.64	13.40 (1.94) 10.42–17.40	11.2	8.6	0.13	0.55
3. MT2 facet dorsoplantar length	19.13 (1.69) 15.67–22.50	12.48 (1.94) 8.74–15.36	15.32 (1.85) 12.54–20.37	13.1*	11.7	**0.04**	0.30
4. MT2 facet mediolateral width	11.90 (1.24) 9.18–14.65	10.98 (1.08) 9.04–13.06	12.87 (1.82) 9.77–15.89	12.4	9.1	1.08	-0.08
5. Dorsal anteroposterior medial width	16.23 (1.80) 11.95–18.79	11.38 (1.25) 9.58–13.68	11.50 (1.42) 9.37–14.83	12.1	11.2	1.02	**–0.95**
6. Dorsal anteroposterior middle width	17.49 (1.58) 13.15–19.92	10.95 (0.99) 9.28–13.41	13.15 (1.57) 11.00–16.17	11.0	10.3	**-0.15**	0.18
7. Dorsal anteroposterior lateral width	15.41 (1.77) 12.52–18.72	8.13 (1.15) 6.21–10.92	9.91 (1.61) 7.98–13.93	10.	8.9	0.35	**–0.18**
8. Plantar anteroposterior width	13.92 (1.99) 10.57 –18.91	10.78 (1.22) 8.40–12.41	13.00 (1.97) 9.40–17.14	11.6	9.8	0.61	**0.03**
9. Medial cuneiform facet anteroposterior (dorsal) width	16.23 (1.80) 11.95–18.79	11.38 (1.25) 9.58–13.68	11.50 (1.42) 9.37–14.83	12.1	9.2	0.00	0.00
10. Lateral cuneifrom facet anteroposterior (dorsal) width	10.34 (1.10) 8.09–12.63	3.96 (0.77) 2.99–5.71	5.62 (1.41) 3.78–8.12	6.4	4.7	0.00	0.00

Measurements in cells of extant taxa correspond to mean and one SD (parenthetical) in the top row and range (i.e., min-max) in the bottom row. Can 1 and Can 2 represent standardized scoring coefficients of the discriminant analysis.
* estimated value

on similarities in linear dimensions of the posterior and anterior talar facets, as well as the cuboid facet. Other studies (Prang, 2015 a,b) also have emphasized African-ape like features in StW 352. Some (Latimer et al., 1982; Stern and Susman, 1983; Deloison, 1985; Latimer and Lovejoy, 1989; Zipfel et al., 2011) have highlighted the large size of the peroneal tubercle as African ape-like, but it is important to note that a recent study of the internal morphology specimen has shown its reconstructed size is artificially inflated to an unknown extent due to an intrusive matrix (McNutt et al., 2017). While we did not include the StW 573h calcaneus in the discriminant function analysis due to its less complete nature, it appears reasonably similar to StW 352 in its overall external dimensions. Trabecular architecture beneath the cuboid facet of StW 352 has been shown to be similar to gorillas and modern humans (Zeininger et al., 2016). The shape of its posterior talar facet is chimpanzee-like, but trabecular architecture beneath this facet is more modern human-like (DeSilva et al., 2018). While trabecular architecture of the StW 573h calcaneus has not been assessed, such an investigation would offer more insight into the extent of its similarities to StW 352.

Tali, being the most numerous tarsal in the Sterkfontein hominin assemblage, offer an opportunity to assess the extent of morphological variation. Tali from Member 4, like other Sterkfontein tarsals, generally combine modern human-like features (shape of the plantar facet and talar head torsional angle) with African ape-like features (talar neck angle and shape of the talar head articulation). In the talar discriminant function analysis, the two most complete specimens, StW 88 and StW 363, plotted in different areas of discriminant space. StW 363 plotted between modern humans and chimpanzees, while StW 88 plotted with modern humans. A modern human-like classification of StW 88 contrasts with results of other multivariate analyses (Harcourt-Smith, 2002; Kidd and Oxnard 2005; Harcourt-Smith et al., 2015), which differed in sampling (e.g., inclusion of *Pongo* or a priori membership of StW 573a as comparative samples), measurement (e.g., present study did not incorporate angular measurements in the discriminant function analysis), and statistical approaches.

The presence of multiple australopith species in Member 4 has been proposed (Lockwood and Tobias, 2002; Clarke, 2013). The StW 88 talus, in having a relatively flat trochlea and narrow talar head angle of torsion, differs from other Sterkfontein Member 4 tali (StW 347, StW 363, and StW 486). Moreover, StW 88 does not exhibit the well-marked groove for the tendon of *m. flexor hallucis longus* or concave trochlea with elevated medial rim that other Sterkfontein tali share with African apes.

Thus, based on Member 4 tali, excluding the possibly subadult StW 363 talus, there appears to be at least one morph comprised of StW 88 and one morph comprised of StW 102, StW 347, and StW486. In this regard, it may

Table 14.12 Comparative lateral cuneiform data. All measurements in mm.

	Modern humans (N = 20)	Pan (N = 22)	Gorilla (N = 20)	StW 573g	A.L.333-79	OH8	Can 1	Can 2
1. Navicular facet dorsoplantar height	14.48 (1.15) 12.19–16.44	8.77 (1.32) 7.25–12.64	11.22 (1.78) 8.8–14.5	9.5	13.4	12.8	**−0.30**	0.05
2. Navicular facet mediolateral width	12.2 (1.25) 10.43–14.4	9.45 (1.46) 7.17–13.06	10.81 (1.49) 8.07–13.28	8.2	10.9	8.7	0.06	**−0.04**
3. Navicular facet dorsoplantar height	21.88 (1.35) 19.85–24.65	17.81 (1.56) 15.37–21.17	20.5 (2.23) 17.78–25.03	20.8	22.8	18.0	0.86	0.35
4. MT3 facet dorsoplantar height	21.51 (1.91) 18.3–24.82	15.29 (1.09) 13.46–17.54	16.83 (1.43) 15.35–20.7	16.3	...	14.8	**−0.37**	**−0.66**
5. MT3 facet dorsal mediolateral width	13.66 (1.37) 11.13–16.87	12.83 (0.98) 10.9–14.4	14.92 (1.63) 12.11–18.17	11.7	...	9.8	0.81	0.17
6. MT3 facet middle mediolateral width	8.4 (0.88) 6.62–10.26	4.98 (0.84) 3.45–6.59	7.13 (1.14) 5.48–9.94	6.8	8.2	5.5	**0.05**	0.57
7. MT3 facet plantar mediolateral width	5.87 (0.9) 3.95–7.35	6.33 (1.04) 4.53–8.32	7.38 (1.17) 4.93–9.52	2.8	5.1	2.2	0.67	0.35
8. Dorsal anteroposterior lateral length	14.61 (1.72) 11.69–19.38	11.12 (1.37) 8.95–14.51	11.51 (1.72) 8.56–14.53	17.8	18.7*	13.2	0.28	**−0.43**
9. Dorsal anteroposterior medial length	19.36 (1.72) 15.47–21.65	11.55 (1.13) 8.98–13.7	12.9 (1.68) 10.59–16.59	13.7	15.3*	12.3	**−0.05**	**−0.43**
10. Plantar anteropsterior length	24.18 (1.56) 21.47–26.9	14.33 (1.39) 11.49–16.76	18.1 (2.1) 14.78–22.72	12.0	15.6	11.3	**−0.32**	0.64
11. Cuboid facet width	15.84 (1.87) 13.12–20.14	7.59 (0.83) 6.24–9.39	11.62 (2.62) 7.33–18.77	7.7	9.5	6.7	**−0.17**	0.61
12. Cuboid facet height	12.6 (2.17) 9.9–17.93	8.91 (2.55) 5.45–18.21	11.23 (1.72) 8.47–14.11	7.1	10.1	7.8	0.72	0.66
13. MT4 facet width	2.16 (2.18) 0–7.8	4.35 (1.18) 1.7–6.08	4.26 (2.76) 0–7.89	4.5	...	2.1	...	...
14. MT4 facet height	3.83 (3.32) 0–8.67	4.97 (0.88) 3.47–7.21	3.67 (2.55) 0–9.1	6.2	...	3.3	...	...
15. Intermediate cuneiform facet width	8.66 (1.41) 4.66–10.65	4.16 (0.99) 2.47–5.99	4.69 (1.33) 2.29–7.18	5.1	...	2.7	0.16	0.06
16. Intermediate cuneiform facet height	13.94 (2.15) 11–18.96	6.06 (1.39) 4.13–8.84	8.17 (1.68) 5–10.68	5.2	...	7.2	**−0.04**	0.42
17. MT2 facet width	4.34 (1.74) 1.78–7.18	2.49 (0.5) 1.95–4.07	2.9 (0.94) 0.92–5.7	4.4	...	3.1	0.65	0.66
18. MT2 facet height	10.66 (3.73) 4.89–16.87	5.7 (1.24) 3.85–9.32	6.83 (1.1) 4.6–8.98	5.2	...	4.4	0.16	0.06

Measurements in cells of extant taxa correspond to mean and one SD (parenthetical) in the top row and range (i.e., minimum-maximum) in the bottom row. Can 1 and Can 2 represent standardized scoring coefficients of the discriminant analysis.

*estimated value.

be instructive to consider the arboreal/terrestrial differences exhibited by lowland and mountain gorillas that have been linked to differences in their talar morphology (Dunn et al., 2014; Knigge et al., 2015). This classification differs from that of Deloison (2003) who proposed four different locomotor groups: arboreal (StW 363), primitive biped (StW 88 and StW 486), biped (StW 347), and unclassified (StW 102). When StW 573a is projected into the talar discriminant space, like the projected StW 88, it plotted with modern humans despite exhibiting a relatively concave trochlear surface and an elevated medial trochlear rim more typical of African apes. Other studies of the external morphology of StW 573a have noted both modern human-like and African ape-like features (Clarke and Tobias, 1995; Harcourt-Smith, 2002; Deloison, 2003; Kidd and Oxnard, 2005; Harcourt-Smith et al., 2015; DeSilva et al., 2019). In an analysis of trabecular architecture, Su and Carlson (2017) noted generally similar trabecular properties beneath the trochleae of StW 102 and StW 486, which suggested similar tibiotalar joint ranges of motion. While trabecular architecture of the StW 573a talus has not been assessed, such an investigation would offer more insight into the extent of its similarities to StW 88.

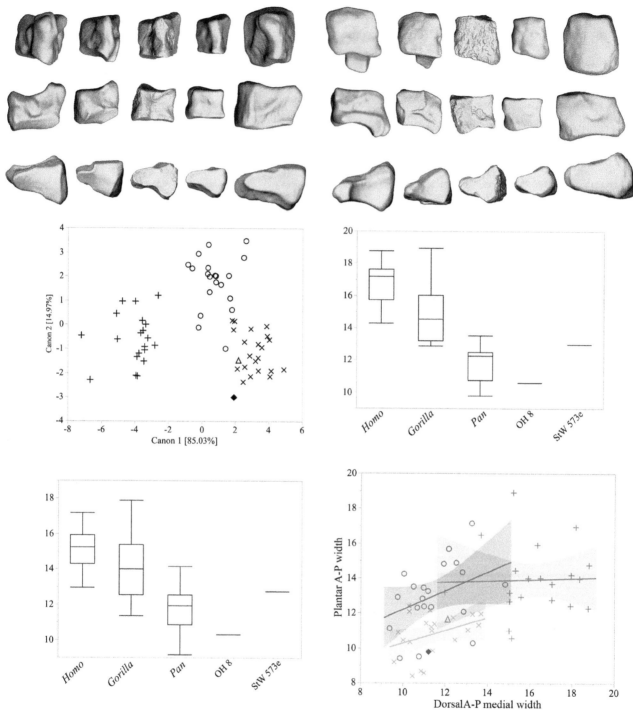

Figure 14.10 Comparative analysis of intermediate cuneiforms. Within the top row, intermediate cuneiform renderings are grouped into plantar (left column) and dorsal (right column) views. Within each column, from left to right, renderings are illustrated for *Gorilla*, *Pan*, StW 573e, OH 8, and modern humans. The second row illustrates lateral (left column) and medial (right column) views, while the third row illustrates anterior (left column) and posterior (right column) views. Discriminant analysis results are illustrated in the fourth row (left) for modern humans (crosses), *Gorilla* (open circles), *Pan* (x-marks), StW 573e (open triangle), and OH 8 (filled diamond). Box plots in the fourth row illustrate navicular facet shape (right) (i.e., calculated as the square root of the product of navicular facet height and width). In the bottom row, the box plots on the left illustrate second metatarsal (MT2) facet shape (i.e., calculated as the square root of the product of MT2 facet height and width), while the bivariate plot with linear 90% confidence and best fit regression line of plantar anteroposterior width vs. dorsal anteroposterior medial width on the right illustrates modern humans (blue), *Gorilla* (red), *Pan* (green), StW 573e (open triangle), and OH 8 (filled diamond).

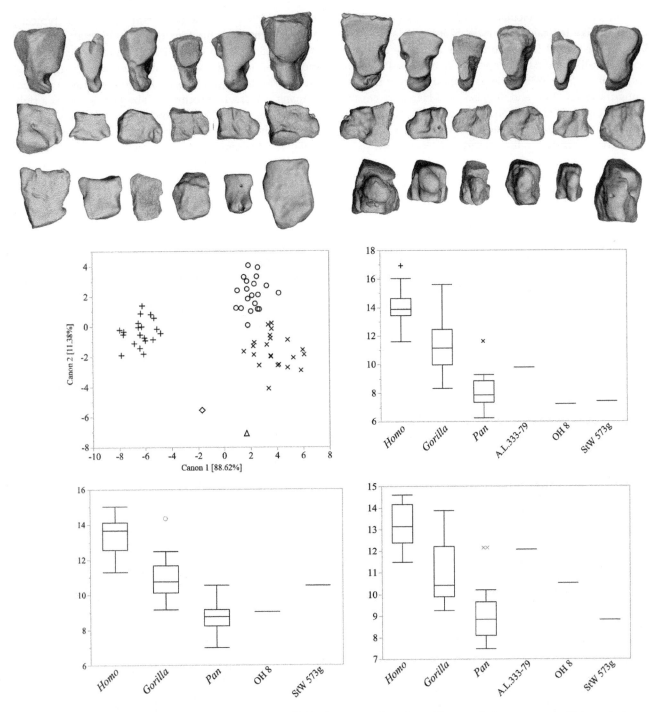

Figure 14.11 Comparative analysis of lateral cuneiforms. Within the top row, lateral cuneiform renderings are grouped into posterior (left column) and anterior (right column) views. Within each column, from left to right, renderings are illustrated for *Gorilla, Pan*, StW 573g, A.L. 333-79, OH 8, and modern humans (right column) and modern humans, OH 8, A.L. 333-79, 573g, Pan and Gorilla (left column) receptively. The second row illustrates medial (left column) and lateral (right column) views, while the third row illustrates dorsal (left column) and plantar (right column) views. Discriminant analysis results are illustrated in the fourth row (left) for modern humans (crosses), *Gorilla* (open circles), *Pan* (x-marks), StW 573g (open triangle), and OH 8 (open diamond). Box plots in the fourth row illustrate cuboid facet shape (right) (i.e., calculated as the square root of the product of cuboid facet height and width), while those in the fifth row illustrate third metatarsal facet shape (left) (i.e., calculated as the square root of the product of third metatarsal facet height and width) and navicular facet shape (right) (i.e., calculated as the square root of the product of navicular facet height and width).

The StW 573b navicular and cuneiforms (StW 573c, e, and g), like the calcaneus (StW 573h) and talus (StW 573a) of the partial foot, also display a mosaic of modern human-like and African ape-like characters. Previous studies of the StW 573b navicular support such a mosaic configuration (Clarke and Tobias, 1995; Harcourt-Smith, 2002; Deloison, 2003; Kidd and Oxnard 2005). The medial cuneiform (StW 573c) exhibits both modern human-like L-shaped facet for the intermediate cuneiform and African ape-like shape of the navicular facet as well as shape of the concave first metatarsal facet. Planes of the navicular and first metatarsal facets on the medial cuneiform are relatively parallel, suggesting the hallux was not notably divergent from the metatarsus (but see Clarke and Tobias, 1995).

By comparison, the intermediate (StW 573e) and lateral (StW 573g) cuneiforms tend to exhibit disproportionately more African-ape like features than modern human-like features (Figures 14.9 and 14.10, Tables 14.11 and 14.12). Specifically, the lateral cuneiform (StW 573g) is shortened anteroposteriorly and has a large, projecting hamulus for insertion of the *m. tibialis posterior* tendon that creates a short, square navicular facet. These findings are consistent with a previous study of the intermediate and lateral cuneiforms (Deloison, 2003).

A recent study of the StW 573 partial foot by DeSilva et al. (2019) questions whether the StW 573h calcaneus is associated with other tarsals. They suggested that the concavity of the cuboid facet on StW 573h is typical of cercopithecoids. This is not consistent with our qualitative and quantitative (Figure 14.6, Table 14.7) analyses, however, which show its mixture of modern human-like and African ape-like features, including concavity of the cuboid facet, and matches the same type of pattern generally exhibited by all of the elements in the StW 573 partial foot, and it articulates perfectly with the talus.

African ape-like characters, indicating selection on adaptive use of arboreal substrates, are persistence throughout Member 2 and Member 4 tarsals, although they are more prevalent in some than in others. A convex posterior talar facet, which is exhibited by StW 352, promotes a greater range of motion in this subtalar joint compared to that exhibited by modern humans (DeSilva, 2009; Venkataraman et al., 2013; Prang, 2016b). The presence of a peroneal tubercle, as also exhibited to some extent by StW 352, has been linked to arboreality, and specifically climbing (Stern and Susman, 1983). A well-marked groove for the tendon of *m. flexor hallucis longus* and a concave trochlea with an elevated medial rim are evident on Sterkfontein tali except StW 88 are typically described as arboreal (climbing) adaptations (Latimer and Lovejoy, 1989; DeSilva, 2009; Dunn et al., 2014; Knigge et al., 2015). Convexity of the first metatarsal facet on the medial cuneiform permits greater dorsoplantar range of

motion in the first tarsometatarsal joint than is typically present in modern humans (Jashashvili et al., 2010; Gill et al., 2015; DeSilva et al., 2018). The anteroposteriorly short intermediate and lateral cuneiforms suggest that the anterior tarsal row has not yet elongated as in modern humans, adversely affecting its ability to serve as a rigid lever (Lovejoy et al., 2009; DeSilva et al., 2019).

In general, calcanei and tali from Sterkfontein and Malapa, attributed to *Australopithecus sediba*, exhibit more morphological similarities to one another than tarsals from either assemblage exhibit to *Australopithecus* tarsals from Hadar. For example, Sterkfontein Member 2 and Member 4 calcanei share convexity and shape of the posterior talar facet with U.W. 88-99 that is quite different from those of A.L. 333-8 and A.L. 333-55 calcanei (Figure 14.5, Table 14.7). The South African calcanei also exhibit a calcaneal robusticity index that is smaller than exhibited by Hadar calcanei (Prang, 2015a, 2016b). An exception to this trend is exhibited by talar head angle of torsion. StW 88, StW 573a, and U.W. 88-98 exhibit a narrow talar head angle of torsion unlike other Sterkfontein hominin tali, which morphologically aligns the former with A.L. 288-1 and A.L. 333-147 (Prang 2015b, 2016b) tali (Figure 14.6). StW 347, StW 363, and StW 486, on the other hand, exhibit similar talar head angles of torsion as KNM-ER-813 and KNM-ER 5428 tali (Figure 14.6).

Additional studies, including investigations of internal structure, are necessary to answer questions about the functional morphology and taxonomy of tarsal bones in the Sterkfontein hominin assemblage. Incorporating all tarsal bones into a stepwise common Principal Component Analysis (PCA) to evaluate ever-increasing similarity among the patterns of morphological variation within tarsals might offer particularly novel and exciting insights.

Acknowledgments

We would like to thank the co-editors of this special volume, Drs. Bernhard Zipfel and Carol Ward, for the invitation to contribute to it. We thank Ms. Stephany Potze for support and access during data collection at the Ditsong (formerly Transvaal) National Museum of Natural History in Pretoria, South Africa. We thank Dr. Bernhard Zipfel for support and access during data collection at the Evolutionary Studies Institute (ESI), University of the Witwatersrand, in Johannesburg, South Africa. We thank Dr. Brendon Billings for support and access during data collection at the Raymond A. Dart Collection of Human Skeletons (*Dart Collection*), which is housed in the School of Anatomical Sciences, University of the Witwatersrand, in Johannesburg, South Africa. We thank Dr. Emmanuel Gilissen, Dr. Wim Wendelen, and Mr. Wim van Neer of the

Royal Museum for Central Africa, in Tervuren, Belgium. We thank Dr. Christine Lefèvre for support and access during data collection of the National Museum of Natural History in Paris, France. We thank Prof. Philippe Grenier and Mrs. Bettina Begenau (Hôpital Pitié-Salpêtrière, Paris) for permitting access to CT facilities. We thank Mr.Walter Coudyzer (UZ Leuven) for permitting access to CT facilities. We thank Dr. Jackie Smilg and Mrs. Queen Letsoalo for assistance in the CT facilities of Charlotte Maxeke Johannesburg Academic Hospital. Finally, we also thank the Virtual Imaging in Palaeosciences (VIP) laboratory for data analysis and Wits micro-CT facility to scan studied fossil materials, ESI at the University of the Witwatersrand.

References

Bruxelles, L., Clarke, R.J., Maire, R., Ortega, R., Stratford, D., 2014. Stratigraphic analysis of the Sterkfontein StW 573 *Australopithecus* skeleton and implications for its age. Journal of Human Evolution 70, 36–48.

Clarke, R.J., 1994. Advances in understanding the craniofacial anatomy of South African early hominids. In: Corruccini, R.S., and Ciochon R.L. (Eds.), Integrative Paths to the Past: Essays in Honor of F. Clark Howell. New Jersey: Prentice-Hall, pp. 205–222.

Clarke, R.J., 2002. Newly revealed information on the Sterkfontein Member 2 *Australopithecus* skeleton: News and views. South African Journal of Science 98, 523–526.

Clarke, R.J., 2006. A deeper understanding of the stratigraphy of Sterkfontein fossil horninid site. Transactions of the Royal Society of South Africa 61, 111–120.

Clarke, R.J., 2007. Taphonomy of Sterkfontein *Australopithecus* skeletons. In: Kuman, K., Pickering, T.R., Schick, K., Toth, N. (Eds.), Breathing life into fossils: taphonomic studies in honour of C.K. (Bob) Brain. Stone Age Institute Press: Bloomington, pp. 195–201.

Clarke, R.J., 2008. Latest information on Sterkfontein's *Australopithecus* skeleton and a new look at *Australopithecus*. South African Journal of Science 104, 443–449.

Clarke, R.J., 2013. *Australopithecus* from Sterkfontein Caves, South Africa. In: Reed, K.E., Fleagle, J.G. (Eds.), The paleobiology of *Australopithecus*. Springer: Dordrecht, pp. 105–123.

Clarke, R.J., Tobias, P.V., 1995. Sterkfontein member 2 foot bones of the oldest South African hominid. Science 269, 521–524.

Dayal, M.R., Kegley, A.D.T., Štrkalj, G., Bidmos, M.A., Kuykendall, K.L., 2009. The history and composition of the Raymond A. Dart Collection of Human Skeletons at the University of the Witwatersrand, Johannesburg, South Africa. American Journal of Physical Anthropology 140, 324–335.

Deloison, Y., 1985. Comparative study of calcanei of primates and *Pan-Australopithecus-Homo* relationship. In: Tobias, P.V. (Ed.), Hominid evolution: past, present and future. Alan R. Liss: New York, pp. 143–147.

Deloison, Y., 1993. Etude des restes fossiles des pieds des premiers hominidés: *Australopithecus* et *Homo habilis*: Essai d'interprétation de leur mode de locomotion. Ph.D. Dissertation, Sciences Naturelles, Paris 5.

Deloison, Y., 2003. Fossil foot bone anatomy from South Africa between 2.4 and 3.5 MY old: interpretation in relation to the kind of locomotion. Biometrie Humaine et Anthropologie 21, 189–230.

Deloison, Y., 2004. A new hypothesis on the origin of hominoid locomotion: from biped to strider. Springer: Boston, pp. 35–47.

DeSilva, J.M., 2009. Functional morphology of the ankle and the likelihood of climbing in early hominins. Proceedings of the National Academy of Science USA 106(16), 6567–6572.

DeSilva, J.M., Devlin, M.J., 2012. A comparative study of the trabecular bony architecture of the talus in humans, non-human primates, and *Australopithecus*. Journal of Human Evolution 63, 536–551.

DeSilva, J.M., Gill, C.M., Prang, T.C, Bredella, M.A., Alemseged, Z., 2018. A nearly complete foot from Dikika, Ethiopia and its implications for the ontogeny and function of *Australopithecus afarensis*. Science Advances 4, eaar7723.

DeSilva, J., McNutt, E., Benoit, J., Zipfel, B., 2019. One small step: a review of Plio-Pleistocene hominin foot evolution. American Journal of Physical Anthropology 168, 63–140.

Dowdeswell, M.R., Jashashvili, T., Patel, B.A., Lebrun, R., Susman, R.L., Lordkipanidze, D., Carlson, K.J. 2017. Adaptation to bipedal gait and fifth metatarsal structural properties in *Australopithecus, Paranthropus*, and *Homo*. Comptes Rendus Palevol 16, 585–599.

Dunn, R.H., Tocheri, M.W., Orr, C.M., Jungers, W.L. 2014. Ecological divergence and talar morphology in gorillas. American Journal of Physical Anthropology 153, 526–541.

Gebo, D.L., Schwartz, G.T., 2006. Foot bones from Omo: implications for hominid evolution. American Journal of Physical Anthropology 129, 499–511.

Gill, C.M., Bredella, M.A., DeSilva, J.M., 2015. Skeletal development of hallucal tarsometatarsal joint curvature and angulation in extant apes and modern humans. Journal of Human Evolution 88, 137–145.

Granger, D.E., Gibbon, R.J., Kuman, K., Clarke, R.J., Bruxelles, L., Caffee, M.W., 2015. New cosmogenic burial ages for Sterkfontein Member 2 *Australopithecus* and Member 5 Oldowan. Nature 522, 85.

Harcourt-Smith, W.E.H., 2002. Form and function in the hominoid tarsal skeleton. Ph.D. Dissertation, University of London.

Harcourt-Smith, W.E.H., Throckmorton, Z., Congdon, K.A., Zipfel, B., Deane, A.S., Drapeau, M.S.M.,

Churchill, S.E., Berger, L.R., DeSilva, J.M. 2015. The foot of *Homo naledi*. Nature Communications 6, 8432.

Jashashvili, T., Dowdeswell, M.R., Lebrun, R., Carlson, K.J., 2015. Cortical structure of hallucal metatarsals and locomotor adaptations in hominoids. PLoS ONE 10, e0117905.

Jashashvili T., Ponce de Leon, M.S., Lordkipanidze, D., Zollikofer, C.P. 2010. First evidence of a bipartite medial cuneiform in the hominin fossil record: a case report from the Early Pleistocene site of Dmanisi. Journal of Anatomy 216, 705–716.

Jungers, W.L., Harcourt-Smith, W.E., Wunderlich, R.E., Tocheri, M.W., Larson, S.G., Sutikna, T., Due, R.A., Morwood, M.J. 2009. The foot of *Homo floresiensis*. Nature 459, 81–84.

Kidd, R., Oxnard, C., 2005. Little foot and big thoughts—a re-evaluation of the StW 573 foot from Sterkfontein, South Africa. Homo 55, 189–212.

Knigge, R.P., Tocheri, M.W., Orr, C.M., McNulty, K.P., 2015. Three-dimensional geometric morphometric analysis of talar morphology in extant gorilla taxa from highland and lowland habitats. Anatomical Record 298, 277–290.

Kuman, K., Clarke, R.J., 2000. Stratigraphy, artefact industries and hominid associations for Sterkfontein, Member 5. Journal of Human Evolution 38, 827–847.

Latimer, B., Lovejoy, C.O., 1989. The calcaneus of *Australopithecus afarensis* and its implications for the evolution of bipedality. American Journal of Physical Anthropology 78, 369–386.

Latimer, B.M., Lovejoy, C.O., Johanson, D.C., Coppens, Y., 1982. Hominid tarsal, metatarsal, and phalangeal bones recovered from the Hadar formation: 1974–1977 collections. American Journal of Physical Anthropology 57, 701–719.

Lockwood, C.A., Tobias, P.V., 2002. Morphology and affinities of new hominin cranial remains from Member 4 of the Sterkfontein Formation, Gauteng Province, South Africa. Journal of Human Evolution 42, 389–450.

Lovejoy, C.O., Latimer, B., Suwa, G., Asfaw, B., White, T.D., 2009. Combining prehension and propulsion: the foot of *Ardipithecus ramidus*. Science 326, 72–72e78.

McHenry, H.M., Jones, A.L., 2006. Hallucial convergence in early hominids. Journal of Human Evolution 50, 534–539.

McNutt, E.J., Claxton, A.G., Carlson, K.J., 2017. Revisiting the peroneal trochlea of the StW 352 calcaneus. South African Journal of Science 113, 1–5.

McNutt, E.J., Zipfel, B., DeSilva, J.M., 2018. The evolution of the human foot. Evolutionary Anthropology 27, 197–217.

Patel, B.A., Jashashvili, T., Bui, S.H., Carlson, K.J., Griffin, N.L., Wallace, I.J., Orr, C.M., Susman R.L., 2018a. Inter-ray variation in metatarsal strength properties in

humans and African apes: implications for inferring bipedal biomechanics in the Olduvai Hominid 8 foot. Journal of Human Evolution 121, 147–165.

Patel, B.A., Organ, J.M., Jashashvili, T., Bui, S.H., Dunsworth, H.M., 2018b. Ontogeny of hallucal metatarsal rigidity and shape in the rhesus monkey (*Macaca mulatta*) and chimpanzee (*Pan troglodytes*). Journal of Anatomy 232, 39–53.

Prang, T.C., 2015a. Calcaneal robusticity in Plio-Pleistocene hominins: implications for locomotor diversity and phylogeny. Journal of Human Evolution 80, 135–146.

Prang, T.C., 2015b. Rearfoot posture of *Australopithecus sediba* and the evolution of the hominin longitudinal arch. Scientific Reports 5:17677.

Prang, T.C., 2016a. Conarticular congruence of the hominoid subtalar joint complex with implications for joint function in Plio-Pleistocene hominins. American Journal of Physical Anthropology 160, 446–457.

Prang, T.C., 2016b. The subtalar joint complex of *Australopithecus sediba*. Journal of Human Evolution 90, 105–119.

Stern, J.T., Susman, R.L., 1983. The locomotor anatomy of *Australopithecus afarensis*. American Journal of Physical Anthropology 60, 279–317.

Su, A., Carlson, K.J., 2017. Comparative analysis of trabecular bone structure and orientation in South African hominin tali. Journal of Human Evolution 106, 1–18.

Susman, R.L., Stern, J.T., Jungers, W.L., 1984. Arboreality and bipedality in the Hadar hominids. Folia Primatologica 43, 113–156.

Tocheri, M.W., 2009. Laser scanning: 3D analysis of biological surfaces. In: Sensen, C.W., Hallgrímsson, B. (Eds.), Advanced imaging in biology and medicine: technology, software environments, applications. Springer: Berlin, pp. 85–101.

Tumarkin-Deratzian, A.R., 2009. Evaluation of long bone surface textures as ontogenetic indicators in centrosaurine ceratopsids. Anatomical Record 292, 1485–1500.

Venkataraman, V.V., Kraft, T.S., Dominy, N.J., 2013. Tree climbing and human evolution. Proceedings of the National Academy of Sciences 110, 1237–1242.

Ward, C.V., Kimbel, W.H., Harmon, E.H., Johanson, D.C., 2012. New postcranial fossils of *Australopithecus afarensis* from Hadar, Ethiopia (1990–2007). Journal of Human Evolution 63, 1–51.

Zeininger, A., Patel, B.A., Zipfel, B., Carlson, K.J., 2016. Trabecular architecture in the StW 352 fossil hominin calcaneus. Journal of Human Evolution 97, 145–158.

Zipfel, B., DeSilva, J.M., Kidd, R.S., Carlson, K.J., Churchill, S.E., Berger, L.R., 2011. The foot and ankle of *Australopithecus sediba*. Science 333, 1417–1420.

15

Metatarsals and pedal phalanges

Bernhard Zipfel and Roshna Wunderlich

The metatarsals and tarsals from most of the foot bones found at Sterkfontein by the time of this monograph have been described in an elegant and thorough publication by Deloison (2003). Our descriptions of the metatarsals are, in large part, in accordance with hers, and we point out only where we deviate from her interpretations. Due to the fragmentary nature of most of the Sterkfontein metatarsals, description and interpretation are limited. Appendix I presents member, locus with grid reference, and date of discovery where known of each specimen.

Descriptions

StW 562—Right first metatarsal

Preservation. StW 562 is a complete right first metatarsal found in an area described as the top of Member 4. It is well preserved except for a small amount of damage on the dorsomedial surface of the head and the medial tubercle (Figure 15.1, Table 15.1).

Morphology. The proximal joint surface is kidney shaped and decidedly concave. The proximal joint is divided into two articular areas by a small ridge that is only present laterally. The concavity is deeper in the dorsal articular area than the ventral area. There is a pit within a raised area at the location of the *m. fibularis longus* attachment. On the lateral side of the base, there is a smooth area for contact with the second metatarsal.

The shaft is triangular in cross-section; dorsolateral, plantar-medial, and plantar-lateral surfaces are separated from one another by smooth ridges. In sagittal view, the shaft expands proximally starting at about three-quarters of the shaft's length from the proximal end. There is a small hypertrophy on the tubercle for origin of the lateral metatarsophalangeal ligament.

The head is broad, and the articular surface extends far dorsally. There is a distinct groove proximal to the head where it meets the shaft. The distal breadth of the articular surface is narrower dorsally (16.5 mm) than plantarly (18 mm). The dorsal portion of the articular surface exhibits a distinct ridge that is sharply slanted, extending further proximally on the lateral side than the medial side. In plantar view, the distal articulation has large distinct tubercles. Two grooves are present for sesamoids and appear worn. The lateral groove runs the entire length of the joint in the sagittal plane. The distal articular surface is oriented medially in the coronal plane relative to the proximal articular surface indicating moderate shaft torsion that Drapeau and Harmon (2012) measured as 4.7°.

StW 573d—Left proximal first metatarsal

The StW 573 foot assemblage was recovered ex situ from Dump 20 left by lime workers within Silberberg Grotto and assigned to Member 2 (Clarke and Tobias, 1995; Tobias, 1998). It was first described together with the associated foot remains with which it was found (Clarke and Tobias, 1995), and subsequently associated with the remainder of the skeleton discovered in situ (Clarke, 1998; Partridge, 2000; see also Jashashvili et al., Chapter 14, this volume, for a description of the tarsals). An associated second metatarsal (StW 573f) is also described together with the Sterkfontein second metatarsals later in this chapter.

Preservation. The left first metatarsal of the StW 573d "Little Foot" assemblage is a partial specimen of which approximately the proximal one-third is preserved (Figure 15.1, Table 15.1). It is broken and repaired across

Bernhard Zipfel and Roshna Wunderlich, *Metatarsals and pedal phalanges*. In: *Hominin Postcranial Remains from Sterkfontein, South Africa, 1936–1995.* Edited by: Bernhard Zipfel, Brian G. Richmond, and Carol V. Ward, Oxford University Press (2020). © Oxford University Press.
DOI: 10.1093/oso/9780197507667.003.0014

StW 562

StW 573d

StW 595

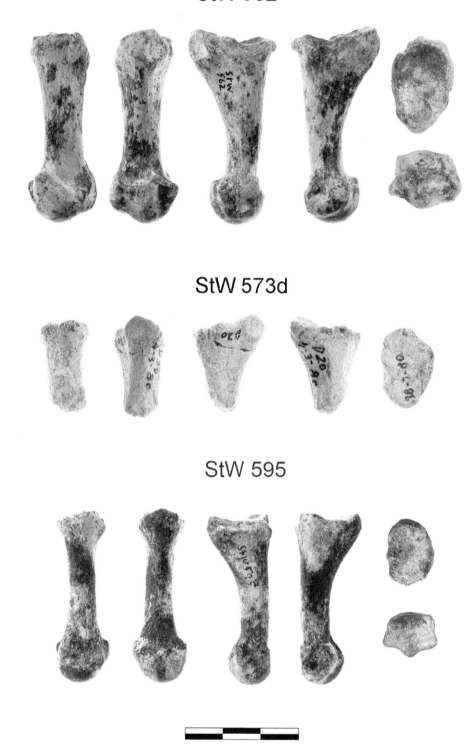

Figure 15.1 The StW 562 right first metatarsal shown in (from left to right) dorsal, plantar, lateral, medial, and proximal (top) and distal (bottom) views. The StW 573d left proximal first metatarsal shown in (from left to right) dorsal plantar, lateral, and proximal views. The StW 595 right first metatarsal shown in (from left to right) dorsal, plantar, lateral, proximal (top), and distal (bottom) views.

Table 15.1 External measurements of fossil metatarsals. All measurement in mm

	MT1			MT2		MT3						MT4	MT5
	StW 595	StW 562	StW 573	StW 89	StW 377	StW 238	StW 387	StW 388	StW 435	StW 496	StW 477	StW 485	StW 114/115
Maximum length	45.7	52.4	–	62.0	–	–	–	–	–	–	–	–	61.1
Midshaft dorsoplantar	8.5	11.4	9.4	8.2	7.1	8.5*	7.9*	8.9*	8.7	7.5*	7.4*	–	7.5
Midshaft mediolateral	7.6	10.4	10.2	5.8	6.1	6.3*	6.1*	5.9*	6.2	5.7*	5.8*	–	8.6
Proximal joint dorsoplantar maximum	18.1	24.2	20.4	13.0	15.2	16.3	16.7	18.5	20.1	16.8	16.5	14.3	11.5
Proximal joint mediolateral maximum	13.0	16.0	12.49	11.7	13.1	12.1	10.9	12.6	13.3	10.9	12.0	9.5	17.0
Distal joint dorsoplantar maximum	12.2	15.2	–	11.4	–	–	–	–	–	–	–	–	11.8
Distal joint mediolateral maximum	15.0	19.0	–	8.3	–	–	–	–	–	–	–	–	9.4

*Estimated dimension

the proximal joint. A small portion of the superolateral portion of the joint is broken.

Morphology. The proximal joint is somewhat kidney shaped with two separate articular areas partially divided by a distinct line laterally. The dorsal portion of this articular facet is deeply concave. The plantar portion of the articular facet is concave but not to the same extent. The tubercle for the *m. fibularis longus* tendon is fairly large and exhibits a large pit. On the lateral side of the base, there is a smooth concave area for contact with the second metatarsal. The proximal shaft is triangular in cross-section with a broad dorsal/dorsomedial side.

StW 595—Right first metatarsal

Preservation. StW 595 is an almost complete right first metatarsal from Member 4 (Deloison, 2003; Zipfel et al., 2010) (Figure 15.1, Table 15.1). The base, including the proximal articular surface, is severely damaged. There is also some damage to the dorsal portion of the distal joint.

Morphology. It is difficult to assess either shape or concavity of the proximal articular surface because it is particularly damaged in its lateral and dorsal aspects. Overall, it seems to be moderately concave but is unlikely to have exhibited the level of concavity present in StW 562.

The shaft has slight longitudinal curvature, is ovoid (proximally) to round (middle) in cross-section, and is flattened dorsoplantarly in its distal portion. The plantar-medial and plantar-lateral aspects are somewhat symmetrical. In sagittal view, the shaft is expanded proximally, but this expansion only starts in the proximal one-third of the shaft.

The head is dorsoplantarly compressed. The articular portion of the head continues onto the dorsal surface and is mediolaterally narrower dorsally than plantarly. It is very difficult to determine the dorsal extent of the joint, but it does not appear to extend as far dorsally as in StW 562 (Zipfel et al., 2010). Despite damage to some of the dorsal aspect, the lateral side preserves the dorsal groove. The joint edge forms an angle dorsally such that the lateral dorsal tubercle extends further proximally than the medial tubercle. Two distinct sesamoid grooves are present on the plantar surface of the head, and neither extends beyond the distal surface of the head (unlike StW 562). Head torsion is minimal (-5.9°; Drapeau and Harmon, 2012), oriented medially about the long axis.

StW 89—Left second metatarsal

Preservation. StW 89 is a complete left second metatarsal catalogued as Member 5 but more recently listed as Member 4 (Clarke, 1985; Deloison, 2003) and has been described and interpreted by DeSilva et al. (2012) (Figure 15.2, Table 15.1).

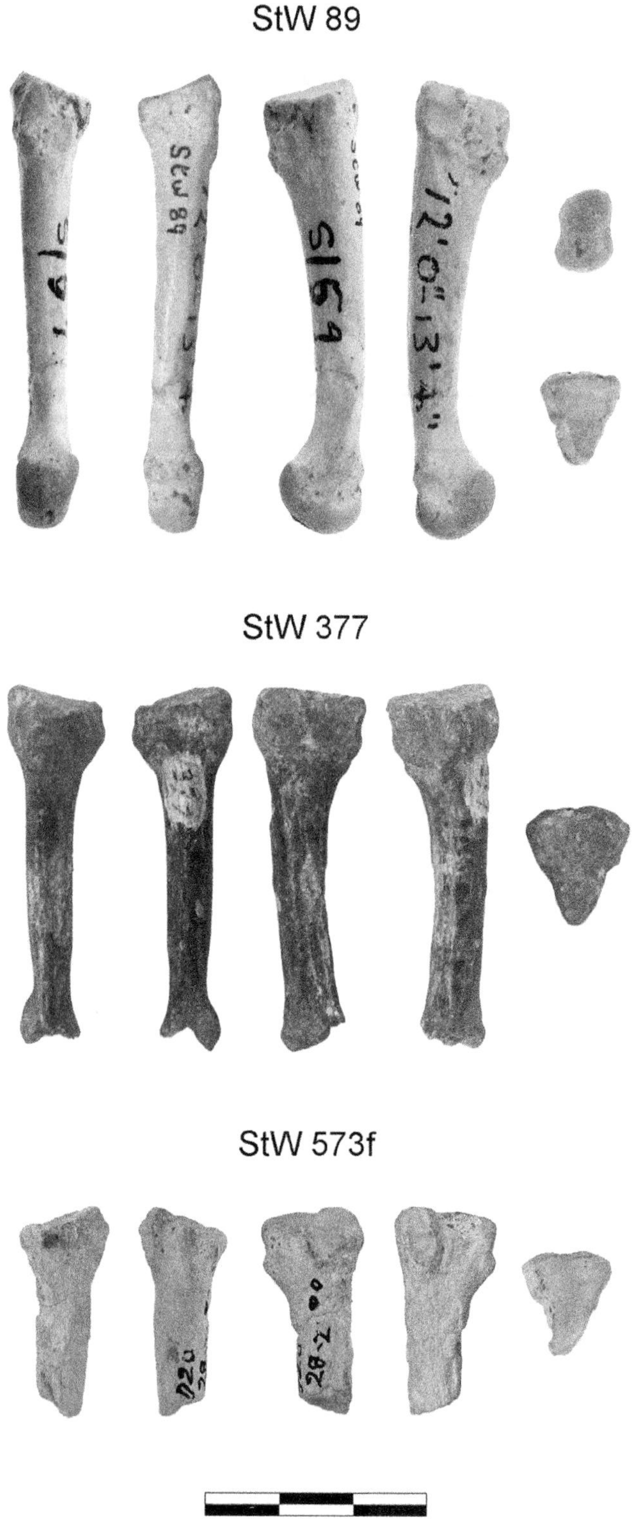

Figure 15.2 The StW 89 left second metatarsal shown in (from left to right) dorsal, plantar, medial, lateral and distal (top), and proximal (bottom) views. The StW 377 left proximal second metatarsal shown in (from left to right) dorsal, plantar, lateral, medial, and proximal views. The StW 573d left proximal second metatarsal shown in (from left to right) plantar, dorsal, medial, lateral, and proximal views.

A transverse break separated the distal one-third, and the two pieces are glued together. In contrast to the other specimens except StW 573d, its preservation is very light in color.

Morphology. The proximal facet faces proximomedially at a distinct angle (medially) relative to the shaft. The proximal articular surface is almost flat but slightly concave mediolaterally in the dorsal half. There is no excavation as is found in *Pan.* On the medial surface of the base the facet for the first metatarsal is semicircular in shape and measures 4.0 mm x 4.2 mm. This facet follows Singh (1964) variation 1. Two facets are present on the lateral side of the base for the second cuneiform and the third metatarsal. These facets are distinct and separated by a ridge. The proximal (cuneiform) facet is smaller (3.3 mm x 5.2 mm) than the distal (metatarsal) facet (4.6 mm x 5.4 mm), and there is an angle between them. They follow most closely Singh (1964) variation 3.

The shaft is fairly straight proximally with some longitudinal curvature (dorsal convex) distally. Some flattening of the dorsal surface is present distally, especially on the medial portion. Dorsal ridges separate the distal shaft from the head.

The head is narrow and exhibits marked torsion (-14.7°; Drapeau and Harmon, 2012) medially relative to the base. The head is narrow mediolaterally but wider plantarly than dorsally. Plantarly, the articular surface presents as two projections, the medial projecting slightly further proximally than the lateral.

StW 377—Left proximal second metatarsal

Preservation. StW 377 is a left second metatarsal recovered from Member 4 (Figure 15.2, Table 15.1). The head and a splinter of the plantar midshaft are missing, but the remainder of the specimen is in good condition. The distal break is asymmetrical, and the ridge extending from the shaft onto part of the dorsal ridge is present. The proximal facets are somewhat worn. The total preserved length is 46.2 mm.

Morphology. The shaft seems straight throughout, unlike the condition in StW 89 (DeSilva et al., 2012). Head torsion, measured just proximal to the metatarsal head, is -11.0° (Drapeau and Harmon, 2012). The base is triangular in shape and angled distomedially. It is mostly flat yet slightly mediolaterally concave in the dorsal aspect of the articular surface. The facet on the medial surface of the base is large (5.2 mm x 5.4 mm) and most closely follows Singh (1964) variation 2, with a beveled facet that would accommodate articulation with the first metatarsal and medial cuneiform. The lateral facet (7.0 mm x 6.3 mm) is fairly worn and does not resemble any of the Singh (1964) variations but might be considered to be closest to facet variation 3.

StW 573f—Left proximal second metatarsal

Preservation. This is a left proximal second metatarsal measuring 30.8 mm (Figure 15.2). The shaft is broken obliquely (from dorsodistolateral to plantarproximomedial) just proximal to what was likely the midshaft.

Morphology. The proximal facet faces proximomedially at a distinct angle (medially) relative to the shaft. The proximal articular surface is roughly triangular in shape, slightly concave in the dorsal half and convex in the plantar half. On the medial surface of the base the facet for the first metatarsal is semicircular in shape and measures 6.5 mm x 5.0 mm. Just below and anterior to this is a smooth area for contact with the first metatarsal following Singh (1964) variation 1 associated with the StW 573d first metatarsal and "Little Foot" assemblage (Clarke and Tobias, 1995; Clarke, 1998). One facet is present on the lateral side of the base, dorsally, and is beveled to articulate with the second cuneiform posteriorly and third metatarsal anteriorly. The proximal (cuneiform) facet is similar in size (5.6 mm x 5.4 mm) to the distal (metatarsal) facet (5.2 mm x 6.3 mm), and there is a distinct angle between them. They follow most closely Singh (1964) variation 3.

The available proximal shaft appears to be fairly straight. The proximal dorsal surface is somewhat flat and becomes round at the shaft tapering slightly. In sagittal view, the base tapers sharply into the shaft.

StW 238—Right proximal third metatarsal

Preservation. StW 238 is a right third metatarsal from Member 4 or 5 (Figure 15.3, Table 15.1). The specimen is very damaged, especially on the medial and plantar aspects of the base; this affects the joint and the plantar tuberosity but not the small portion of proximal shaft that is present. The total preserved length is 31.2 mm.

Morphology. The facet on the lateral side of the proximal joint is large and oval (8.1 mm x 7.8 mm) and most closely resembles the Singh (1964) variation 2. The medial side of the base is too damaged to identify the articular facets.

StW 387—Left proximal third metatarsal

Preservation. StW 387 is a left third metatarsal recovered from Member 4 (Figure 15.3, Table 15.1). Only the base and less than one-half of the proximal end of the shaft are present, and there is damage to the inferolateral base. The total preserved length is 33.3 mm.

Morphology. The base faces proximally and slightly medially and is flat. The facet for the second metatarsal is small and semicircular (4.4 mm x 5.9 mm), but very worn, making measurement difficult, and it most closely resembles a small version of Singh (1964) variation 2. The facet for metatarsal 4 is large and oval shaped (8.4 mm x 7.7 mm).

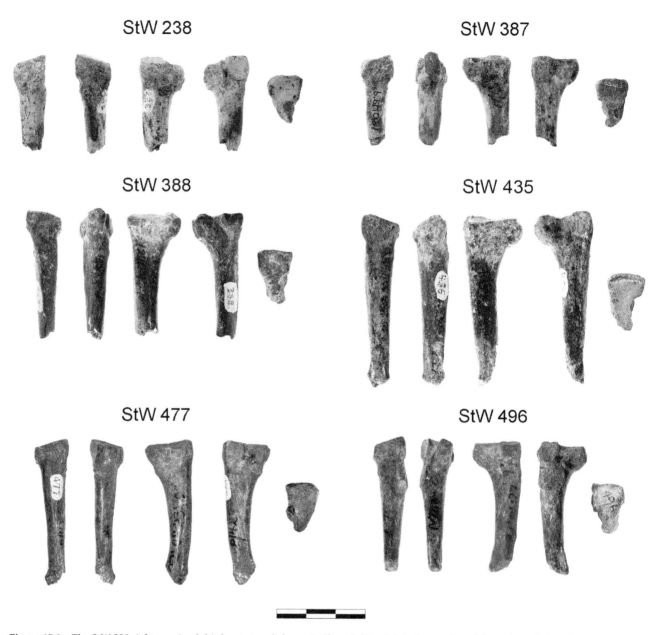

Figure 15.3 The StW 238 right proximal third metatarsal shown in (from left to right) plantar, dorsal, lateral, medial, and proximal views. The StW 387 left proximal third metatarsal shown (from left to right), dorsal, plantar, medial, lateral, and proximal views. The StW 388 right proximal third metatarsal shown (from left to right), dorsal, plantar, lateral, medial, and proximal views. The StW 435 right proximal third metatarsal shown (from left to right), lateral, medial, and proximal views. The StW 477 left proximal third metatarsal shown (from left to right), dorsal, plantar, medial, lateral, and proximal views. The StW 496 left proximal third metatarsal shown (from left to right), dorsal, plantar, lateral, and proximal views.

The lateral facets most resemble Singh (1964) variation 1. A very large prominent plantar tubercle is present. The shaft is only preserved in a small portion. It is straight proximally and flattened dorsally at the proximal part.

StW 388—Right proximal third metatarsal

Preservation. StW 388 is a right third metatarsal recovered from Member 4 (Figure 15.3, Table 15.1). The head and

distal shaft are missing, and the inferior margin of the base is abraded, exposing the underlying trabecular bone. Otherwise, it is well preserved. The total preserved length is 44 mm.

Morphology. Because the bone is broken neatly just distal to the midshaft, linear measurements of cortical thickness can be taken and are as follows: dorsal 2.6 mm, plantar 3.2 mm, medial 2.1 mm, and lateral 1.6 mm. The proximal articular surface is largely flat but slightly dorsoplantarly concave on the distal end, and it faces proximally and slightly medially. The facet for the second metatarsal, as in StW 387, is small and semicircular (4.4 mm x 6.6 mm), most closely resembles Singh (1964) variation 2, and becomes continuous with the proximal articular surface. The facet for metatarsal 4 is oval shaped (7.9 mm x 6.9 mm) with a slight bevel distally. The lateral facet most resembles Singh (1964) variation 1. Neutral axis torsion was measured at 65% the length of the specimen to be 14.7° medially (Harmon and Drapeau, 2012). There is a prominent plantar tubercle.

StW 435—Right proximal third metatarsal

Preservation. StW 435 is a right third metatarsal previously catalogued as a second metatarsal. It is a large specimen recovered from Member 4 (Figure 15.3, Table 15.1). The proximal joint and shaft are preserved entirely up to the dorsal ridges proximal to the head. The head distal to the dorsal ridge and some of the plantar distal shaft are missing. The total preserved length is 57.8 mm.

Morphology. The shaft is straight dorsoplantarly although somewhat twisted mediolaterally, resulting in medial torsion (inversion) of the distal shaft in relationship to the base. The proximal articulation is flat, slightly concave dorsally, and faces medially relative to the axis of the shaft. The medial proximal facet for the second metatarsal is semicircular and small; it measures 3.3 mm x 5.2 mm and most closely resembles Singh (1964) variation 2 but is much smaller. The lateral facet for metatarsal 4 is large and oval-shaped. It measures 8.3 mm x 7.1 mm and mostly closely resembles Singh (1964) variation 2. There is a prominent plantar tubercle.

StW 477—Left proximal third metatarsal

Presrvation. StW 477 is a left third metatarsal recovered from Member 4. The base, shaft, and some of the dorsal ridge posterior to the head are preserved (Figure 15.3, Table 15.1). The head is missing entirely. The surface preservation is excellent and total preserved length is 48.7 mm.

Morphology. The shaft is slightly curved dorsoplantarly. The base faces proximally and slightly medially and is flat. Shaft torsion at 80% the length of the shaft measured 28.8° inverted (Drapeau and Harmon, 2012). The medial facet for the second metatarsal is semicircular and measures 5.0 mm x 5.7 mm. It most closely resembles a shortened version of Singh (1964) variation 2. The lateral facet is large and broken distally. It most closely resembles Singh (1964) variation 2. A plantar tubercle is present, but it is not as large as in StW 387. On the lateral side of the base is a small prominence.

StW 496—Left proximal third metatarsal

Preservation. StW 496 is a left third metatarsal recovered from Member 4 (Deloison, 2003). The proximal joint and most of the shaft are preserved but broken proximal to the metaphysis for the head (Figure 15.3, Table 15.1). The total preserved length is 46.4 mm.

Morphology. The shaft is only slightly curved dorsoplantarly. Neutral axis torsion measured 18.7° at 70% the length of the bone (Drapeau and Harmon, 2012). Proximally the articular facet is very slightly concave dorsally but otherwise flat. The proximal articulation is directed proximomedially. The medial facet for metatarsal 2 is raised, is semicircular, and measures 4.6 mm x 5.5 mm. It most closely resembles Singh (1964) variation 2. The lateral facet is large and oval shaped and measures 7.7 mm x 6.3 mm and most closely resembles Singh (1964) pattern 2. Laterally there is a hypertrophied, roughened area 10.5 mm long indicating contact between the third and fourth metatarsals in the most proximal part of the shaft.

StW 485—Left proximal fourth metatarsal

Preservation. StW 485 is a left fourth metatarsal recovered from Member 4. Only the base and most proximal part of the shaft are present (Figure 15.4, Table 15.1). The base is damaged medially and has a small abraded area dorsally. The total preserved length is 27.2 mm.

Morphology. The shaft appears to be fairly straight, at least in the proximal portion that is present. The break across the shaft provides the opportunity for cortical measurements, although as the total length of the specimen is only 27.3 mm, these measurements are likely very proximally positioned, at about 25% of the bone's length. The cortical measurements are as follows: dorsal 2.4 mm, plantar 1.7 mm, medial 2.0 mm, and lateral 2.1 mm.

StW 114/115—Right fifth metatarsal

Preservation. StW 114/115 is a complete right fifth metatarsal in excellent condition (Figure 15.5, Table 15.1). The fossil is composed of a right distal fifth metatarsal bone (StW 114) recovered from grid square W/44 depth 8'9"–9'10" and a right proximal fifth metatarsal bone (StW 115) recovered from grid square W/45 depth 7'10"– 8'10" originally assigned to Member 5 (Deloison, 2003) but subsequently thought to come from Member 4 (Pickering et al., 2004; see also Moggi-Cecchi et al., 2006; Zipfel et al., 2009).

Morphology. This specimen has been described previously and interpreted by Zipfel et al. (2009). The base is expanded, and the articular surface is curved. The proximal articular surface for the cuboid is separated laterally from the tuberosity by a groove. The cuboid facet is mildly convex mediolaterally and separated from the fourth

Figure 15.4 The StW 485 left proximal fourth metatarsal shown (bottom from left to right), plantar, dorsal, medial, lateral, and proximal views.

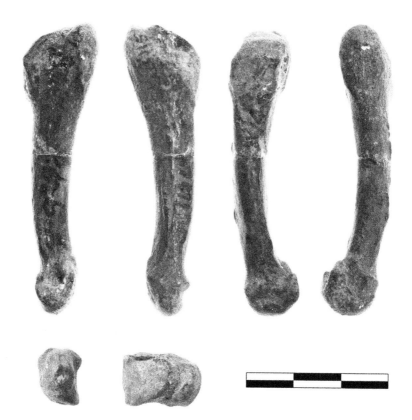

Figure 15.5 The StW 114/115 right fifth metatarsal shown (from top left to right), dorsal, plantar, medial, lateral, and (from bottom left to right) distal and proximal views.

metatarsal facet by a faint ridge running dorsoplantarly. Medially the proximal facet sits at an angle relative to the shaft such that the angle between the cuboid and fourth metatarsal facets is approximately 120°. The fourth metatarsal facet is semiovoid in shape.

The shaft is curved both mediolaterally and dorsoplantarly, such that it presents concavities laterally and plantarly. It is robust, especially proximally, expanding toward the tuberosity. Ridges are present dorsally on the shaft. Zipfel et al. (2009) suggest cautiously that these may represent attachments for *m. fibularis tertius* or perhaps even part of the *abductor digiti minimi*.

The head is narrow, especially dorsally. Plantarly, the articular surface of the head extends further proximally on the lateral side and further plantarly on the medial side. The head is considerably narrower dorsally than plantarly. Dorsally, a groove is present posterior to the articular surface, and a prominent ridge is present proximal to the groove. The groove is not continuous mediolaterally but rather is interrupted by joint surface in the middle.

StW 470—Right proximal hallucal phalanx

Preservation. This is a complete right proximal hallucal phalanx recovered from Sterkfontein Member 4. Aside from a chip of bone missing from the dorsal margin of the base, the preservation is excellent (Figure 15.6, Table 15.2). *Morphology.* Comparing the length (23.1 mm) to body breath (10.4 mm) suggests that it is quite robust. The proximal articular surface is concave mediolaterally and dorsoplantarly. The dorsal and plantar articular margins are concave. In the sagittal plane, the entire proximal articular surface is dorsally canted, angled dorsoproximally. The dorsal surface is slightly convex from the proximal superior articular margin to just posterior to the head, where it flattens to the distal superior articular margin. On the plantar surface, the body is flat, tapering distally. The medial and lateral condyles of the head are prominent and broader than the shaft. The condyles are slightly asymmetric.

StW 355—Right proximal non-hallucal phalanx

Preservation. This is a right complete proximal pedal non-hallucal phalanx recovered from Sterkfontein Member 4 (Figure 15.7, Table 15.2). Preservation is excellent.

Morphology. The bone is gracile and markedly curved with a distinct dorsal convexity from just anterior to the base to just posterior to the distal articular margin. The plantar aspect is concave but less curved. In the sagittal plane, the shaft tapers from a dorsoplantar depth of 6.7 mm proximally to 3.9 mm distally. Thin flexor sheath ridges occupy the sides of the middle one-third of plantar shaft. They do not project plantarly beyond the shaft. The proximal articular surface is concave and canted dorsally. The dorsal and plantar articular margins are concave. In plantar view, the distal articular surface is slightly asymmetric. In lateral view, the condyles are round and slightly flattened distoplantarly.

This phalanx has been listed by some to be manual (e.g., Susman et al., 2001; Kivell et al., Chapter 9, this volume).

StW 617—Right distal hallucal phalanx

Preservation. This specimen is a complete right hallucal distal phalanx of an adult individual (Figure 15.8, Table 15.2). Other than a small area of bone missing from the dorsal aspect and lateral side of the distal end, the surface preservation is excellent. This specimen was thought to be a pollical distal phalanx, but comparisons with the

Figure 15.6 The StW 470 right proximal hallucal phalanx shown (from left to right), dorsal, plantar, medial, distal (top), and proximal (bottom) views.

Table 15.2 External measurements of fossil phalanges. All measurements in mm.

	Hallucal proximal phalanx	Nonhallucal proximal phalanx	Hallucal distal phalanx
	StW 470	StW 355	StW 617
Maximum length	23.2	24.6	15.4
Midbody dorsoplantar	7.8	5.0	4.4
Midbody mediolateral	10.6	5.9	9.6
Proximal joint dorsoplantar maximum	11.6	7.8	6.4
Proximal joint mediolateral maximum	14.9	9.1	13.8
Distal joint/tuft dorsoplantar maximum	7.3	4.5	5.1
Distal joint/tuft mediolateral maximum	13.0	7.1	9.5

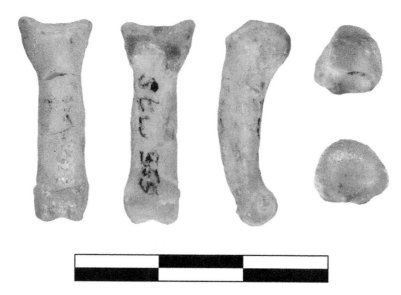

Figure 15.7 The StW 355 right proximal non-hallucal phalanx shown (from left to right), dorsal, plantar, medial and distal (top), and proximal (bottom) views.

Figure 15.8 The StW 617 right distal hallucal phalanx shown (from left to right), plantar, dorsal, lateral, and proximal views.

pollical distal phalanx StW 294 (Ricklan, 1990; Kivell et al., Chapter 9, this volume) and with pollical and non-pollical distal phalanges from Hadar (Bush et al., 1982; Ward et al., 2012) show that it is a distal hallucal phalanx. *Morphology.* This is a short (15.4 mm), broad (13.8 mm) distal hallucal phalanx. The ungual tuberosity is almost as wide (9.5 mm estimated) as the base (13.8 mm). The plantar surface of the tuft is dorsoplantarly deep (5.1 mm) and has no ungual fossa or spines. The ridge for the attachment of the *m. flexor hallucis longus* lies just proximal to the tuft at about two-thirds of the length from the proximal end. Proximal to that is a large volar fossa (7.5 mm proximodistally by 9 mm mediolaterally) occupying the plantar shaft for more than half of the proximal end. The shaft and tuberosity angle laterally. The basal tubercles are pronounced. The margins of the fossa are slightly asymmetric with a more prominent fibular margin.

The center of the dorsal margin is raised in the region of the attachment of the *m. extensor digitorum longus* tendon. The articular surface is dorsoplantarly short

(4.1 mm) compared to its breadth (12 mm), with shallow concavities for the trochlea of the proximal phalanx.

Comparative morphology

First metatarsal

Two complete first metatarsals from Member 4, StW 562 and StW 595 (Deloison 2003), vary morphologically from each other to an extent unlikely to fall within the range of intraspecific variability. StW 562 is more robust than StW 595, the former having a distal articular surface extending onto the dorsal surface of the metatarsal head, possibly allowing for human-like first metatarsophalangeal dorsiflexion, and the later having a more ape-like distal articular surface (Figure 15.9). Principal Components Analysis (PCA) and Canonical Variates Analysis (CVA) comparing the fossils with humans, chimpanzees, gorillas, and orangutans place StW 562 with an affinity closer

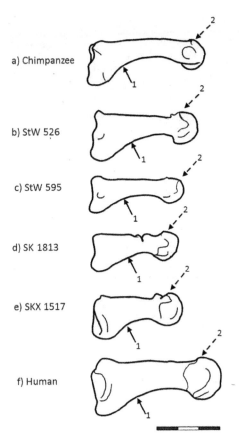

a) Chimpanzee

b) StW 526

c) StW 595

d) SK 1813

e) SKX 1517

f) Human

Figure 15.9 Profiles of hallucal metatarsals. The enhanced cross-section near the base ("1") is seen in the fossils (b, d, and e) and in humans (f). Chimpanzees have a reduced basal dorsoplantar diameter (a). StW 562, SK 1813, SKX 5017, and humans all have an extension of the distal phalanx onto the dorsum. This is not present in chimpanzees or StW 595 (c).

to humans and chimpanzees and StW 595 with an affinity closer to orangutans (Zipfel et al., 2010). Both these metatarsals have a head that is angled laterally in the transverse plane, similar to that of apes, suggesting some degree of abduction of the hallux, possibly for grasping, similar to that of chimpanzees (Figure 15.10). In contrast, the Swartkrans first metatarsals SK 1813 and SKX 5017 (Susman and Brain, 1988; Susman and deRuiter, 2004; Zipfel and Kidd, 2006) have a more human-like head (Figure 15.10). Proctor (2010b), found that the proximal articular surfaces of StW 562, StW 573 ("Little Foot") and A.L. 333-54 were intermediate in shape between humans and apes, while O.H. 8 is indistinguishable from modern *Homo*. However, the base of the first metatarsal of *Homo naledi* is oriented perpendicularly to the long axis of the shaft and is more narrow mediolaterally than StW 562, StW 595, and O.H. 8 (Harcourt-Smith et al., 2015). The Swartkrans metatarsals, in comparison, have a more ape-like proximal articular surface (Proctor, 2010b).

In StW 573d, the tubercle for the *m. fibularis longus* tendon is similar in size to that StW 562. The tubercle exhibits a large pit similar to that seen in StW 562 and SKX 5017 (Susman and Brain, 1988). Both StW 562 and 573 have a contact area on the lateral side of the base for articulation with the second metatarsal and is also present in SKX 5017. This is a feature not found in any of the apes (Le Minor and Winter, 2003) but is variable in humans (Singh, 1960; Le Minor and Winter 2003; Zipfel, 2004), and its absence may indicate more mobility in the tarsometatarsal joint (Fritz and Prieskorn, 1995).

From the available evidence, even though all three of these Sterkfontein individuals were almost certainly bipedal, it is unlikely that they had the same same pedal biomechanics neither of which would have been identical to modern humans. The morphology suggests different functions which may result in an earlier heel lift to compensate for the possible limited first metatarsophalangeal dorsiflexion in StW 595, and hyperpronation of the foot at toe-off with a more laterally deviated hallux.

Second metatarsal

Unfortunately, with the exception of StW 89, all the Sterkfontein second metatarsals are incomplete, missing their distal ends to varying degrees. In a detailed study by DeSilva et al. (2012), StW 89 was shown to display a mosaic of human-like and African ape-like features. Like human second metatarsals, the dorsal head is domed, and there is a prominent dorsal sulcus just proximal to the metatarsal head, indicative of phalangeal dorsiflexion during the toe-off phase of bipedal locomotion. This bone also possesses some ape-like features. Multivariate analyses show that the biggest differences between humans and apes are the relative proximal and distal dimensions. In these dimensions, StW 89 clusters with the modern apes (DeSilva et al., 2012). Additionally, the base of StW 89 is quite gorilla-like in overall shape (Proctor, 2010a). The base of StW 89 lacks the dorsoplantar depth typical for human second metatarsals and falls within the range of chimpanzees and gorillas, entirely outside the range of humans and fossils StW 377, *Homo habilis* (OH 8), and *Ardipithecus ramidus* (ARA-VP-6/1000) (DeSilva et al., 2012). The base morphology of StW 573 and StW 377 is more human-like than that of StW 89 (Proctor, 2010a; DeSilva et al., 2012). Lovejoy et al. (2009) suggest that a robust metatarsal base is an important adaptation to stiffen the midfoot and limit midfoot laxity during the propulsive phase of bipedalism. This suggests that StW 89 may have had a less stable midfoot when compared to the other Sterkfontein second metatarsals.

The StW 355 proximal phalanx is possibly associated with StW 89 (Kuman and Clarke, 2000; DeSilva et al.,

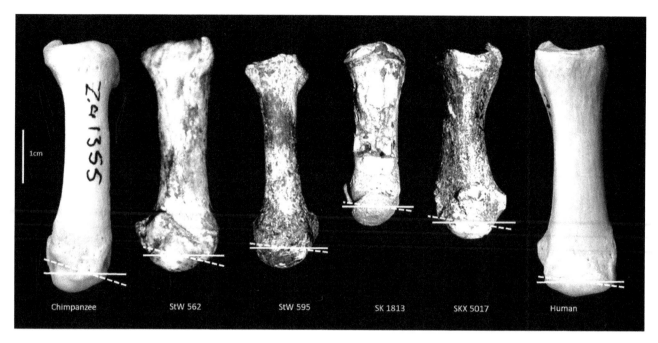

Figure 15.10 The lateral angulation of the first metatarsal head in chimpanzees, fossil hominins, and humans. Compared to the other metatarsals, StW 562 has a distinctly laterally deviated proximal articular set angle similar to that of apes.

2012) and when articulated using the technique of Latimer and Lovejoy (1990), the range of dorsiflexion is similar to that seen in modern humans and fossil *Australopithecus*. However, StW 89 possesses an ape-like range of plantarflexion, quite unlike that found in the associated foot bones of *A. afarensis*. If StW 355 and StW 89 are associated, more advanced pedal-grasping abilities may have been present in this hominin than for the East African australopiths (DeSilva et al., 2012). The internal torsion of StW 89 functionally places the second digit in opposition to the first, and is a critical adaptation for climbing in the apes (Morton, 1922) and may have had the same function in the fossil hominin.

Third metatarsal

As only the proximal third metatarsals are available, it is not possible to carry out any extensive assessment of the total bone morphology. Of their preserved portions, the six Sterkfontein third metatarsals look similar to each other, in particular the proximal articular surface of StW 387, 388, 435, 477, and 496, which are virtually indistinguishable from each other, being human-like in appearance (Figure 15.11). StW 477 and 496 appear to be more curved than humans in the sagittal plane of the shaft with StW 238 and 387 being more flat. On the lateral side of the base of both StW 477 and 494 is a small prominence, similar to one seen on O.H. 8, that might have contacted the fourth metatarsal which is a common variation in humans.

Deloison (2003) suggests that these morphologies are consistent with a dominantly bipedal foot, and from the available evidence, we concur.

Fourth and fifth metatarsals

Given the incompleteness of the StW 485 fourth metatarsal, there is little to distinguish it from that of modern humans and *Homo naledi*, having a dorsoplantarly flat articular surface, unlike that of *Australopithecus sediba* (U.W. 88-22) (Zipfel et al., 2011) and *A. afarensis* (A.L. 333-160) (Ward et al., 2011). This flat cuboid facet suggests that the lateral midfoot was human-like and rigid during heel lift (DeSilva et al., 2013).

Comparisons of the fifth metatarsal, StW 114/115 to those of modern humans, extant apes, and partial hominin fifth metatarsals AL 333-13, AL 333-78, SKX 33380, O.H. 8, KNM-ER 803f, U.W. 88-33, and *H. naledi* reveal a similar morphology in all eight fossils consistent with habitual bipedality (Zipfel et al., 2009; Zipfel et al., 2013; Harcourt-Smith et al., 2015). This includes a humanlike proximal articular morphology and internal torsion of the head suggesting a stable lateral column and the likely presence of lateral longitudinal and transverse tarsal arches. This may support the hypothesis of an increased calcaneocuboid stability having been an early evolutionary event in the history of terrestrial bipedalism. This is to an extent also supported by Dowdeswell et al. (2017) examining internal properties found that human-like functional loading

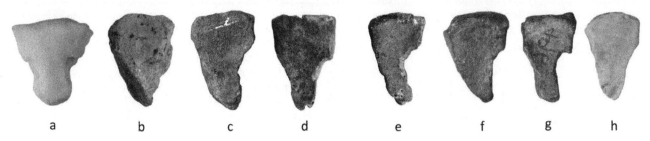

Figure 15.11 The proximal articular surface of the third metatarsal, a. chimpanzee, b. StW 238, c. StW 387, d. StW 388, e. StW 435, f. StW 477, g. 496 and human. StW 387 (c), StW 388 (d), StW 477 (f), and StW 496 (g) appear particularly human in morphology.

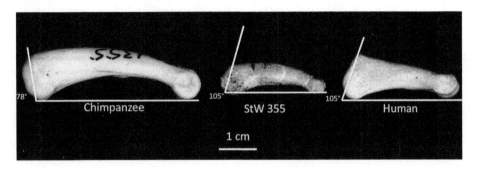

Figure 15.12 Proximal non-hallucal phalanges in profile view. A pin was held against the proximal facet to visualize the orientation of the facet relative to a horizontal. This angle has been redrawn for clarity (after McNutt et al., 2018; see also DeSilva et al., 2019), and because the proximal facet is obscured in lateral view by the rim of bone along the edge of the metatarsal facet. The chimpanzee proximal facet has a plantar cant and the StW 470 proximal phalanx has a human-like dorsal cant.

of the lateral column existed in at least some fossil hominins, although perhaps surprisingly not in hominins from Dmanisi, Georgia, or Dinaledi, South Africa.

Phalanges

Deloison (2003) had difficulty in interpreting the StW 470 proximal hallucal phalanx; nevertheless, she thought the morphology to be more suited to arboreal rather than bipedal locomotion. Our view is that it is human-like and more robust than that of the apes. It also has an indentation on the dorsal margin of the proximal articular surface as found in humans, associated with extreme dorsiflexion of the first metatarsophalangeal joint, reflecting habitual dorsiflexion that would occur in bipedality.

The StW 355 proximal phalanx may be associated with the StW 89 second metatarsal (DeSilva et al., 2012). It is curved in the sagittal plane, much like that of the apes, but its proximal articular surface is canted dorsally as found in humans associated with dorsiflexion (Figure 15.12) (see also McNutt et al., 2018; DeSilva et al. 2019). If StW 355 is associated with StW 89, then this explains the apelike curvature of the bone associated with grasping (e.g. Stern and Susman, 1983; Congdon, 2012).

The StW 617 hallucal distal phalanx has no discernable axial torsion as characterized in humans, positioning its plantar surface in contact with the ground. However, it is slightly deflected laterally toward the other toes—a derived feature important for bipedal function in which weight is shifted laterally during toe-off (Aiello and Dean, 1990).

Conclusions

The hominin forefoot, comprised of the metatarsus and phalanges, represents an important functional component of the foot that can potentially answer a number of questions regarding the evolution of the foot, more specifically the advent of bipedalism in the human lineage. In the forefront is the question as to the degree to which hominins had an opposable hallux and its function in arboreal and terrestrial locomotion. Unfortunately, the Sterkfontein forefoot pedal elements are mostly fragmentary, with hardly any association with each other. Nevertheless, some important information could be obtained from them. Broadly, it appears that the lateral column of the Sterkfontein hominin feet was human-like, and similar to that of virtually all other known hominins, suggesting a stable lateral column early in the evolution of the foot. In contrast, the medial column shows a mosaic of derived and ape-like characters consistent with bipedalism but also with grasping in arboreal behavior.

There is arguably also evidence of two morphs in foot morphology and function, suggesting a second species of *Australopithecus* within Member 4 of Sterkfontein (Zipfel et al., 2010; McNutt et al., 2018; DeSilva et al., 2019).

Acknowledgments

The authors would like to thank Stephanie Potze and Lazerus Kgasi of the Ditsong Museum of Natural History and the Fossil Access Advisory Committee, University of the Witwatersrand for access to specimens. Funding to conduct this work was provided by the Wenner Gren Foundation, Ford Foundation, and Palaeontological Scientific Trust. We thank the two reviewers for their helpful comments and suggestions.

References

Aiello, L., Dean, C., 1990. An introduction to human evolutionary Anatomy. Academic Press: London, pp. 507–538.

Clarke, R.J.,1985. *Australopithecus* and early *Homo* in southern Africa. In: Delson, E. (Ed.), Ancestors, the hard evidence. Alan R. Liss: NewYork, pp. 171–177.

Clarke R.J., 1998. First ever discovery of a well-preserved skull and associated skeleton of *Australopithecus*. South African Journal of Science 94, 460–463.

Clarke R.J., Tobias P.V., 1995. Sterkfontein Member 2 foot bones of the oldest South African hominid. Science 269, 521–524.

Congdon, K.A., 2012. Interspecific and ontogenetic variation in proximal pedal phalangeal curvature of great apes (*Gorilla gorilla, Pan troglodyts* and *Pongo pygmaeus*). International Journal of Primatology 33, 418–427.

DeLoison, Y., 2003. Anatomie des fossils de pieds des hominidés d'Afrique du Sud dates entre 2,4 et 3,5 millions d'ánnées. Interprétation quand à leur mode de locomotion. Biométrie Humaine et Anthropologie 21, 189–230.

DeSilva, J.M., Holt, K.G., Churchill, S., Carlson, K., Walker, C., Zipfel, B., Berger, L., 2013. The lower limb and the mechanics of walking in *Australopithecus sediba*. Science 340, 1232999-1-5.

DeSilva, J., McNutt, E., Benoit, J., Zipfel, B., 2019. One small step: a review of Plio-Pleistocene hominin foot evolution. Yearbook of Physical Anthropology S67, 63–140.

DeSilva, J.M., Proctor, D.J., Zipfel, B., 2012. A complete second metatarsal (StW 89) from Sterkfontein Member 4, South Africa. Journal of Human Evolution 63, 487–489.

Dowdeswell, M.R., Jashashvili, T., Patel, B.A., Lebrun, R., Susman, R.L., Lordkipanidze, D., Carlson, K.J., 2017. Adaptation to bipedal gait and fifth metatarsal structural properties in *Australopithecus, Paranthropus,* and *Homo.* Comptes Rendues Paleovol 16, 585–599.

Drapeau, M.S.M., Harmon, E.H., 2012. Metatarsal torsion in monkeys, apes, humans and Australopiths. Journal of Human Evolution 64, 93–108.

Fritz, G.R., Prieskorn, D., 1995. First metatarsocuneiform motion: a radiographic and statistical analysis. Foot and Ankle International 16, 117–123.

Harcourt-Smith W., Aiello L.C., 2004. Fossils, feet and the evolution of human bipedal locomotion. Journal of Anatomy 204, 403–416.

Harcourt-Smith, W.E.H., Throckmorton, Z., Condon, K., Zipfel, B., Dean, A., Drapeau, M., Churchill, S.E., Berger, L.R., DeSilva, J.M., 2015. The foot of *Homo naledi*. Nature Communications 6, 8432. https://doi.org/10.1038/ncomms9432. Retried 6 October 2015.

Kuman, K., Clarke, R.J., 2000. Stratigraphy, artifact industries and hominin associations for Sterkfontein, Member 5. Journal of Human Evolution 38, 827–847.

Latimer, B.M., Lovejoy, C.O., 1990. Metatarsophalangeal joints of *Australopithecus afarensis*. American Journal of Physical Anthropology 83, 13–23.

Le Minor, J.-M., Winter, M., 2003 The intermetatarsal articular facet of the first metatarsal bone in humans: a derived trait unique with primates. Annals of Anatomy, Anatomischer Anzeiger 185, 359–365.

Lovejoy, C.O., Latimer, B., Suwa, G., Asfaw, B., White, T.D., 2009. Combining prehension and propulsion: the foot of *Ardipithecus ramidus*. Science 326, 72e1e72e8.

McNutt, E.J., Zipfel, B., DeSilva, J.M., 2018. The evolution of the human foot. Evolutionary Anthropology 27(5), 197–217. https://doi.org/10.1002/evan.21713.

Moggi-Cecchi, J., Grine, F.E., Tobias, P.V., 2006. Early hominid dental remains from Members 4 and 5 of the Sterkfontein Formation (1966–1996 excavations): catalogue, individual associations, morphological descriptions and initial metrical analysis. Journal of Human Evolution 50, 239–328.

Morton, D.J., 1922. Evolution of the human foot. American Journal of Physical Anthropology 5, 305–336.

Partridge, T.C., 2000. Hominid-bearing cave and tufa deposits. In: Partridge, T.C., Maud, R.R. (Eds.), The Cenozoic of southern Africa. New York: Oxford University Press, pp. 100–125.

Partridge, T.C., Shaw, J., Helsop, D., Clarke, R.J., The new hominid skeleton from Sterkfontein, South Africa: age and preliminary assessment. Journal of Quaternary Science 14, 293–298.

Pickering, T.R., Clarke, R.J., Moggi-Cecchi, J., 2004. Role of carnivores in the accumulation of the Sterkfontein Member 4 hominid assemblage: a taphanomic reassessment of the complete hominid fossil sample (1936–1999). American Journal of Physical Anthropology 125, 1–15.

Proctor, D.J., 2010a. Three-dimensional morphometrics of the proximal metatarsal articular surfaces of *Gorilla, Pan, Hylobates,* and shod and unshod humans. Ph.D. Dissertation, University of Iowa.

Proctor, D. J., 2010b. Brief communication: Shape analysis of the MT 1 proximal articular surface in fossil hominins and shod and unshod *Homo*. American Journal of Physical Anthropology 143, 631–637.

Singh, I., 1960. Variations in the metatarsal bones. Journal of Anatomy 94, 345–350.

Stern, J.T., Susman, R.L., 1983. The locomotor anatomy of *Australopithecus afarensis*. American Journal of Physical Anthropology 60, 279–317.

Susman, R.L., Brain, T.M., 1988. New first metatarsal (SKX 5017) from Swartkrans and the gait of *Paranthropus robustus*. American Journal of Physical Anthropology 77, 7–15.

Susman, R.L., de Ruiter, D.J., 2004. New hominin first metatarsal (SK 1813) from Swartkrans. Journal of Human Evolution 47, 171–181.

Susman, R.L., de Ruiter, D.J., Brain, C.K., 2001. Recently identified postcranial remains of *Paranthropus* and early *Homo* from Swartkrans Cave, South Africa. Journal of Human Evolution 41, 607–629.

Tobias, P.V., 1998. History of the discovery of a fossilised Little Foot at Sterkfontein, South Africa, and the light that it sheds on the origins of hominin bipedalism. Mitteilungen der Berliner Geselschaft für Anthropologische, Ethnologische und Urgeschichte 19, 47–56.

Zipfel, B., 2004. Morphological variation in the metatarsal bones of selected recent and pre-pastoral humans from South Africa. Ph.D. Dissertation, University of the Witwatersrand.

Zipfel, B., De Silva, J.M., Kidd, R.S., 2009. Earliest complete hominin fifth metatarsal—implications for the evolution of the lateral column of the foot. American Journal of Physical Anthropology 140, 532–545.

Zipfel, B., DeSilva, J., Kidd, R.S., Carlson, C., Churchill, S., Berger, L.R., 2011. The foot and ankle of *Australopithecus sediba*. *Science* 333, 1417–1420.

Zipfel, B., Kidd, R., 2006. Hominin first metatarsals (SKX 5017 and SK 1813) from Swartkrans: a morphometric analysis. Homo: Journal of Comparative Human Biology 57, 117–131.

Zipfel, B., Kidd, R.S., Clarke, R.J., 2010. The "second australopithecine species hypothesis" in Sterkfontein Member 4: the post-cranial evidence. Proceedings of the 16th Conference of the Palaeontological Society of Southern Africa (Howick, 5-8 August). Interpak Books: Pietermaritzburg, pp. 124–125.

Section 3

Functional anatomy and biology

16

Long bone cross-sectional geometry

Christopher B. Ruff, Ryan W. Higgins, and Kristian J. Carlson

Structural analyses of long bone cortices have become an increasingly important means to evaluate mechanical properties and reconstruct locomotor behavior among fossil hominins (Trinkaus and Ruff, 1989; Ruff et al., 1993; Ruff, 1995; Trinkaus et al., 1996; Ruff, 1998; Ruff et al., 1999; Lovejoy et al., 2002; Pickford et al., 2002; Holt, 2003; Shackelford, 2007; Ruff, 2008; Kuperavage and Eckhardt, 2009; Ruff, 2009; Trinkaus, 2009; Berger et al., 2010; Kuperavage and Eckhardt, 2010; Puymerail et al., 2012; Trinkaus and Ruff, 2012; Ruff and Higgins, 2013; Di Vincenzo et al., 2015; Ruff et al., 2016; Rodriguez et al., 2018). Early hominin specimens from South Africa have been included in several previous investigations, including a study of the *A. robustus* SK 82 and 97 proximal femora (Ruff et al., 1999), a preliminary study of the *A. sediba* MH1 femur and humerus (Berger et al., 2010), a comparative analysis of femoral neck structure in East and South African specimens (Ruff and Higgins, 2013), and in comparisons with A.L. 288-1 (Ruff et al., 2016). We report here on all available long bone diaphyseal and femoral neck cross sections of specimens from Sterkfontein (Member 4). We also carry out comparisons of these section properties and other cross sectional dimensions with those of other early hominins as well as modern samples.

In terms of general locomotor reconstructions among primates (e.g., degree of arboreality versus terrestriality, bipedalism versus quadrupedalism), some of the most informative types of comparisons involve assessments of upper (fore) versus lower (hind) limb bone strengths (Schaffler et al., 1985; Ruff, 2002, 2003; Ruff et al., 2016, 2018a). Such analyses have shown, for example, marked differences in limb use, and by implication, locomotor behavior in *Australopithecus afarensis*, *Homo habilis* (*sensu stricto*), and early *H. erectus* (Ruff, 2009; Ruff et al., 2016). Unfortunately, paired upper and lower long bone elements are not available for any of the Sterkfontein specimens, although some comparisons between upper limb and pelvic articular dimensions for StW 431 were carried out by McHenry and Berger (1998a), and we include an analysis of upper limb bone strength relative to body mass calculated from acetabular size for this specimen. However, structural analyses of isolated long bones can also provide useful information regarding locomotor behavior and mechanical loadings (e.g., Ruff, 1995; Trinkaus et al., 1996; Ohman et al., 1997; Ruff et al., 1999; Pickford et al., 2002). Because of the number of femoral sections available from Sterkfontein, as well as the diagnostic value of this element in locomotor studies (see references above), the main focus here is on this bone, although this chapter and Appendix II also present data for other bones.

Methods

Computed tomography (CT) scans were taken of all available long bone diaphyses from Sterkfontein on a Phillips Tomoscan SR 7000 CT scanner located in Johannesburg Hospital in July and August 1998. Slice thickness was 1.5. Images were displayed at a window center of 1000 H with a window width of 1300 H and recorded as hard copies. (These settings are somewhat different from those used previously on the same scanner for the SK 82 and 97 femora (Ruff et al., 1999), but the effect on measurement parameters is minimal.) External breadth measurements taken with calipers at the same section locations were used to confirm dimensions as recorded on the hard copies. Only sections requiring little or no reconstruction were included in the analyses, with the exception of the mid-femoral neck sections from StW 99 and StW 311 for which reconstruction of missing portions were made. All

Christopher B. Ruff, Ryan W. Higgins, and Kristian J. Carlson, *Long bone cross-sectional geometry.* In: *Hominin Postcranial Remains from Sterkfontein, South Africa, 1936–1995.* Edited by: Bernhard Zipfel, Brian G. Richmond, and Carol V. Ward, Oxford University Press (2020). © Oxford University Press. DOI: 10.1093/oso/9780197507667.003.0016

other sections were intact except for minor chipping or cracks, which were digitally filled in using existing contours as a guide. Internal trabecular bone or matrix was also removed, again using the existing visible endosteal surface for guidance (see Ruff and Higgins, 2013, for an example).

Prior to scanning, all bones were set up relative to standardized anatomical axes as described in detail in Ruff (2002). Diaphyseal section locations were determined as percentages of bone length', with length' as defined in Ruff (2002) (note that for the femur this does not include the femoral head): at 20%, 35%, 50%, 65%, and 80% of length', measured from the distal end and taken perpendicular to the long axis of the diaphysis. In the humerus, an additional section at 40% of length' was taken to facilitate comparisons with non-human primates (see Ruff, 2002, 2008, 2009). Bone lengths and section locations for incomplete specimens were estimated using available anatomical landmarks. Femoral neck sections were taken at the base of the neck (immediately adjacent to the trochanters) and at the midpoint of the neck (midway between the trochanters and femoral head), in both cases perpendicular to the neck longitudinal axis, correcting for any anteversion (Ruff and Higgins, 2013). Sections from the left side were reversed to maintain constant anatomical orientation.

Hard copies of section images were scanned using a flatbed scanner, and images imported into ImageJ (rsb.info.nih.gov/ij/). A custom macro (MomentMacro; www.hopkinsmedicine.org/fae/mmacro.html) was used within ImageJ to calculate mechanical section properties: total subperiosteal (TA) and cortical (CA) areas, percent cortical area (CA/TA), second moments of area (I_x, I_y, and J), and section moduli (Z_x, Z_y, and Z_p). I_x and I_y measure bending rigidities in the A-P and M-L planes, respectively, and Z_x and Z_y measure bending strengths in the same planes. J and Z_p measure torsional and (twice) average bending rigidity and strength, respectively. (This assumes equivalent material properties, which is reasonable—see Ruff, 2018, for further discussion.) Section moduli can be calculated in a number of ways. MomentMacro includes calculation of true section moduli, derived as second moments of area (SMAs) divided by the maximum perpendicular distance from the neutral axis to the outermost fiber of the section (or, for Z_p, from the section centroid). However, many of the available comparative section moduli data were derived in a different way, by either dividing SMAs by half the external breadth, or by taking SMAs to the 0.73 power (see Ruff et al., 1999; Ruff, 2008, 2009). To use these data, and for consistency with earlier studies, alternative methods of calculating section moduli from SMAs are used for some comparisons and are identified in the specific analyses.

Past studies of femoral neck cortical bone distribution in early hominins have concentrated on the relative thickness of the inferior versus superior cortex (Ohman et al., 1997; Lovejoy et al., 2002; Pickford et al., 2002; Kuperavage and Eckhardt, 2010), because of the possible significance of this ratio in deciphering load transmission across the hip (Lovejoy, 1988; Stern and Susman, 1991; also see Rafferty, 1998). Therefore, we also provide here measurements of superior and inferior cortical thicknesses of the available femoral necks from Sterkfontein, measured directly on CT scans using ImageJ (for methodological details, see Ruff and Higgins, 2013).

Comparative samples were derived from previous studies of extant hominoid and fossil hominin taxa (Ruff, 1995; Ohman et al., 1997; Rafferty, 1998; Ruff et al., 1999; Puymerail et al., 2012; Ruff et al., 2016). Modern humans (n = 100) are represented by two samples of very different body proportions: East Africans (generally tall and linear) and Pecos Pueblo Native Americans (shorter and relatively broader). For plotting in graphs, taxonomic assignments of fossil hominins follow Ruff et al. (2016) except that here KNM-ER 1472 and 1481a are grouped together with *Homo erectus/ergaster*. KNM-ER 5881 is tentatively grouped with OH 62 as *Homo habilis* (sensu stricto), based on their morphological similarity (Ward et al., 2015) and following arguments in Ruff et al. (2018b). Several possibly *Paranthropus boisei* specimens (KNM-ER 738, 815, 1465a, and 1503; OH 20) are grouped with KNM-ER 1500d and OH 80-12 within this taxon.

Results

A total of 29 long bones (femur, tibia, fibula, humerus, radius, ulna) from Sterkfontein preserved at least one usable diaphyseal and/or femoral neck section. A complete list of these bones and sections is given in Appendix II. Analyses here concentrate on the two more complete femora (StW 99 and 121, the only femora that include the subtrochanteric and/or midshaft regions), the most complete humerus (StW 431, the only humerus with a mid-region section), midshaft sections from three radii and three ulnae, and femoral neck sections from six specimens. Section outlines are derived from CT scans for all of these are shown Figure 16.1. Diaphyseal sections are shown with medial to the left and anterior above; femoral neck sections with anterior to the left and superior above. Table 16.1 lists section properties, and Table 16.2 presents femoral neck superior and inferior cortical breadths. Section outlines and associated geometric properties for all other sections are included in Appendix II, Table 1 and Appendix II Figures 1 and 2. The tables give true section moduli.

Table 16.1 Cross-sectional properties of selected sections illustrated in Figure 16.1.

Specimen	Bone	Section	TA	CA	%CA	Ix	Iy	Imax	Imin	J	Zx	Zy	Zp
StW 99	Femur	50%	484	356	73.6	15121	20339	20724	14736	35460	1168	1439	2518
		65%	525	450	85.6	19049	25550	26108	18492	44600	1375	1720	2922
		80%	634	552	86.9	27805	39253	44671	22386	67057	2023	2372	3743
		base-nk	458	249	54.4	22785	8003	24216	6573	30789	1358	801	1733
		mid-nk	404	168	41.5	10838	6797	11125	6510	17634	813	666	1329
StW 121	Femur	20%	459	226	49.3	10220	15076	15350	9947	25297	929	1094	1819
		35%	384	266	69.2	10278	10595	10597	10276	20873	917	944	1676
		50%	363	292	80.5	10982	9346	11052	9276	20328	892	852	1653
		65%	376	314	83.4	11336	10534	11364	10507	21870	960	950	1842
StW 311	Femur	mid-nk	412	155	37.6	10939	6121	11149	5911	17060	757	578	1457
StW 403	Femur	mid-nk	325	151	46.5	8608	4159	8738	4029	12767	702	448	1207
StW 479	Femur	base-nk	337	130	38.5	7281	3836	7362	3754	11116	535	429	797
		mid-nk	326	101	31.0	5408	3447	5418	3436	8855	410	376	673
StW 501	Femur	mid-nk	297	95	32.2	4364	3497	4973	2889	7861	408	370	677
StW 522	Femur	base-nk	274	101	36.9	5072	2079	5072	2079	7151	358	246	506
		mid-nk	269	78	28.9	3638	2149	3662	2124	5787	326	231	516
StW 431	Humerus	20%	309	207	66.9	5336	8923	9001	5258	14260	552	647	997
		35%	309	213	69.1	7026	6772	7168	6630	13798	655	680	1286
		40%	318	217	68.1	7546	6887	7857	6577	14433	708	695	1327
StW 348	Radius	50%	126	111	88.8	963	1668	1683	947	2630	161	199	314
StW 354	Radius	50%	103	95	91.6	759	999	1017	742	1759	131	130	229
StW 431	Radius	50%	132	103	77.9	1282	1343	1346	1278	2625	190	198	380
StW 326	Ulna	50%	84	71	84.5	529	583	593	519	1113	100	101	195
StW 349	Ulna	50%	109	100	91.4	1225	767	1262	731	1992	169	138	274
StW 431	Ulna	50%	119	104	87.8	1343	975	1424	893	2317	184	153	312

TA: total subperiosteal area (mm²); CA: cortical area (mm²); %CA: percent cortical area (CA/TA · 100); I_x, I_y: second moments of area about M-L and A-P axes (mm⁴); I_{max}, I_{min}: maximum and minimum second moments of area (mm⁴); J: Polar second moment of area (mm⁴); Z_x, Z_y: section moduli about M-L and A-P axes (mm³); Z_p: polar section modulus (mm³).

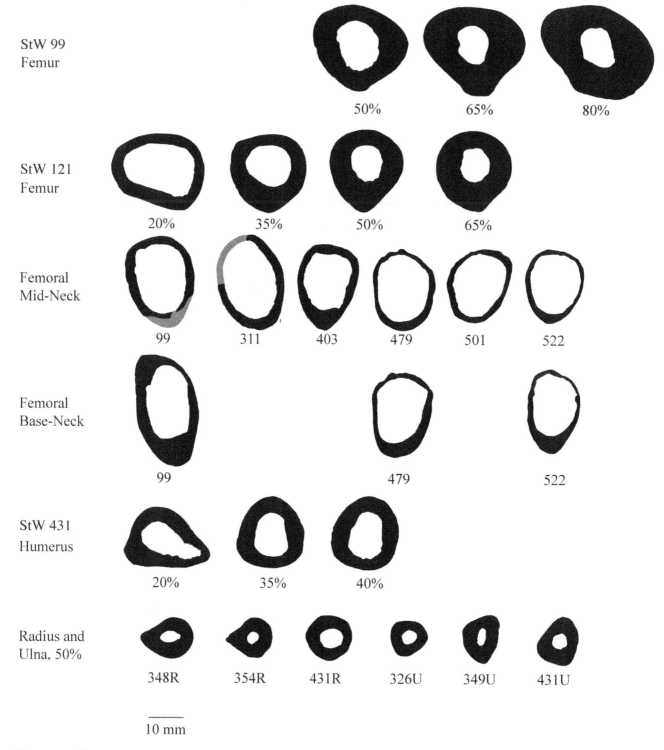

Figure 16.1 Cross-sectional outlines for primary Sterkfontein specimens, derived from CT scans. Diaphyseal sections oriented with medial to the left, anterior above, femoral neck sections with anterior to the left, superior above.

Femoral diaphysis

General section contours for the two more complete Sterkfontein femora are typical for hominins in general (see Ruff, 1995: Figure 16.1; Ruff et al., 1999: Figure 16.2; Ruff, 2009: Figure 16.4). However, subtle differences in diaphyseal shape also characterize different hominin taxa. One characteristic that has been shown to vary systematically among early hominin femora, and between

Table 16.2. Superior and inferior cortical breadths of femoral neck sections.

Specimen	Base of neck			Mid-neck		
	Superior	Inferior	Superior/Inferior	Superior	Inferior	Sup/Inf
StW 99	3.9	8.2	0.48	2.4	4.3	0.56
StW 311				1.8	3.4	0.53
StW 403				2.9	5.8	0.50
StW 479	1.7	4.4	0.39	1.8	2.8	0.62
StW 501				1.6	1.9	0.85
StW 522	1.6	5.1	0.31	1.6	3.0	0.54
SK 82	1.9	4.1	0.47	3.0	2.9	1.05
SK 97	1.2	2.8	0.42	1.5	2.2	0.68
A.L. 288-1	1.2	3.5	0.34	1.5	2.6	0.58

Dimensions in mm. Note that due to rounding of cortical breadths in the table, ratios may be slightly different than those calculated from the values given here. All measurements in mm.

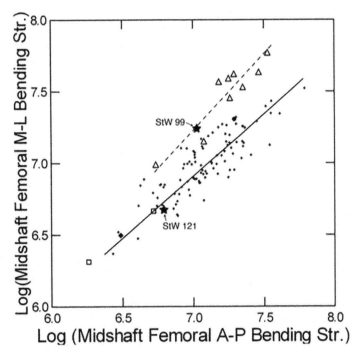

Figure 16.2 Midshaft femoral M-L (Z_y) vs. A-P (Z_x) bending strength. Section moduli calculated as second moments of area to the 0.73 power, following Ruff (1995). Small + : modern humans; open triangles: *H. erectus/ergaster* (Kresna 11; OH 28; BOU-VP-2/15 and 19/63; KNM-ER 737, 803a, 1472, 1481a, and 1808m); open squares: *H. habilis* (OH 62, KNM-ER 5881); filled circle: *P. boisei* (OH 80-2); filled diamond: *A. afarensis* (A.L. 288-1); filled stars: Sterkfontein specimens (StW 99 and 121). Reduced major axis regression lines plotted through modern humans (solid), and *H. erectus* (dashes). Comparative data from Ruff, 1995, Puymerail et al., 2012, and Ruff et al., 2016.

early hominin and modern femora, is relative mediolateral (M-L) to anteroposterior (A-P) bending strength (or breadth) of the diaphysis (Ruff, 1995, and references therein; 1998; Ruff et al., 1999; Haeusler and McHenry, 2004). Compared to modern humans, early *Homo erectus* (or *ergaster*) shows a very large increase in relative M-L bending strength of the mid-proximal femur, which is consistent with a long femoral neck and an inferred wide biacetabular breadth (Ruff, 1995; Ward et al., 2015; Ruff, 2017). Despite also possessing a long femoral neck

and wide biacetabular breadth, australopiths show only a moderate increase in M-L bending strength of the femur (Ruff, 1995, 1998; Ruff et al., 1999; Haeusler and McHenry, 2004; Ruff et al., 2016). This, together with other morphological features, has been argued to indicate a possible alteration in the gait of australopiths (Ruff, 1998; Ruff et al., 1999; Ruff and Higgins, 2013; Ruff et al., 2016).

The StW 99 femur shows evidence of increased mediolateral relative to anteroposterior dimensions in the mid-proximal diaphysis, while this is less apparent in StW 121 (which also does not preserve the most proximal, subtrochanteric section) (Figure 16.1, Table 16.1). Figures 16.2 and 16.3 are plots of M-L to A-P section moduli in the femoral midshaft and subtrochanteric (80%) sections, respectively, for modern humans, *H. erectus/ergaster*, *H. habilis*, australopiths, and StW 99 and 121. Reduced major axis (RMA) regressions are plotted through modern human and *H. erectus* samples. All data are logged to preserve proportionality across size ranges.

The relatively increased mediolateral bending strength of *H. erectus* femora compared to modern humans is clearly apparent in both sections, as observed previously (Ruff, 1995; Ward et al., 2015). Not including the Sterkfontein specimens, at midshaft (Figure 16.2), australopiths fall well within the modern human range (although OH 80-2 is near its upper end), while in the subtrochanteric section (Figure 16.3) they consistently fall at the upper edge of or just beyond the modern human distribution but still below that for *H. erectus* (with the exception of one modern human outlier discussed in Ruff, 1995). Thus, compared to modern humans, australopiths show evidence for some increase in relative M-L bending strength proximally, but not to the same degree as *H. erectus*. The midshaft sections available for *H. habilis* are similar to those of modern humans and australopiths and not *H. erectus*.

In the femoral midshaft comparisons (Figure 16.2), StW 121 is similar to australopiths and modern humans, while StW 99 falls almost directly on the *H. erectus* regression line and above modern humans and australopiths. However, in the proximal femur (Figure 16.3), StW 99 falls below all *H. erectus* specimens and almost directly on the australopith line; that is, it does not exhibit the extreme relative M-L strengthening characteristic of *H. erectus*. In this respect it is similar in proportions to two other South African specimens, SK 82 and 97, attributed to *A. robustus*, as well as East African *P. boisei* and *A. afarensis*. The proximal femoral shaft is likely to be more informative regarding load transmission across the hip (Ruff, 1995, 1998), and in this region StW 99 clearly groups with australopiths rather than with *H. erectus*. The "*erectus*-like" morphology of the midshaft of StW 99 is interesting, though, and may have functional implications (see "Discussion and conclusions").

Femoral head and neck

Another characteristic that has been shown to be variable in early hominin femora is the size of the femoral head relative to the breadth of the proximal shaft, with australopiths showing relatively smaller femoral heads than *H. erectus* or modern humans (e.g., Napier, 1964; Wood, 1976; Ruff et al., 1999, 2018b). Figure 16.4 is a plot of subtrochanteric average bending strength against femoral head superoinferior breadth in modern humans, African apes, three *H. erectus*, four australopiths, and StW 99. A femoral head breadth for StW 99 of about 38 was obtained from its partially preserved surface using laser scans and a sphere-fitting procedure (following Hammond et al., 2013). Other fossil femoral head data were derived from Ruff et al. (2018b).

As noted previously (Ruff et al., 1999), in terms of relative femoral head size to shaft strength, australopiths group with African apes, while early *H. erectus* groups with modern humans. In this respect, StW 99 clearly falls with australopiths and African apes; that is, it has a relatively small femoral head. As with proximal femoral M-L and A-P bending strength (Figure 16.3), the shaft of StW 99 is somewhat larger (stronger) overall than either of the two *P. robustus* Swartkrans specimens (SK 82 and 97). In previous comparisons of femoral head breadth to distal humeral articular breadth, StW 431 fell with chimpanzees and below humans, as did A.L. 288-1 (Ruff et al., 2016), again indicating a relatively small head. (Femoral head breadth of StW 431 was derived from acetabular breadth in these analyses; note that acetabular breadth (42.0) was mistakenly listed as femoral head breadth in Table 2 of Ruff et al., 2016.)

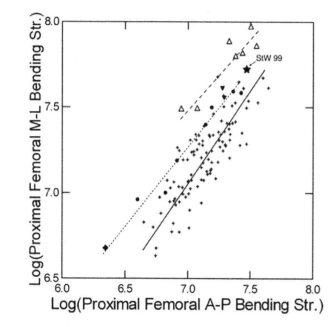

Figure 16.3 Subtrochanteric femoral M-L (Z_y) vs. A-P (Z_x) bending strength. Section moduli calculated as second moments of area to the 0.73 power, following Ruff (1995) and Ruff et al., 2016. Small + : modern humans; open triangles: *H. erectus/ergaster* (Kresna 11; OH 28; KNM-ER 737, 803a, 1472, 1481a, and 1808m); filled circles: *P. boisei* (OH 20 and 80-2; KNM-ER 738, 815, 1465a, 1500d, 1503); filled triangles: *P. robustus* (SK 82 and 97); filled diamond: *A. afarensis* (A.L. 288-1); filled star: Sterkfontein specimen (StW 99). Reduced major axis regression lines plotted through modern humans (solid), australopiths except Sterkfontein specimens (dotted), and *H. erectus* (dashes). Comparative data from Ruff, 1995, Puymerail et al., 2012, and Ruff et al., 2016.

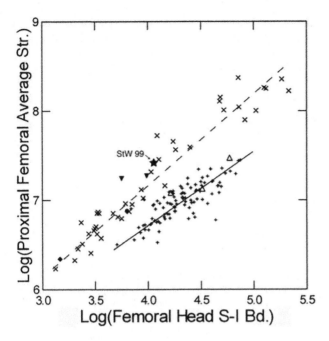

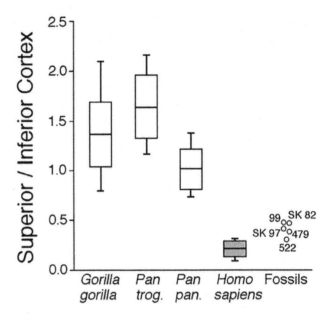

Figure 16.4 Subtrochanteric femoral average strength (one-half polar second moment of area to the 0.73 power, following Ruff et al., 1999) vs. superoinferior femoral head breadth. Small crosses: modern humans; Xs: African apes; open triangles: *H. erectus/ergaster* (OH 28 [femoral head calculated from acetabulum]; KNM-ER 1472 and 1481a); filled circle: *P. boisei* (KNM-ER 738); filled triangles: *P. robustus* (SK 82 and 97); filled diamond: *A. afarensis* (A.L. 288-1); filled star: Sterkfontein specimen (StW 99). Reduced major axis regressions plotted through humans (solid line) and apes (long dashes). Comparative data from Ruff, 1995, and Ruff et al., 1999, 2016, and 2018.

Figure 16.5 Box plots of femoral neck superior to inferior cortical breadths measured at the base of the femoral neck in African apes (*Gorilla gorilla, Pan troglodytes,* and *P. paniscus*), modern humans, SK 82 and 97, A.L. 288-1, and three Sterkfontein specimens (filled circles; numbers only given). Comparative data from Ohman et al., 1997. Modified from Ruff and Higgins (2013) and Ruff et al. (2016).

Table 16.2 lists femoral neck superior and inferior cortical breadths for the Sterkfontein specimens. StW 99 is missing a large chunk of cortex inferiorly at mid-neck (Figure 16.1), but adjacent cortices are intact (e.g., see base of neck section in Figure 16.1), permitting an accurate filling of this defect (see Ruff and Higgins, 2013: Figure 2). StW 311 is missing a large section of the superior-anterior cortex but enough of the remaining cortical contours are present so that a reasonable reconstruction of the missing portion can be made (Figure 16.1). However, resulting geometric properties for this specimen should be considered approximate. In addition to the Sterkfontein specimens, femoral neck cortical breadths were also measured on CT scans of the SK 82 and 97 specimens and A.L. 288-1 (Ruff et al., 2016), with data included in Table 16.2. Comparative data for the base of the femoral neck were taken from Ohman et al. (1997), and for the mid-femoral neck from Rafferty (1998).

All of the Sterkfontein specimens, as well as SK 82 and 97 and A.L. 288-1, fall closer to modern humans than to African apes in superior to inferior cortical ratios at the base of the neck (Figure 16.5); that is, they have relatively asymmetric cortices (Figure 16.1). However, they also all fall near or above the limit of variation in the human sample (n = 10) of Ohman et al. (1997), indicating that their cortices are not as asymmetric on average as those of modern humans. Mid-femoral neck cortical ratios for australopiths are in the upper range or above those of the human sample (n = 24) of Rafferty (1998) and overlap extensively with her non-human hominoids (Figure 16.6). SK 82 and StW 501 are particularly high, falling in the upper range for chimpanzees. Thus, cortical distribution in the mid-femoral neck in these South African hominins appears less human-like on average than that in the base of the neck, although both sections are less asymmetric than those of modern humans.

In terms of overall cross-sectional shape of the mid-femoral neck, most specimens from Sterkfontein appear somewhat "A-P flattened" compared to modern humans (e.g., see Ruff and Hayes, 1983: Figure 6), although this feature is variable, with some specimens such as StW 479, 501, and 522 showing a somewhat more rounded contour (Figure 16.1). Elsewhere we have demonstrated that the general morphology of australopith femoral necks is actually better characterized as superoinferiorly expanded rather than A-P compressed, compared to femoral head size and probable hip joint loading (Ruff and Higgins, 2013). Figure 16.7 demonstrates this relationship between S-I breadth of the mid-neck

313

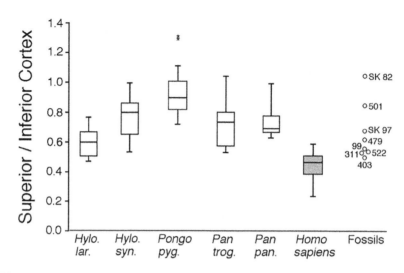

Figure 16.6 Box plots of femoral neck superior to inferior cortical breadths measured at the midpoint of the femoral neck in apes (*Pan troglodytes, P. paniscus, Pongo pygmaeus, Hylobates lar, H. syndactylus*), modern humans, SK 82 and 97, and six Sterkfontein specimens (filled circles; numbers only given). Comparative data from Rafferty (1998). Modified from Ruff and Higgins (2013) and Ruff et al. (2016).

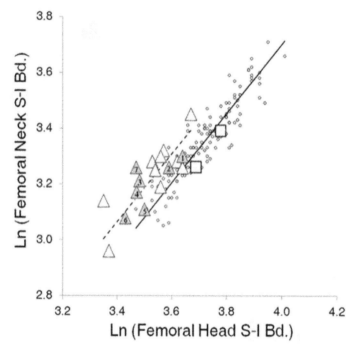

Figure 16.7 Femoral mid-neck superoinferior breadth vs. femoral head superoinferior breadth in australopiths (triangles), early *H. erectus* (KNM-ER 1472 and 1481a; squares), and modern humans. Specimens from Sterkfontein shaded: (1) StW 99; (2) StW 311; (3) StW 403; (4) StW 479; (5) StW 501; (6) StW 598. Data for StW 598 (*A.* sp.) from Pickering et al. (2012); data for other Sterkfontein specimens from personal measurements and Plavcan, pers. comm. (femoral head breadth estimates for StW 99, 403, and 479). Comparative modern human data from Richmond and Jungers (2008, and pers. comm.). Solid line: RMA regression through modern humans; dashed line: RMA regression through australopiths. Modified from Ruff and Higgins (2013), with the addition of StW 403 and 479. See Ruff and Higgins (2013) for identification of other australopith data points.

and head in australopiths, early *H. erectus*, and modern humans (modern comparative data from Richmond and Jungers, 2008). The Sterkfontein specimens conform to the typical australopith pattern (although there is also much overlap with modern humans).

Relative upper limb strength

Upper limb cross sections, in isolation, are less informative regarding locomotor and postural adaptations in hominins. However, the relatively great strength of the StW 431 upper limb bones is noteworthy. The polar section modulus (Z_p) of the mid-distal (40%) humerus of 1306 mm³ (Table 16.1) is greater than that of the large-bodied adult early *H. erectus* KNM-ER 1808 (1041 mm³, Ruff, 2008) as well the *H. erectus* Zhoukoudian III humerus (1009 mm³, Xing et al., 2018: Table S3). Figure 16.8 is a plot of average mid-distal humeral bending strength against estimated body mass in the modern human Pecos and East African samples, 15 chimpanzees with known body masses (Burgess et al., 2018), StW 431, A.L. 288-1, and KNM-WT 15000. Body mass in the modern human samples was estimated from femoral head superoinferior breadth (Ruff et al., 2018b), or from stature and bi-iliac breadth when possible (Ruff et al., 2005). Body masses for StW 431 (45.0 kg) and A.L. 288-1 (28.9 kg) were calculated from femoral head S-I breadth, adjusted for australopiths as described in Ruff et al. (2018b). The femoral head breadth of StW 431 (34.1, Ruff et al., 2018b) was derived from its acetabular breadth using a conversion formula (Ruff, 2010).[1] The body mass for KNM-WT 15000 (as a juvenile, i.e., at his time of death) of 51.5 kg is derived from Ruff (2007).

As expected, chimpanzees have much stronger humeri for their body masses than modern humans (Figure 16.8). Both StW 431 and A.L. 288-1 fall within the chimpanzee distribution (although in its lower half), well above modern humans. Thus, both specimens have very strong humeri compared to modern humans of similar body size. It should be noted that the body mass estimate for StW 431 is considerably higher than that recently proposed by Grabowski et al. (2015) of 34.5 kg. For reasons presented elsewhere (Ruff et al., 2018b), we feel that the Grabowski et al. values for StW 431 as well as A.L. 288-1 (26.0 kg) are underestimates. This also applies to earlier estimates for these specimens based on unadjusted femoral head breadths (Ruff, 2010). However, lowering body mass estimates for the fossils would only increase the difference in proportions from those of modern humans and make them appear even more chimpanzee-like. KNM-WT 15000 falls within the modern human data scatter, alhough within its upper half.

Midshaft radial and ulnar polar section moduli (Z_p) could also be calculated for StW 431 (Table 16.1). These are plotted against its body mass estimate in Figures 16.9a,b, together with modern humans and chimpanzees (radial and ulnar data were only available for the Pecos sample [N = 60]). Again, StW 431 has bone strength proportions

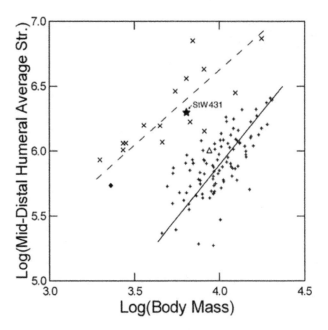

Figure 16.8 Mid-distal (40%) humeral average strength (one-half polar second moment of area to the 0.73 power) vs. body mass. Small crosses: modern humans; X's: *Pan*; open triangle: KNM-WT 15000; filled diamond: *A. afarensis* (A.L. 288-1); filled star: Sterkfontein specimen (StW 431). Reduced major axis regressions plotted through modern humans (solid line) and apes (long dashes). *Pan* data for individuals of known body mass; body mass in modern humans calculated from stature and bi-iliac breadth or femoral head breadth (see text). Body masses for StW 431 and A.L. 288-1 from Ruff et al., 2018, and for KNM-WT 15000 (as a juvenile) from Ruff (2007). Cross-sectional data for A.L. 288-1 from Ruff et al. (2016) and for KNM-WT 15000 from Ruff (2007). Human (n = 100) and *Pan* (n = 15) unpublished data from samples described in Ruff, 1995 and Burgess et al., 2018.

more like chimpanzees than like modern humans, although this is clearer for the radius (Figure 16.9a) than for the ulna (Figure 16.9b). Radial strength proportions are more distinct in chimpanzees and humans than are ulnar proportions (with one low chimpanzee outlier— also a low outlier in Figure 16.8). Among African apes, greater arboreality is associated with increasing radial to ulnar strength, i.e, chimpanzees>western lowland gorillas>mountain gorillas, possibly associated with different load-sharing during knuckle-walking versus arboreal climbing and suspensory behavior (Ruff, 2002; Ruff et al., 2013; Ruff et al., 2018a). While the knuckle-walking versus climbing distinction would not apply to StW 431, the fact that it exhibits a particularly strong radius

---

1 The current best estimate of average acetabular breadth of StW 431, using a sphere-fitting model (Hauesler and Ruff, Chapter 11, this volume), is 41.1. Assuming that average breadth is midway between S-I and A-P breadths, and rearranging terms in the formula given in Table 11.1 (Hauesler and Ruff, Chapter 11, this volume), gives an estimate for S-I acetabular breadth of 42.0, which is slightly lower than the estimate of 42.5 used in Ruff et al. (2018). However, given the various uncertainties inherent in this process and the minimal effects on femoral head breadth and thus body mass, the values used in Ruff et al. (2018) are employed here.

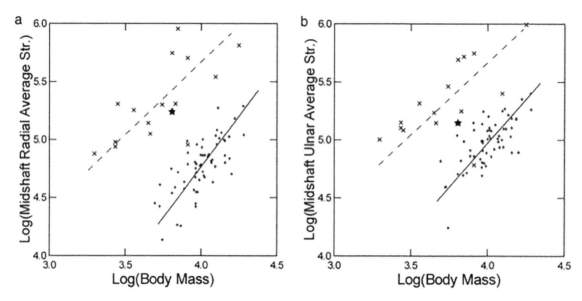

Figure 16.9 Midshaft radial (a) and ulnar (b) average strength (one-half polar second moment of area to the .73 power) vs. body mass. Small crosses: modern humans; Xs: *Pan*; filled star: Sterkfontein specimen (StW 431). Reduced major axis regressions plotted through humans (solid line) and apes (long dashes). Body masses as in Figure 8. Human (n = 60) and *Pan* (n = 15) unpublished data from samples described in Ruff, 1995 and Burgess et al., 2018.

Discussion and conclusions

Morphology of the proximal femur in StW 99—M-L to A-P bending strength of the shaft and size of the head relative to shaft strength—allies it with other australopith femora from East and South Africa. Its midshaft femoral shape is more similar to that of early *H. erectus* femora from East Africa. However, the femoral midshaft is less informative regarding load transfer around the hip, and by extension, pelvic shape, and bipedal locomotion (Ruff, 1995, 1998), so this observation carries less weight in terms of biomechanical inferences. Cortical bone distribution in the femoral neck of StW 99 is also consistent with other australopiths. Following from previous analyses (Ruff, 1995, 1998; Ruff et al., 1999; Ruff and Higgins, 2013), the proximal morphology of StW 99 may thus indicate a slightly altered gait relative to that of modern humans involving greater lateral transfer of body mass over the stance leg, resulting in a reduction in hip joint reaction force (and thus smaller femoral head), a more vertical hip joint reaction force (and thus increase in superoinferior bending of the femoral neck), and only moderately increased M-L bending of the proximal shaft (and thus less extreme M-L strengthening of the proximal shaft than in *H. erectus*). These models assume a relatively long femoral neck, which has been observed in all early hominin femora where it can be evaluated (Ruff, 1995; Ruff and Higgins, 2013; Ward et al., 2015; Ruff et al., 2016). Femoral neck length of StW 99 was not specifically evaluated here, but it appears quite long—at least as long as those of SK 82 and 97 (also see DeSilva and Grabowski, Chapter 12,

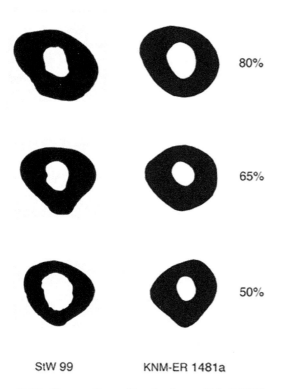

StW 99   KNM-ER 1481a

Figure 16.10 Cross section outlines for three midshaft (50%) through subtrochanteric (80%) sections in StW 99 and KNM-ER 1481a. Sections for KNM-ER 1481a derived from Figure 18 in Ruff (1995). Drawn to approximately the same scale.

(Figure 16.9a) may be another indication of significant forelimb use during arboreal locomotion.

this volume). StW 121 is smaller than StW 99, and shows less M-L expansion in the midshaft area. Its most proximal section, at 65% of bone length from the distal end, also shows much less M-L expansion than StW 99 (Figure 16.1), and thus in terms of proximal femoral morphology would also certainly not group with *H. erectus* (Figure 16.3; also compare with Figure 16.10).

Still, despite falling within the general australopith morphological, and by implication, biomechanical pattern more proximally, the early *H. erectus*-like morphology of the femoral midshaft of StW 99 is striking. This indicates that mediolateral bending loads remained fairly high throughout at least the proximal half of the diaphysis. Figure 16.10 shows a comparison of cross sectional outlines for 50%, 65%, and 80% locations in StW 99 and KNM-ER 1481a, an early (1.965 Ma, Joordens et al., 2013) *H. erectus* specimen from East Africa (sections for the latter were reported in Ruff, 1995). The overall similarity in morphology between the two specimens is apparent. StW 99 is also quite large for an australopith—A-P and M-L "subtrochanteric" (i.e., 80% location) external breadths are 25.7 and 33.6, respectively, which are as large as or larger than those for other large femora attributed to *Australopithecus* or *Paranthropus*, e.g., OH 20, KNM-ER 993, A.L. 827-1 (Walker, 1973; Ward et al., 2012; also see Ruff et al., 2018b), as well as SK 82 and 97 (A.L. 333-3 may be larger, but is broken too proximally to obtain comparable measurements; Lovejoy et al., 1982). The original complete length of StW 99 is difficult to determine, given that it is broken somewhere in the middle shaft region and has suffered some erosion of the distal posterior surface (i.e., around the linea aspera) (see Desilva and Grabowski, this volume). Pontzer (2012) used a maximum length estimate of 433.5, obtained from T. Holliday and based on the relative position of minimum M-L shaft breadth in modern femora (Sládek et al., 2010)—this appears to occur just proximal to the distal break in StW 99. However, compared to modern humans, minimum M-L breadth is more distally placed in early *Homo* (Weidenreich, 1938; Day, 1971; Ruff, 1995) and likely also *Australopithecus* (Ward et al., 2012), so this procedure may overestimate its length to some degree. Other authors have derived shorter length estimates for StW 99: 313 and 382, based on regressions from femoral head size in modern apes and humans, respectively (McHenry and Berger, 1998b); and StW 346, based on a photographic reconstruction using other Sterkfontein femora as references for the missing parts of StW 99 (Haeusler, unpublished data). We used an estimate of 400 for the length used here to locate cross-sections, which is equivalent to about 420 for maximum length, but we consider this to be very approximate. The longest nearly complete *Australopithecus* femur—A.L. 827-1—is about 368 in reconstructed maximum length (Ward et al., 2012). Walker (1973) estimated the length of KNM-ER 993, a possible *Australopithecus*, to be about 360. Based on femoral head size, length estimates

of 373–391 have been cited for A.L. 333-3 (Pontzer, 2012; Ward et al., 2012), but this technique is obviously very approximate and possibly biased because of the relatively small size of the femoral head in australopiths. Depending on which length estimate is used for StW 99, it appears to be among the longest or possibly the longest australopith femur in the fossil record.

Thus, StW 99 is unusual in both middle shaft morphology and overall size for an australopith. Although its stratigraphic position in Member 4 (and thus, its attribution to *Australopithecus*) was questioned by McHenry and Berger (1998a), more recent work at Sterkfontein has placed it clearly within that level (Stratford, this volume). It is possible that its large size is in some way related to its partial deviation from "typical" australopith morphology in the midshaft region (e.g., by affecting body breadth to height proportions). Another large australopith femur, A.L. 827-1, is also relatively M-L broad throughout its shaft (Ward et al., 2012), with M-L/A-P external breadth ratios in the subtrochanteric and midshaft regions similar to those for StW 99. StW 121, a smaller specimen, is less mediolaterally expanded at midshaft and the 65% section.

It should also be noted that the biomechanical model of the hip used to interpret proximal femoral morphology (Ruff, 1998; Ruff and Higgins, 2013) would predict less variation in mediolateral bending loads along the length of the shaft in australopiths, because a more vertically oriented hip joint reaction force (i.e., one more parallel to the shaft) would create more equal M-L bending moment arms down the shaft (also see Ruff, 1995). The smaller difference in M-L/A-P bending strength between 80% and 50% sections in StW 99 than in early *Homo* (or modern humans) implied by its relative positions in Figures 16.2 and 16.3 may be a product of this difference in joint force orientation.

Cortical bone distribution in the femoral necks from Sterkfontein is similar to, but subtly different from, that of modern humans in showing less superoinferior asymmetry. Theoretical and experimental studies of the modern human femoral neck have shown that a combination of hip joint and gluteal abductor forces produces high compressive loads on the inferior cortex and very low loads on the superior cortex, explaining the asymmetric distribution of cortical bone in the neck (Frankel, 1960; Pauwels, 1980; Lovejoy, 1988). However, a more vertical hip joint reaction force produces relatively more bending of the neck and thus higher loads on the superior cortex (Frankel, 1960; Carter et al., 1989). Thus, the less asymmetric cortical bone distribution in the Sterkfontein femoral necks are consistent with other evidence above indicating an altered gait pattern involving lateral displacement of the body center of mass over the stance limb. The superoinferiorly expanded femoral neck, relative to femoral head size, of australopiths—including the Sterkfontein specimens—is also consistent with the same biomechanical mechanism. As discussed elsewhere (Ruff and Higgins, 2013; Ruff et al., 2016) such an alteration in gait would have involved an increase in energy costs of terrestrial bipedal locomotion. This likely would have limited terrestrial mobility (e.g., range size) in both South and East African australopiths.

The extremely strong humerus and radius of StW 431 relative to its probable body size agrees with observations for *A. africanus* based on relative articular size (McHenry and Berger, 1998a), and is another indication that locomotor behavior may have been altered relative to that of modern humans. Like A.L. 288-1, it groups with chimpanzees in this regard (also see Ruff et al., 2016). Unlike results by McHenry and Berger (1998a) based on articular size, StW 431 is not more ape-like than A.L. 288-1—if anything, StW 431 is lightly closer to the human distribution than A.L. 288-1, as are larger chimpanzees (this is in part due to more positive allometry in humeral dimensions among humans than among nonhuman apes, which is a general phenomenon (Zelazny, 2018). It should be noted that the modern human comparative sample consists of populations that engaged in manual labor (Ruff, 1995), although some decrease in relative upper limb bone strength likely occurred between the Pleistocene and Holocene (Chirchir et al., 2015; Ruff et al., 2015). The relatively strong humerus of KNM-WT 15000, even as a juvenile, compared to modern humans, is consistent with this temporal trend. However, the relative humeral strength of KNM-WT 15000 is still far below that of StW 431 or A.L. 288-1.

These observations are similar to those made earlier for the *H. habilis* OH 62 (Ruff, 2009) and *A. afarensis* A.L. 288-1 specimens (Ruff et al., 2016), although unlike in these specimens, direct upper/lower limb comparisons (of diaphyseal strength) cannot be carried out for StW 431. Still, the relatively very strong upper limb of StW 431 implies that arboreal behavior formed a significant component of its locomotor repertoire. Its particularly strong radius may also imply use of the upper limb in climbing (Ruff et al., 2013; Ruff et al., 2018a). This complements much other evidence for increased robusticity in the upper limb skeleton of earlier australopiths (White et al., 1993; Kimbel et al., 1994; Haile-Selassie et al., 2010; Ward et al., 2012; Ruff et al., 2016) as well as *H. habilis* (*sensu stricto*) (Hartwig-Scherer and Martin, 1991; Ruff, 2009), and suggests retention of climbing behavior among at least some early hominin taxa for several million years following the initial acquisition of terrestrial bipedality (Senut et al., 2001; Zollikofer et al., 2005; Richmond and Jungers, 2008).

In sum, the cross-sectional geometry of the Sterkfontein Member 4 long bone specimens suggests some similarities to, but also interesting differences, in mechanical loading of these elements relative to modern humans. Bipedal gait may have been less efficient and arboreal climbing more prevalent in these hominins.

Acknowledgments

We would like to thank Francis Thackeray for helping to arrange CT scanning, Kathy Rafferty for her original femoral neck cortical thickness data, Mike Plavcan for his unpublished femoral head breadth estimate for StW 99, Carol Ward for providing measurements of the A.L. 827-1 femur, Loring Burgess for providing chimpanzee comparative data, and Kinley Russell for providing technical support. Support for scanning of the fossil specimens was provided by the L.S.B. Leakey Foundation, and comparative data were collected with funding from the National Science Foundation (BSC 1316104) and Wenner-Gren Foundation for Anthropological Research (8657).

References

Berger, L.R., de Ruiter, D.J., Churchill, S.E., Schmid, P., Carlson, K.J., Dirks, P.H., Kibii, J.M., 2010. *Australopithecus sediba*: a new species of *Homo*-like australopith from South Africa. Science 328, 195–204.

Burgess, M.L., McFarlin, S.C., Mudakikwa, A., Cranfield, M.R., Ruff, C.B., 2018. Body mass estimation in hominoids: age and locomotor effects. Journal of Human Evolution 115, 36–46.

Carter, D.R., Orr, T.E., Fyhrie, D.P., 1989. Relationships between loading history and femoral cancellous bone architecture. Journal of Biomechanics 22, 231–244.

Chirchir, H., Kivell, T.L., Ruff, C.B., Hublin, J.-J., Carlson, K.J., Zipfel, B., Richmond, B.G., 2015. Recent origin of low trabecular bone density in modern humans. Proceedings of the National Academy of Sciences 112, 366–371.

Day, M.H., 1971. Postcranial remains of *Homo erectus* from Bed IV, Olduvai Gorge, Tanzania. Nature 232, 383–387.

Di Vincenzo, F., Rodríguez, L., Carretero, J.M., Collina, C., Geraads, D., Piperno, M., Manzi, G., 2015. The massive fossil humerus from the Oldowan horizon of Gombore I, Melka Kunturé (Ethiopia, > 1.39 Ma). Quaternary Science Reviews 122, 207–221.

Frankel, V.H., 1960. The femoral neck. Almqvist and Wiksells Boktryckeri AB: Uppsala.

Grabowski, M., Hatala, K.G., Jungers, W.L., Richmond, B.G., 2015. Body mass estimates of hominin fossils and the evolution of human body size. Journal of Human Evolution 85, 75–93.

Haeusler, M., McHenry, H.M., 2004. Body proportions of *Homo habilis* reviewed. Journal of Human Evolution 46, 433–465.

Haile-Selassie, Y., Latimer, B.M., Alene, M., Deino, A.L., Gibert, L., Melillo, S.M., Saylor, B.Z., Scott, G.R., Lovejoy, C.O., 2010. An early *Australopithecus afarensis* postcranium from Woranso-Mille, Ethiopia. Proceedings of the National Academy of Sciences 107, 12121–12126.

Hammond, A.S., Plavcan, J.M., Ward, C.V., 2013. Precision and accuracy of acetabular size measures in fragmentary hominin pelves obtained using sphere-fitting techniques. American Journal of Physical Anthropology 150, 565–578.

Hartwig-Scherer, S., Martin, R.D., 1991. Was "Lucy" more human than her "child"? Observations on early hominid postcranial skeletons. Journal of Human Evolution 21, 439–450.

Holt, B.M., 2003. Mobility in Upper Paleolithic and Mesolithic Europe: evidence from the lower limb. American Journal of Physical Anthropology 122, 200–215.

Joordens, J.C., Dupont-Nivet, G., Feibel, C.S., Spoor, F., Sier, M.J., van der Lubbe, J.H., Nielsen, T.K., Knul, M.V., Davies, G.R., Vonhof, H.B., 2013. Improved age control on early Homo fossils from the upper Burgi Member at Koobi Fora, Kenya. Journal of Human Evolution 65, 731–745.

Kimbel, W.H., Johanson, D.C., Rak, Y., 1994. The first skull and other new discoveries of Australopithecus afarensis at Hadar, Ethiopia. Nature 368, 449–451.

Kuperavage, A.J., Eckhardt, R.B., 2009. Biomechanical inferences about the origins of bipedal locomotion from ancient African femora. Journal of Engineering Mechanics ASME 135, 479–484.

Kuperavage, A.J., Eckhardt, R.B., 2010. Moment coefficients of skewness in the femoral neck cortical bone distribution of BAR 1002'00. Homo 61, 244–252.

Lovejoy, C.O., 1988. Evolution of human walking. Scientific American 259, 118–125.

Lovejoy, C.O., Johanson, D.C., Coppens, Y., 1982. Hominid lower limb bones recovered from the Hadar Formation: 1974–1977 Collections. American Journal of Physical Anthropology 57, 679–700.

Lovejoy, C.O., Meindl, R.S., Ohman, J.C., Heiple, K.G., White, T.D., 2002. The Maka femur and its bearing on the antiquity of human walking: applying contemporary concepts of morphogenesis to the human fossil record. American Journal of Physical Anthropology 119, 97–133.

McHenry, H.M., Berger, L.R., 1998a. Body proportions in Australopithecus afarensis and africanus and the origin of the genus Homo. Journal of Human Evolution 35, 1–22.

McHenry, H.M., Berger, L.R., 1998b. Limb lengths in Australopithecus and the origin of the genus Homo. South African Journal of Science 94, 447–450.

Napier, J.R., 1964. The evolution of bipedal walking in the hominids. Archives Biologique (Leige) 75 (Suppl.), 673–708.

Ohman, J.C., Krochta, T.J., Lovejoy, C.O., Mensforth, R.P., Latimer, B., 1997. Cortical bone distribution in the femoral neck of hominoids: implications for the locomotion of Australopithecus afarensis. American Journal of Physical Anthropology 104, 117–131.

Pauwels, F., 1980. On the distribution of the density of cancellous bone in the upper end of the femur and its significance for the theory of the functional structure of bone. In: Pauwels, F. (Ed.), Biomechanics of the Locomotor Apparatus. Springer-Verlag, Berlin, pp. 299–309.

Pickford, M., Senut, B., Gommery, D., Treil, J., 2002. Bipedalism in Orrorin tugenesis revealed by its femora. Comptes Rendu Palevol 1, 191–203.

Pontzer, H., 2012. Ecological energetics in early Homo. Current Anthropology 53, Suppl. 6, S346–S348.

Puymerail, L., Ruff, C.B., Bondioli, L., Widianto, H., Trinkaus, E., Macchiarelli, R., 2012. Structural analysis of the Kresna 11 Homo erectus femoral shaft (Sangiran, Java). Journal of Human Evolution 63, 741–749.

Rafferty, K.L., 1998. Structural design of the femoral neck in primates. Journal of Human Evolution 34, 361–383.

Richmond, B.G., Jungers, W.L., 2008. Orrorin tugenensis femoral morphology and the evolution of hominin bipedalism. Science 319, 1662–1665.

Rodriguez, L., Carretero, J.M., Garcia-Gonzalez, R., Arsuaga, J.L., 2018. Cross-sectional properties of the lower limb long bones in the Middle Pleistocene Sima de los Huesos sample (Sierra de Atapuerca, Spain). Journal of Human Evolution 117, 1–12.

Ruff, C.B., 1995. Biomechanics of the hip and birth in early Homo. American Journal of Physical Anthropology 98, 527–574.

Ruff, C.B., 1998. Evolution of the hominid hip. In: Strasser, E., Rosenberger, A., McHenry, H., Fleagle, J. (Eds.), Primate Locomotion: Recent Advances. Plenum Press: Davis, pp. 449–469.

Ruff, C.B., 2002. Long bone articular and diaphyseal structure in Old World monkeys and apes, I: Locomotor effects. American Journal of Physical Anthropology 119, 305–342.

Ruff, C.B., 2003. Ontogenetic adaptation to bipedalism: Age changes in femoral to humeral length and strength proportions in humans, with a comparison to baboons. Journal of Human Evolution 45, 317–349.

Ruff, C.B., 2007. Body size prediction from juvenile skeletal remains. American Journal of Physical Anthropology 133, 698–716.

Ruff, C.B., 2008. Femoral/humeral strength in early African Homo erectus. Journal of Human Evolution 54, 383–390.

Ruff, C.B., 2009. Relative limb strength and locomotion in Homo habilis American Journal of Physical Anthropology 138, 90–100.

Ruff, C.B., 2010. Body size and body shape in early hominins—implications of the Gona pelvis. Journal of Human Evolution 58, 166–178.

Ruff, C.B., 2017. Mechanical constraints on the hominin pelvis and the "obstetrical dilemma." Anatomical Record 30, 946–955.

Ruff, C.B., 2018. Biomechanical analyses of archaeological human skeletal samples. In: Katzenburg, M.A., Grauer, A.L. (Eds.), Biological Anthropology of the Human Skeleton, 3rd ed. John Wiley and Sons: New York, pp. 189–224.

Ruff, C.B., Burgess, M.L., Bromage, T.G., Mudakikwa, A., McFarlin, S.C., 2013. Ontogenetic changes in limb bone structural proportions in mountain gorillas (*Gorilla beringei beringei*). Journal of Human Evolution 65, 693–703.

Ruff, C.B., Burgess, M.L., Junno, J.-A., Mudakikwa, A., Zollikofer, P.E., Ponce de Leon, M.S., McFarlin, S.C., 2018a. Phylogenetic and environmental effects on limb bone structure in gorillas. American Journal of Physical Anthropology 166, 353–372.

Ruff, C.B., Burgess, M.L., Squyres, N., Junno, J.-A., Trinkaus, E., 2018b. Lower limb articular scaling and body mass estimation in Pliocene and Pleistocene hominins. Journal of Human Evolution 115, 85–111.

Ruff, C.B., Burgess, M.L., Ketcham, R.A., Kappelman, J., 2016. Limb bone structural proportions and locomotor behavior in A.L. 288–1 ("Lucy"). PLoS One 11, e0166095.

Ruff, C.B., Hayes, W.C., 1983. Cross-sectional geometry of Pecos Pueblo femora and tibiae—a biomechanical investigation. I. Method and general patterns of variation. American Journal of Physical Anthropology 60, 359–381.

Ruff, C.B., Higgins, R., 2013. Femoral neck structure and function in early hominins. American Journal of Physical Anthropology 150, 512–525.

Ruff, C.B., Holt, B.M., Niskanen, M., Sládek, V., Berner, M., Garofalo, E., Garvin, H.M., Hora, M., Junno, J.-A., Schuplerova, E., Vilkama, R., Whittey, E., 2015. Gradual decline in mobility with the adoption of food production in Europe. Proceedings of the National Academy of Science 112, 7147–7152.

Ruff, C.B., McHenry, H.M., Thackeray, J.F., 1999. Cross-sectional morphology of the SK 82 and 97 proximal femora. American Journal of Physical Anthropology 109, 509–521.

Ruff, C.B., Niskanen, M., Junno, J.A., Jamison, P., 2005. Body mass prediction from stature and bi-iliac breadth in two high latitude populations, with application to earlier higher latitude humans. Journal of Human Evolution 48, 381–392.

Ruff, C.B., Trinkaus, E., Walker, A., Larsen, C.S., 1993. Postcranial robusticity in *Homo*, I: Temporal trends and mechanical interpretation. American Journal of Physical Anthropology 91, 21–53.

Schaffler, M.B., Burr, D.B., Jungers, W.L., Ruff, C.B., 1985. Structural and mechanical indicators of limb specialization in primates. Folia Primatologica 45, 61–75.

Senut, B., Pickford, M., Gommery, D., Mein, P., Cheboi, K., Coppens, Y., 2001. First hominid from the Miocene (Lukeino Formation, Kenya). Comptes Rendus De L'Academie Des Sciences Serie Ii Fascicule a-Sciences De La Terre Et Des Planetes 332, 137–144.

Shackelford, L.L., 2007. Regional variation in the postcranial robusticity of Late Upper Paleolithic humans. American Journal of Physical Anthropology 133, 655–668.

Sládek, V., Berner, M., Galeta, P., Friedl, L., Kudrnova, S., 2010. Technical note: The effect of midshaft location

on the error ranges of femoral and tibial cross-sectional parameters. American Journal of Physical Anthropology 141, 325–332.

Stern, J.T., Susman, R.L., 1991. "Total morphological pattern" versus the "magic trait": Conflicting approaches to the study of early hominid bipedalism. In: Coppens, Y., Senut, B. (Eds.), Origine(s) de la Bipédie chez les Hominidés. Centre National de la Recherche Scientifique: Paris, pp. 99–111.

Trinkaus, E.T., 2009. The human tibia from Broken Hill, Kabwe, Zambia. Paleoanthropology 2009, 145–165.

Trinkaus, E., Ruff, C.B., 1989. Cross-sectional geometry of Neandertal femoral and tibial diaphyses: implications for locomotion. American Journal of Physical Anthropology 78, 315–316.

Trinkaus, E., Ruff, C.B., 2012. Femoral and tibial diaphyseal cross-sectional geometry in Pleistocene *Homo*. PaleoAnthropology 2012, 13–62.

Trinkaus, E., Stringer, C.B., Ruff, C.B., Hennessy, R.J., Roberts, M.B., Parfitt, S.A., 1996. The Boxgrove tibia. American Journal of Physical Anthropology Suppl. 22, 230–231.

Walker, A., 1973. New *Australopithecus* femora from East Rudolf, Kenya. Journal of Human Evolution 2, 545–555.

Ward, C.V., Feibel, C.S., Hammond, A.S., Leakey, L.N., Moffett, E.A., Plavcan, J.M., Skinner, M.M., Spoor, F., Leakey, M.G., 2015. Associated ilium and femur from Koobi Fora, Kenya, and postcranial diversity in early *Homo*. Journal of Human Evolution 81, 48–67.

Ward, C.V., Kimbel, W.H., Harmon, E.H., Johanson, D.C., 2012. New postcranial fossils of *Australopithecus afarensis* from Hadar, Ethiopia (1990-2007). Journal of Human Evolution 63, 1–51.

Weidenreich, F., 1938. The discovery of the femur and the humerus of *Sinanthropus pekinensis*. Nature 141, 614–617.

White, T.D., Suwa, G., Hart, W.K., Walter, R.C., Woldegabriel, G., Deheinzelin, J., Clark, J.D., Asfaw, B., Vrba, E., 1993. New Discoveries of *Australopithecus* at Maka in Ethiopia. Nature 366, 261–265.

Wood, B., 1976. Remains attributable to *Homo* in East Rudolf succession. In: Coppens, Y., Howell, F.C., Issac, G.L., Leakey, R.E.F. (Eds.), Earliest Man and Environments in the Lake Rudolf Basin. University of Chicago Press: Chicago, pp. 490–506.

Xing, S., Carlson, K.J., Wei, P., He, J., Liu, W., 2018. Morphology and structure of *Homo erectus* humeri from Zhoukoudian, Locality 1. PeerJ 6, e4279.

Zelazny, K.G., 2018. Morphological variation in the distal humerus of extant hominids and fossil hominins. Ph.D. Thesis, The Johns Hopkins University.

Zollikofer, C.P., Ponce de Leon, M.S., Lieberman, D.E., Guy, F., Pilbeam, D., Likius, A., Mackaye, H.T., Vignaud, P., Brunet, M., 2005. Virtual cranial reconstruction of *Sahelanthropus tchadensis*. Nature 434, 755–759.

17

Limb proportions and positional behavior

Revisiting the theoretical and empirical underpinnings for locomotor reconstruction in Australopithecus africanus

Adam D. Gordon, David J. Green, William L. Jungers, and Brian G. Richmond

Introduction

Many previous studies have addressed the question of whether limb proportions in various fossil hominins were human-like, ape-like, or intermediate. While many studies have focused on the limb length proportions available for the few fossil hominin specimens that preserve them (Jungers, 1982, 1991, 1994, 2009; Jungers and Stern, 1983; Franciscus and Holliday, 1992; Richmond et al., 2002; Haeusler and McHenry, 2004; Reno et al., 2005; Holliday and Franciscus, 2009, 2012), long bones which are complete enough to measure overall length have historically been relatively rare in the hominin fossil record, including at Sterkfontein (with the notable exception of StW 573, a.k.a. "Little Foot"; Clarke, 2013). As a result, other work has focused on the proportions of articular and/or diaphyseal long bone measurements (McHenry, 1978; McHenry and Berger, 1998; Dobson, 2005; Green et al., 2007). However, as Jungers (2009) notes, interlimb proportionality as measured by bone lengths may differ from that measured by diaphyseal and articular dimensions, raising the question: what exactly do these various limb proportions represent?

For various length-based limb proportions (e.g., intermembral index), there are well-established empirical relationships between length proportions and various broad locomotor categories, and such relationships typically have at least plausible conceptual grounding if not always explicitly articulated and tested hypotheses. For example, specialized leapers, habitual bipeds, and suspensory animals tend to exhibit greater length in the limbs that are most involved in locomotion, and quadrupeds tend to have relatively similar lengths for fore- and hindlimbs (Fleagle, 2013). Among extant hominoids, the forearm is longer relative to the upper arm in more highly suspensory apes such as gibbons and orangutans as compared to the African apes and humans, although there is not unequivocal support for any particular biomechanical theoretical model that might drive this relationship (Takahashi, 1991), and it might also result from the downstream effects of genes responsible for long manual digits in suspensory primates (Reno et al., 2008). However, what is less clear is which aspects of an individual's positional repertoire affect articular-diaphyseal limb proportions, and how they do so. The relative sizes of these dimensions have been argued to reflect the loading regime and magnitude of loads applied to limbs (Jungers, 1988; Godfrey et al., 1995; Currey, 2002), but they may also be constrained by the demands of joint mobility (Rafferty and Ruff, 1994; Hamrick, 1996), or some combination of the two.

In the present study we reexamine articular-diaphyseal limb proportions in the Sterkfontein Member 4 hominins using the resampling approach developed

Adam D. Gordon, David J. Green, William L. Jungers, and Brian G. Richmond, *Limb proportions and positional behavior*. In: *Hominin Postcranial Remains from Sterkfontein, South Africa, 1936–1995*. Edited by: Bernhard Zipfel, Brian G. Richmond, and Carol V. Ward, Oxford University Press (2020). © Oxford University Press.
DOI: 10.1093/oso/9780197507667.003.0016

Table 17.1   Skeletal sample used in this study.[a]

Species	HUMHEAD	ELBOW[0.5]	RADTV	FEMHEAD	FEMSHAFT[0.5]	DISTFEM[0.5]	PROXTIB[0.5]	DISTTIB[0.5]
Australopithecus afarensis	2	3	2	4	8	3	5	4
Australopithecus africanus	3	1	4	13	2	2	1	3
Gorilla gorilla	24F, 24M	*	*	*	*	*	*	*
Homo sapiens	24F, 24M	*	*	*	*	*	*	*
Pan troglodytes	24F, 24M	*	*	*	*	*	*	*
Pongo abelii	7F, 3M	*	*	*	*	*	*	*
Pongo pygmaeus	9F, 8M	*	*	*	*	*	*	*

[a] Abbreviations: F, female; M, male. *Sample sizes are the same across all measurements within each extant species.

by Green et al. (2007) as applied to a slightly expanded fossil sample for *Australopithecus africanus* and *A. afarensis*. However, before we do so we will address two questions: (1) how does the scaling of various postcranial measurements affect the value of limb proportion indices and thus comparisons between taxa of different size, and (2) how well do various limb proportion indices correspond with various components of locomotor repertoire among extant hominids (the Ponginae and Homininae)? We then explore how the answers to these two questions affect the interpretation of the results of the limb proportion resampling analysis.

Materials and methods

Skeletal sample

A sample of postcranial measurements from Sterkfontein was compared to a fossil sample drawn from *Australopithecus afarensis*, as well as extant comparative samples from *Homo sapiens, Gorilla gorilla, Pan troglodytes, Pongo abelii*, and *P. pygmaeus*.

The vast majority of hominin postcranial specimens at Sterkfontein are found in Member 4, one particularly notable exception being the nearly complete skeleton StW 573 ("Little Foot") from Member 2 (see Clarke, 2013, for a recent list of hominin postcranial fossils from Sterkfontein). Most authors attribute all of the Member 4 (and Member 2) hominin material to *A. africanus*, although it has been debated whether two or more hominin species might be represented in these deposits (Clarke, 1988; Kimbel and Rak, 1993; Lockwood, 1997; Moggi-Cecchi et al., 1998; Lockwood and Tobias, 2002; Clarke, 2013; Grine, 2013, 2019). Here we treat all of the postcranial material from Member 4 as representatives of a single species, *A. africanus*, although we recognize that the results of limb proportion comparisons with other fossil and extant taxa might differ if the Member 4 hominins represent multiple species rather than only one.

All Member 4 specimens that preserve the measurements of interest (see section "Morphometric

variables") were included in this analysis (Table 17.1, Appendix III). For the majority of specimens, we collected measurements from the original specimens and compared them to previously published data from McHenry (1992) and McHenry and Berger (1998). In most cases, new and published measurements were in close agreement, in which case previously published measurements were used to facilitate comparison with measurements from *A. afarensis* in those same studies. In cases of substantial difference or for specimens not included in previously published analyses, we provide new measurements (Appendix III). Similarly, we attempted to minimize interobserver error between our measurements on extant hominoids and previously published measurements by also collecting measurements from a high-quality research cast of A.L. 288-1 at the Smithsonian and found good agreement between our measurements and those of McHenry and Berger (1998). *A. afarensis* specimens were included as a comparative sample, and data from McHenry (1992) and McHenry and Berger (1998) were supplemented by published data from Ward et al. (2012) (noted in Appendix III).

The extant comparative sample was drawn from the collections of three museums: American Museum of Natural History (AMNH), Cleveland Museum of Natural History (CMNH), and National Museum of Natural History (Smithsonian; NMNH). All specimens were identified to the species level using current taxonomic designations (Groves, 2001) (e.g., distinguishing between *G. gorilla* and *G. beringei*, and between *P. abelii* and *P. pygmaeus*). All non-human specimens were wild-caught individuals; all specimens were adult. As much as possible, extant comparative samples were constrained to be composed of equal numbers of males and females and equal sample sizes across taxa to minimize the effect of sample size difference on results. For all non-orangutan species, 24 males and 24 females were included (Table 17.1). Due to the relative scarcity of adult orangutan postcranial material in these museums, samples for the two *Pongo* species were smaller with a combined genus-level sample of 16 males and 11 females (Table 17.1). The two *Pongo* species are distinguished with different symbols in figures but are treated as a single taxon due to overall similarity in limb proportions. Appendix IV provides specimen IDs and all measurements included in this analysis.

Morphometric variables

For all extant specimens, we collected standard measures of long bone length for the humerus, radius, femur, and tibia (some of the length measurements for the Cleveland specimens were provided by Lyman Jellema) (see Appendix V). We additionally collected articular and diaphyseal measurements from these elements that are measureable in at least one specimen in the Sterkfontein Member 4 sample, following previously published protocols (McHenry and Corruccini, 1978; McHenry, 1992; McHenry and Berger, 1998; Green et al., 2007). These include proximal articular measurements from the humerus (HUMHEAD), radius (RADTV), femur (FEMHEAD), and tibia (PROXTIB); distal articular measurements from the humerus (ELBOW), femur (DISTFEM), and tibia (DISTTIB); and a diaphyseal measurement from the femur (FEMSHAFT). Appendix V provides detailed definitions. This same set of skeletal variables has been used in previous analyses of body proportions in australopiths (McHenry, 1992; McHenry and Berger, 1998; Green et al., 2007) due to the number of fossil specimens that preserve the anatomy necessary to collect these measurements. In some cases, variables are the product of two orthogonal measurements; in those cases, the square root of the original variable is used here so that all included variables are linear.

Four limb proportion metrics are constructed from these variables, two based on lengths and two based on articular and diaphyseal measurements, as follows.

1. *Length Intermembral Index (LII)*. This is the standard intermembral index, measured as:

$$LII = \frac{\text{forelimb length}}{\text{hindlimb length}} \times 100$$

where forelimb length = humerus length + radius length, and hindlimb length = femur length + tibia length.

2. *Length Brachial Index (LBI)*. The standard measure of forearm to upper arm length, measured as:

$$LBI = \frac{\text{radius length}}{\text{humerus length}} \times 100$$

3. *Articular-diaphyseal Intermembral Index (AII)*. When logged, this is very similar to the limb proportion index used by Green et al. (2007), differing only in the inclusion of ELBOW and PROXTIB. It is a ratio of the geometric mean of forelimb measurements to the geometric mean of hindlimb measurements, such that

$$AII = \frac{GM_F}{GM_H} = \frac{\left(\begin{array}{c} \text{HUMHEAD} \times \text{ELBOW}^{0.5} \\ \times \text{RADTV} \end{array}\right)^{1/3}}{\left(\begin{array}{c} \text{FEMHEAD} \times \text{FEMSHAFT}^{0.5} \\ \times \text{DISTFEM}^{0.5} \times \text{PROXTIB}^{0.5} \\ \times \text{DISTTIB}^{0.5} \end{array}\right)^{1/5}}$$

4. *Articular Brachial Index (ABI)*. The ratio of the size of the radial head to the geometric mean of humeral measurements:

$$ABI = \frac{\text{RADTV}}{GM_{\text{Humerus}}} = \frac{\text{RADTV}}{\left(\text{HUMHEAD} \times \text{ELBOW}^{0.5}\right)^{1/2}}$$

In all analyses, these indices are log-transformed in order to render distributions symmetrical and unbounded (Smith, 1999).

Locomotor/positional variables

Among the extant great apes, although there is significant variation in proportional use of locomotor behavior between populations within any taxon (Doran, 1996), there is general agreement that orangutans are almost entirely arboreal in the wild and thus more arboreal than the African apes, and common chimpanzees are more arboreal in general than Western lowland gorillas (Rodman, 1979, 1984; Doran, 1996). In order to quantitatively compare various limb proportion indices to locomotor and positional repertoires in adults within the extant taxa, estimates of time spent in various locomotor categories were drawn from published sources. Table 17.2 provides estimates taken from a recent compilation by Crompton et al. (2010) drawing on the work of several field researchers (Rodman, 1979; Hunt, 1992; Doran, 1996). Two different locomotor variables were analyzed in this study: (1) overall proportion of time spent arboreal, and (2) the ratio of the proportion of time spent brachiating[1] to the proportion of time spent bipedal, representing a coarse representation of the relative loading of the fore- and hindlimbs during locomotion.

---

1 Here we use "brachiation" as a general term to mean movement using forelimb suspension underneath a branch at any speed, as opposed to the ricochetal brachiation used by hylobatids in which the body is propelled quickly using a rhythmic bimanual sequence of arm swings.

Table 17.2 Approximate percentage of time spent in various positional behaviors by great apes.

	Percentage of locomotion spent brachiating[a]	Percent of locomotion spent bipedal[a]	Percent of locomotion spent arboreal
Gorilla gorilla	3.6	6.1	41[b]
Homo sapiens	0	100	0
Pan troglodytes	0.8	0.7	50[c]
Pongo sp.	13.0	7.0	95[d]

[a] Taken from Crompton et al. (2010); see footnote in text regarding our use of the term "brachiating."
[b] From Hunt (1992) and Crompton et al. (2010).
[c] Mean of males and females at three sites from Doran (1996).
[d] Mean of male (80%) and female (100%) for Bornean orangutans (Rodman, 1979) and both sexes (100%) for Sumatran orangutans (Rodman, 1984).

Analytical methods

Scaling relationships among postcranial variables are evaluated using the standardized major axis equivalent of analysis of covariance as described by Warton et al. (2006) through the use of their package "smatr" for *R* (Warton et al., 2012; R Core Team, 2013). Resampling analyses of AII values follow the same procedure used in Green et al. (2007), using the *R* code developed for that study. The resampling procedure is an iterative process; in each iteration and for each taxon, a random sample is drawn with replacement for each measurement in Table 17.1, where the sample is equal in sample size to the minimum number of measurements for that variable across all taxa. Geometric means are calculated for each variable, and then AII is calculated following the equation provided in the section "Morphometric variables." This iterative procedure is repeated 10,000 times for each taxon (extant and extinct) and then the means of the resulting distributions are compared among taxa. For a fuller description of this method, please refer to Green et al. (2007).

Results and discussion

Limb proportions and limb scaling

Due to the small sample sizes available for the two *Pongo* species, limb proportions were compared to determine whether the two samples could be combined into a single genus-level sample. Student's t-tests show no significant difference in mean log-transformed limb proportions between *P. abelii* and *P. pygmaeus* for any of the four metrics considered here (LII, p = 0.613; LBI, p = 0.490; AII, p = 0.508; ABI, p = 0.421). Consequently, we have combined the two species to analyze *Pongo* as a single taxon in all subsequent analyses.

Within each extant taxon, the scaling relationship between the log-transformed numerator and denominator of LII (forelimb length and hindlimb length, respectively) does not differ significantly from isometry, and scaling slopes do not differ significantly among taxa (Figure 17.1A, Table 17.3). This indicates that LII is effectively independent of size within species, such that comparisons of LII between individuals or samples need not take into account size differences. However, while LII performs well at distinguishing among extant hominoids between species with an arboreal locomotor component and those without (i.e., between great apes and humans), it does not correctly recover the differences in arboreality among extant great apes: although western gorillas are widely recognized to be less arboreal than chimpanzees, they have higher LII values (Figure 17.1C). The same statements are true of humerofemoral index (not shown here). In contrast, AII recovers the correct ranking of degree of arboreality in extant great apes and humans (Figure 17.1D). However, while scaling slopes of the numerator and denominator of AII do not differ among taxa, the shared slope shows slight positive allometry of GM_F with respect to GM_H (Figure 17.1B, Table 17.3), indicating that values of AII increase slightly as individuals increase in size.

Note that allometry in and of itself is not an explanatory relationship but simply a description of a pattern, and that shape differences between differently sized individuals may reflect real functional differences rather than something which needs to be corrected for (Fleagle, 1985). However, in this case the positive allometry present in AII is unlikely to reflect an underlying locomotor signal because larger individual apes are typically less arboreal than smaller conspecifics, not more arboreal (e.g., compare adult male western gorillas to adult females, or adult male orangutans to adult females, Rodman, 1979; Doran, 1996). Instead, this slight positive allometry reflects a confounding influence that probably should be taken into account when inferring a locomotor signal from AII.

If limb proportions are instead restricted to the forelimb, LBI exhibits a pattern similar to that described for AII above. Although scaling slopes for the log-transformed numerator (radius length) and denominator (humerus length) do not differ among taxa, radius length does scale

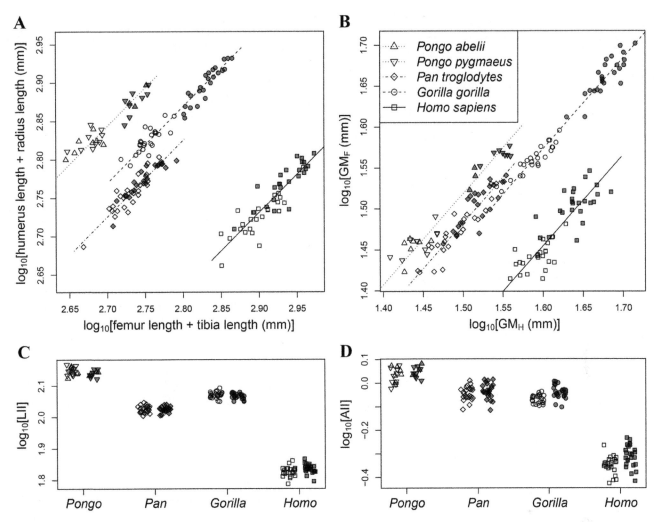

Figure 17.1 Scaling relationships of forelimb and hindlimb variables (A&B) and logged intermembral ratios (C&D) by taxon. Each data point represents an individual specimen; open symbols are female, filled symbols are male. *P. abelii* and *P. pygmaeus* are analyzed as a single taxon (see text for further detail). Forelimb length scales isometrically with hindlimb length within each taxon (A), while GM_F scales slightly positively allometrically with GM_H (B) (slopes given in Table 17.3). Logged ratios corresponding to A and B are shown by taxon in C and D (position on the X-axis is jittered in C and D for better visibility of individual data points). Note that on average chimpanzees have lower intermembral indices (LII) than gorillas, although higher GM_F:GM_H ratios (AII).

Table 17.3 Standardized major axis regression parameters for scaling of forelimb and hindlimb variables.

	interaction effect p	species effect p[a]	common slope[a]	slope 95% CI[a]
$\log_{10}$[forelimb length] vs. $\log_{10}$[hindlimb length]	0.149	< 0.001	0.995	(0.947–1.047)
$\log_{10}$[GM_F] vs. $\log_{10}$[GM_H]	0.396	< 0.001	1.114	(1.067–1.165)
$\log_{10}$[radius length] vs. $\log_{10}$[humerus length]	0.080	< 0.001	1.091	(1.023–1.166)
$\log_{10}$[RADTV] vs. $\log_{10}$[$GM_{Humerus}$]	0.195	< 0.001	1.038	(0.983–1.097)

[a] Because the interaction effect is not significant at alpha = 0.05 in all cases, models are refit without the interaction effect, such that the slope is the shared in common for all species but the intercept may differ among species (compare lines in Figure 17.1A for an example).

slightly positively allometrically in relation to humerus length (Figure 17.2A, Table 17.3). That said, the correct rank order of arboreality is recovered using LBI, although there is considerable overlap in individual human and gorilla LBI values while there was no overlap between humans and

gorillas in LII or AII (compare Figure 17.2C to Figure 17.1C and 1D). Within-taxon variability relative to between-taxon variability is considerably higher in LBI than LII, resulting in lower discriminatory power for LBI when comparing samples. This pattern of variation may reflect the influence

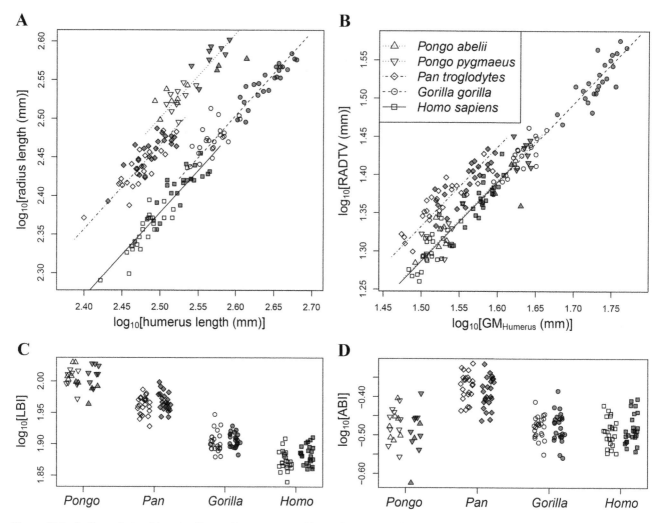

Figure 17.2 Scaling relationships of radius and humerus variables and logged brachial ratios by taxon. Symbols and conventions are as in Figure 17.1. Radius length scales slightly positively allometrically with humerus length within each taxon (A), while RADTV scales isometrically with $GM_{Humerus}$ (B) (slopes given in Table 17.3). Logged brachial ratios corresponding to A and B are shown by taxon in C and D. Note that length brachial indices track rank order of degree of arboreality across all taxa, while there is little variation in $RADTV:GM_{Humerus}$ ratios (ABI) between humans, gorillas, and orangutans.

of non-biomechanical factors on the relative length of proximal and distal limb segments such as thermoregulatory concerns (Trinkaus, 1981; Holliday, 1999). If the analogous articular limb proportion metric is considered (ABI), the numerator (RADTV) and denominator ($GM_{Humerus}$) scale with the same slope across taxa, and that slope does not differ significantly from isometry (Figure 17.2B, Table 17.3). Unfortunately, the intercept does not differ significantly between most taxa, such that orangutans, gorillas, and humans have ABI values that are indistinguishable from each other, while chimpanzees have ratio values that are slightly higher on average but overlap considerably with the other taxa (Figure 17.2D). Relative size of the radial head does not appear to vary with locomotor differences among the taxa included in this sample (Patel, 2005); thus locomotor pattern is not recovered using this metric.

Choice of limb proportion metric

Leaving aside for the moment whether certain measurements such as lengths might be available for a fossil sample, we compare the various metrics with the goal of linking limb proportions to locomotor patterns. Because several previous studies have indicated ape-like limb proportions for *A. africanus* (however one might interpret what that means), we will focus here on those metrics that were shown above to recover the correct rank ordering of degree of arboreality among the great apes, namely, LBI and AII.

A bivariate comparison of logged LBI and AII shows a general positive correlation between the two measures across the extant sample (species means, r = 0.831; individual data points, r = 0.750), with a generally linear trend among humans, chimpanzee, and orangutans (Figure

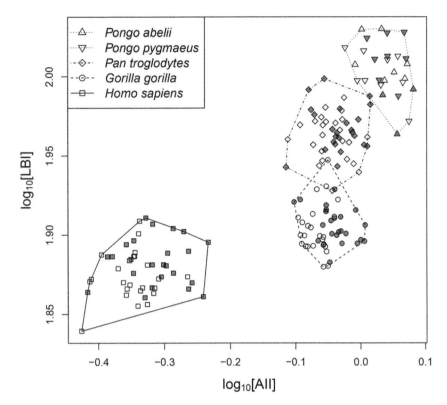

Figure 17.3 Log-transformed length brachial index (LBI) plotted against log-transformed articular-diaphyseal intermembral index (AII). Convex polygons delineate the extent of each taxon. There is a general positive association between the two variables, but gorillas deviate from the overall trend.

17.3). However, gorillas deviate from this overall trend. While both metrics preserve the correct rank ordering of species means based on relative degree of arboreality (Figures 17.1D and 17.2C), AII shows considerable overlap between chimpanzees and gorillas, whereas LBI shows marked overlap between humans and gorillas (Figure 17.3). Two important points should be noted here: (1) while limb proportion indices may distinguish between species based on their means, any individual measurement may overlap with the range of multiple species that in some cases exhibit dramatically different locomotor modes (e.g., humans and gorillas), and (2) while LBI and AII preserve the same rank ordering, they are clearly measuring something different from each other.

In order to get at that difference, we have quantified two different measures of locomotor repertoire: proportion of time spent arboreal and the ratio of proportion of time spent brachiating to proportion of time spent bipedal (regardless of substrate). These indices are admittedly coarse and highly variable among populations and individuals within a single taxon, but they provide a means to illustrate two of the many ways that "degree of arboreality" might be parsed and to show that any given limb proportion index will have different relationships with different measures of arboreality. Figure 17.4 illustrates

the relationship between these two metrics and the two limb proportion indices under consideration. What follows is not a statistical analysis; the effective sample size for each plot in Figure 17.4 is four, the number of taxa, and such a small sample does not allow for a meaningful significance test of difference in correlation. However, visual inspection demonstrates that LBI appears to have a more linear relationship with the ratio of time spent brachiating to time spent bipedal (Figure 17.4C) than it does with the proportion of time spent arboreal (Figure 17.4A). Regardless of whether or not this reflects an underlying biomechanical mechanism regarding the biomechanical importance of distal arm elongation in brachiation or is simply a coincidence resulting from the coarse nature of the positional behavior metrics, it remains the case that LBI has different relationships with these two measures of "degree of arboreality." Similarly, AII also differs with respect to its relationship with these two metrics (Figures 17.4B and 17.4D).

It should be noted that direct comparison across taxa in all of these cases may be complicated by the allometric scaling of numerator and denominator in both LBI and AII, which would effectively mean that the gorilla and human data points sit higher on the y-axis in all four plots of Figure 17.4 than they would if they were the

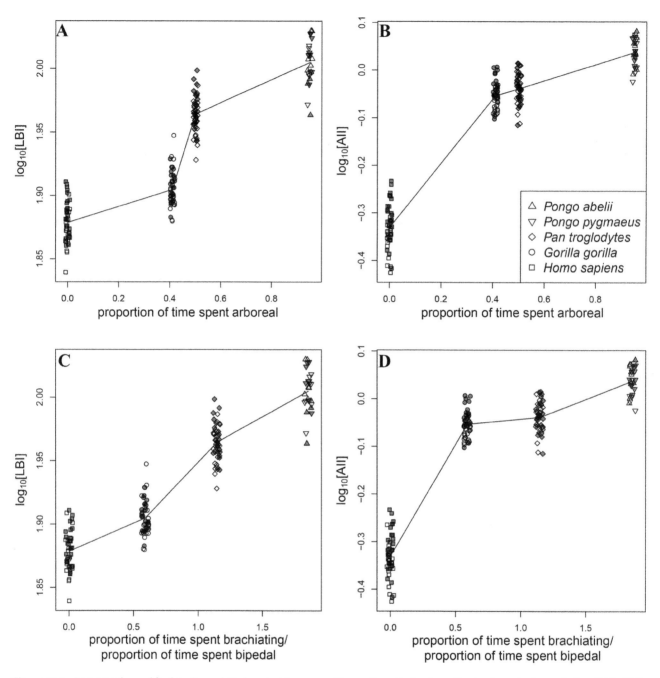

Figure 17.4 Log-transformed limb indices plotted against taxon-specific positional behavior ratios for the extant sample (see Table 17.2). Ideally positional behavior ratios would also be logged since they are also ratios (Smith, 1999), but are not logged here to allow the inclusion of humans which have values of zero for both metrics and thus have an infinitely negative logarithm. Jitter has been added to the X-axis of all plots for greater visibility of individual data points; in all cases, the same positional behavior value is assigned to all individuals for each taxon. Lines connect taxon means. Length brachial index (LBI) increases monotonically with the proportion of time spent arboreal (A) and the ratio of time spent brachiating to time spent bipedal (C), as does the articular-diaphyseal intermembral index (AII) (B and D), although the specific nature of the relationship differs between limb indices.

same size as chimpanzees or orangutans. However, such a shift would apply equally to comparisons of AII with either locomotor metric, so the point still stands that limb proportions will have different relationships with different measures of arboreality. As mentioned above, it is possible that differences in the relative lengths of proximal and distal limb segments measured by LBI may be less reflective of differences in substrate and more reflective of locomotor differences within a particular substrate setting (i.e., differences among arboreal taxa). Similarly, it might be argued that a greater theoretical justification could be proposed for taxon differences in AII resulting from differential loading of the fore- and hindlimb or different mobility requirements, regardless of substrate, rather than a simple differentiation between arboreality versus terrestriality. Thus, in ideal situations where complete long bones are available for study, different limb proportion indices may be selected to measure different components of substrate use and locomotor mode.

Then there is the practical question of whether long bone lengths are available for a particular fossil sample. The majority of *A. africanus* postcranial elements are incomplete and do not allow for length measurements, with the notable exception of StW 573, whose limb proportions will be of great interest when the long bone lengths are published. But in the case of the Sterkfontein Member 4 sample, it is not possible to calculate LII or LBI.

However, as Figure 17.1 illustrates, AII is probably preferable to LII when attempting to distinguish between various "degrees of arboreality" within the range exhibited by extant great apes, although the potential effect of size differences between taxa must be considered when interpreting the results of analyses of AII.

Limb proportions in *A. africanus* (and *A. afarensis*) revisited

This analysis repeats the resampling analysis of Green et al. (2007) with a slightly expanded sample and produces highly similar results. Comparisons between extant and fossil taxa indicate significantly higher mean resampled log AII for *A. africanus* as compared to *H. sapiens* and *A. afarensis*, but no significant difference between *A. africanus* and any of the great apes (Figure 17.5, Table

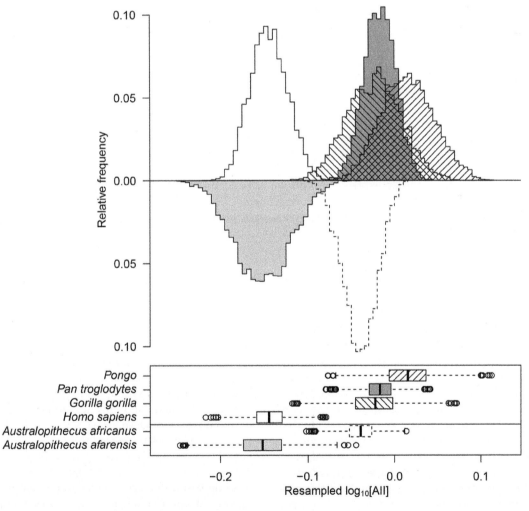

Figure 17.5 Histograms and stem and leaf plots for the distributions of resampled log AII for extant and extinct samples. Note that resampled values for *A. africanus* are most similar to those for *G. gorilla* and *P. troglodytes*, although the mean value for *A. africanus* is slightly lower than that for all extant great apes. Mean resampled log AII for *A. africanus* is significantly higher than that of *A. afarensis* and *H. sapiens* (Table 17.4).

Table 17.4 Comparison of mean resampled log-transformed AII for six-taxa comparison.[a]

	A. africanus	A. afarensis	Pongo	Pan troglodytes	Gorilla gorilla	Homo sapiens
A. africanus	–	**−0.112**	0.055	0.023	0.016	**−0.105**
A. afarensis	**0.002**	–	**0.167**	**0.135**	**0.129**	0.008
Pongo	0.123	**< 0.001**	–	−0.033	−0.039	**−0.160**
Pan troglodytes	0.394	**< 0.001**	0.366	–	−0.006	**−0.127**
Gorilla gorilla	0.638	**0.002**	0.367	0.850	–	**−0.121**
Homo sapiens	**< 0.001**	0.838	**< 0.001**	**< 0.001**	**< 0.001**	–

[a] Values above the diagonal are mean resampled values for column taxa minus means for row taxa. Values below the diagonal are two-sided p-values associated with those differences. Differences significant at alpha = 0.05 highlighted in bold.

17.4). One reviewer noted the unusually small radial head of A.L. 288-1 and wondered whether it had any effect on these results, but exclusion of RADTV from the analysis produces qualitatively the same results (i.e., all of the differences that are significant at alpha = 0.05 in Table 17.4 remain significant, and those that are not significantly different also remain so). There is also no significant difference among the great apes, although all great ape means differ significantly from humans (Table 17.4). If the analysis is rerun excluding A. afarensis to allow for a larger sample size during the resampling procedure, results are similar although resampled distributions are slightly narrower, and thus p values for comparison of means are slightly lower (Figure 17.6, Table 17.5). In this analysis, all relationships follow the same pattern of significance described above, although the difference in mean resampled log AII between A. africanus and Pongo approaches significance (p = 0.073).

These results reaffirm earlier work demonstrating relatively large forelimb articular dimensions with respect to hindlimb articular and diaphyseal dimensions in A. africanus when compared to H. sapiens and A. afarensis (McHenry, 1978; McHenry and Berger, 1998; Green et al., 2007), and replicates results of Green et al. (2007) with an expanded sample. Our intent here is not to rehash those studies in full but to provide a more nuanced interpretation of the significance of that particular finding.

In particular, this study illustrates two important points that bear on the interpretation of the resampling analysis: (1) the positive allometric scaling of GM_F with GM_H indicates a positive correlation between AII and size within species even though larger individuals tend to be less arboreal than smaller conspecifics, not more, suggesting that size differences between taxa should be taken into consideration when comparing AII values, and (2) AII may not be tracking differences in arboreality per se, but may instead reflect the relative constraints on fore- and hindlimb size due to loading and mobility requirements, regardless of substrate.

With regard to the first point, comparison of resampled GM_F and GM_H among taxa demonstrates substantial

size differences between taxa that may affect AII values. For example, there is considerable overlap in resampled GM_F values among A. africanus, humans, chimpanzees, and orangutans, while the majority of gorilla values are considerably higher and the majority of A. afarensis values are considerably lower (Figure 17.7A). For resampled GM_H, the ranges of both australopith species overlap substantially with chimpanzees and orangutans, while all values for all four of these taxa are less than all values for both gorillas and humans (Figure 17.7B). As a consequence, the positions of the distributions for human and gorillas AII values in Figures 17.5 and 17.6 relative to those of the other apes and australopiths are potentially misleading, as the AII values for humans and gorillas may be elevated relative to what they would be if those two species were smaller in size but unchanged in locomotor repertoire. If that were the case, it would affect the overall similarity of A. africanus and G. gorilla on the one hand and A. afarensis and H. sapiens on the other.

In relation to the second point, the question that arises is what, exactly, is indicated by similarity in AII between two taxa. For example, while the range of resampled AII values for humans is completely contained within the range for A. afarensis, multiple avenues of evidence indicate that A. afarensis was adapted for arboreal locomotion in addition to terrestrial bipedality. This is supported not only by many anatomical features such as the cranial orientation of the shoulder girdle and digital curvature (Stern and Susman, 1983; Susman et al., 1984; Duncan et al., 1994; Stern, 2000; however, see Ward, 2002) but by other limb proportions as well. A.L. 288-1 ("Lucy") preserves both a humerus and femur for which length can be measured, and the humerofemoral index falls between that of modern humans and extant African apes (Jungers, 1982, 1991; Jungers and Stern, 1983; Richmond et al., 2002). It has been argued that such a relatively high humerofemoral index in A.L. 288-1 may simply "result" from negative allometry of this index with body size in bipeds (Wolpoff, 1983; Franciscus and Holliday, 1992; Lovejoy, 1993), although it has been shown that this argument rests on the flawed underlying assumption that two

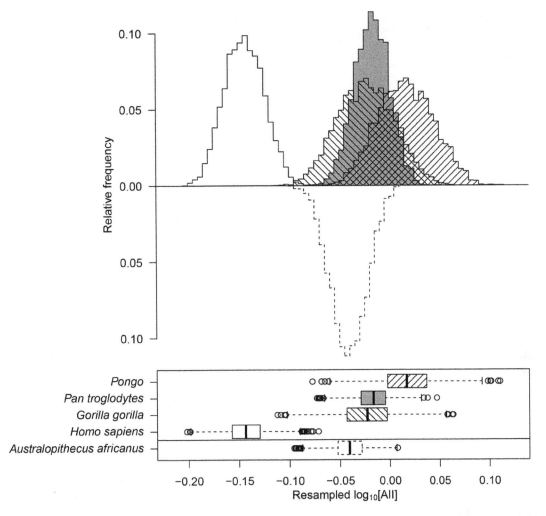

Figure 17.6 Distributions of resampled log AII for extant samples and *A. africanus* using the sampling regime based on *A. africanus* alone (see text for details). Note that resampled values are very similar to those shown in Figure 17.5. Mean resampled log AII for *A. africanus* falls within the range of all great ape ratios but is significantly higher than that of *H. sapiens* (Table 17.5).

Table 17.5 Comparison of mean resampled log-transformed AII for five-taxa comparison.[a]

	A. africanus	Pongo	Pan troglodytes	Gorilla gorilla	Homo sapiens
A. africanus	–	0.058	0.023	0.018	**−0.103**
Pongo	0.073	–	−0.034	−0.040	**−0.161**
Pan troglodytes	0.348	0.302	–	−0.006	**−0.127**
Gorilla gorilla	0.606	0.316	0.859	–	**−0.121**
Homo sapiens	**< 0.001**	**< 0.001**	**< 0.001**	**< 0.001**	–

[a] Conventions follow Table 17.4.

individuals who sit on the same allometric scaling line represent the same functional adaptation despite having markedly different limb proportions that have real biomechanical consequences (Jungers and Stern, 1983; Jungers, 1991). More to the point, Jungers (2009) recently showed that the humerofemoral index scales isometrically with respect to size, not negatively allometrically, and that A.L. 288-1 has a substantially higher value for

this index than any real or predicted value based on the smallest-bodied modern humans. He also showed that humerofemoral index has a high intraspecific correlation with intermembral index in humans (Jungers, 2009). So while one possible interpretation of the results of the resampling analysis of AII is that they are in conflict with previous work on the humerofemoral index, an alternative interpretation is that humerofemoral index and AII

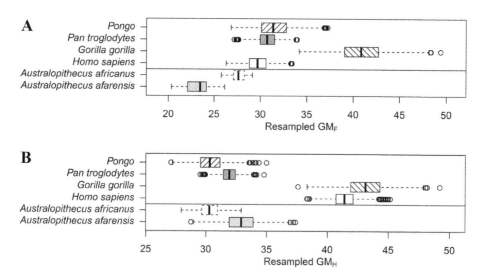

Figure 17.7 Comparisons of resampled GM$_F$ and GM$_H$ among extant and extinct samples. Note that resampled values for *A. africanus* are considerably smaller than those of *G. gorilla* for both the forelimb and the hindlimb, and resampled values for *A. afarensis* are likewise much smaller than those of *H. sapiens* in both limbs, with no overlap between taxa in either case. In both of these situations, deviation from isometric scaling becomes a factor to consider when comparing ratios (see text).

relate to different components of a species' positional repertoire.

The implication of this latter interpretation is that AII and other similar ratios of forelimb to hindlimb articular and diaphyseal measurements do not simply map onto the distinction between arboreal and non-arboreal taxa. Rather, it is more likely that differences in AII among hominids that incorporate both terrestrial and arboreal components into their locomotor repertoire (e.g., australopiths, chimpanzees, and gorillas) reflect the differential loads applied to the fore- and hindlimb as well as different articular size requirements imposed by range of motion considerations (e.g., Rafferty and Ruff, 1994) in both terrestrial and arboreal settings. In this context, AII may be less informative in terms of direct comparison between australopiths and the knuckle-walking African apes (with regular terrestrial loading of forelimbs in addition to hindlimbs) or the nearly exclusively suspensory orangutans. Instead, AII may be most useful in examining the diversity of locomotor behavior within the australopiths. While the functional morphology of *A. africanus* and *A. afarensis* indicates that both species are clearly committed bipeds when terrestrial and both also have substantial arboreal adaptations (Stern and Susman, 1983; Susman et al., 1984; Clarke and Tobias, 1995; Ward, 2002; Harcourt-Smith and Aiello, 2004; Ward et al., 2011; Green and Alemseged, 2012), our results reaffirm that from a locomotor perspective, *A. africanus* was not simply a younger southern version of *A. afarensis*. Instead, while acknowledging the overall similarities in the two australopith species, if we also examine them in the context of the parallel comparison of AII values in the African apes, *A. africanus* represents a species much more distinct from *A. afarensis* in its overall locomotor repertoire than the difference between chimpanzees and Western lowland gorillas.

Acknowledgments

We would like to thank The Wenner-Gren Foundation, Ford Foundation, and NSF-IGERT (DGE-0801634) for supporting this research, and the reviewers for their time and efforts. Special thanks to Carol Ward, Bernhard Zipfel, and Charles Lockwood for organizing the workshop for this monograph. We would also like to thank Eileen Westwig (AMNH), Jennifer Clark and Richard Potts (Human Origins Program at the Smithsonian Institution), Lyman Jellema and Yohannes Haile-Selassie (CMNH), Linda Gordon and David Hunt (NMNH), Stephany Potze (Ditsong National Museum of Natural History, Pretoria), and Bernhard Zipfel (University of the Witwatersrand) for access to extant specimens, fossil casts, and fossils under their care. Heather Dingwall kindly assisted ADG in measuring human skeletal material at the Smithsonian, and Lyman Jellema generously contributed length measurements for some CMNH specimens.

References

Clarke, R.J., 1988. A new *Australopithecus* cranium from Sterkfontein and its bearing on the ancestry of *Paranthropus*. In: Grine, F.E. (Ed.), Evolutionary

History of the "Robust" Australopithecines. New York: Aldine de Gruyter, pp. 285–292.

Clarke, R.J., 2013. *Australopithecus* from Sterkfontein Caves, South Africa. In: Reed, K.E., Fleagle, J.G., Leakey, R.E. (Eds.), The Paleobiology of *Australopithecus*: Springer: New York, pp. 105–123.

Clarke, R.J., Tobias, P.V., 1995. Sterkfontein Member 2 foot bones of the oldest South African hominid. Science 269, 521–524.

Crompton, R.H., Sellers, W.I., Thorpe, S.K.S., 2010. Arboreality, terrestriality and bipedalism. Philosophical Transactions of the Royal Society B 365, 3301–3314.

Currey, J.D., 2002. Bones: Structure and Mechanics. Princeton University Press: Princeton.

Dobson, S.D., 2005. Are the differences between StW 431 (*Australopithecus africanus*) and A.L. 288–1 (*A. afarensis*) significant? Journal of Human Evolution 49, 143–154.

Doran. D.M., 1996. The comparative positional behaviour of the African Apes. In: McGrew, W., Nishida, T. (Eds.), Great Ape Societies. Cambridge University Press: Cambridge, pp. 213–224.

Duncan, A.S., Kappelman, J., Shapiro, L.J., 1994. Metatarsophalangeal joint function and positional behavior in *Australopithecus afarensis*. American Journal of Physical Anthropology 93, 67–81.

Fleagle, J.G., 1985. Size and adaptation in primates. In: Jungers W.L. (Ed.), Size and Scaling in Primate biology. New York: Plenum Press, pp. 1–19.

Fleagle, J.G., 2013. Primate Adaptation and Evolution, 3rd ed. Elsevier: San Diego.

Franciscus, R.G., Holliday, T.W., 1992. Hindlimb skeletal allometry in Plio-Pleistocene hominids with special reference to AL-288-1 ("Lucy"). Bulletins et mémoires de la Société d'anthropologie de Paris 22, 1–16.

Godfrey, L.R., Sutherland, M.R., Paine, R.R., Williams, F.L., Boy, D.S., Vuillaume-Randriamanantena, M., 1995. Limb joint surface areas and their ratios in Malagasy lemurs and other mammals. American Journal of Physical Anthropology 97, 11–36.

Green, D.J., Alemseged, Z., 2012. *Australopithecus afarensis* scapular ontogeny, function, and the role of climbing in human evolution. Science 338, 514–517.

Green, D.J., Gordon, A.D., Richmond, B.G., 2007. Limb-size proportions in *Australopithecus afarensis* and *Australopithecus africanus*. Journal of Human Evolution 52, 187–200.

Grine, F.E., 2013. The alpha taxonomy of *Australopithecus africanus*. In: Reed, K.E., Fleagle, J.G., Leakey, R.E. (Eds.), The paleobiology of *Australopithecus*. Springer: New York, pp. 73–104.

Grine, F.E., 2019. The alpha taxonomy of *Australopithecus* at Sterkfontein: The postcranial evidence. Comptes Rendues Paleovol 18, 335–352.

Groves, C., 2001. Primate Taxonomy. Smithsonian Institution Press: Washington.

Haeusler, M., McHenry, H.M., 2004. Body proportions of *Homo habilis* reviewed. Journal of Human Evolution 46, 433–465.

Hamrick, M.W., 1996. Articular size and curvature as determinants of carpal joint mobility and stability in strepsirhine primates. Journal of Morphology 230, 113–127.

Harcourt-Smith, W.E.H., Aiello, L.C., 2004. Fossils, feet and the evolution of human bipedal locomotion. Journal of Anatomy 204, 403–416.

Holliday, T.W., 1999. Brachial and crural indices of European Late Upper Paleolithic and Mesolithic humans. Journal of Human Evolution 36, 549–566.

Holliday, T.W., Franciscus, R.G., 2009. Body size and its consequences: allometry and the lower limb length of Liang Bua 1 (*Homo floresiensis*). Journal of Human Evolution 57, 253–257.

Holliday, T.W., Franciscus, R.G., 2012. Humeral length allometry in African hominids (*sensu lato*) with special reference to A.L. 288–1 and Liang Bua 1. PaleoAnthropology 2012, 1–12.

Hunt, K.D., 1992. Positional behavior of *Pan troglodytes* in the Mahale Mountains and Gombe Stream National Parks, Tanzania. American Journal of Physical Anthropology 87, 83–105.

Jungers, W.L., 1982. Lucy's limbs: skeletal allometry and locomotion in *Australopithecus afarensis*. Nature 297, 676–678.

Jungers, W.L., 1988. Relative joint size and hominoid locomotor adaptations with implications for the evolution of hominid bipedalism. Journal of Human Evolution 17, 247–265.

Jungers, W.L., 1991. A pygmy perspective on body size and shape in *Australopithecus afarensis* (AL 288-1, "Lucy"). In: Senut, B., Coppens, Y. (Eds.), Origine(s) de la Bipedie chez les Hominides. Cahiers de Paleoanthropologie: Paris, pp. 215–224.

Jungers, W.L., 1994. Ape and hominid limb length. Nature 369, 194.

Jungers, W.L., 2009. Interlimb proportions in humans and fossil hominins: variability and scaling. In: Grine, F.E., Fleagle, J.G., Leakey, R.E. (Eds.), The First Humans: Origin and Early Evolution of the Genus *Homo*. Springer: New York, pp. 93–98.

Jungers, W.L., Stern, J.T., 1983. Body proportions, skeletal allometry and locomotion in Hadar hominids: a reply to Wolpoff. Journal of Human Evolution 12, 673–684.

Kimbel, W.H., Rak, Y. 1993. The importance of species taxa in paleoanthropology and anargument for the phylogenetic concept of the species category. In: Kimbel, W.H., Martin, L.B. (Eds.), Species, Species Concepts, and Primate Evolution. Plenum Press: New York, pp. 461–484.

Lockwood, C.A., 1997. Variation in the face of *Australopithecus africanus* and other African hominoids. Ph.D. Dissertation, University of the Witwatersrand.

Lockwood, C.A., Tobias, P.V., 2002. Morphology and affinities of new hominin cranial remains from Member 4 of the Sterkfontein Formation, Gauteng Province, South Africa. Journal of Human Evolution 42, 389–450.

Lovejoy, C.O., 1993. Modeling human origins: are we sexy because we're smart, or smart because we're sexy? In: Rasmussen, D.T. (Ed.), The Origin and Evolution of Humans and Humanness. Jones and Bartlett: Boston, pp. 1–28.

McHenry, H.M., 1978. Fore- and hindlimb proportions in Plio-Pleistocene hominids. American Journal of Physical Anthropology 49, 15–22.

McHenry, H.M., 1992. Body size and proportions in early hominids. American Journal of Physical Anthropology 87, 407–431.

McHenry, H.M., Berger, L.R., 1998. Body proportions in Australopithecus afarensis and A. africanus and the origin of the genus Homo. Journal Human Evolution 35, 1–22.

McHenry, H.M., Corruccini, R.S., 1978. The femur in early human evolution. American Journal of Physical Anthropology 49, 473–488.

Moggi-Cecchi, J., Tobias, P.V., Beynon, A.D., 1998. The mixed dentition and associated skull fragments of a juvenile fossil hominid from Sterkfontein, South Africa. American Journal of Physical Anthropology 106, 425–466.

Patel, B.A., 2005. The hominoid proximal radius: re-interpreting locomotor behaviors in early hominins. Journal of Human Evolution 48, 415–432.

R Core Team, 2013. R: A language and environment for statistical computing. R Foundation for Statistical Computing: Vienna.

Rafferty, K.L., Ruff, C.B., 1994. Articular structure and function in Hylobates, Colobus, and Papio. American Journal of Physical Anthropology 94, 395–408.

Reno, P.L., DeGusta, D., Serrat, M.A., Meindl, R.S., White, T.D., Eckhardt, R.B., Kuperavage, A.J., Galik, K., Lovejoy, C.O., 2005. Plio-Pleistocene hominid limb proportions: evolutionary reversals or estimation errors? Current Anthropology 46, 575–588.

Reno, P.L., McCollum, M.A., Cohn, M.J., Meindl, R.S., Hamrick, M., Lovejoy, C.O., 2008. Patterns of correlation and covariation of anthropoid distal forelimb segments correspond to Hoxd expression territories. Journal of Experimental Zoology B 310B, 240–258.

Richmond, B.G., Aiello, L.C., Wood, B.A., 2002. Early hominin limb proportions. Journal of Human Evolution 43, 529–548.

Rodman, P.S., 1979. Individual activity pattern and the solitary nature of orangutans. In: Hamburg, D.A., McCown, E.R. (Eds.), The Great Apes. Benjamin/Cumings: Menlo Park, pp. 234–255.

Rodman, P.S., 1984. Foraging and social systems of orangutans and chimpanzees. In: Rodman, P.S., Cant, J.G.H. (Eds.), Adaptations for Foraging in Non-Human Primates. New York: Columbia University Press, pp. 134–160.

Smith, R.J. 1999. Statistics of sexual size dimorphism. Journal of Human Evolution 36, 423–459.

Stern Jr., J.T., 2000. Climbing to the top: a personal memoir of Australopithecus afarensis. Evolutionary Anthropology 9, 113–133.

Stern Jr., J.T., Susman, R.L., 1983. The locomotor anatomy of Australopithecus afarensis. American Journal of Physical Anthropology 60, 279–317.

Susman, R.L., Stern, Jr., J.T., Jungers, W.L., 1984. Arboreality and bipedality in the Hadar hominids. Folia Primatologia 43, 113–156.

Takahashi, L.K., 1991. Forearm elongation in gibbons: hypothesis and preliminary results. International Journal of Primatology 12, 599–614.

Trinkaus, E., 1981. Neanderthal limb proportions and cold adaptation. In: Stringer, C.B. (Ed.), Aspects of Human Evolution. London: Taylor and Francis, pp. 187–224.

Ward, C.V., 2002. Interpreting the posture and locomotion of Australopithecus afarensis: where do we stand? Yearbook of Physical Anthropology 35, 185–215.

Ward, C.V., Kimbel, W.H., Harmon, E.H., Johanson, D.C., 2012. New postcranial fossils of Australopithecus afarensis from Hadar, Ethiopia (1990–2007). Journal of Human Evolution 63, 1–51.

Ward, C.V., Kimbel, W.H., Johanson, D.C., 2011. Complete fourth metatarsal and arches in the foot of Australopithecus afarensis. Science 331, 750–753.

Warton, D.I., Duursma, R.A., Falster, D.S., Taskinen, S. 2012. smatr 3—an R package for estimation and inference about allometric lines. Methods in Ecology and Evolution 3, 257–259.

Warton, D.I., Wright, I.J., Falster, D.S., Westoby, M., 2006. Bivariate line-fitting methods for allometry. Biology Review 81, 259–291.

Wolpoff, M.H., 1983. Lucy's lower limbs: long enough for Lucy to be fully bipedal? Nature 304, 59–61.

18

Summary and synthesis

Carol V. Ward and Bernhard Zipfel

The fossil record is the only evidence we have of the order and timing of the acquisition of the key characteristics that have shaped hominin evolution. Every fossil has a chapter of the story to tell, so it is imperative that all fossils be considered in analyses of the biology and evolution of hominin species. For this reason, we are pleased to present the complete collection of postcranial fossils recovered through 1995 and available for study as of 2009 from the historic site of Sterkfontein, Gauteng Province, South Africa.

These fossils have been recovered thanks to the painstaking work of many scientists and their teams over many years (reviewed in Thackeray, Chapter 1). A number of the fossils included in this volume have been discussed in the literature for many years. Not the least of these is the iconic partial skeleton Sts 14 which, combined with some isolated specimens from Makapansgat, along with the *Australopithecus robustus* sites of Swartkrans and Kromdraai, provided the foundation for analysis and interpretation of all australopith postcranial fossils recovered since the 1960s, including fossils attributed to *Australopithecus afarensis*, *A. anamensis*, *A. sediba*, and the "robust" australopith species often attributed to *Paranthropus*. More recent discoveries of the partial skeleton StW 431 and a wealth of other isolated bones made available for study through the time of the Sterkfontein Hominin Postcranial Workshop have continued to enrich our picture of the Sterkfontein hominins, but many of these specimens have received little or no attention in the literature.

To our knowledge, this volume represents the first source presenting all these fossils together in one collection. The chapters in this book provide photographs, measurements, and basic descriptions of each postcranial fossil. They present comparative observations and

analyses and statistical analysis of taxonomically and functionally relevant aspects of morphology, and in some cases they provide reconstructions of the bones.

In addition to the materials presented here, three-dimensional polygonal models of most of the bones can be accessed at www.morphosource.org. These freely available models were created from surface scan files, collected primarily by J.M. Plavcan with contributions from other scholars. The models supplement the photographs, descriptions, and metrics presented in this book. We encourage any researchers to download, print, and study them. There is no substitute, though, for the detail and quality of studying the original fossils. We urge those who avail themselves of these models, and any other interested scholars, to examine the original fossil material. The fossils are housed at the Evolutionary Studies Institute, University of the Witswatersrand, and Ditsong Museum of Natural History, and they are all available for study following the application procedures of each institution.

Each of the chapters in this volume addresses key questions surrounding the Sterkfontein hominins. Is more than one hominin species represented at Sterkfontein, and if so how might their biology or behavior have differed? How are the Sterkfontein hominins similar to or different from hominins at other sites, particularly other australopiths? What do the postcranial fossils from Sterkfontein reveal about the posture and locomotion of *Australopithecus africanus* (or other species)?

Answers to these questions depend in part on understanding the geologic context of the fossils and stratigraphic interpretation of the Sterkfontein deposits (Stratford, Chapter 2) and geochronology of these deposits (Pickering and Herries, Chapter 3). The various members present at Sterkfontein are complicated, and preserve different temporal records of the hominins

Carol V. Ward and Bernhard Zipfel, *Summary and synthesis*. In: *Hominin Postcranial Remains from Sterkfontein, South Africa, 1936–1995*. Edited by: Bernhard Zipfel, Brian G. Richmond, and Carol V. Ward, Oxford University Press (2020). © Oxford University Press.
DOI: 10.1093/oso/9780197507667.003.0018

and their environments. These chapters present the most recent understanding of the site, and set the stage for the interpretation of the fossils within.

The most prominent theme running through the chapters in this volume is that of variation, which underlies all interpretations of taxonomy and functional interpretation.

The Member 5 deposits, which are more recent, have yielded fossils attributable to *Homo* in the craniodental sample, as it does for many postcrania. The postcranial elements that can be clearly attributed to Member 5 also tend to resemble those of *Homo* specimens at other sites. Within Member 4, however, although there is substantial morphological variation in the craniodental material, this variation has been difficult to sort into two or more distinct clusters that would clearly distinguish more than one species. Confounding the interpretation of the significance of morphological variation within the sample is, of course, possible effects of time averaging, given that even Member 4 appears to span a time frame, from the bottom of the deposit at 2.6 Ma and to 1.8 Ma (Pickering and Herries, Chapter 3). Even without considering possible time-averaging, statistical analyses of dental remains have failed to falsify the hypothesis that more than one species is represented.

Each chapter in this volume comes to a similar conclusion, that although variation in size or morphology may exceed that found in other australopith species, or even among mixed samples of other fossil hominins, the hypothesis that only one species is represented by the Sterkfontein Member 4 hominins cannot be falsified. Given the fact that at least most of the craniodental fossils at Sterkfontein are attributed to *Australopithecus africanus*, the Member 4 postcrania remain attributable to this taxon as well despite clear association between cranial and postcranial remains.

Another theme running through the chapters in this volume is whether the Sterkfontein hominins differed in functionally relevant ways from other hominins. The *Australopithecus africanus* postcrania resemble those of *A. afarensis* and *A. sediba*, the two other most well-represented australopith species, in many aspects of their size and morphology. All of these taxa share features of the vertebral column, thorax, hip, knee, ankle, and foot indicating fully upright, human-like posture. Their upper limb morphology, is intermediate between that of humans and apes in many ways, with a glenohumeral joint not as laterally facing in humans compared with African apes, conformation of the elbow and wrist, a slightly thicker cortex along the superior surface of the femoral neck, and possibly slightly more mobility along the medial column of the foot. Femoral proximal morphology may hint at a kinematically different and less efficient gait than characterized *Homo*. *Australopithecus africanus*, as with these

other australopith species, had relatively long and strong upper limbs relative to lower ones, especially in terms of joint proportions or overall limb size (Gordon et al., Chapter 17), as well as in measure of bone strength as measured by long bone cross-sectional geometry (Ruff et al., Chapter 16). These morphologies are derived relative to almost any interpretation of the primitive morphology from which hominins originated, although are not as derived as seen in *Homo*. The significance of these primitive retentions continues to be debated. Although the adaptive significance of primitive retentions in light of the strong directional signal toward committed terrestrial bipedality and away from highly effective arboreality, they are most commonly interpreted as representing retention of climbing behaviors in addition to terrestrial bipedal locomotion. Some of the insights and considerations presented by Gordon et al. (Chapter 17) may provide a fruitful approach for sorting out the significance of shape variations and proportions in discriminating between major differences in mode of locomotion versus variations in emphasis between taxa with similar overall repertoires.

Among many anatomical regions, though, the *A. africanus* fossils appear distinct from those attributed to *A. afarensis*, yet the variation between these samples often does not exceed that observed within single species of extant hominoids. Statistical analysis is confounded by the fact that most bones are represented by only a small handful of fossils, if that, making it difficult to rule out sampling bias in any such interpretations. Still, there is often a directionality in these differences. The glenoid fossa may have been oriented slightly further cranially in *A. africanus* than in *A. afarensis*, a feature characterizing apes and often linked to overhead arm movements during climbing or suspension. The radial head appears slightly more beveled, and forearm flexor muscle attachments often more prominent. There may be more mobility of the hallucal metatarsal joint and more mobility through the medial portion of the foot. Despite a potential signal of more arboreality in *A. africanus*, the less keeled and anteriorly oriented humeroulnar joint, rounder ulnar diaphysis, shorter manual phalanges, Kivell et al. (Chapter 9) suggest the possibility of a greater emphasis on manual manipulatory behaviors in *A. africanus* than in *A. afarensis*, which may or may not be related to climbing. Although these differences may or may not be borne out with further study and the recovery of more fossils, the chapters in this volume open up intriguing hypotheses for further investigation.

These ideas are but a sample of the results presented and discussed throughout the book. Conclusions presented in the chapters within this volume occasionally differ from previously published interpretations. The studies presented here are the most thorough and extensive published to date, and so can be considered to be the most reasonable hypotheses at the present time.

The research presented in this volume, combined with previous work on some of the Sterkfontein fossils, is only a start. The scholars who have contributed to this work have made significant advances in our understanding of the paleobiology of *Australopithecus africanus*, but, as with all good research, these chapters raise as many questions as they answer. It is our sincere hope that making the descriptions, models, and basic comparative observations on this amazing collection of fossils accessible and available in a single volume will inspire renewed interest, new perspectives, and approaches by the authors of the chapters in this volume and even more importantly by other paleobiologists in the future. These fossils are particularly significant in light of more recent discovery and publication of the nearly complete skeleton of StW 573 from the Silderberg Grotto of Member 2 at Sterkfontein, known until very recently from only a handful of pedal elements. The postcranial fossils presented in this volume provide the framework for understanding this stunning skeleton, and this skeleton, in turn, will be pivotal for addressing the many questions raised by the fossils considered herein. Together, these fossils will serve as the comparative basis for new fossils that will be recovered at Sterkfontein in the future.

Our understanding of hominin evolution continues to be enriched by the wealth of new fossils being recovered across Africa, including at Sterkfontein, but also from careful consideration and reconsideration of the collection of postcranial fossils already recovered. The australopith species that "started it all" back in 1925 still has more to teach us, and we all look forward to learning what the future holds for *Australopithecus africanus*.

Section 4

Appendices

Appendix I

Table of post-cranial remains

Table 1 Table of specimens, locus (grid and depth, identification, and date of discovery).
List of the historic Sterkfontein post-cranial fossil hominin collection recovered from 1936–1995. A few relevant, more recently discovered specimens are also included.

Accession No.[a]	Member	Locus[b]	Element[c]	Discovery Date
Sts 7a	Member 4		R partial Scapula	June 24, 1947
Sts 7b	Member 4		R proximal Humerus	June 24, 1947
Sts 34	Member 5		R distal Femur	June 29, 1948
Sts 65	Unknown		Partial Hipbone	1949
Sts 68	Unknown		L proximal Radius	Unknown
Sts 73	Member 4		Vertebra	1936
TM 1513	Member 4		L distal Femur	August 1, 1947
TM 1526	Unknown		R Capitate	Late 1930's
Sts 14a	Member 4		Vertebra	August 1, 1947
Sts 14b	Member 4		Vertebra	August 1, 1947
Sts 14c	Member 4		Vertebra	August 1, 1947
Sts 14d	Member 4		Vertebra	August 1, 1947
Sts14e	Member 4		Vertebra	August 1, 1947
Sts14f	Member 4		Vertebra	August 1, 1947
Sts 14g	Member 4		Vertebra	August 1, 1947
Sts 14w	Member 4		Vertebra	August 1, 1947
Sts 14j	Member 4		Vertebra	August 1, 1947
Sts 14k-n	Member 4		Vertebra	August 1, 1947
Sts 14o-p	Member 4		Bodies of Vertebrae	August 1, 1947
Sts 14q	Member 4		First two sacral Vertebrae	August 1, 1947
Sts 14r	Member 4		L Hipbone	August 1, 1947
Sts 14s	Member 4		R Hipbone	August 1, 1947
Sts 14t	Member 4		R proximal Femur	August 1, 1947
Sts 14u	Member 4		Long bone fragment	August 1, 1947
Sts 14w	Member 4		3 partial Ribs	August 1, 1947
Sts 14x	Member 4		Rib fragment	
Sts 14y	Member 4		Rib fragment	August 1, 1947
Sts 14z	Member 4		Rib fragment	August 1, 1947
Sts 14aa	Member 4		Rib fragment	August 1, 1947
Sts 14ab	Member 4		Rib fragment	August 1, 1947
Sts 2198b	Member 4		R (?) proximal Radius	
StW 8a	Member 4	From red brown breccia	Lumbar Vertebra	November 25, 1969
StW 8b	Member 4	From red brown breccia	Lumbar Vertebra	November 25, 1969
StW 8c	Member 4	From red brown breccia	Lumbar Vertebra	November 25, 1969
StW 8d	Member 4	From red brown breccia	Lumbar Vertebra	November 25, 1969

Accession No.[a]	Member	Locus[b]	Element[c]	Discovery Date
StW 25	Member 4	From D/18	R proximal Femur	January 7, 1974
StW 26	Member 4	From D/18	Distal fourth Metacarpal	January 5, 1974
StW 27	Member 5	From D/3	R (?) third Metacarpal	April 11, 1973
StW 28	Member 4	From D/18	Proximal manual Phalanx	May 28, 1974
StW 29	Member 4	From D/18	Proximal manual Phalanx	November 9, 1973
StW 30a	Member 4	From D/18	Proximal Femur (felid?)	July 8, 1974
StW 30b	Member 4	From D/18	Proximal Femur (felid?)	July 8, 1974
StW 30bb	Member 4	From D/18	Proximal Femur (felid?)	February 22, 1978
StW 31	Member 4	From D/18	Proximal Femur (felid?)	August 21, 1974
StW 38	Member 4	From D/18	R Humerus shaft fragment	August 15, 1974
StW 41a	Member 4	From D/18	Thoracic Vertebra	January 10, 1975
StW 41b	Member 4	From D/18	Thoracic Vertebra	January 10, 1975
StW 46	Member 4	From D/18	L distal Radius	March 12, 1976
StW 63	Member 4	From D/18	L fifth Metacarpal	November 17, 1976
StW 64	Member 4	From D/18	L third Metacarpal	February 22, 1977
StW 65	Member 4	From D/18	R fourth Metacarpal	March 1, 1977
StW 68	Member 4	From D/18	R third Metacarpal	April 12, 1977
StW 88	Member 4	From R/59 12'3"-13'3" at breccia	R Talus	September 19, 1980
StW 89	Member 4	From S/59 12'0"-13'4" at breccia	L second Metatarsal	September 26, 1980
StW 99a/100	Member 4	From W/45 6'10"-7'10"	R proximal Femur	July 1, 1982
StW 99b	Member 4	From X/44 5'9"-6'9"	R proximal Femur	July 5, 1982
StW 99c	Member 4	From W/46 6'11"-7'11"	R proximal Femur	July 6, 1982
StW 102	Member 4	From X/47 3'11"-4'11"	R Talus	July 6, 1982
StW 105	Member 4	From S/48 8'8"-9'8"	R proximal juvenile Radius	July 16,1982
StW 108	Member 4	From T/47 8'4"- 9'4"	L proximal Ulna	Unknown
StW 113	Member 4	From U/44 7'10"-8'10"	Left proximal Ulna	August 9, 1982
StW 114/115	Member 4	From W/44 8'9"-9'10"	R fifth Metatarsal	August 10, 1982
StW 121	Member 4	From W/46 9'11"-10'11"	R Femoral shaft	August 12, 1982
StW 122	Member 4/5	From R/49 14'1"-15'1"	Proximal Phalanx	August 26, 1982
StW 124	Member 4	From T/49 15'8"-16'8"	L distal Humerus shaft	September 15, 1982
StW 125	Member 4	From T/49 16'8"-17'8"	R Radius shaft	September 15, 1982
StW 129	Member 4/5	From T/50 18'10"-19'10"	L distal Femur	September 20, 1982
StW 139	Member 4	From O/46 10'2"-11'2"	Radius head & partial neck	January 27, 1983
StW 150	Member 4	From O/45 13'3"-14'3"	L distal Humerus shaft	March 26, 1983
StW 162	Member 4/5	From Aa/49 5'2"-6'2"	L partial Scapula	May 16, 1983
StW 181	Member 4	From R/45 10'8"-11'8"	L distal Tibia	January 23, 1984
StW 182	Member 4	From R/46 13'11"-14'11"	R distal Humerus shaft	January 25, 1984
StW 238	Member 4/5	From T/46 21'5"-22'7"	R proximal third Metatarsal	May 2, 1984
StW 292	Member 4/5	From T/48 25'9"-26'9"	Distal fourth Metacarpal	August 1, 1984
StW 293	Member 4/5	From U/48 24'9"-25'9"	Proximal manual Phalanx	August 1, 1984
StW 294	Member 4/5	From U/48 24'9"-25'9"	Pollical distal Phalanx	August 1, 1984
StW 311	Member 5	From R/53 14'5"-15'5"	R proximal Femur	August 27, 1980
StW 318	Member 4/5	From U/49 24'0"-25'0"	R distal Femur	November 6, 1984
StW 326a	Member 4	From S/45 17'9"-18'9"	R distal Ulna	May 9, 1984
StW 326b	Member 4	From S/45 26'6"-27'6"	R distal Ulna	July 16, 1984
StW 326c	Member 4	From S/45 23'6"-24'6"	R distal Ulna	July 11, 1984
StW 328	Member 4	Unknown	R proximal Humerus	Unknown
StW 330	Member 4	From U/42 9'0"-10'4"	L fourth Metacarpal	April 30, 1985
StW 331	Member 4	From T/40 8'2"-9'2"	Manual intermed. Phalanx	May 2, 1985
StW 339	Member 4	From V/43 6'7"-7'7"	R distal Humerus shaft	June 18, 1985
StW 340	Member 4	From U/44 7'10"-8'10"	Distal Ulna	August 9, 1982
StW 347	Member 4	From R/43 11'2"-12'3"	L Talus	July 9, 1985
StW 348	Member 4	From W/42 8'7"-9'7"	R Radius shaft	July 17, 1985
StW 349	Member 4	From W/42 8'7"-9'7"	R Ulna shaft	July 17, 1985
StW 352	Member 4	From V/41 3'10"-4'10"	R Calcaneus	August 23, 1985
StW 354	Member 4/5	From V/57 12'5"-13'3" and S/57 14'10"-15'10"	L Radius shaft fragment	Unknown

Accession No.[a]	Member	Locus[b]	Element[c]	Discovery Date
StW 355	Member 4/5	From T/59 10'7"-11'7"	R proximal Phalanx	January 16, 1986
StW 356	Member 4	From P/44 14'5"-15'1"	L Fibula shaft	March 7, 1986
StW 358	Member 4	From P/43 13'6"-14'6"	L distal Tibia	March 7, 1986
StW 361	Member 4	From S/43 14'10"-15'2"	Proximal Femur	March 7, 1986
StW 363	Member 4	From Q/43 13'4"-14'4"	L Talus	March 14, 1986
StW 366	Member 4	From R/44 15'10"-16'11"	L Scapula fragment	April 11, 1986
StW 367	Member 4	From R/43 16'0"-17'0"	Distal Femur	April 11, 1986
StW 377	Member 4	From N/44 14'8"-15'8"	L proximal Metatarsal	May 9, 1986
StW 380	Member 4/5	From O/46 16'3"-17'3"	R proximal Ulna	July 4, 1986
StW 382	Member 4/5	From P/47 14'4"-15'4"	L second Metacarpal	May4, 1986
StW 387	Member 4/5	From N/46 18'0"-19'9"	L proximal third Metatarsal	July 21, 1986
StW 486	Member 5	From S/40 9'11"-10'10"	R Talus	April 29, 1988
StW 388	Member 4	From T/43 12'10"-13'10"	R proximal third Metatarsal	July 21, 1986
StW 389	Member 4/5	From O/47 18'10"-19'9"	L distal Tibia	July 21, 1986
StW 390	Member 4/5	From O/46 18'3"-19'8"	R proximal Ulna	July 21, 1986
StW 392	Member 4	From N/45 16'8"-17'8"	R proximal Femur	July 25, 1986
StW 394	Member 4	From O/45 18'4"-19'7"	L distal third Metacarpal	July 28, 1986
StW 396	Member 4	From O/45 18'4"-19'7"	R proximal Tibia	August 4, 1986
StW 398a	Member 4	From P/44 18'11"-19'10"	L proximal Ulna	August 18, 1986
StW 398b	Member 4	From P/44 18'11"-19'10"	L distal Ulna	August 18, 1986
StW 400	Member 4	From Q/44 16'8"-17'8"	Proximal manual Phalanx	August 18, 1986
StW 403	Member 4	From P/44 19'10"-20'10"	R proximal Femur	September 7, 1986
StW 418	Member 4	From Q/44 25'8"-26'8"	L first Metacarpal	March 23, 1987
StW 431a	Member 4	From R/45 22'6"-23'6"	R proximal Radius	February 9, 1987
StW 431ba	Member 4	From R/45 22'6"-23'6"	R proximal Ulna	February 9, 1987
StW 431bb	Member 4	From Q/45 22'11"-23'11"	R proximal Ulna	February 23, 1987
StW 431bc	Member 4	From P/44 13'5"-14'5"	R proximal Ulna	March 7, 1987
StW 431ca	Member 4	From R/45 22'6"-23'6"	R distal Humerus	February 9, 1987
StW 431cb	Member 4	From R/45 23'6"-24'6"	R distal Humerus	February 9, 1987
StW 431da	Member 4	From R/45 22'6"-23'6"	R partial Scapula	February 9, 1987
StW 431db	Member 4	From R/45 24'6"-25'6"	R partial Scapula	February 9, 1987
StW 431dc	Member 4	From Q/45 21'11"-22'11"	R partial Scapula	February 9, 1987
StW 431ea	Member 4	From R/45 25'6"-26'6"	R Ilium fragment	February 9, 1987
StW 431eb	Member 4	From Q/45 21'11"-22'11"	R Ilium fragment	February 23, 1987
StW 431ec	Member 4	From Q/45 22'11"-23'11"	R Ilium fragment	February 23, 1987
StW 431ed	Member 4	From Q/45 22'11"-23'11"	R Ilium fragment	February 23, 1987
StW 431ee	Member 4	From Q/45 23'11"-24'11"	R Ilium fragment	February 25, 1987
StW 431ef	Member 4	From Q/45 22'11"-23'11"	R Ilium fragment	February 23, 1987
StW 431f	Member 4	From Q/46 28'6"-29'6"	R Acetabulum fragment	February 23, 1987
StW 431g	Member 4	From Q/45 21'11"-22'11"	R partial Clavicle	February 23, 1987
StW 431ha	Member 4	From Q/45 22'11"-23'11"	Sacrum	February 23, 1987
StW 431hb	Member 4	From Q/45 24'11"-25'11"	Sacrum	February 25, 1987
StW 431i	Member 4	From Q/46 20'10"-21'10"	L superior Pubic ramus	September 12, 1986
StW 431j	Member 4	From Q/45 24'11"-25'11"	L Ilium fragment	February 23, 1987
StW 431k	Member 4	From Q/45 24'11"-25'11"	L Acetabulum fragment	February 25, 1987
StW 431l	Member 4	From R/45 24'6"-25'6"	Vertebra	February 9, 1987
StW 431ma	Member 4	From R/45 25'6"-26'6"	Vertebra	February 9, 1987
StW 431mb	Member 4	From Q/45 23'11"-24'11"	Vertebra	February 25, 1987
StW 431na	Member 4	From Q/45 25'6"-26'6"	Vertebra	February 9, 1987
StW 431nb	Member 4	From Q/45 24'11"-25'11"	Vertebra	February 25, 1987
StW 431o	Member 4	From Q/45 21'11"-22'11"	Vertebra	February 23, 1987
StW 431p	Member 4	From Q/45 24'11"-25'11"	Proximal middle Rib	February 27, 1987
StW 431qa	Member 4	From R/45 21'6"-22'6"	Vertebra	February 9, 1987
StW 431qb	Member 4	From Q/45 24'11"-25'11"	Vertebra	February 25, 1987
StW 431r	Member 4	From R/45 21'6"-22'6"	Vertebra	February 9, 1987
StW 431s	Member 4	From R/45 21'6"-22'6"	Vertebra	February 9, 1987
StW 431t	Member 4	From R/45 21'6"-22'6"	Vertebra	February 9, 1987

Accession No.[a]	Member	Locus[b]	Element[c]	Discovery Date
StW 431ua	Member 4	From R/45 21'6"-22'6"	Vertebra	February 9, 1987
StW 431ub	Member 4	From R/45 21'6"-22'6"	Vertebra	February 9, 1987
StW 431v	Member 4	From R/45 21'6"-22'6"	Vertebra	February 9, 1987
StW 431v1	Member 4	From Q/45 21'11"-22'11"	Vertebra	February 23, 1987
StW 431w	Member 4	From R/43 20'11"-21'10"	L Auricular fragment	February 26, 1987
StW 435	Member 4	From R/45 24'6"-25'6"	R proximal third Metatarsal	February 9, 1987
StW 448	Member 4	From S/44 30'8"-31'6"	R femoral shaft	April 24, 1987
StW 470	Member 4	From V/43 21'2"-22'2"	R proximal hallucal Phalanx	August 14, 1987
StW 477	Member 4	From P/42 22'0"-23'0"	L proximal Metatarsal	August 31, 1987
StW 478	Member 4	From P/43 23'2"-24'2"	L pollical proximal Phalanx	August 31, 1987
StW 479	Member 4	From V/43 23'2"-24'2"	R proximal Femur	September 7, 1987
StW 485	Member 4	From S/40 8'11"-9'11"	L fourth Metatarsal	April 29, 1988
StW 486	Member 5	From S/40 9'11"-10'10"	R Talus	April 29, 1988
StW 496	Member 4	From N/43 21'9"-22'9"	L proximal third Metatarsal	September 19, 1988
StW 501	Member 4	From P/42 25'0"-26'0"	L proximal Femur	October 11, 1988
StW 514	Member 4	From J/42 20'4"-21'5"	R proximal Tibia	August 18, 1989
StW 515	Member 4	From J/42 20'4"-21'5"	R distal Tibia	August 18, 1989
StW 516	Member 4	From K/40 20'0"-21'4"	L proximal Radius	August 18, 1989
StW 517	Member 4	From K/40 20'0"-21'4"	L proximal Humerus	August 18, 1989
StW 522	Member 4	From P/45 26'1"-27'1"	L proximal Femur	November 10, 1989
StW 527	Member 4	From P/44 23'7"-24'7"	L (?) proximal Femur	November 10, 1989
StW 528	Member 4	From O/46 26'10"-27'10"	R Radius shaft	January 15, 1990
StW 552	Member 4	From T/44 32'5"-33'5"	Distal fourth Metacarpal	June 18, 1990
StW 556	Member 4	From Q/45 30'6"-31'6"	Phalanx	July 20, 1990
StW 562	Unknown	From R/49 19'11"-20'11"	R first Metatarsal	March 26, 1992
StW 567	Unknown	From P/58 19'10"-20'10"	R distal Tibia	February 23, 1993
StW 568	Member 5 E	From Q/56 27'2"-28'2"	R proximal Ulna shaft	May 6, 1993
StW 571	StW 53 Infill	From Bb/52 6'1"-7'1"	R proximal Ulna	September 29, 1981
StW 572	Member 5 E	From Q/59 14'10"-15'10"	Lumbar Vertebra	October 16, 1980
StW 573a	Member 2	From D/20 Silverberg Grotto	L Talus	April 29, 1988
StW 573b	Member 2	From D/20 Silverberg Grotto	L Navicular	April 29, 1988
StW 573c	Member 2	From D/20 Silverberg Grotto	L Cuneiform	April 29, 1988
StW 573d	Member 2	From D/20 Silverberg Grotto	L proximal first Metatarsal	April 29, 1988
StW 573e	Member 2	From D/20 Silverberg Grotto	L intermediate Cuneiform	April 29, 1988
StW 573f	Member 2	From D/20 Silverberg Grotto	L second Metatarsal	April 29, 1988
StW 573g	Member 2	From D/20 Silverberg Grotto	L lateral Cuneiform	April 29, 1988
StW 573h	Member 2	From D/20 Silverberg Grotto	L Calcaneus	April 29, 1988
StW 575	Unknown	From U/54 6'3"-7'3"	R pollical proximal Phalanx	April 10, 1988
StW 582	Member 4	From D/8	R partial Clavicle	July 15, 1976
StW 583	Member 4	From D/8	L distal Metacarpal	March 7, 1977
StW 595	Member 4	Collapsed Member	R first Metatarsal	March 10, 1998
StW 617	Member 4	From V/46 29'8"-30'8"	R hallucal distal Phalanx	April 2, 2003

[a] Fossils with prefixes Sts and TM are housed at the Ditsong Museum of Natural History. Fossils with prefix StW are housed at the Evolutionary Studies Institute, University of the Witwatersrand.

[b] Letter/number codes refer to excavation grids. Grid location of fossils where known are listed in feet and inches.

[c] L = left, R = right

Appendix II

Cross sectional properties of long bones

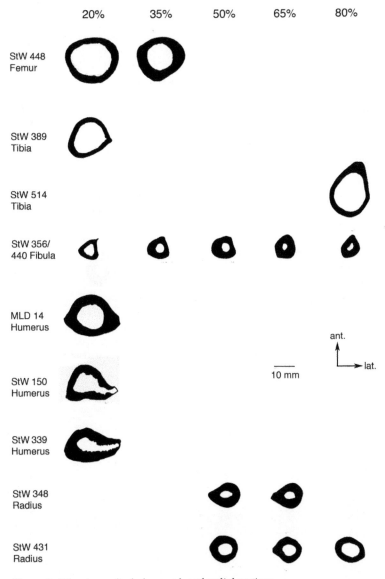

Figure 1 Other lower limb, humeral, and radial sections.

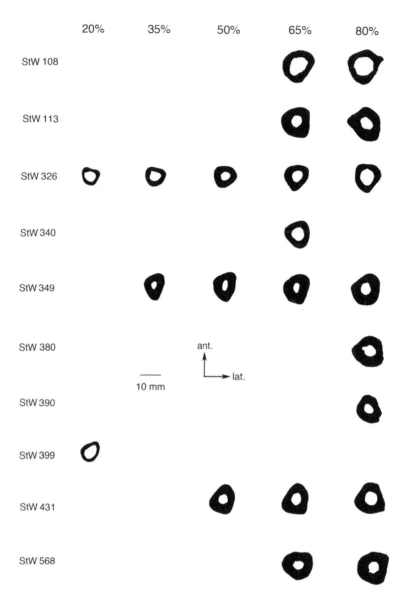

Figure 2 Other ulna sections.

Table 1 Cross-sectional properties of all long bone cross sections.

	Bone	Section	TA	CA	%CA	Ix	Iy	Imax	Imin	J	Zx	Zy	Zp
StW 99	Femur	50%	484	356	73.6	15121	20339	20724	14736	35460	1168	1439	2518
		65%	525	450	85.6	19049	25550	26108	18492	44600	1375	1720	2922
		80%	634	552	86.9	27805	39253	44671	22386	67057	2023	2372	3743
		base-nk	458	249	54.4	22785	8003	24216	6573	30789	1358	801	1733
		mid-nk	404	168	41.5	10838	6797	11125	6510	17634	813	666	1329
StW 121	Femur	20%	459	226	49.3	10220	15076	15350	9947	25297	929	1094	1819
		35%	384	266	69.2	10278	10595	10597	10276	20873	917	944	1676
		50%	363	292	80.5	10982	9346	11052	9276	20328	892	852	1653
		65%	376	314	83.4	11336	10534	11364	10507	21870	960	950	1842
StW 311	Femur	mid-nk	412	155	37.6	10939	6121	11149	5911	17060	757	578	1177
StW 403	Femur	mid-nk	325	151	46.5	8608	4159	8738	4029	12767	702	448	1028
StW 448	Femur	20%*	441	205	46.5	9643	12664	12732	9574	22307	879	945	1641
		35%*	360	240	66.7	8059	10525	10865	7719	18584	751	888	1565
StW 479	Femur	base-nk	337	130	38.5	7281	3836	7362	3754	11116	535	429	797
		mid-nk	326	101	31.0	5408	3447	5418	3436	8855	410	376	673
StW 501	Femur	mid-nk	297	95	32.2	4364	3497	4973	2889	7861	408	370	677
StW 522	Femur	base-nk	274	101	36.9	5072	2079	5072	2079	7151	358	246	506
		mid-nk	269	78	28.9	3638	2149	3662	2124	5787	326	231	516
StW 389	Tibia	20%*	285	126	44.1	4253	4891	5434	3711	9145	430	417	781
StW 514	Tibia	80%*	373	163	43.7	10616	5697	11222	5092	16313	791	550	1200
StW 356/440	Fibula	20%*	68	48	71.1	368	356	398	327	725	68	66	107
		35%*	83	70	83.5	537	559	559	537	1096	95	103	179
		50%*	95	83	87.1	667	761	765	663	1427	118	129	224
		65%*	89	83	93.8	704	562	717	549	1266	116	108	206
		80%*	70	63	90.9	466	346	513	299	812	78	74	134
MLD 14	Humerus	20%*	373	260	69.6	7742	13513	13589	7665	21254	743	998	1529
StW 150	Humerus	20%*	287	185	64.5	4716	7213	7232	4697	11929	492	498	810
StW 339	Humerus	20%*	309	238	77.0	4352	11457	11465	4343	15808	493	759	1035
StW 431	Humerus	20%	309	207	66.9	5336	8923	9001	5258	14260	552	647	997
		35%	309	213	69.1	7026	6772	7168	6630	13798	655	680	1286
		40%	318	217	68.1	7546	6887	7857	6577	14433	708	695	1327
StW 348	Radius	50%	126	111	88.8	963	1668	1683	947	2630	161	199	314
		65%*	126	112	88.8	1007	1668	1700	975	2675	164	193	307
StW 354	Radius	50%	103	95	91.6	759	999	1017	742	1759	131	130	229
StW 431	Radius	50%	132	103	77.9	1282	1343	1346	1278	2625	190	198	380
		65%*	121	94	77.8	992	1265	1276	980	2256	158	175	312
		80%*	135	90	67.0	1087	1557	1644	1000	2645	176	208	331

(continued)

Table 1 Continued

	Bone	Section	TA	CA	%CA	Ix	Iy	Imax	Imin	J	Zx	Zy	Zp
StW 108	Ulna	65%*	166	115	69.4	2106	1970	2358	1718	4076	262	253	496
		80%*	180	119	65.8	2478	2261	2671	2068	4739	288	253	501
StW 113	Ulna	65%*	154	133	86.9	1934	1812	1935	1811	3747	274	250	478
		80%*	170	147	86.0	2435	2263	2724	1974	4698	303	262	535
StW 326	Ulna	20%*	54	35	63.8	200	218	226	192	418	46	48	87
		35%*	69	51	73.6	330	385	386	329	715	70	74	134
		50%	84	71	84.5	529	583	593	519	1113	100	101	195
		65%*	102	82	80.7	840	782	909	714	1622	127	132	243
		80%*	122	85	69.9	1190	1009	1213	986	2200	162	154	304
StW 340	Ulna	65%*	112	85	75.7	1018	905	1027	896	1923	156	143	287
StW 349	Ulna	35%*	89	82	92.2	857	499	876	480	1356	129	97	204
		50%	109	100	91.4	1225	767	1262	731	1992	169	138	274
		65%*	130	121	92.5	1568	1206	1639	1135	2774	217	189	374
		80%*	152	132	86.8	1999	1676	2131	1543	3674	260	230	461
StW 380	Ulna	80%*	159	136	85.3	1968	2031	2085	1914	3999	263	254	496
StW 390	Ulna	80%*	112	98	87.4	1176	866	1228	813	2042	171	145	293
StW 399	Ulna	20%*	64	38	59.6	283	275	344	214	558	58	59	101
StW 431	Ulna	50%	119	104	87.8	1343	975	1424	893	2317	184	153	312
		65%*	135	116	85.8	1720	1252	1792	1181	2972	217	187	373
		80%*	153	124	81.5	1962	1722	1970	1714	3684	250	235	442
StW 568	Ulna	65%*	147	132	89.8	1535	1932	1933	1533	3466	220	254	454
		80%*	172	152	88.5	2456	2300	2562	2193	4755	315	308	550

* Sections illustrated in Appendix I, Figure I, and Appendix II, Figure 1; all other sections shown in Figure 16.1. (Radial and ulnar 50% sections shown in both main text and Appendix figures.)

NK: neck; TA: total subperiosteal area (mm²); CA: cortical area (mm²); %CA: percent cortical area (CA/TA · 100); I_x, I_y: second moments of area about M-L and A-P axes (mm⁴); I_{max}, I_{min}: maximum and minimum second moments of area (mm⁴); J: Polar second moment of area (mm⁴); Z_x, Z_y: section moduli about M-L and A-P axes (mm³) (true—see text). Z_p: polar section modulus (mm³).

Appendix III

Fossil sample used in Chapter 17

ID	Species	HUMHEAD	ELBOW$^{0.5}$	RADTV	FEMHEAD	FEMSHAFT$^{0.5}$	DISTFEM$^{0.5}$	PROXTIB$^{0.5}$	DISTTIB$^{0.5}$	Comments
Sts 7	A. africanus	39.7	–	–	–	–	–	–	–	a
Sts 14	A. africanus	–	–	–	30.0	20.12	–	–	–	a
Sts 34	A. africanus	–	–	–	–	–	43.97	–	–	a
Sts 68	A. africanus	–	–	21.1	32.4	–	–	–	–	b
StW 25	A. africanus	–	–	–	32.4	–	–	–	–	a
StW 30a	A. africanus	–	–	–	33.1	–	–	–	–	b
StW 30b	A. africanus	–	–	–	33.4	–	–	–	–	b,c
StW 31	A. africanus	–	–	–	29.8	–	–	–	–	b,d
StW 99	A. africanus	–	–	–	38.0	29.40	–	–	–	a
StW 139	A. africanus	–	–	22.5	–	–	–	–	–	b
StW 328	A. africanus	29.2	–	–	–	–	–	–	–	b,e
StW 358	A. africanus	–	–	–	–	–	–	–	17.85	a
StW 361	A. africanus	–	–	–	29.7	–	–	–	–	b,f
StW 389	A. africanus	–	–	–	–	–	–	–	23.23	a
StW 392	A. africanus	–	–	–	31.5	–	–	–	–	a
StW 431	A. africanus	–	27.74	22.2	36.0	–	–	–	–	a,g
StW 479	A. africanus	–	–	–	28.6	–	–	–	–	b,f
StW 501	A. africanus	–	–	–	33.2	–	–	–	–	a
StW 514	A. africanus	–	–	–	–	–	–	38.68	–	b,h
StW 515	A. africanus	–	–	–	–	–	–	–	19.39	a
StW 516	A. africanus	–	–	22.5	–	–	–	–	–	a
StW 517	A. africanus	35.3	–	–	–	–	–	–	–	a
StW 522	A. africanus	–	–	–	31.1	–	–	–	–	a
StW 527	A. africanus	–	–	–	32.5	–	–	–	–	a
TM 1513	A. africanus	–	–	–	–	–	40.44	–	–	a
A.L. 128-1/129-1	A. afarensis	–	–	–	–	21.59	37.50	39.94	–	a
A.L. 137-48a	A. afarensis	–	22.93	–	–	–	–	–	–	a
A.L. 152-2	A. afarensis	–	–	–	33.1	19.93	–	–	–	i
A.L. 211-1	A. afarensis	–	–	–	–	28.20	–	–	–	j
A.L. 288-1	A. afarensis	27.3	20.51	15.0	28.6	20.88	–	40.31	18.15	a
A.L. 322-1	A. afarensis	–	22.94	–	–	–	–	–	–	a
A.L. 330-6	A. afarensis	–	–	–	–	–	–	52.90	–	i
A.L. 333-3	A. afarensis	–	–	–	40.2	31.25	–	–	–	a

Specimen	Species								Note
A.L. 333-4	*A. afarensis*	–	–	–	–	45.63	–	–	a
A.L. 333-6	*A. afarensis*	–	–	–	–	–	–	21.70	a
A.L. 333-7	*A. afarensis*	–	–	–	–	–	–	24.75	a
A.L. 333-42	*A. afarensis*	–	–	–	–	–	50.60	–	a
A.L. 333-96	*A. afarensis*	–	–	–	–	–	–	21.00	a
A.L. 333-107	*A. afarensis*	35.1	–	–	–	–	–	–	a
A.L. 333-131	*A. afarensis*	–	–	–	33.93	–	–	–	i
A.L. 333w-40	*A. afarensis*	–	–	–	30.83	–	–	–	j
A.L. 333w-56	*A. afarensis*	–	–	–	–	45.01	–	–	a
A.L. 333x-14	*A. afarensis*	–	22.2	–	–	–	–	–	a
A.L. 333x-26	*A. afarensis*	–	–	–	–	–	52.25	–	a
A.L. 827-1	*A. afarensis*	–	–	38.0	23.56	–	–	–	i

Note: All measurements are in mm; measurement descriptions are provided in Appendix V. All *A. africanus* specimens included in this analysis are from Sterkfontein Member 4.

[a] Data from McHenry and Berger (1998).

[b] Previously unpublished data collected by ADG.

[c] Measurement for this partial specimen taken from the anteroposterior dimension.

[d] Although the femoral head is broken and much of it is missing; this appears to be a very good estimate.

[e] McHenry and Berger (1998) provide a measurement of 34.2 mm for this specimen but appear to have accidentally swapped the anteroposterior and superoinferior dimensions, so we provide a newly measured value.

[f] Although the femoral head is broken and much is missing, this measurement represents a complete diameter measurable on the specimen.

[g] McHenry and Berger (1998) note that although the femoral head is missing, this is an estimate based on the acetabulum.

[h] The square root of the value provided by McHenry and Berger (1998) is 42.15 mm, but this is clearly too large for this specimen, so we provide a newly measured value.

[i] Data from Ward et al. (2012).

[j] Data from McHenry (1992).

Appendix IV

Extant sample used in Chapter 17

ID	Species	Sex	HUMHEAD	ELBOW$^{0.5}$	RADTV	FEMHEAD	FEMSHAFT$^{0.5}$	DISTFEM$^{0.5}$	PROXTIB$^{0.5}$	DISTTIB$^{0.5}$	Humerus length	Radius length	Femur length	Tibia length
AMNH 54327	*Gorilla gorilla*	F	51.77	38.81	28.04	44.79	31.28	49.63	57.70	27.53	346	274	291	247
AMNH 54356	*Gorilla gorilla*	F	51.44	35.97	27.16	40.56	31.25	45.91	58.54	26.78	382	299	325	262
AMNH 81652	*Gorilla gorilla*	F	50.67	37.74	28.94	43.64	30.97	52.23	58.91	29.04	370	289	310	251
AMNH 167339	*Gorilla gorilla*	F	45.70	33.81	23.68	38.25	27.42	43.87	50.45	23.79	372	295	306	249
AMNH 167340	*Gorilla gorilla*	F	52.75	36.62	27.47	41.50	32.35	50.04	59.22	28.72	384	300	320	260
CMNH HTB 1419	*Gorilla gorilla*	F	46.61	35.72	25.06	39.93	28.32	47.80	53.06	26.24	379	305	331	277
CMNH HTB 1423	*Gorilla gorilla*	F	51.57	37.60	27.32	41.64	29.83	52.26	59.53	29.79	402	319	327	271
CMNH HTB 1710	*Gorilla gorilla*	F	48.00	33.96	24.61	39.65	27.13	45.06	51.42	25.31	384	303	324	256
CMNH HTB 1743	*Gorilla gorilla*	F	49.83	35.92	27.04	40.50	30.77	48.80	54.17	28.21	364	298	312	261
CMNH HTB 1756	*Gorilla gorilla*	F	48.72	36.64	25.66	40.10	28.93	47.36	50.99	27.24	356	276	294	229
CMNH HTB 1794	*Gorilla gorilla*	F	49.27	35.36	26.45	39.96	28.04	48.56	53.24	25.05	363	296	316	259
CMNH HTB 1798	*Gorilla gorilla*	F	51.61	38.70	26.72	40.91	31.28	50.26	55.09	27.49	370	281	306	253
CMNH HTB 1801	*Gorilla gorilla*	F	48.53	36.13	25.92	40.27	30.98	48.06	55.50	28.11	356	290	301	251
CMNH HTB 1806	*Gorilla gorilla*	F	50.34	36.39	26.61	40.14	31.90	49.68	56.06	28.97	379	301	312	262
CMNH HTB 1849	*Gorilla gorilla*	F	46.51	38.36	25.65	39.73	30.01	47.39	55.40	27.42	355	285	296	251
CMNH HTB 1851	*Gorilla gorilla*	F	51.31	38.81	28.27	40.14	31.62	50.34	56.71	29.98	370	289	306	248
CMNH HTB 1854	*Gorilla gorilla*	F	52.77	37.92	25.79	41.75	33.51	51.38	56.31	27.80	403	316	331	284
CMNH HTB 1856	*Gorilla gorilla*	F	53.75	39.67	28.76	45.05	32.14	50.15	57.79	30.32	389	295	322	260
CMNH HTB 1997	*Gorilla gorilla*	F	49.77	37.21	25.61	39.28	30.11	49.50	53.31	25.62	368	326	316	256
CMNH HTB 2069	*Gorilla gorilla*	F	49.87	35.35	26.20	38.59	30.34	52.47	57.06	26.54	366	296	306	256
CMNH HTB 2784	*Gorilla gorilla*	F	48.57	36.20	26.70	39.62	29.82	49.69	54.22	26.88	377	320	313	250
CMNH HTB 2787	*Gorilla gorilla*	F	47.17	37.89	27.09	40.56	27.04	50.33	55.99	27.12	359	306	308	248
CMNH HTB 3393	*Gorilla gorilla*	F	46.26	36.35	24.82	39.07	27.94	47.83	51.77	26.77	371	290	312	242
NMNH 174698	*Gorilla gorilla*	F	45.08	34.34	24.60	36.78	27.19	42.99	50.07	23.85	348	291	314	255
AMNH 54355	*Gorilla gorilla*	M	63.31	47.18	33.29	49.60	37.50	62.81	71.52	33.38	449	369	380	323
AMNH 69398	*Gorilla gorilla*	M	55.07	41.26	30.16	45.38	33.76	53.21	57.61	29.93	408	332	349	287
AMNH 81651	*Gorilla gorilla*	M	65.96	49.39	36.27	51.77	36.10	66.65	71.29	35.85	475	378	396	320
AMNH 90289	*Gorilla gorilla*	M	62.49	44.96	30.31	48.20	34.59	59.55	66.95	31.49	410	313	345	288
AMNH 90290	*Gorilla gorilla*	M	61.17	44.70	31.26	48.06	36.71	58.60	63.47	29.94	425	339	355	292
AMNH 167335	*Gorilla gorilla*	M	62.35	47.01	33.97	49.23	36.26	59.06	68.10	30.92	453	373	372	320
AMNH 167336	*Gorilla gorilla*	M	58.78	43.19	32.78	44.80	35.71	56.11	66.35	30.75	413	324	353	293
AMNH 167338	*Gorilla gorilla*	M	66.16	48.99	32.80	50.63	39.78	61.86	71.02	36.11	458	369	383	334
AMNH 201459	*Gorilla gorilla*	M	64.65	49.22	34.87	50.91	38.36	67.38	73.45	33.33	455	369	384	328
AMNH 201460	*Gorilla gorilla*	M	65.43	47.98	35.48	49.39	34.85	62.08	69.75	33.44	453	365	371	312

ID	Species	Sex												
AMNH 201471	*Gorilla gorilla*	M	63.56	45.92	32.86	48.18	35.39	60.42	68.44	33.16	447	357	373	309
AMNH 202932	*Gorilla gorilla*	M	69.61	49.72	36.84	56.58	41.41	65.60	74.31	32.99	471	384	396	337
AMNH 239597	*Gorilla gorilla*	M	65.09	46.53	35.02	52.07	40.04	62.92	71.44	30.96	410	348	366	310
CMNH HTB 1795	*Gorilla gorilla*	M	61.35	46.22	32.11	48.92	35.59	58.87	67.02	34.99	426	351	367	305
CMNH HTB 1857	*Gorilla gorilla*	M	63.81	48.04	33.60	52.33	39.14	65.52	68.95	32.50	439	344	371	295
CMNH HTB 1859	*Gorilla gorilla*	M	62.99	46.02	36.64	50.58	36.52	62.14	69.37	30.69	455	357	376	301
CMNH HTB 1954	*Gorilla gorilla*	M	64.10	49.43	36.09	50.72	35.90	62.59	71.26	33.10	478	377	399	325
CMNH HTB 1991	*Gorilla gorilla*	M	62.07	46.27	32.12	48.51	37.32	58.52	64.82	34.26	450	358	383	310
CMNH HTB 1994	*Gorilla gorilla*	M	62.48	43.93	33.11	49.34	40.12	65.47	74.40	31.76	420	350	350	292
CMNH HTB 1995	*Gorilla gorilla*	M	62.91	46.04	32.51	48.82	35.40	61.74	69.39	31.67	432	343	370	294
NMNH 174722	*Gorilla gorilla*	M	57.61	46.67	31.66	49.27	34.87	58.12	67.30	34.02	400	315	345	290
NMNH 174723	*Gorilla gorilla*	M	54.45	43.08	29.25	47.70	33.88	57.06	67.28	29.47	403	337	359	301
NMNH 176225	*Gorilla gorilla*	M	68.01	48.90	37.62	51.70	35.68	64.98	73.66	34.46	465	366	386	311
NMNH 220325	*Gorilla gorilla*	M	59.65	44.53	32.36	50.01	34.73	57.36	67.59	32.44	434	360	375	312
NMNH 11R	*Homo sapiens*	F	37.45	26.95	19.62	41.17	24.29	48.70	50.27	27.72	310	231	386	359
NMNH 18R	*Homo sapiens*	F	36.30	25.56	18.87	42.84	27.16	51.76	55.87	29.48	288	199	427	368
NMNH 41R	*Homo sapiens*	F	36.95	27.12	18.72	40.84	28.35	53.02	56.90	29.59	290	216	430	348
NMNH 135R	*Homo sapiens*	F	37.95	27.34	19.38	43.07	27.37	54.92	58.48	30.99	291	216	416	346
NMNH 162R	*Homo sapiens*	F	36.87	27.51	19.67	42.73	25.41	49.96	53.51	25.79	298	214	431	342
NMNH 255	*Homo sapiens*	F	42.76	29.72	21.84	45.42	30.06	55.63	59.26	30.81	343	273	488	421
NMNH 304	*Homo sapiens*	F	37.44	27.46	19.69	42.65	28.16	53.22	55.08	30.81	320	247	446	387
NMNH 405R	*Homo sapiens*	F	36.64	27.01	18.20	40.92	24.31	50.50	53.14	26.98	264	195	394	315
NMNH 680	*Homo sapiens*	F	41.85	27.28	20.72	42.74	28.83	52.68	54.21	30.35	328	235	465	382
NMNH 723	*Homo sapiens*	F	38.12	27.64	20.72	41.02	27.66	51.96	54.27	29.92	305	235	468	376
NMNH 745R	*Homo sapiens*	F	39.73	30.35	20.26	43.49	28.15	54.53	53.13	28.19	311	227	430	353
NMNH 880	*Homo sapiens*	F	39.47	28.60	22.02	47.18	27.34	52.28	54.88	30.73	329	241	469	374
NMNH 891	*Homo sapiens*	F	40.87	28.69	21.26	43.78	27.38	54.33	56.77	31.10	306	248	461	397
NMNH 906	*Homo sapiens*	F	37.94	27.23	19.46	40.60	25.50	48.71	51.46	27.64	291	214	386	322
NMNH 929	*Homo sapiens*	F	39.06	27.33	19.64	40.48	25.70	55.49	55.64	28.98	307	235	445	371
NMNH 970	*Homo sapiens*	F	38.40	28.75	19.44	40.86	26.10	53.03	55.32	29.65	306	237	435	381
NMNH 983	*Homo sapiens*	F	35.87	26.39	18.50	38.11	24.80	48.15	51.85	26.47	295	217	409	342
NMNH 1032	*Homo sapiens*	F	37.48	27.36	20.73	41.67	26.62	53.12	54.92	28.48	300	231	443	359
NMNH 1105	*Homo sapiens*	F	36.78	26.58	18.58	41.08	26.61	49.97	53.07	27.53	300	227	436	352
NMNH 1120	*Homo sapiens*	F	37.11	27.88	20.51	43.04	27.23	51.25	55.69	29.73	286	212	403	324
NMNH 1482	*Homo sapiens*	F	38.26	28.06	20.53	42.04	29.05	51.93	53.92	30.31	305	222	429	346
NMNH 1561	*Homo sapiens*	F	36.84	28.90	19.80	40.10	27.03	50.07	55.09	27.43	309	235	430	363

ID	Species	Sex	HUMHEAD	ELBOW$^{0.5}$	RADTV	FEMHEAD	FEMSHAFT$^{0.5}$	DISTFEM$^{0.5}$	PROXTIB$^{0.5}$	DISTTIB$^{0.5}$	Humerus length	Radius length	Femur length	Tibia length
NMNH 1566	Homo sapiens	F	39.07	29.03	19.50	43.00	26.77	54.56	58.71	28.10	320	235	458	383
NMNH 1617	Homo sapiens	F	38.39	27.84	20.98	38.64	23.12	50.72	52.99	27.90	297	222	416	340
NMNH 4R	Homo sapiens	M	45.48	35.74	26.76	51.12	30.19	58.99	61.80	34.73	359	266	505	406
NMNH 13R	Homo sapiens	M	38.94	28.63	22.26	43.93	25.30	55.45	59.31	29.76	323	263	458	388
NMNH 30R	Homo sapiens	M	44.99	31.44	22.05	48.55	27.27	57.08	62.36	32.20	347	264	491	407
NMNH 31R	Homo sapiens	M	45.93	33.72	24.17	45.17	31.46	57.86	61.43	30.92	339	263	449	399
NMNH 34R	Homo sapiens	M	45.23	31.28	23.78	49.73	28.47	58.81	64.49	33.92	313	227	456	376
NMNH 89R	Homo sapiens	M	44.48	32.67	22.76	48.25	27.57	57.25	63.87	31.59	347	264	469	394
NMNH 203	Homo sapiens	M	40.65	31.86	23.60	44.25	28.64	56.20	60.02	30.87	333	253	462	383
NMNH 210	Homo sapiens	M	42.63	33.87	22.96	44.14	30.52	57.93	66.77	32.13	342	276	488	434
NMNH 222	Homo sapiens	M	42.10	32.99	22.21	48.64	29.57	59.05	64.28	31.38	330	260	474	396
NMNH 303R	Homo sapiens	M	43.48	33.42	24.55	50.23	31.18	61.92	66.83	33.80	350	263	496	420
NMNH 311	Homo sapiens	M	42.27	30.35	23.00	44.29	25.44	50.69	54.54	29.01	313	246	446	385
NMNH 343	Homo sapiens	M	42.05	32.50	22.67	45.08	28.41	54.80	59.68	30.51	310	233	457	413
NMNH 591	Homo sapiens	M	44.68	34.30	24.01	47.30	27.15	53.68	61.86	31.82	336	244	465	383
NMNH 594	Homo sapiens	M	45.73	35.93	25.11	48.27	30.38	60.04	63.79	32.71	365	268	489	416
NMNH 595	Homo sapiens	M	43.46	33.47	25.02	46.01	28.59	54.75	61.12	35.11	357	285	502	437
NMNH 606	Homo sapiens	M	44.40	33.26	23.24	47.23	29.97	57.82	61.35	30.24	323	259	435	356
NMNH 645	Homo sapiens	M	43.38	32.99	23.19	50.78	29.46	59.21	66.14	33.77	343	263	490	421
NMNH 696	Homo sapiens	M	46.39	33.11	23.72	55.27	31.61	62.50	67.63	36.02	339	261	494	402
NMNH 989	Homo sapiens	M	40.50	29.39	20.30	48.24	29.13	58.35	61.80	31.98	316	231	450	363
NMNH 1126	Homo sapiens	M	45.85	34.72	24.23	51.03	31.19	61.83	64.36	34.23	367	270	491	401
NMNH 1158RR	Homo sapiens	M	41.03	29.66	20.29	45.33	24.91	50.86	57.08	29.29	293	219	404	333
NMNH 1183	Homo sapiens	M	43.20	33.57	23.10	47.17	28.31	56.47	61.24	32.92	327	253	456	377
NMNH 1187	Homo sapiens	M	44.47	32.16	23.38	48.17	33.02	61.55	64.70	32.75	351	275	471	435
NMNH 1564	Homo sapiens	M	42.70	30.49	21.43	46.64	30.93	58.65	63.13	33.13	304	234	422	359
AMNH 51376	Pan troglodytes	F	37.77	29.53	22.52	32.62	23.62	39.58	44.73	21.16	282	259	279	230
AMNH 89351	Pan troglodytes	F	38.65	27.52	23.46	30.76	22.31	39.14	43.25	21.09	281	264	274	230
AMNH 89354	Pan troglodytes	F	36.11	28.61	22.50	30.08	22.38	36.96	41.45	21.21	290	276	284	238
AMNH 89355	Pan troglodytes	F	36.56	30.35	23.92	32.27	23.04	41.46	46.88	21.85	279	260	287	247
AMNH 89426	Pan troglodytes	F	34.14	25.97	20.99	30.17	22.03	36.28	41.71	20.66	280	261	282	235
AMNH 90292	Pan troglodytes	F	35.11	27.05	19.92	31.24	22.74	37.27	41.59	21.35	286	261	285	243
AMNH 174860	Pan troglodytes	F	41.69	31.54	23.67	34.23	26.28	43.60	45.52	24.45	302	283	295	259
AMNH 201469	Pan troglodytes	F	38.52	30.17	24.91	33.03	25.32	39.57	44.25	21.84	294	272	292	240

CMNH HTB 1720	Pan troglodytes	F	36.07	27.93	22.16	29.82	23.53	36.92	41.53	22.19	301	270	297	246
CMNH HTB 1755	Pan troglodytes	F	36.08	28.63	23.23	30.80	23.31	38.32	40.49	21.74	306	286	306	261
CMNH HTB 1766	Pan troglodytes	F	44.80	32.55	26.75	36.09	25.02	43.74	50.14	23.89	302	266	303	248
CMNH HTB 1843	Pan troglodytes	F	40.48	29.74	24.29	33.65	25.72	41.84	45.86	22.43	321	284	299	252
CMNH HTB 1853	Pan troglodytes	F	40.06	29.98	24.02	33.73	25.42	40.87	47.58	23.42	317	287	303	262
CMNH HTB 1880	Pan troglodytes	F	45.47	33.84	25.41	36.63	24.73	45.39	49.21	25.24	297	288	307	259
CMNH HTB 2771	Pan troglodytes	F	36.61	28.70	21.89	29.87	24.15	38.39	42.75	21.61	296	260	299	243
CMNH HTB 3551	Pan troglodytes	F	37.15	28.26	22.14	28.83	22.16	35.74	39.72	21.15	308	268	294	239
NMNH 176226	Pan troglodytes	F	33.99	26.53	20.79	28.23	20.28	34.86	38.93	21.19	288	244	277	236
NMNH 176227	Pan troglodytes	F	36.39	27.96	22.45	30.84	22.02	36.51	40.78	19.85	295	268	287	240
NMNH 176229	Pan troglodytes	F	42.68	32.95	24.46	34.95	26.84	42.77	47.89	24.81	319	305	302	265
NMNH 220063	Pan troglodytes	F	39.56	28.78	22.75	31.81	23.05	39.81	45.24	23.21	296	279	279	237
NMNH 220064	Pan troglodytes	F	40.08	31.75	23.79	32.94	22.89	39.17	44.70	21.15	332	304	314	273
NMNH 236971	Pan troglodytes	F	35.60	25.46	20.41	29.23	19.94	35.27	40.97	20.06	251	235	252	214
NMNH 477333	Pan troglodytes	F	38.39	28.62	23.35	31.31	23.14	37.34	42.81	21.37	291	258	282	228
NMNH 481803	Pan troglodytes	F	39.68	31.19	24.30	32.19	24.58	40.56	45.38	22.92	301	274	300	251
AMNH 51202	Pan troglodytes	M	37.43	29.21	21.51	30.50	23.47	37.88	43.41	20.20	308	280	307	253
AMNH 51377	Pan troglodytes	M	39.53	29.56	21.99	34.55	25.16	42.55	46.27	23.21	332	291	319	274
AMNH 51379	Pan troglodytes	M	44.11	29.60	22.74	36.41	24.75	43.28	47.02	22.70	318	303	318	268
AMNH 51381	Pan troglodytes	M	40.44	30.67	22.64	32.04	23.50	41.30	45.16	22.37	299	274	294	246
AMNH 51393	Pan troglodytes	M	42.75	32.22	26.17	34.11	24.72	40.53	47.13	23.22	311	281	304	256
AMNH 51394	Pan troglodytes	M	45.09	35.04	25.09	38.34	26.85	44.86	48.55	25.01	330	299	338	281
AMNH 54330	Pan troglodytes	M	42.58	32.41	24.89	31.25	24.90	39.21	46.45	23.88	283	275	276	240
AMNH 89353	Pan troglodytes	M	40.32	33.01	25.40	32.25	24.52	43.04	46.61	23.32	327	300	319	272
AMNH 89406	Pan troglodytes	M	38.19	30.15	23.66	31.26	22.41	40.37	45.03	23.24	306	282	308	256
AMNH 89407	Pan troglodytes	M	38.04	28.75	23.23	32.00	23.19	40.86	44.54	21.37	297	267	295	241
AMNH 90190	Pan troglodytes	M	45.76	34.09	26.84	37.25	26.37	46.13	51.72	25.25	305	291	314	252
AMNH 174861	Pan troglodytes	M	44.47	34.27	26.06	36.66	26.99	44.77	51.56	25.15	312	289	304	260
CMNH HTB 1056	Pan troglodytes	M	45.01	33.22	25.54	35.41	24.39	44.98	47.52	25.13	320	296	326	273
CMNH HTB 1718	Pan troglodytes	M	41.69	32.46	26.01	32.84	25.32	39.60	46.15	23.09	301	289	295	249
CMNH HTB 1722	Pan troglodytes	M	43.13	34.62	25.90	36.83	25.39	43.10	47.98	25.44	321	300	310	270
CMNH HTB 1758	Pan troglodytes	M	41.91	31.29	23.94	36.10	26.73	42.62	47.77	23.68	313	307	319	265
CMNH HTB 1761	Pan troglodytes	M	43.08	33.01	25.08	34.31	24.18	41.31	47.53	24.12	314	295	299	257
CMNH HTB 3552	Pan troglodytes	M	41.92	32.56	25.40	34.69	26.37	44.89	50.37	25.41	296	280	294	250
NMNH 176228	Pan troglodytes	M	44.07	33.89	27.15	34.98	25.56	41.91	49.29	25.23	327	296	318	273
NMNH 220065	Pan troglodytes	M	42.90	32.87	25.74	34.01	24.17	43.39	48.70	24.37	295	272	288	243

ID	Species	Sex	HUMHEAD	ELBOW$^{0.5}$	RADTV	FEMHEAD	FEMSHAFT$^{0.5}$	DISTFEM$^{0.5}$	PROXTIB$^{0.5}$	DISTTIB$^{0.5}$	Humerus length	Radius length	Femur length	Tibia length
NMNH 220326	Pan troglodytes	M	41.48	32.29	24.42	34.47	24.89	39.58	46.72	24.56	327	293	309	256
NMNH 220327	Pan troglodytes	M	37.41	29.56	22.91	32.15	24.15	39.32	42.75	22.14	293	292	292	256
NMNH 395820	Pan troglodytes	M	37.00	31.23	24.25	32.63	23.95	40.62	44.50	21.62	270	247	281	229
NMNH 481804	Pan troglodytes	M	37.66	28.84	24.04	32.10	22.62	39.31	43.80	21.98	302	265	294	235
CMNH HTB 1030	Pongo abelii	F	37.52	29.88	21.30	30.61	19.63	35.40	39.68	19.99	322	345	247	213
CMNH HTB 1055	Pongo abelii	F	39.07	27.97	22.06	32.99	19.74	35.12	41.26	21.13	311	333	242	212
CMNH HTB 1443	Pongo abelii	F	33.41	28.89	19.26	29.61	17.33	34.86	38.71	19.62	343	349	258	223
NMNH 143597-A49853	Pongo abelii	F	38.65	30.20	20.58	31.52	18.33	35.41	39.19	19.24	333	335	264	222
NMNH 143602-A49861	Pongo abelii	F	36.76	30.28	20.20	30.10	18.65	34.85	40.29	18.91	333	331	256	224
NMNH 270807	Pongo abelii	F	37.61	30.90	20.38	29.26	17.68	34.12	38.80	19.83	327	333	264	232
NMNH A22937	Pongo abelii	F	38.84	29.54	21.43	30.76	17.93	33.35	39.60	20.22	316	315	241	201
CMNH HTB 1444	Pongo abelii	M	45.05	34.62	23.75	37.43	21.71	39.34	46.34	23.46	375	365	296	250
NMNH 143590-A49855	Pongo abelii	M	45.52	37.84	25.92	37.33	21.67	41.38	48.34	23.05	377	370	293	244
NMNH 143593-A49859	Pongo abelii	M	48.06	37.93	22.86	35.12	21.57	41.31	49.22	24.48	411	378	301	264
AMNH 200898	Pongo pygmaeus	F	36.13	27.74	21.01	29.85	17.14	31.26	36.38	18.95	334	313	243	201
AMNH 200900	Pongo pygmaeus	F	38.91	30.78	20.39	31.37	17.32	34.40	39.87	19.85	352	349	255	222
NMNH 142169-A49768	Pongo pygmaeus	F	38.83	31.04	21.36	31.52	19.89	35.76	42.99	22.52	341	339	257	237
NMNH 142170-A49769	Pongo pygmaeus	F	37.59	30.67	19.46	31.64	17.89	36.23	40.87	21.82	332	330	256	228
NMNH 145300-A49957	Pongo pygmaeus	F	34.92	29.67	19.79	29.37	17.38	33.34	39.50	19.57	327	323	244	219
NMNH 145302-A49959	Pongo pygmaeus	F	38.46	29.33	21.35	31.03	19.14	32.47	39.18	21.63	322	330	253	227
NMNH 145306-A49963	Pongo pygmaeus	F	37.66	31.33	21.69	32.51	20.75	35.04	41.91	21.90	339	352	259	234
NMNH 145308-A49965	Pongo pygmaeus	F	39.22	32.73	23.06	31.37	20.27	35.90	42.67	21.77	333	343	247	227
NMNH 145309-A49966	Pongo pygmaeus	F	34.76	29.74	20.85	30.19	18.89	35.68	42.15	21.91	326	340	260	230

AMNH 140426	*Pongo pygmaeus*	M	47.22	37.60	25.50	37.56	22.43	42.03	48.64	23.43	380	390	288	254
AMNH 61586	*Pongo pygmaeus*	M	49.31	38.50	26.64	40.19	23.35	45.25	50.17	26.40	366	377	281	256
CMNH HTB 0172	*Pongo pygmaeus*	M	46.73	40.22	27.49	40.58	24.13	42.62	51.23	25.00	355	345	283	246
NMNH 145301-A49958	*Pongo pygmaeus*	M	49.07	39.47	25.68	40.95	24.16	44.51	53.01	25.86	390	400	298	271
NMNH 145304-A49961	*Pongo pygmaeus*	M	46.74	37.33	28.23	38.14	24.29	44.99	53.92	28.06	366	387	277	253
NMNH 145305-A49962	*Pongo pygmaeus*	M	48.03	40.16	27.86	40.36	25.30	44.86	50.14	27.16	368	392	292	266
NMNH 145310-A49967	*Pongo pygmaeus*	M	47.14	39.45	27.25	39.51	22.35	43.05	50.84	25.56	347	370	284	253
NMNH 153823	*Pongo pygmaeus*	M	47.94	39.78	26.05	39.21	23.90	44.99	49.36	27.99	385	383	304	263

Note: All measurements are in mm; measurement descriptions are provided in the main text and Appendix V.

Appendix V

Measurement definitions used in Chapter 17 (reproduced from Green et al. [2007], pp. 189–190)

1. HUMHEAD is the maximum anteroposterior (AP) diameter of the humeral head taken perpendicular to the shaft axis.
2. ELBOW is the product of capitular height and articular width of the distal humerus. Capitular height was taken from the anteroproximal border of capitulum to the distoposterior border along the midline. Articular width was taken across the anterior aspect of the articular surface from the lateral border of the capitulum to the medial edge of the articular surface.
3. RADTV is the mediolateral (ML) diameter of the radial head.
4. FEMHEAD is the maximum superoinferior (SI) diameter of the femoral head.
5. FEMSHAFT is the product of the AP and transverse diameters of the femoral shaft, taken just inferior to the lesser trochanter.
6. DISTFEM is the product of the biepicondylar and shaft AP diameters of the distal femur.
7. PROXTIB is the product of the AP and transverse diameters of the proximal tibia. The AP diameter was taken with one jaw of the calipers on the line connecting the posterior surfaces of the medial and lateral condyles and the other jaw on the most distant point on the medial condyle. Transverse diameter was the distance between the most lateral point on the lateral condyle to the most medial point on the medial condyle (perpendicular to the AP diameter).
8. DISTTIB is the product of the AP and transverse diameters of the distal tibia. The AP diameter is the distance between the most anterior and posterior points of the talar facet in the AP plane. Transverse diameter is the distance between the midline of the medial malleolus and the midline of the most medial point of the talar facet before the fibular facet begins.

Index